Isovector methods for equations of balance

Monographs and textbooks on mechanics of solids and fluids
editor-in-chief: G. Æ. Oravas

Mechanics: Analysis
editor: V. J. Mizel

1. M. A. KRASNOSELSKII *et al.*
   Integral operators in spaces of summable functions

2. V. V. IVANOV
   The theory of approximate methods and their application to the numerical solution of singular integral equations

3. A. KUFNER *et al.*
   Function spaces

4. S. G. MIKHLIN
   Approximation on a rectangular grid

5. D. G. B. EDELEN
   Isovector methods for equations of balance

6. R. S. ANDERSSEN, F. R. DE HOOG, M. A. LUKAS
   The application and numerical solution of integral equations

Isovector methods for equations of balance

with programs for computer assistance in operator calculations and an exposition of practical topics of the exterior calculus

Dominic G. B. Edelen

Center for the application of mathematics Lehigh University

SIJTHOFF & NOORDHOFF 1980
Alphen aan den Rijn, The Netherlands
Germantown, Maryland, USA

ISBN 90 286 0420 0

Printed in The Netherlands

Marybeth Bowling Edelen
1902-1974

Edmund Nalle Bowling
1899-1976

John Dominic Bowling
1895-1971

in gentle recollection

PREFACE

Although the basic laws of mechanics and physics are conceptually quite simple, stating as they do the conservation or balance of various collections of fields such as mass, momentum, energy, charge, magnetic flux, and so forth, the actual field equations often prove to be of great difficulty when it comes to obtaining exact or even good approximate solutions of specific nonlinear problems. This state of affairs can be credited primarily to the loss of superposition principles and the inapplicability of linear methods, yet it would also appear fair to say that some of the difficulty can be traced to the innate inertia associated with the acquisition of new mathematical techniques and methods that are more naturally attuned to nonlinear problems. A striking example of what can be achieved with new techniques is the marked success with which geometric methods have come to grips with certain classes of problems associated with nonlinear, 1-dimensional wave phenomena. These techniques also provide access to classes of problems of significantly greater breadth and generality than 1-dimensional wave phenomena, and, in fact, they will be shown to be naturally associated with any system of laws of balance. There are, however, certain well-known problems, such as $Q(T)\partial T/\partial t = \partial^2 T/\partial x^2$, for which the techniques provide only a very restricted amount of information. The modern geometric approach is thus not a panacea, even though it provides a uniform framework that often proves to be of great intrinsic utility.

The basic subject matter of this monograph is the study of systems of laws of balance that determine the fundamental physical field quantities as functions of position and time. Such laws have the generic structure

$$\sum_{i=1}^{4} \frac{\partial}{\partial x^i} W^i_\beta \left(x^j, \phi^\alpha, \partial\phi^\alpha/\partial x^j\right) = W_\beta\left(x^j, \phi^\alpha, \partial\phi^\alpha/\partial x^j\right) ,$$

with x^i, $i=1,2,3$ as spatial variables, x^4 as the time variable, and $\phi^\alpha(x^j)$, $\alpha=1,\ldots,N$ as the N physical field quantities. We thus deal with second order, nonlinear partial differential equations that are determined by N 4-dimensional flux quantities W^i_β and N 4-dimensional source quantities W_β that are specified functions of the independent variables, the field quantities, ϕ^α, and the first derivatives of these field quantities. Now, as is well known, difficulties arise primarily because of the occurrence of second rather than first order derivatives and the concomitant requirement of two quadratures rather than one in order to secure a solution. The power and scope of the geometric approach is that it replaces the second order partial differential system by an appropriately structured system of first order partial differential equations and associates them in a natural and computationally useful manner with exterior differential forms. This usefulness and computational efficacy is achieved through use of the theory of exterior differential equations, linear homotopy operators, and isovectors of ideals that are generated by the exterior differential forms.

Lest the reader be led astray, we hasten to point out that this monograph is not a text in modern differential geometry. A perusal of the text shows that only a limited part of the vast machinery of modern differential geometry [1-6] is used. Most of the considerations are strictly local in nature, occur in a single coordinate neighborhood, and use only those aspects of the exterior calculus that can be formulated without the explicit introduction of the concept of a manifold. In this regard, the text is self-contained, for the five chapters of the Appendix present the exterior calculus from an applications point of view. Here, the emphasis is on the practical aspects that aid solving partial differential equations rather than the more customary treatment via tangent and cotangent bundles and the theory of jets.

After collecting together the essential aspects of exterior

differential equations, linear homotopy operators and ideal theory in Chapter 1, the basis for the decomposition of the given system of second order equations into equivalent systems of first order equations is studied in Chapter 2. The idea here is quite simple. We introduce a collection of new variables y_i^α so that we have a space K with coordinates $(x^i, q^\alpha, y_i^\alpha)$. The new variables (y_i^α) are then correlated with the first derivatives of the field variables through a system of exterior differential forms

$$C^\alpha = dq^\alpha - y_i^\alpha \, dx^i ,$$

called contact forms, and the requirement that $q^\alpha = \phi^\alpha(x^j)$ yields $\Phi^* C^\alpha = d\phi^\alpha(x^j) - \Phi^*(y_i^\alpha)\, dx^i = 0$; that is, $\Phi^*(y_i^\alpha) = \partial\phi^\alpha(x^j)/\partial x^i$ for any map Φ from the space of independent variables (x^i) into the space K. Since the quantities C^α are exterior differential forms of degree one, their intrinsic structure is determined through the study of the ideal of the algebra of exterior differential forms on K that is generated by $\{C^\alpha\}$ and $\{dC^\alpha\}$. In other words, we study all exterior forms that are sums of exterior multiples of $\{C^\alpha\}$ and $\{dC^\alpha\}$. This study is accomplished through a characterization of the isovectors of this contact ideal. The reason for this is that the isovectors are those vector fields that transport elements of the contact ideal into new elements of the same ideal; they specify those properties of the ideal that are intrinsic, as opposed to those that are associated with particular elements of the ideal. A complete characterization of all isovectors is obtained in closed form and this, to a large extent, provides the explicit results upon which much of the theory depends. The characterization also allows us to study the orbits of the isovector fields and to obtain relations between the equations for the orbits and systems of partial differential equations of the first order that admit the isovector fields as characteristics. Direct correlations with classic well-known results are thereby achieved, as well as certain additional structural results that prove to be instrumental in the construction and study of similarity solutions of the original system of second order partial differential equations.

The contact forms and their associated ideal are used in Chapter 3 to reduce the original system of second order partial differential equations to a new system of first order partial differential equations. This is achieved by adjoining to the contact forms an additional collection $\{E_\alpha\}$ of differential forms whose degree is equal to the dimension of the space of original independent variables (x^i) . Now, not all systems $\{E_\alpha\}$ are of equal importance, and, in particular, many different systems $\{E_\alpha\}$ can result in the same system of second order partial differential equations. A natural equivalence relation is shown to exist on the collection of all possible $\{E_\alpha\}$; two systems being equivalent if and only if they possess exactly the same collection of solutions. This equivalence is used in conjunction with the closed ideal generated by $\{C^\alpha\}$ and $\{E_\alpha\}$ in order to obtain canonical forms for the possible systems $\{E_\alpha\}$, which we refer to as balance systems. The canonical forms thus arrived at turn out to be systems of laws of balance of the form stated at the beginning. The mathematics thus contains no more and no less than the original physics and states that any arbitrary system, that fancy might dictate one study, can always be reduced to an equivalent canonical form that has the structure of a system of laws of balance.

Again, it is the isovector fields of the closed ideal generated by $\{E_\alpha\}$ and $\{C^\alpha\}$ that provide the basic computational tool. The calculation of such isovector fields has been considerably simplified, for we have already obtained the isovectors of the closed ideal generated by $\{C^\alpha\}$ and these isovector fields contain all isovector fields of $\{C^\alpha\}$ and $\{E_\alpha\}$ combined. Now, the isovectors of the closed ideal generated by $\{E_\alpha\}$ and $\{C^\alpha\}$

transport the ideal into itself and form a Lie algebra. This Lie algebra thus generates mappings of K to K with the important property that it always transports solutions of the original system into solutions of the original system. We are thus able to imbed any given solution in a whole Lie group of nontrivial solutions of the given system of partial differential equations. In most cases, this imbedding amounts to the construction of a Lie group of completely integrated contact transformations. Obvious modifications lead to a characterization of the similarity solutions of the system and their Bäcklund mappings. The isovectors of the closed balance ideal and the linear homotopy operators are also shown to lead to systems of laws of conservation and balance that every solution of the given balance system will satisfy. Infinitely many such systems of laws of conservation are explicitly exhibited for a wide class of systems, thus dispelling the "mystery" of their occurrence in certain classic problems. In addition, similarity solutions of $E_\alpha = W_\alpha \mu - d(W_\alpha^i \mu_i)$ are shown to satisfy laws of conservation in all instances.

The more recent notions of pseudopotentials and prolongations are considered in the fourth Chapter. These provide natural generalizations of the classic results on potential-like representations of solutions. In particular, we establish the existence of pseudopotentials and their associated prolongations for a wide class of problems with a finite but arbitrary number of independent variables, rather than only two independent variables. These include systems of types $E_\alpha = (d\xi_\beta) R_\alpha^\beta$ and $E_\alpha = W_\alpha \mu - d(W_\alpha^i \mu_i)$. A more general prolongation structure is also obtained. This turns out to be particularly useful, for the isovectors of the prolonged closed ideal allow us to imbed solutions in Lie groups of solutions that are much larger than those previously obtained. It also turns out that this new imbedding leads to a large collection of Bäcklund transformations, and these transformations are also imbedded in a Lie group structure, although they do not themselves form a Lie group necessarily.

Chapter 5 makes use of the concept of prolongation in quite a different fashion. After obtaining necessary and sufficient conditions in order that a balance system arise from a variational principle (constitute a variational system), we prove that any balance system admits a prolongation to a variational system. Of particular importance is the fact that the Lagrangian function for the prolonged variational system is linear in the fiber variables of the prolongation. Further, the structure of the prolonged variational system is such that a simple reordering of the indices allows us to carry over all of the results of the previous four chapters. We thus obtain immediate results concerning isovectors of the prolonged contact ideal, isovectors of the prolonged balance ideal, and mappings that generate new solutions by transport along the orbits of the isovector fields.

A significant amount of additional information can be obtained directly from the new aspect of the system, namely, the existence of a variational principle. The exterior differential structure already in evidence is used to construct an equivalent "free variational principle" for which there is no requirement that the variations vanish on the boundary. This leads in a natural way to the definition and complete characterization of both finite and infinitesimal variations and the variation process is shown to be generated by isovector fields of the prolonged contact ideal. We show that all finite variations of the action functional are evaluated in terms of surface integrals whenever the action itself is evaluated at a solution of the prolonged balance system.

The ability to evaluate finite variations of the action functional provides direct access to the Lie algebra of Noetherian vector fields. This is a Lie subalgebra of the Lie algebra of isovector fields of the contact ideal that generates identically zero finite variations of the action functional (global symmetries). Two nontrivial generalizations of Noetherian vector fields are defined by use of the contact ideal and the balance ideal of the system. Both generalizations yield vector fields that form Lie subalgebras. Each of the three kinds of Neotherian vector fields are shown to lead to systems of current densities that are conserved for any solution of the balance system. They also provide a means of constructing new (n-1)-forms whose exterior derivatives belong to the balance ideal of the system. This result, in conjunction with those established in Chapter 4, shows that there exist as many independent systems of pseudopotentials as there are independent Noetherian vector fields: prolongation to a variational system provides an automatic pseudopotential calculator.

E-transformations of variational systems are defined and given explicit characterization. They are shown to define mappings from one variational system to another such that all solutions of one map onto solutions of the other. In particular, the equations include a Lagrangian formulation of the multidimensional Hamilton-Jacobi equation and generalized Bäcklund transformations between the old the new y's of the two systems. These results, although useful for certain classes of problems, are not universally applicable to all systems in terms of feasible computational methods. A study of methods of solution that obtain from transport of the variational system for discrete values of the orbital parameter is given. Again, the results are useful, but not universally applicable.

The concept of pointwise transversality is used to show that the tangent space to graph space is the direct sum of the contact subspace and the transversality subspace at all points of graph space for which the Lagrangian function is different from zero. This allows us to complete the contact 1-forms to a basis for the dual of the tangent space to graph space by adjoining a system of *transversality* 1-forms. A direct sum decomposition of

$\Lambda^1(K_4)$ is thereby obtained and this leads to a simple characterization of the null class of the Euler-Lagrange operator and to the equivalence of action integrals. A study of the possible linear transformations of the system of transversality 1-forms is shown to lead to generalizations of the complete figure of Carathéodory and to a family of Hamilton-Jacobi equations and canonical variables. This reduces the problem of solving a variational system to that of solving the corresponding Hamilton-Jacobi equation together with the constraints $S^*C^a = 0$.

The study of variational systems ends with an examination of problems with given differential constraints of arbitrary exterior degree. A Lagrange multiplier method is exhibited in all cases, but differs from the classic method in that the number of multipliers is equal to the number of independent constraint equations only when the constraint forms are completely integrable.

The sixth Chapter gives the essential aspects of what we have termed the method of explicit extension. This method combines the more useful aspects of several of the methods studied in previous chapters and appears to be of use for certain classes of problems. There is some difficulty in giving a concise summary of this method, so we simply refer the reader to Chapter 6 for the details.

Anyone who has dealt with isovector methods knows that an inordinate amount of labor is involved for even the simplest balance systems. In fact, the restriction to only two independent variables that is so prevalent in the literature is a direct consequence of the need to make the calculations manageable. Chapter 7 gives program listings in the commerically available REDUCE 2 system that implement the basic operations of the exterior calculus. These programs provide direct computer assistance in the calculation of exterior products, exterior derivatives, inner products, Lie derivatives, Lie products, and linear homotopy operators. Since the REDUCE 2 system is interactive and implements a general class of global equivalence statements for operators, a significant part of the calculations for isovectors can be done automatically and with perfect accuracy the first time around. Use of these programs would seem to provide the last ingredient whereby isovector methods may be applied to problems as a general rule rather than as a drudgerous last resort. In this vein, we note that problems that used to take a week or more may now be finished in a morning's work at a computer terminal.

The monograph concludes with a rather long appendix that naturally breaks into two parts. The first part, consisting of Chapter I through IV, provides a detailed development of the exterior calculus without introducing the usual manifold theory since the applications deal with the local properties of solutions of partial differential equations. This material has been included so that the text is complete in itself. The reader who is not overly conversant with the exterior calculus is strongly urged to master these chapters before proceeding to the text proper. The second part, consisting of Chapter V, develops the theory of linear homotopy operators, antiexact exterior differential forms, and their application to problems associated with systems of exterior differential equations. Much of the material in this chapter is not currently available in the literature, but essential to a full understanding of the text. The general reader will thus need to familiarize himself with Chapter V of the appendix before turning to the text.

The author owes a significant debt to a number of people in connection with the research, writing, and publication of this work. First to Professor Eric Varley for the many useful discussions of specific problems and for pointing out the basic questions associated with Bäcklund transformations. The computer programs were significantly simplified and improved through implementation of several suggestions of Professor Gerhard Rayna, which are gratefully acknowledged. This work would still be in its infancy without the encouragement and advice of Professor Hanno Rund. A major alteration of the text resulted from the substantive comments of Professor V. Mizel in the course of review for publication (a task often thankless and ignored). Discussions with numerous colleagues and graduate students at Lehigh University are gratefully acknowledged, together with the forbearance of the students of my Math 302 classes who acted as a testing ground for much of the material in the Appendix. Finally, particular thanks are due to Mrs. Fern Sotzing for the patience and care with which she typed the camera-ready typescript and to Miss Aida Kadić for her help and diligence in the proofreading.

Bethlehem, PA
December, 1979

Dominic G. B. Edelen

CONTENTS

CHAPTER 1

PRELIMINARY CONSIDERATIONS

The approach taken in this monograph is primarily a geometric one. Full use is therefore made of E. Cartan's exterior calculus. We use the following standard notation [1-6]:

$\wedge$ for exterior multiplication,

d for exterior differentiation,

$\lrcorner$ for inner multiplication,

£ for Lie differentiation.

These operations often lead to very remarkable simplifications, both in theory and in actual calculations, for they exhibit the following interrelations.[†] Let α and β be exterior differential forms, u and v be vector fields, f a function, and a = degree of α, b = degree of β, then

$$\alpha \wedge \beta = (-1)^{ab} \beta \wedge \alpha ,$$

$$d(\alpha+\beta) = d\alpha + d\beta ,$$

$$d(\alpha \wedge \beta) = d\alpha \wedge \beta + (-1)^{a} \alpha \wedge d\beta ,$$

[†]Those not overly familiar with the exterior calculus are urged to consult the Appendix before proceeding further.

$$dd\alpha = 0 ,$$

$$(u+v) \rfloor \alpha = u \rfloor \alpha + v \rfloor \alpha ,$$

$$u \rfloor (\alpha+\beta) = u \rfloor \alpha + u \rfloor \beta ,$$

$$u \rfloor f = 0 ,$$

$$u \rfloor (\alpha \wedge \beta) = (u \rfloor \alpha) \wedge \beta + (-1)^a \alpha \wedge (u \rfloor \beta) ,$$

$$\pounds_u \alpha = u \rfloor d\alpha + d(u \rfloor \alpha) ,$$

$$\pounds_u (\alpha+\beta) = \pounds_u \alpha + \pounds_u \beta ,$$

$$\pounds_u (\alpha \wedge \beta) = (\pounds_u \alpha) \wedge \beta + \alpha \wedge (\pounds_u \beta) ,$$

$$\pounds_u d\alpha = d\pounds_u \alpha ,$$

$$\pounds_{fu} \alpha = f\pounds_u \alpha + df \wedge (u \rfloor d\alpha) ,$$

$$\pounds_u (v \rfloor \alpha) = [u,v] \rfloor \alpha + v \rfloor (\pounds_u \alpha) ,$$

$$\pounds_{[u,v]} \alpha = (\pounds_u \pounds_v - \pounds_v \pounds_u) \alpha ,$$

$$\int_{S_{a+1}} d\alpha = \int_{\partial S_{a+1}} \alpha ,$$

where $[u,v]$ is the commutater or Lie product of the vectors u and v that is defined by $[u,v]f = u(vf) - v(uf)$ and where S_{a+1} is a region of dimension $a+1$ with boundary ∂S_{a+1} (remember a=degree of α).

Let $\Lambda(M)$ denote the graded algebra of exterior differential forms on a space M and let $\Lambda^k(M)$ denote the exterior forms of degree k. If $\Phi: N \to M$ is a differentiable mapping, we write $\Phi^*: \Lambda(M) \to \Lambda(N)$ for the induced mapping of the graded algebras of exterior forms. Thus, if $\Phi: N \to M$, then Φ^* maps any form of degree k on M onto a form of degree k on N; that is, it maps in the direction opposite to that of Φ. We note in particular that

$$\Phi^*(d\alpha) = d(\Phi^*\alpha)$$

$$\Phi^*[(\Phi_*v)\rfloor\alpha] = v\rfloor\Phi^*\alpha ,$$

where (Φ_*v) denotes the image of v on M that is induced by the map Φ from the vector field v on N. Φ^* is really quite simple, for if Φ is realized by $y^i = f^i(x^1,\dots,x^n)$ then $\Phi^*(\alpha_i(y^k)dy^i) = \alpha_i(f^k(x^1,\dots,x^n))\,\frac{\partial f^i}{\partial x^j}\,dx^j$; in other words, the chain rule.

1.1 THE SPACE OF INDEPENDENT VARIABLES

It is sufficient for our purposes to take the space of independent variables to be n-dimensional and flat. We therefore identify this space with n-dimensional number space, E_n , with a global Cartesian coordinate cover $(x^i) = (x^1, x^2, \dots, x^n)$. Any other coordinate cover can be introduced by regular coordinate mappings. By a domain set, B_n , we mean the closure of an open, simply connected, arcwise simply connected point set in E_n . The volume element of E_n is denoted by μ , that is

(1.1) $$\mu = dx^1 \wedge dx^2 \wedge \dots \wedge dx^n$$

is a basis for $\Lambda^n(E_n)$ in the coordinate cover (x^i) .

The basis for the tangent space, $T(E_n)$, of E_n is denoted by

(1.2) $$t_i = \partial/\partial x^i , \quad i = 1,\dots,n ,$$

in the coordinate cover (x^i) , so that any $v(x)\varepsilon T(E_n)$ may be written as $v(x) = v^i(x)\ t_i$, where the standard summation convention is adopted. It is then immediate that

(1.3) $$\mu_i = t_i\rfloor\mu , \quad i = 1,\dots,n ,$$

is a basis for $\Lambda^{n-1}(E_n)$ and exhibits the following properties (see Appendix, Section 3-5)

$$(1.4) \qquad d\mu_i = 0 , \quad dx^j \wedge \mu_i = \delta^j_i \, \mu .$$

Similarly,

$$(1.5) \qquad \mu_{ji} = t_j \rfloor \mu_i , \quad i,j = 1,\ldots,n ,$$

exhibit the properties

$$(1.6) \qquad d\mu_{ji} = 0 , \quad \mu_{ji} = -\mu_{ij} , \quad dx^k \wedge \mu_{ji} = \delta^k_j \mu_i - \delta^k_i \mu_j$$

and hence $\{\mu_{ij} | i < j\}$ is a basis for $\Lambda^{n-2}(E_n)$.

Since $\{\mu_i\}$ constitutes a basis for $\Lambda^{n-1}(E_n)$, any $\omega \in \Lambda^{n-1}(E_n)$ is uniquely expressible as

$$(1.7) \qquad \omega = W^i(x) \, \mu_i .$$

The properties given by (1.4) show that

$$(1.8) \qquad d\omega = (\partial_i W^i) \mu .$$

It is thus clear that use of the bases $\{\mu_i\}$, $\{\mu_{ij} | i < j\}$, provides a simple and direct means of modeling "divergence" type structures such as $\partial_i W^i$ without introducing a "metric".

1.2 HOMOTOPY OPERATORS AND ANTIEXACT FORMS

Let B be a star-like region (Appendix, Section 5-2) of an m-dimensional space M that is contained in a single coordinate patch of M and let this coordinate patch assign the coordinates $z^a = 0$, $a = 1,\ldots,m$ to the center of B . If $\omega(z) \in \Lambda^r(B)$, then define $\omega(\lambda z)$ to be the 1-parameter family of r-forms on B that obtains by evaluation of the coefficients of $\omega(z)$ at (λz^a) for $0<\lambda<1$, and define $\chi(z) \in T(B)$ by

$$(1.9) \qquad \chi(z) = z^a \, \partial/\partial z^a .$$

These quantities serve to define the linear homotopy operator H by (Appendix, Section 5-3)

(1.10) $$H\omega(z) = \int_0^1 \lambda^{r-1}\, \chi(z) \rfloor \omega(\lambda z)\, d\lambda \; .$$

The operator H clearly generates a mapping of $\Lambda^r(B)$ to $\Lambda^{r-1}(B)$ and is such that (Appendix, Section 5-3)

(1.11) $$Hd\omega + dH\omega = \omega \; , \qquad HH\omega \equiv 0$$

throughout B. Thus, each $\omega \varepsilon \Lambda(B)$ has a uniquely associated exact part, $\omega_e = dH\omega$, and a uniquely defined antiexact part, $\omega_a = Hd\omega$, that belongs to $\ker(H)$. In fact, H maps $\Lambda(B)$ onto the graded submodule $A(B)$ of *antiexact* forms (Appendix, Section 5-5). The collection of all antiexact forms of degree r is defined by

(1.12) $$A^r(B) = \{\omega \varepsilon \Lambda^r(B) \,|\, \chi(z) \rfloor \omega(z) = 0 \; , \;\; \omega(0) = 0 \;\; \text{for} \;\; r>0\} \; ,$$

and $(1.11)_1$ can be rewritten as the equivalent statement

(1.13) $$\omega = d\sigma + U \; ,$$

with

(1.14) $$\sigma = H\omega \,\varepsilon\, A^{r-1}(B) \; , \qquad U = Hd\omega \,\varepsilon\, A^r(B) \; .$$

A number of other properties of H and of $A(B)$ are given in the Appendix. We mention only one other that is of considerable practical importance (Appendix, Section 5-5): *the homotopy operator* H *inverts the exterior derivative operator* d *on the submodule* $A(B)$.

1.3 DIFFERENTIAL SYSTEMS OF DEGREE r AND ORDER k

The properties of antiexact forms and the homotopy operator H are very useful in studying systems of exterior differential equations. Since such systems deal with collections of differential forms, a significant simplification is achieved through use of matrix notation. Let $\Lambda^r_{p,q}(B)$ denote the collection of all p-by-q matrix valued differential forms on B of degree r. If $\underset{\sim}{\Omega} \in \Lambda^r_{a,b}(B)$, $\underset{\sim}{\Sigma} \in \Lambda^s_{b,c}(B)$, then $\underset{\sim}{\Omega} \wedge \underset{\sim}{\Sigma}$ will belong to $\Lambda^{r+s}_{a,c}(B)$ and is to be computed by matrix multiplication with the exterior product; that is,

$$(\underset{\sim}{\Omega} \wedge \underset{\sim}{\Sigma})^i_j = \sum_{\ell=1}^{b} \Omega^i_\ell \wedge \Sigma^\ell_j .$$

Although there can be no meaning attached to the corresponding inverse operation in matrix algebra, the transpose operation is well defined and useful. In particular, we have

$$(1.15) \qquad (\underset{\sim}{\Omega} \wedge \underset{\sim}{\Sigma})^T = (-1)^{rs} \underset{\sim}{\Sigma}^T \wedge \underset{\sim}{\Omega}^T .$$

A collection $\underset{\sim}{\Omega} \in \Lambda^r_{k,1}(B)$, $\underset{\sim}{\Sigma} \in \Lambda^{r+1}_{k,1}(B)$, $\underset{\sim}{\Gamma} \in \Lambda^1_{k,k}(B)$, $\Theta \in \Lambda^2_{k,k}(B)$ is said to form a *differential system* of *degree* r and *order* k on B if and only if (Appendix, Section 5-11)

$$(1.16) \qquad d\underset{\sim}{\Omega} = \underset{\sim}{\Gamma} \wedge \underset{\sim}{\Omega} + \underset{\sim}{\Sigma} , \qquad d\underset{\sim}{\Sigma} = \underset{\sim}{\Gamma} \wedge \underset{\sim}{\Sigma} - \underset{\sim}{\Theta} \wedge \underset{\sim}{\Omega}$$

$$(1.17) \qquad d\underset{\sim}{\Gamma} = \underset{\sim}{\Gamma} \wedge \underset{\sim}{\Gamma} + \underset{\sim}{\Theta} , \qquad d\underset{\sim}{\Theta} = \underset{\sim}{\Gamma} \wedge \underset{\sim}{\Theta} - \underset{\sim}{\Theta} \wedge \underset{\sim}{\Gamma}$$

hold throughout B. In view of the similarity between these equations and the structure equations of E. Cartan, we refer to $\underset{\sim}{\Gamma}$ as the connection 1-forms, $\underset{\sim}{\Theta}$ as the curvature 2-forms and $\underset{\sim}{\Sigma}$ as the torsion (r+1)-forms of the differential system.

General solutions of such systems are given in Section 5-16 of the Appendix for star-like regions B of a differentiable manifold, and prove to be most useful in the study of systems of partial differential equations given in this monograph. Let $A \in \Lambda^0_{k,k}(B)$ satisfy the matrix Riemann-Graves integral equation

$$(1.18) \qquad \underset{\sim}{A} = \underset{\sim}{A}_o + H(\underset{\sim}{\Gamma}\ \underset{\sim}{A})$$

with

(1.19) $\quad \det(\underset{\sim}{A}_o) \neq 0 , \quad \underset{\sim}{A}_o = $ constant-valued matrix.

The general solution of the differential system (1.16)-(1.17) is given by

(1.20) $$\underset{\sim}{\Omega} = \underset{\sim}{A}(d\underset{\sim}{\phi} + \underset{\sim}{\eta} + H(\underset{\sim}{\theta} \wedge d\underset{\sim}{\phi}))$$

(1.21) $$\underset{\sim}{\Sigma} = \underset{\sim}{A}(d\underset{\sim}{\eta} - \underset{\sim}{\theta} \wedge \underset{\sim}{\eta} - \underset{\sim}{\theta} \wedge H(\underset{\sim}{\theta} \wedge d\underset{\sim}{\phi}) - H(d\underset{\sim}{\theta} \wedge d\underset{\sim}{\phi}))$$

(1.22) $$\underset{\sim}{\Gamma} = (d\underset{\sim}{A} + \underset{\sim}{A}\,\underset{\sim}{\theta})\underset{\sim}{A}^{-1} , \quad \underset{\sim}{\Theta} = \underset{\sim}{A}(d\underset{\sim}{\theta} - \underset{\sim}{\theta} \wedge \underset{\sim}{\theta})\underset{\sim}{A}^{-1}$$

where $\underset{\sim}{\eta}$ and $\underset{\sim}{\theta}$ are matrix-valued antiexact forms and

(1.23) $$\underset{\sim}{\phi} = H(\underset{\sim}{A}^{-1}\underset{\sim}{\Omega}) + d\sigma , \quad \underset{\sim}{\eta} = H(\underset{\sim}{A}^{-1}\underset{\sim}{\Sigma}) , \quad \underset{\sim}{\theta} = H(\underset{\sim}{A}^{-1}\underset{\sim}{\Gamma}\underset{\sim}{A} - \underset{\sim}{A}^{-1}d\underset{\sim}{A}) .$$

Studies of systems of partial differential equations require not just column matrices of forms, $\underset{\sim}{\Omega}$, but row matrices of forms as well. If we take the transpose of the system (1.16)-(1.17) and set

(1.24) $$\hat{\underset{\sim}{\Gamma}} = -\underset{\sim}{\Gamma}^T , \quad \hat{\underset{\sim}{\Theta}} = -\underset{\sim}{\Theta}^T ,$$

we obtain

(1.25) $$d\underset{\sim}{\Omega}^T = (-1)^{r+1}\underset{\sim}{\Omega}^T \wedge \hat{\underset{\sim}{\Gamma}} + \underset{\sim}{\Sigma}^T , \quad d\underset{\sim}{\Sigma}^T = (-1)^{r+2}\underset{\sim}{\Sigma}^T \wedge \hat{\underset{\sim}{\Gamma}} + \underset{\sim}{\Omega}^T \wedge \hat{\underset{\sim}{\Theta}}$$

(1.26) $$d\hat{\underset{\sim}{\Gamma}} = \hat{\underset{\sim}{\Gamma}} \wedge \hat{\underset{\sim}{\Gamma}} + \hat{\underset{\sim}{\Theta}} , \quad d\hat{\underset{\sim}{\Theta}} = \hat{\underset{\sim}{\Gamma}} \wedge \hat{\underset{\sim}{\Theta}} - \hat{\underset{\sim}{\Theta}} \wedge \hat{\underset{\sim}{\Gamma}} .$$

The substitutions

(1.27) $$\underset{\sim}{B} = \underset{\sim}{A}^{T-1} , \quad \hat{\underset{\sim}{\theta}} = -\underset{\sim}{\theta}^T$$

then yield

(1.28) $$\underset{\sim}{B} = \underset{\sim}{B}_o + H(\hat{\underset{\sim}{\Gamma}}\,\underset{\sim}{B})$$

(1.29) $$\underset{\sim}{\Omega}^T = (d\underset{\sim}{\phi}^T + \underset{\sim}{\eta}^T + (-1)^{r+1}H(d\underset{\sim}{\phi}^T \wedge \underset{\sim}{\hat{\theta}}))\underset{\sim}{B}^{-1}$$

(1.30) $$\underset{\sim}{\Sigma}^T = (d\underset{\sim}{\eta}^T + (-1)^r\underset{\sim}{\eta}^T \wedge \underset{\sim}{\hat{\theta}} + H(d\underset{\sim}{\phi}^T \wedge \underset{\sim}{\hat{\theta}}) \wedge \underset{\sim}{\hat{\theta}} + H(d\underset{\sim}{\phi}^T \wedge d\underset{\sim}{\hat{\theta}}))\underset{\sim}{B}^{-1}$$

(1.31) $$\underset{\sim}{\hat{\Gamma}} = (d\underset{\sim}{B} + \underset{\sim}{B}\underset{\sim}{\hat{\theta}})\underset{\sim}{B}^{-1} , \quad \underset{\sim}{\hat{\Theta}} = \underset{\sim}{B}(d\underset{\sim}{\hat{\theta}} - \underset{\sim}{\hat{\theta}} \wedge \underset{\sim}{\hat{\theta}})\underset{\sim}{B}^{-1} ,$$

with

$$\underset{\sim}{\phi}^T = H(\underset{\sim}{\Omega}^T\underset{\sim}{B}) + d\underset{\sim}{\sigma}^T , \quad \underset{\sim}{\eta}^T = H(\underset{\sim}{\Sigma}^T\underset{\sim}{B}) .$$

1.4 CLOSED IDEALS

An idea essential to the study of partial differential equations is that of an ideal of $\Lambda(M)$. Amongst all such ideals, the closed ideals (Appendix, Section 4-2) are of particular interest. Let $A^1, A^2, \ldots$ be given elements of $\Lambda(M)$. The ideal generated by $A^1, A^2, \ldots$ will be denoted by $I\{A^1, A^2, \ldots\}$. Accordingly, any $\eta \in I\{A^1, A^2, \ldots\}$ is expressible as

(1.32) $$\eta = \alpha_1 \wedge A^1 + \alpha_2 \wedge A^2 + \ldots$$

for some $\alpha_1, \alpha_2, \ldots$ belonging to $\Lambda(M)$. The closed ideal generated by $A^1, A^2, \ldots$ is denoted by $\bar{I}\{A^1, A^2, \ldots\}$, so that

(1.33) $$\bar{I}\{A^1, A^2, \ldots\} = I\{A^1, dA^1, A^2, dA^2, \ldots\} .$$

The importance of closed ideals is that $\eta \in \bar{I}\{A^1, A^2, \ldots\}$ implies that $d\eta \in \bar{I}\{A^1, A^2, \ldots\}$. In this regard, it is customary to write $\eta \in \bar{I}\{A^1, A^2, \ldots\}$ as

(1.34) $$\eta = 0 \bmod \bar{I}\{A^1, A^2, \ldots\} ,$$

in which case

(1.35) $$d\eta = 0 \bmod \bar{I}\{A^1, A^2, \ldots\} .$$

Let us assume that M is starlike, in which case the

Poincaré lemma holds. The relations (1.34) and (1.35) clearly generalize in the following way:

$$(1.36) \qquad \eta = d\rho \bmod \bar{I}\{A^1, A^2, \ldots\}$$

implies

$$(1.37) \qquad d\eta = 0 \bmod \bar{I}\{A^1, A^2, \ldots\} .$$

On the other hand, the simple example $A^1 = dx^1 + x^2 dx^3$, $d\eta = 7\, A^1 \wedge dx^3 = d(7x^1 dx^3)$ yields $\eta = d\rho + 7x^1 dx^3$ and hence $d\eta = 0 \bmod \bar{I}\{A^1, A^2, \ldots\}$ does not imply $\eta = d\beta \bmod \bar{I}\{A^1, A^2, \ldots\}$. The question thus arises as to what additional conditions must be imposed in order to obtain the converse of (1.36), (1.37).

<u>LEMMA 1.1</u>. *If* $\bar{I}\{A^1, A^2, \ldots\}$ *is a closed ideal of* $\Lambda(M)$ *and* η *is a differential form of degree* r, *then* η *can satisfy*

$$(1.38) \qquad d\eta = 0 \bmod \bar{I}\{A^1, A^2, \ldots\}$$

if and only if there exists a $P \in I\{A^1, A^2, \ldots\}$ *of degree* r, *and a closed element* Q *of* $I\{A^1, A^2, \ldots\}$ *of degree* $r+1$ *such that*

$$(1.39) \qquad d(\eta + P) = Q .$$

<u>Proof</u>. If η satisfies (1.38), there exist forms M_α and N_α, with $\text{degree}(M_\alpha) = r - \text{degree}(A_\alpha) + 1 \overset{\text{def}}{=} a^\alpha$ and $\text{degree}(N_\alpha) = r - \text{degree}(A^\alpha)$, such that

$$(1.40) \qquad d\eta = M_\alpha \wedge A^\alpha + N_\alpha \wedge dA^\alpha .$$

Since the left-hand side of (1.40) is a closed form, the right-hand side must likewise be a closed form, and we obtain the requirement

$$(1.41) \qquad 0 = dM_\alpha \wedge A^\alpha + (dN_\alpha + (-1)^{a^\alpha} M_\alpha) \wedge dA^\alpha .$$

Now, define the forms U_α by

$$(1.42) \qquad (-1)^{a^\alpha} U_\alpha = dN_\alpha + (-1)^{a^\alpha} M_\alpha ,$$

so that $\text{degree}(U_\alpha) = \text{degree}(M_\alpha) = a^\alpha$. When (1.42) is solved for M_α and the results are substituted back into (1.40) and (1.41), we obtain

$$(1.43) \qquad d\eta = U_\alpha \wedge A^\alpha + (-1)^{a^\alpha - 1} d(N_\alpha \wedge A^\alpha)$$

and

$$0 = d(U_\alpha \wedge A^\alpha) .$$

The result then follows with the identifications $P = (-1)^{a^\alpha} N_\alpha \wedge A^\alpha \varepsilon I\{A^1, A^2, \ldots\}$ and $Q = U_\alpha \wedge A^\alpha \varepsilon I\{A^1, A^2, \ldots\}$.

THEOREM 1.1. *If* M *is star-like and* $\bar{I}\{A^1, A^2, \ldots\}$ *is a closed ideal of* $\Lambda(M)$, *then*

$$(1.44) \qquad d\eta = 0 \bmod \bar{I}\{A^1, A^2, \ldots\}$$

implies

$$(1.45) \qquad \eta = d\beta \bmod \bar{I}\{A^1, A^2, \ldots\}$$

if and only if there exists at least one form $\rho \,\varepsilon\, \bar{I}\{A^1, A^2, \ldots\}$ *for each closed* $Q \,\varepsilon\, I\{A^1, A^2, \ldots\}$ *such that*

$$(1.46) \qquad Q = d(\rho + d\phi) .$$

Proof. A direct application of Lemma 1.1 reduces (1.44) to the equivalent problem

$$d(\eta + P) = Q , \quad dQ = 0$$

with P and Q belonging to $I\{A^1, A^2, \ldots\}$, for otherwise there could exist no η that satisfies (1.44). Since Q is closed and M is star-like, the Poincaré lemma yields the existence of forms ρ and ϕ such that $Q = d(\rho + d\phi)$. In this event, the condition becomes $d(\eta + P - \rho - d\phi) = 0$, and hence we obtain

$$(1.47) \qquad \eta = d(\alpha + \phi) + \rho - P .$$

Now, set $\alpha + \phi = \beta$. The relation (1.47) then gives

$$(1.48) \qquad \eta = d\beta + \rho \bmod \bar{I}\{A^1, A^2, \ldots\}$$

because $P \in I\{A^1, A^2, \ldots\}$. It thus follows that $\eta = d\beta \bmod \bar{I}\{A^1, A^2, \ldots\}$ if and only if $\rho \in \bar{I}\{A^1, A^2, \ldots\}$.

The results of this theorem show that it is useful to know the form of all possible $\rho \in \bar{I}\{A^1, A^2, \ldots\}$ such that $d\rho \in I\{A^1, A^2, \ldots\}$. The following Lemma gives an answer to this question for a class of generators of the ideals that will feature prominently in the remainder of this monograph.

LEMMA 1.2. *If* $A^1, A^2, \ldots, A^N$ *are of degree one and satisfy*

$$(1.49) \qquad A^1 \wedge A^2 \wedge \ldots \wedge A^N \overset{\text{def}}{=} A \neq 0 ,$$

$$(1.50) \qquad dA^\alpha \wedge A \neq 0 , \quad \alpha = 1, \ldots, N ,$$

then any $\rho \in \bar{I}\{A^1, A^2, \ldots\}$ *of degree* r *such that* $d\rho \in I\{A^1, A^2, \ldots\}$ *has the form*

$$(1.51) \qquad \rho = d(N_\alpha \wedge A^\alpha) + (R_{\alpha\beta} \wedge A^\beta + S_{\alpha\beta} \wedge dA^\beta) \wedge A^\alpha ,$$

with

$$(1.52) \qquad R_{\alpha\beta} = - R_{\beta\alpha} , \quad S_{\alpha\beta} = - S_{\beta\alpha} .$$

Under these conditions

$$(1.53) \qquad d\rho = U_\alpha \wedge A^\alpha$$

with

$$(1.54) \qquad U_\alpha = dR_{\alpha\beta} \wedge A^\beta + (dS_{\alpha\beta} + 2(-1)^r R_{\alpha\beta}) \wedge dA^\beta .$$

Proof. If $\rho \in \bar{I}\{A^1, A^2, \ldots\}$, then

$$\rho = P_\alpha \wedge A^\alpha + T_\alpha \wedge dA^\alpha$$

and a rearrangement of terms yields

$$(1.55) \qquad \rho = d(N_\alpha \wedge A^\alpha) + M_\alpha \wedge A^\alpha .$$

Taking the exterior derivative of (1.55) thus gives

$$(1.56) \qquad d\rho = dM_\alpha \wedge A^\alpha + (-1)^r M_\alpha \wedge dA^\alpha .$$

Since $A \neq 0$, A spans the domain of $I\{A^1, A^2, \ldots\}$ and hence $d\rho \in I\{A^1, A^2, \ldots\}$ if and only if

$$(1.57) \qquad 0 = d\rho \wedge A = (-1)^r M_\alpha \wedge dA^\alpha \wedge A .$$

By hypothesis, $dA^\alpha \wedge A \neq 0$ for all α, and hence (1.57) can be satisfied if and only if

$$(1.58) \qquad M_\alpha = R_{\alpha\beta} \wedge A^\beta + S_{\alpha\beta} \wedge dA^\beta , \qquad S_{\alpha\beta} = - S_{\beta\alpha} .$$

When (1.58) is substituted into (1.55), we obtain (1.51), from which it is clear that we may require $R_{\alpha\beta} = - R_{\beta\alpha}$ for the symmetric part of $R_{\alpha\beta}$ makes no contribution to $R_{\alpha\beta} \wedge A^\alpha \wedge A^\beta$. Exterior differentiation of (1.51) and use of the relations (1.55) then gives

$$\begin{aligned} d\rho &= dR_{\alpha\beta} \wedge A^\beta \wedge A^\alpha + (-1)^{r-2} R_{\alpha\beta} \wedge dA^\beta \wedge A^\alpha \\ &\quad + (-1)^{r-1} R_{\alpha\beta} \wedge A^\beta \wedge dA^\alpha + dS_{\alpha\beta} \wedge dA^\beta \wedge A^\alpha \\ &\quad + (-1)^{r-3} S_{\alpha\beta} \wedge dA^\beta \wedge dA^\alpha \\ &= \{dR_{\alpha\beta} \wedge A^\beta + (dS_{\alpha\beta} + 2(-1)^r R_{\alpha\beta}) \wedge dA^\beta\} \wedge A^\alpha , \end{aligned}$$

so that (1.53) and (1.54) are established.

CHAPTER 2

CONTACT STRUCTURES

The principal topic of this monograph is the structure that can be associated with the solution sets of systems of nonlinear, second order partial differential equations with n independent variables. Since it is the structure of the solution set, rather than the solutions themselves that is of interest, realization of the solution set in terms of geometric constructs proves to be most useful. These geometric constructs should, however, represent the intrinsic structure of the solution set of the system of partial differential equations rather than an induced geometric structure that arises from the intrinsic geometry of the space of independent variables. It is therefore sufficient for our purposes to take the space of independent variables to be an n-dimensional Euclidean space E_n with a fixed Cartesian coordinate cover (x^i), $i = 1,\ldots,n$. Other coordinate covers and other manifold structures of the independent variable space can be introduced subsequently by standard techniques of differential geometry.

2.1 GRAPH SPACE

The simplest geometric structure that can be associated with the solution set of a system of partial differential equations is that of the graph of the solution set. We therefore introduce a *graph space* $G = E_{n+N}$ of $n+N$ dimensions with a global Cartesian coordinate cover (x^i, q^α), $i = 1,\ldots,n$, $\alpha = 1,\ldots,N$, and realize the solution set in terms of mappings from E_n into G.

Let Φ denote a map from an n-dimensional point set B_n of E_n into G. Such a map is said to be *regular* if

$$(2.1) \qquad \Phi^*\mu = \Phi^*(dx^1 \wedge dx^2 \wedge \ldots \wedge dx^n) \neq 0$$

throughout B_n. The collection of all regular maps of B_n into G is denoted by $R(B_n)$, so that

$$(2.2) \qquad R(B_n) = \{\Phi: B_n \to G \mid \Phi^*\mu \neq 0\} .$$

Regular maps Φ are realized by

$$(2.3) \qquad \Phi: x^i = \phi^i(\tau^j) , \quad q^\alpha = \phi^\alpha(\tau^j)$$

where (τ^j) denote the coordinates of points in B_n relative to a fixed coordinate cover of E_n. Since Φ is regular, $\Phi^*\mu = (\partial(x)/\partial(\tau))\, d\tau^1 \wedge d\tau^2 \wedge \ldots \wedge d\tau^n \neq 0$, we may solve for the parameters τ^j in terms of the x's, at least locally, so as to obtain $\tau^j = m^j(x^i)$. Composition of this with the second of (2.3) yields the relations

$$(2.4) \qquad q^\alpha = \phi^\alpha(m^j(x^k)) = \Phi^\alpha(x^k) .$$

Thus, every regular map Φ of B_n to G yields the graph $(x^i = \phi^i(\tau^j) , \; q^\alpha = \phi^\alpha(\tau^j))$ that constitutes an n-dimensional differentiable manifold contained in G. Of course, if we are given the solution set $q^\alpha = \Phi^\alpha(x^j)$, then the choice $\phi^i(\tau^j) = \tau^i$, $q^\alpha = \Phi^\alpha(\tau^j)$ gives the graph of such a solution directly as a map from B_n to G. From this point of view, the parameters (τ^j) are superfluous. They are introduced, however, in order to keep the distinction between E_n and G quite clear. Otherwise, there can be a difficulty in that G has the structure $G = E_{n+N} = E_n \times E_N$, and hence regular maps could become confused with mappings from the first factor E_n of G onto the second factor E_N.

If $f = f(x^i, q^\alpha)$ is a function defined on G, then composition of f with a map $\Phi: B_n \to G \mid x^i = \tau^i$, $q^\alpha = \phi^\alpha(\tau^j)$ yields $F(x^j) = f \circ \Phi = f(x^j, \phi^\alpha(x^j))$. The total x^i-derivative of $F = f \circ \Phi$ is denoted by $D_i F$ and we have

$$D_i F = \partial_i f(x^j, \phi^\alpha(x^j)) + \frac{\partial f(x^j, \phi^\alpha(x^j))}{\partial \phi^\beta} \partial_i \phi^\beta$$

The operator ∂_i is thus used throughout these discussions to denote *explicit* x^i-wise differentiation; i.e., differentiation of functions of several variables with all variables but the explicit occurrence of the x's held fixed.

2.2 CONTACT SPACE AND CONTACT FORMS

The representation of systems of second order partial differential equations by means of systems of first order partial differential equations demands that first partial derivatives of the dependent variables appear as a new system of dependent variables. This is most easily done by imbedding the graph space G in a larger space. Let $K = G \times E_{nN} = E_{n+N+nN}$ be an $(n+N+nN)$-dimensional Euclidean space with the global coordinate cover $(x^i, q^\alpha, y_i^\alpha)$, $i = 1,\ldots,n$, $\alpha = 1,\ldots,N$. Since K has the product structure $G \times E_{nN}$, K can be viewed as a trivial fiber space with projection

$$(2.5) \qquad \pi: (x^i, q^\alpha, y_i^\alpha) \longrightarrow (x^i, q^\alpha)$$

and fibers E_{nN}. The new variables (y_i^α) of K allow us to introduce N nontrivial 1-forms

$$(2.6) \qquad C^\alpha = dq^\alpha - y_i^\alpha \, dx^i , \quad \alpha = 1,\ldots,N ,$$

which are referred to as the *contact forms* of K . Since

$$(2.7) \qquad C^1 \wedge C^2 \wedge \ldots \wedge C^N = dq^1 \wedge dq^2 \wedge \ldots \wedge dq^N + \ldots \neq 0 ,$$

the N contact forms of K are independent. Further, since

$$(2.8) \qquad dC^\alpha = - dy_i^\alpha \wedge dx^i \neq 0 , \quad C^1 \wedge \ldots \wedge C^N \wedge dC^\alpha \neq 0 ,$$

it follows that $C^\alpha \wedge (dC^\alpha)^{(n)} \neq 0$, $(dC^\alpha)^{(n+1)} = 0$ where $(\ )^{(k)}$ denotes the kth exterior power. Thus, each of the N contact forms has Darboux class = 2n + 1 (Appendix, Section 4-4).

Up to this point, the coordinates (y_i^α) in the space E_{nN} of fibers have been quite arbitrary. We now give them a precise identification. Let $\Phi : B_n \to G$, $x^i = \phi^i(\tau^j)$, $q^\alpha = \phi^\alpha(\tau^j)$ be regular. We extend Φ to a map of $B_n \to K$ by the requirements

$$(2.9) \qquad \Phi^* C^\alpha = 0 , \quad \alpha = 1,\ldots,N ;$$

that is, *the section of* K *that is generated by any regular section of* G *annihilates the contact forms of* K . A substitution of (2.6) into (2.9) yields

$$(2.10) \qquad 0 = \Phi^*(dq^\alpha - y_i^\alpha\, dx^i) = d\Phi^* q^\alpha - \Phi^*(y_i^\alpha)\, d\Phi^* x^i$$

$$= d\phi^\alpha(\tau^j) - \Phi^*(y_i^\alpha)\, d\phi^i(\tau^j) .$$

Since $\Phi : B_n \to G$ is regular, the lift of Φ to K yields

$$(2.11) \qquad \Phi^*(y_i^\alpha) = \partial\phi^\alpha(\tau^j)/\partial\phi^i(\tau^j) = \partial\Phi^\alpha(x^k)/\partial x^i$$

when (2.4) is used (i.e., $q^\alpha = \phi^\alpha(m^j(\tau^k)) = \Phi^\alpha(x^k)$). Thus, although the coordinates (y_i^α) of K are quite arbitrary and independent of the other coordinates (x^i, q^α) , the values of the y's on any section Φ of K such that $\Phi^*\mu \neq 0$, $\Phi^* C^\alpha = 0$ are given by $y_i^\alpha = \partial\Phi^\alpha(x^k)/\partial x^i$. This, however, is exactly what is required in order that the y's shall serve as place holders that assign the first derivatives of the dependent variables on any regular section of G. We therefore give the following specific delineation of regular maps from B_n to K.

Definition 2.1. A map $\Phi : B_n \longrightarrow K$ is *regular* if and only if

$$(2.12) \qquad \Phi^* \mu \neq 0 ,$$

$$(2.13) \qquad \Phi^* C^\alpha = 0 , \quad \alpha = 1, \ldots, N .$$

The collection of all regular maps of B_n to K is denoted by $R(B_n)$.

The following view thus emerges. The space K is an arbitrary $(n+N+nN)$-dimensional Euclidean space with coordinate cover $(x^i, q^\alpha, y_i^\alpha)$, so that the x's, the q's and the y's can take on arbitrary values and can be incremented independently of one another. The restriction of maps from B_n to K to be regular maps then defines all possible sections of K for which the y's become the derivatives of the q's with respect to the x's. Thus, if we use π to project any regular section of K onto a section of G, then the image of this section of G under π^{-1} will intersect the original section of K in such a fashion that the y's become the derivatives of the q's with respect to the x's.

2.3 THE CONTACT IDEAL AND ITS ISOVECTORS

Since any regular map $\Phi : B_n \longrightarrow K$ annihilates all N contact 1-forms of K, it will annihilate the ideal of $\Lambda(K)$ that is generated by the contact forms. Further, from $dC^\alpha = -dy_i^\alpha \wedge dx^i$, and $\Phi^*(y_i^\alpha) = \partial_i \Phi^\alpha(x^j)$, we have

$$(2.14) \qquad \Phi^* dC^\alpha = -d(\partial_i \Phi^\alpha(x^k)) \wedge dx^i = -(\partial_j \partial_i \Phi^\alpha(x^k)) dx^j \wedge dx^i = 0 .$$

Thus, any regular map $\Phi : B_n \longrightarrow K$ annihilates the closed ideal of $\Lambda(K)$ that is generated by the contact forms. If we denote the closed ideal generated by the contact forms by $\bar{C}$, so that

(2.15) $$\bar{C} = I\{C^1,\ldots,C^N,\ dC^1,\ldots,dC^N\}\ ,$$

we have the following result.

LEMMA 2.1. *A map* $\Phi : B_n \to K$ *is regular if and only if*

$$\Phi^*\mu \neq 0\ ,\qquad \Phi^*\bar{C} = 0\ .$$

This lemma allows us to replace the conditions $\Phi^* C^\alpha = 0$ by the annihilation of the closed ideal $\bar{C}$ of $\Lambda(K)$, which is referred to as the *contact ideal*. We are thus led in a natural manner to the study of the closed ideal $\bar{C}$, and its associated collection $R(B_n)$ of *regular maps*.

Let H be the linear homotopy operator of K that is defined by (1.9) and (1.10). It follows directly from (2.6) that

$$HC^\alpha = q^\alpha - \frac{1}{2} y_i^\alpha x^i\ .$$

Thus, we have

$$C_\alpha = C_e^\alpha + C_a^\alpha$$

where

(2.16) $$C_e^\alpha = d(q^\alpha - \frac{1}{2} y_i^\alpha x^i)\ ,\qquad C_a^\alpha = \frac{1}{2}(x^i\, dy_i^\alpha - y_i^\alpha\, dx^i)$$

are the exact and antiexact parts of C^α, respectively. Thus, in particular, H does not map $\bar{C}$ either into or onto $\bar{C}$. We shall find this to be of particular importance in the next section.

The most useful structure associated with the contact ideal, $\bar{C}$, is its behavior under deformations. Since deformations are defined in terms of changes that result from transport along the orbits of vector fields, it is useful to introduce the following operators

(2.17) $$\partial_i \equiv \frac{\partial}{\partial x^i}\ ,\qquad \partial_\alpha \equiv \frac{\partial}{\partial q^\alpha}\ ,\qquad \partial_\alpha^i \equiv \frac{\partial}{\partial y_i^\alpha}\ .$$

Thus, if v is any vector field on K, v can be uniquely expressed by

$$(2.18) \qquad v = v^i \partial_i + v^\alpha \partial_\alpha + v_i^\alpha \partial_\alpha^i$$

where the (n+N+nN)-tuple $(v^i, v^\alpha, v_i^\alpha)$ are functions defined on K. The collection of all vector fields on K is denoted by $T(K)$. We remind the reader that $T(K)$ is a Lie algebra under the Lie product operation $[v_1,v_2]f = v_1(v_2f) - v_2(v_1f)$ (Appendix, Section 2-6). The fundamental notion here is that of an isovector of $\bar{C}$, a concept that appears to have originated with Estabrook and Harrison [8].

The Lie Subalgebra of Isovectors of $\bar{C}$

Definition 2.2. A $v \in T(K)$ is an *isovector* of $\bar{C}$ if and only if

$$(2.19) \qquad \pounds_v \eta = 0 \bmod \bar{C}$$

for all $\eta \in \bar{C}$; that is, $\pounds_v \bar{C} \subset \bar{C}$.

Thus, $\bar{C}$ is a closed subset of $\Lambda(K)$ under Lie transport by any of its isovectors. We use $TC(K)$ to denote the collection of all isovectors of $\bar{C}$, so that

$$(2.20) \qquad TC(K) = \{v \in T(K) \mid \pounds_v \bar{C} \subset \bar{C}\} .$$

The underlying utility of isovectors of $\bar{C}$ comes about as follows. If $v \in TC(K)$ then $\pounds_v \bar{C} \subset \bar{C}$ and hence any $\Phi \in R(B_n)$ yields

$$\Phi^*(\pounds_v \bar{C}) = 0$$

because any such Φ annihilates $\bar{C}$. Thus, in particular

$$\Phi^*(\exp(s\pounds_v)\bar{C}) = 0 ,$$

which will prove to be fundamental in what follows.

LEMMA 2.2. *A* $v \in T(K)$ *is an isovector of* $\bar{C}$ *if and only if* v *is such that*

$$(2.21) \qquad £_v C^\alpha = A^\alpha_\beta \, C^\beta , \qquad \alpha = 1,\dots,N$$

holds throughout K *for some* N^2*-tuple of functions* $(A^\alpha_\beta ; \alpha,\beta = 1,\dots,N)$.

Proof. Since $\bar{C}$ is a closed ideal of $\Lambda(K)$ that is generated by C^α , $\alpha = 1,\dots,N$, it is necessary that (2.21) hold throughout K for some N^2-tuple of functions (A^α_β) defined on K . If (2.21) holds, then $£_v dC^\alpha = d£_v C^\alpha = d(A^\alpha_\beta \, C^\beta) = dA^\alpha_\beta \wedge C^\beta + A^\alpha_\beta \, dC^\beta$ $= 0 \bmod \bar{C}$ because $\bar{C}$ is the closed ideal generated by C^α , $\alpha = 1,\dots,N$. If η is any element of $\bar{C}$, then $\eta = B_\alpha \wedge C^\alpha + b_\alpha \wedge dC^\alpha$ and $£_v \eta = (£_v B_\alpha) \wedge C^\alpha + B_\alpha \wedge £_v C^\alpha + (£_v b_\alpha) \wedge dC^\alpha + b_\alpha \wedge £_v dC^\alpha$. Satisfaction of (2.21) thus gives $£_v \eta = 0 \bmod \bar{C}$, and hence (2.21) is both necessary and sufficient for $v \in TC(K)$.

THEOREM 2.1. $TC(K)$ *is a Lie subalgebra of* $T(K)$.

Proof. Let u and v belong to $TC(K)$. If b and c are real numbers, $b\,u + c\,v = w \in TC(K)$ because $£_w = b\,£_u + c\,£_v$. Thus, $TC(K)$ is a vector space over the real number field. If we define W by $W = [u,v]$, then $£_W C^\alpha = (£_u £_v - £_v £_u) C^\alpha = 0 \bmod \bar{C}$, and hence $W = [u,v] \in TC(K)$. Thus, $TC(K)$ is closed under the vector space operations and under the Lie product operation [,] of $T(K)$, so that $TC(K)$ is a Lie subalgebra of $T(K)$.

Explicit Characterization of TC(K)

We now turn to the problem of obtaining explicit characterization of the isovectors of $\bar{C}$. Lemma 2.2 shows that $v \in TC(K)$ if and only if $£_v C^\alpha = A^\alpha_\beta \, C^\beta$ holds for some collection of functions (A^α_β) . Since

$$v = v^i \, \partial_i + v^\alpha \, \partial_\alpha + v^\alpha_i \, \partial^i_\alpha ,$$

use of (2.6) to calculate $£_v C^\alpha$ leads to the conditions

(2.22) $$dv^\alpha - y^\alpha_j\, dv^j - v^\alpha_i\, dx^i = A^\alpha_\beta (dq^\beta - y^\beta_i\, dx^i) .$$

Thus, since v^i and v^α are functions of the arguments $(x^i, q^\beta, y^\beta_i)$, an expansion of dv^α and dv^j in terms of the basis of elements dx^i, dq^β, dy^β_i shows that (2.22) can be satisfied if and only if

(2.23) $$\partial_i v^\alpha - y^\alpha_j\, \partial_i v^j - v^\alpha_i = - A^\alpha_\beta\, y^\beta_i ,$$

(2.24) $$\partial_\beta v^\alpha - y^\alpha_j\, \partial_\beta v^j = A^\alpha_\beta ,$$

(2.25) $$\partial^k_\beta v^\alpha - y^\alpha_j\, \partial^k_\beta v^j = 0 .$$

If we note that

(2.26) $$v \rfloor C^\alpha = v^\alpha - y^\alpha_i\, v^i ,$$

then (2.24) yields

(2.27) $$A^\alpha_\beta = \partial_\beta (v \rfloor C^\alpha) .$$

Now that $((A^\alpha_\beta))$ is known, we may solve (2.23) for v^α_i. This yields

(2.28) $$v^\alpha_i = Z_i (v \rfloor C^\alpha) ,$$

where Z_i is the linear differential operator

(2.29) $$Z_i \equiv \partial_i + y^\beta_i\, \partial_\beta .$$

Note that $Z_i \rfloor C^\alpha = 0$, so that $\{Z_i\}$ are characteristic vectors of the ideal $\{C^1, \ldots, C^N\}$. The only conditions that remain to be satisfied are those given by (2.25). Let us write these conditions in the trivially equivalent form

(2.30) $$\partial^k_\beta v^\alpha = y^\alpha_j\, \partial^k_\beta v^j .$$

The integrability conditions of these equations, namely

$$(2.31)\qquad (\partial^m_\gamma \partial^k_\beta - \partial^k_\beta \partial^m_\gamma) v^\alpha = 0 ,$$

yield

$$(2.32)\qquad \delta^\alpha_\gamma \, \partial^k_\beta v^m = \delta^\alpha_\beta \, \partial^m_\gamma v^k .$$

Since these equations must hold for all $(\alpha,\beta,\gamma,k,m)$, setting $\alpha=\gamma$ and summing shows that $N \, \partial^k_\beta v^m = \partial^m_\beta v^k$, and hence (2.32) become

$$(2.33)\qquad N^{-1} \, \delta^\alpha_\gamma \, \partial^m_\beta v^k = \delta^\alpha_\beta \, \partial^m_\gamma v^k .$$

If we now set $\alpha=\beta$ and sum again, we obtain the condition

$$(2.34)\qquad (1 - N^2) \partial^m_\gamma v^k = 0 .$$

There are thus two cases according to whether $N = 1$ or $N > 1$. For the case $N > 1$, we obtain $\partial^m_\gamma v^k = 0$, in which case (2.30) yields $\partial^k_\beta v^\alpha = 0$, and hence $v^i = f^i(x^j, q^\beta)$, $v^\alpha = f^\alpha(x^j, q^\beta)$. For the case $N = 1$, (2.32) reduces to

$$(2.35)\qquad \partial^k_1 v^m = \partial^m_1 v^k ,$$

from which we conclude that there exists a function $\xi(x^j, q^\beta, y^\beta_j)$ such that

$$(2.36)\qquad v^m = \partial^m_1 \xi .$$

A substitution of this result back into (2.30) gives

$$\partial^k_1 v^1 = y^1_j \, \partial^k_1 \partial^j_1 \xi = \partial^k_1 (y^1_j \, \partial^j_1 \xi - \xi) ,$$

and hence an integration yields

$$(2.37)\qquad v^1 = f(x^i, q^1) + (y^1_j \, \partial^j_1 - 1)\xi(x^i, q^1 y^1_i) .$$

THEOREM 2.2. *A vector field*

$$(2.38) \qquad v = v^i \partial_i + v^\alpha \partial_\alpha + v^\alpha_i \partial^i_\alpha$$

belongs to TC(K) *if and only if*

$$(2.39) \qquad v^\alpha_i = Z_i(v \rfloor C^\alpha) = Z_i(v^\alpha - y^\alpha_j v^j) ,$$

$$(2.40) \qquad \pounds_v C^\alpha = \partial_\beta(v \rfloor C^\alpha) C^\beta ,$$

where Z_i *is the linear differential operator*

$$Z_i \equiv \partial_i + y^\beta_i \partial_\beta ,$$

and, for $N > 1$,

$$(2.41) \qquad v^i = f^i(x^j, q^\beta) , \qquad v^\alpha = f^\alpha(x^j, q^\beta) ,$$

while, for $N = 1$,

$$(2.42) \qquad v^i = \partial^i_1 \xi(x^j, q^1, y^1_j) ,$$

$$(2.43) \qquad v^1 = f(x^j, q^1) + (y^1_i \partial^i_1 - 1)\xi(x^j, q^1, y^1_j) ,$$

in which case

$$(2.44) \qquad v \rfloor C^1 = f(x^j, q^1) - \xi(x^j, q^1, y^1_j) .$$

A short recapitulation would seem useful at this point. We have established an (n+N+nN)-dimensional Euclidean space K with coordinates $(x^i, q^\alpha, y^\alpha_i)$. This space has the property that the regular maps from B_n to K yield sections of K on which the y's are the x-wise derivatives of the q's. This comes about from the fact that a regular map must satisfy the conditions $\Phi^*\mu \neq 0$, $\Phi^*C^\alpha = 0$, and these latter requirements are equivalent to the requirement that Φ^* annihilate the closed contact ideal $\bar{C}$ of $\Lambda(K)$. Lastly, we have obtained a complete characterization of the isovectors of the closed ideal $\bar{C}$. Since these constructions form the basis for our subsequent considerations, we label them a *contact structure*. A contact structure is thus a quadruplicate

$$(2.45) \qquad K = \{K, R(B_n), \bar{C}, TC(K)\} ,$$

and consists of a space K , the collection of regular maps $R(B_n)$ from B_n to K , the closed ideal $\bar{C}$ of $\Lambda(K)$ that is generated by the contact forms C^α , and the collection $TC(K)$ of all isovectors of the closed ideal $\bar{C}$.

2.4 ORBITS, CHARACTERISTICS, AND EXTENSION STRUCTURES

The simplest way of obtaining a full understanding of the properties of the isovectors of a contact structure K is to study the orbits of these vector fields. Let v be a fixed element of $TC(K)$. For $N > 1$, the characterization given by Theorem 2.2 yields the orbital equations

$$(2.46) \qquad d\overset{*}{x}{}^i/ds = v^i = f^i(\overset{*}{x}{}^j,\overset{*}{q}{}^\beta) , \quad d\overset{*}{q}{}^\alpha/ds = v^\alpha = f^\alpha(\overset{*}{x}{}^j,\overset{*}{q}{}^\beta)$$

$$(2.47) \qquad d\overset{*}{y}{}^\alpha_i/ds = \overset{*}{Z}_i(f^\alpha - \overset{*}{y}{}^\alpha_j f^j) , \quad \overset{*}{Z}_i \equiv \overset{*}{\partial}_i + \overset{*}{y}{}^\beta_i \overset{*}{\partial}_\beta , \quad \overset{*}{\partial} \equiv \partial/\partial(\overset{*}{\ }) ,$$

with orbital parameter s , and the solutions of these equations may be required to satisfy initial data

$$(2.48) \qquad \overset{*}{x}{}^i(0) = x^i , \quad \overset{*}{q}{}^\alpha(0) = q^\alpha , \quad \overset{*}{y}{}^\alpha_i(0) = y^\alpha_i .$$

As is well known (Appendix, Sections 1-5, 2-4), the general solution of such a system can be written as

$$(2.49) \qquad \begin{aligned} \overset{*}{x}{}^i &= H^i(s;x^j,q^\beta,y^\beta_j) = \exp(s\pounds_v)x^i , \\ \overset{*}{q}{}^\alpha &= H^\alpha(s;x^j,q^\beta,y^\beta_j) = \exp(s\pounds_v)q^\alpha , \\ \overset{*}{y}{}^\alpha_i &= H^\alpha_i(s;x^j,q^\beta,y^\beta_j) = \exp(s\pounds_v)y^\alpha_i . \end{aligned}$$

These relations describe a 1-parameter (s) family (1-parameter group) of diffeomorphisms of K into itself; that is, a 1-parameter group of point transformations acting on K ,

$$(2.50)\qquad \exp(s,v) : (x^i, q^\alpha; y_i^\alpha) \longrightarrow (\overset{*}{x}{}^i, \overset{*}{q}{}^\alpha, \overset{*}{y}{}_i^\alpha) .$$

The Associated First Order System

If we start with the N functions

$$(2.51)\qquad \overset{*}{F}{}^\alpha = -f^\alpha(\overset{*}{x},\overset{*}{q}) + \overset{*}{y}{}_i^\alpha f^i(\overset{*}{x},\overset{*}{q}) = -\overset{*}{v}\rfloor \overset{*}{C}{}^\alpha ,$$

where the superimposed * denotes evaluation at $(\overset{*}{x}{}^i, \overset{*}{q}{}^\alpha, \overset{*}{y}{}_i^\alpha)$, then

$$\partial\overset{*}{F}{}^1/\partial\overset{*}{y}{}_i^1 = \partial\overset{*}{F}{}^2/\partial\overset{*}{y}{}_i^2 = \dots = \partial\overset{*}{F}{}^N/\partial\overset{*}{y}{}_i^N = f^i$$

$$(2.52)\qquad \overset{*}{y}{}_i^\beta\, \partial\overset{*}{F}{}^\alpha/\partial\overset{*}{y}{}_i^\beta = \overset{*}{y}{}_i^\alpha f^i = \overset{*}{F}{}^\alpha + f^\alpha$$

$$-\overset{*}{Z}_i\overset{*}{F}{}^\alpha = \overset{*}{Z}_i(f^\alpha - \overset{*}{y}{}_j^\alpha f^j)$$

and the orbital equations (2.46), (2.47) become

$$(2.53)\qquad d\overset{*}{x}{}^i/ds = f^i = \partial\overset{*}{F}{}^1/\partial\overset{*}{y}{}_i^1 = \dots = \partial\overset{*}{F}{}^N/\partial\overset{*}{y}{}_i^N ,$$

$$(2.54)\qquad d\overset{*}{q}{}^\alpha/ds = f^\alpha = \overset{*}{y}{}_i^\beta\, \partial\overset{*}{F}{}^\alpha/\partial\overset{*}{y}{}_i^\beta - \overset{*}{F}{}^\alpha ,$$

$$(2.55)\qquad d\overset{*}{y}{}_i^\alpha/ds = \overset{*}{Z}_i(f^\alpha - \overset{*}{y}{}_j^\alpha f^j) = -\overset{*}{Z}_i\overset{*}{F}{}^\alpha .$$

These are, however, nothing more than the equations for the characteristics [12] of the system of *quasilinear* first order partial differential equations

$$(2.56)\qquad 0 = \overset{*}{F}{}^\alpha = -\overset{*}{v}{}^\alpha\rfloor \overset{*}{C}{}^\alpha , \quad \overset{*}{y}{}_i^\alpha = \partial\overset{*}{q}{}^\alpha(\overset{*}{x}{}^j)/\partial\overset{*}{x}{}^i$$

for the determination of the N functions $\overset{*}{q}{}^\alpha(\overset{*}{x}{}^j)$. Thus, if the initial data (2.49) (s=0) is assigned in such a fashion that

(2.57) $$F^\alpha = - v \rfloor C^\alpha = 0 \, , \quad y^\alpha_i = \partial q^\alpha(x^j)/\partial x^i \, ,$$

then the solutions of the orbital equations of $v \varepsilon TC(K)$ will be solutions of the system of quasilinear first order partial differential equations (2.56). Comparison of (2.56) and (2.57) show them to be form invariant images of each other under the mapping defined by (2.49). Further, (2.57) may equally well be written as

(2.58) $$\Phi^* F^\alpha = - \Phi^*(v \rfloor C^\alpha) = 0 \, , \quad \Phi^* C^\alpha = 0 \, ,$$

while (2.56) may be written as

(2.59) $$\Phi_v(s)^* \overset{*}{F}{}^\alpha = - \Phi_v(s)^*(\overset{*}{v} \rfloor \overset{*}{C}{}^\alpha) = 0 \, , \quad \Phi_v(s)^* \overset{*}{C}{}^\alpha = 0 \, ,$$

where

(2.60) $$\Phi_v(s) = \exp(s,v) \circ \Phi$$

is the section of K that is generated by transport of the section $\Phi : B_n \longrightarrow K$ along the orbits of $v \varepsilon TC(K)$. Now, (2.60) leads directly to

$$\Phi_v(s)^* = \Phi^* \circ (\exp(s,v))^* \, ,$$

and hence

$$\Phi_v(s)^* \Omega = \Phi^*(\exp(s,v)^* \overset{*}{\Omega}) \, ,$$

where the superimposed $*$ designates evaluation at $(\overset{*}{x}{}^i, \overset{*}{q}{}^\alpha, \overset{*}{y}{}^\alpha_i)$. However, the definition of the Lie derivative with respect to v gives the finite transport relation (Appendix, Section 4-5 with $\exp(s,v)^* \overset{*}{\Omega} = \overset{m}{\Omega}(s)$)

(2.61) $$\exp(s,v)^* \overset{*}{\Omega} = \exp(s \pounds_v) \Omega \; ;$$

that is, the old field at the new point, $\overset{*}{\Omega}$, dragged back to the old point by $\exp(s,v)^*$ is equal to the new field $\exp(s \pounds_v) \Omega$ at the old point. Hence,

$$(2.62) \qquad \Phi_v(s)^*\mu = \Phi^*(\exp(s\pounds_v)\mu) = f(x^j,s)\,\Phi^*\mu$$

so that $\Phi_v(s)^*\mu \neq 0$ for all values of s for which $f(x^j,s) \neq 0$. Thus, since $f(x^j,0) = 1$ and $f(x^j,s)$ is analytic in s in some open interval $I(\Phi,v)$ that contains $s=0$ for each $\Phi \in R(B_n)$, it follows that $\Phi^*\mu \neq 0$ implies $\Phi_v(s)^*\mu \neq 0$ for all $s \in I$. Similarly $\Phi_v(s)^*C^\alpha = \Phi^*(\exp(s\pounds_v)C^\alpha)$. In this case, however, $v \in TC(K)$ implies $\pounds_v C^\alpha = \partial_\beta(v \rfloor C^\alpha)C^\beta$, by Theorem 2.2 and hence there exists a matrix $((A^\alpha_\beta))$ that is analytic in s such that

$$(2.63) \qquad \Phi_v(s)^*C^\alpha = \Phi^*(A^\alpha_\beta)\,\Phi^*C^\beta .$$

This result shows that $\Phi^*C^\alpha = 0$ implies $\Phi_v(s)^*C^\alpha = 0$ for all s. Accordingly, $\Phi \in R(B_n)$ implies that $\Phi_v(s) \in R(B_n)$ for all s in the open interval $I(\Phi,v)$ for which $f(x^j,s) \neq 0$.

<u>THEOREM 2.3</u>. *The orbital equations of any* $v \in TC(K)$, *for* $N > 1$, *are the characteristic equations of the system of first order quasilinear partial differential equations*

$$(2.64) \qquad \overset{*}{F}{}^\alpha = -\overset{*}{v} \rfloor \overset{*}{C}{}^\alpha = 0 , \quad \overset{*}{y}{}^\alpha_i = \partial \overset{*}{q}{}^\alpha(\overset{*}{x}{}^j)/\partial \overset{*}{x}{}^i , \quad \alpha = 1,\ldots,N ,$$

and generate a 1-parameter family of maps

$$(2.65) \qquad \Phi_v(s) = \exp(s,v)\circ\Phi$$

that maps $R(B_n)$ *into itself for all* s *in some open interval* $I(\Phi,v)$ *that contains* $s=0$. *Further, there is associated with each* $v \in TC(K)$, *for* $N > 1$, *a collection*

$$(2.66) \qquad R(v,B_n) = \{\Phi \in R(B_n) \mid \Phi^*F^\alpha = -\Phi^*(v \rfloor C^\alpha) = 0\}$$

of regular maps with the property that $\Phi \in R(v,B_n)$ *implies that the 1-parameter family of regular maps* $\Phi_v(s) = \exp(s,v)\circ\Phi$ *belongs to* $R(v,B_n)$ *for all* s , *and the family* $R(v,B_n)$ *is the maximal proper subset of* $R(B_n)$ *with this property.*

Proof. The only specific not already established is the maximality of the proper subset $R(v,B_n)$. This, however, follows immediately from the fact that the general solution of the system of partial differential equations that admits the orbital equations as characteristic equations defines the maximal invariant varieties admitted by the orbital equations.

COROLLARY 2.1. *If* $v \in TC(K)$, *then the orbital equations for* $\overset{*}{y}{}^{\alpha}_{i}$ *are*

$$\frac{d\overset{*}{y}{}^{\alpha}_{i}}{ds} = \overset{*}{\partial}_{i}\overset{*}{v}{}^{\alpha} + \overset{*}{y}{}^{\beta}_{i}\overset{*}{\partial}_{\beta}\overset{*}{v}{}^{\alpha} - \overset{*}{y}{}^{\alpha}_{j}\overset{*}{\partial}_{i}\overset{*}{v}{}^{j} - \overset{*}{y}{}^{\beta}_{i}\overset{*}{y}{}^{\alpha}_{j}\overset{*}{\partial}_{\beta}\overset{*}{v}{}^{j}$$

and these equations are linear differential equations whenever $\overset{*}{\partial}_{\beta}\overset{*}{v}{}^{j} = 0$ $(\partial_{\beta}v^{j} = 0)$. *If the solutions*

$$\overset{*}{y}{}^{\alpha}_{i} = H^{\alpha}_{i}(s;\ x^{j},\ q^{\beta},\ y^{\beta}_{j}) = \exp(s\pounds_{v})y^{\alpha}_{i}$$

are such that $\partial_{\beta}H^{\alpha}_{i} \neq 0$, *for* $s \neq 0$, *then they represent generalized contact transformations, for any fixed value of* $s \neq 0$, *when* K *is sectioned by any* $\Phi \in R(B_n)$ *; that is, if* $\Phi:(\tau^{j}) \longrightarrow (x^{j},\ \phi^{\alpha})$ *then* $\Phi_{v}(s):(\tau^{j}) \longrightarrow (\overset{*}{x}{}^{j},\ \overset{*}{\phi}{}^{\alpha})$ *for any fixed* s *such that*

$$\frac{\partial\overset{*}{\phi}{}^{\alpha}}{\partial\overset{*}{x}{}^{i}} = H^{\alpha}_{i}(s;\ x^{j},\ \phi^{\beta},\ \partial\phi^{\beta}/\partial x^{j}) .$$

Proof. The expression for $d\overset{*}{y}{}^{\alpha}_{i}/ds$ follows directly from (2.47) and these equations are linear in the variables $(\overset{*}{y}{}^{\alpha}_{i})$ whenever $\overset{*}{\partial}_{\beta}\overset{*}{v}{}^{j} = 0$. The given representation of the solutions is simply a statement of the last of the system (2.49). Since $\Phi_{v}(s) = \exp(s,v)\circ\Phi$, by (2.60), a section of K by any $\Phi \in R(B_n)$ yields a section of K by $\Phi_{v}(s)$ and $\Phi_{v}(s) \in R(B_n)$. Thus $\Phi^{*}C^{\alpha} = 0$ and $\Phi^{*}_{v}(s)\overset{*}{C}{}^{\alpha} = 0$ imply that

$$\Phi:(\tau^{j}) \longrightarrow (x^{j},\ \phi^{\alpha},\ \partial\phi^{\alpha}/\partial x^{j})$$

and

$$\Phi_v(s):(\tau^j) \longrightarrow (\overset{*}{x}{}^j,\ \overset{*}{\phi}{}^\alpha,\ \partial\overset{*}{\phi}{}^\alpha/\partial\overset{*}{x}{}^j)\ ,$$

in which case we obtain

$$\frac{\partial\overset{*}{\phi}{}^\alpha}{\partial\overset{*}{x}{}^i} = H_i^\alpha(s;\ x^j,\ \phi^\beta,\ \partial\phi^\beta/\partial x^j)\ ,$$

for any fixed value of s. For $s = 0$, $H_i^\alpha(0;\ x^j,\ \phi^\beta,\ \partial\phi^\beta/\partial x^j) = \partial\phi^\alpha/\partial x^i$ and hence the result is trivial in this case. If $\partial H_i^\alpha/\partial\phi^\beta \not\equiv 0$ for $s \neq 0$, then the new derivatives, $\partial\overset{*}{\phi}{}^\alpha/\partial\overset{*}{x}{}^i$, become functions of the old fields (ϕ^β) and their first derivatives and these functions depend on both (ϕ^β) and on $(\partial\phi^\beta/\partial x^j)$ in a nontrivial manner. This, however, is exactly the classic definition of a contact transformation [13].

This result should not be completely unexpected, for $\Phi^* C^\alpha = 0$ correlates the $\Phi^*(y_i^\alpha)$ with the first derivatives of the $\Phi^*(q^\alpha)$ while $\pounds_v \bar{C} \subset \bar{C}$ preserves this correlation under transport along the orbits of v; that is, $\Phi_v^*(s)\overset{*}{C}{}^\alpha = 0$ correlates the $\Phi_v^*(s)(\overset{*}{y}{}_i^\alpha)$ with the first derivatives of $\Phi_v^*(s)(\overset{*}{q}{}^\alpha)$. We thus have mappings that map tangent planes to solutions onto tangent planes to new solutions, and this is exactly what classic contact transformations perform.

From yet another point of view, completely integrable contact transformations are naturally associated with the contact ideal $\bar{C}$. In order to see this, we consider an invertible map

$$\rho:\ G \longrightarrow {}^{\backprime}G \mid\ {}^{\backprime}x^i = \rho^i(x^j, q^\beta)\ ,\quad {}^{\backprime}q^\alpha = \rho^\alpha(x^j, q^\beta)$$

from one graph space to another. If we extend G to K and ${}^{\backprime}G$ to ${}^{\backprime}K$, we have the contact forms $C^\alpha = dq^\alpha - y_i^\alpha\, dx^i$ and

$${}^{\backprime}C^\alpha = d{}^{\backprime}q^\alpha - {}^{\backprime}y_i^\alpha\, d{}^{\backprime}x^i\ .$$

A direct substitution from the transformation equations then gives

$$`c^\alpha = (\partial_\beta \rho^\alpha - `y_i^\alpha \partial_\beta \rho^i) c^\beta$$
$$+ \{Z_i(\rho^\alpha) - `y_j^\alpha Z_i(\rho^j)\} dx^i \ .$$

Hence ρ extends to a map of K to $`K$ such that $\overset{-1}{\rho}{}^*$ maps $\bar{C}$ onto $`\bar{C}$ if and only if

$$`y_j^\alpha Z_i(\rho^j) = Z_i(\rho^\alpha) \ .$$

When this is written out, we have

$$`y_j^\alpha (\partial_i \rho^j + y_i^\beta \partial_\beta \rho^j) = \partial_i \rho^\alpha + y_i^\beta \partial_\beta \rho^\alpha$$

which is clearly a contact transformation. Further, $v = v^i \partial_i + v^\alpha \partial_\alpha + Z_i(v^\alpha - y_j^\alpha v^j) \partial_\alpha^i$ is seen to be a generator of an infinitesimal contact transformation $(x^i, q^\alpha, y_i^\alpha) \to (x^i + \varepsilon v \rfloor dx^i,$ $q^\alpha + \varepsilon v \rfloor dq^\alpha,\ y_i^\alpha + \varepsilon v \rfloor dy_i^\alpha)$.

First Group Extension Structure

For $N > 1$, the orbital equations (2.46), (2.47) decompose naturally into two parts. The first part, consisting of $d\overset{*}{x}{}^i/ds = f^i(\overset{*}{x},\overset{*}{q})$, $d\overset{*}{q}{}^\alpha/ds = f^\alpha(\overset{*}{x},\overset{*}{q})$, is a system of ordinary differential equations on the graph space G with coordinates (x^i, q^α) ; that is, the orbital equations (2.46) project onto the graph space G . Once these projected equations are solved, their solutions can be substituted into the remaining orbital equations (2.47) for the determination of the y's. Thus, if we use the projection map

$$\pi : K \to G \mid \ (x^i, q^\alpha, y_i^\alpha) \longrightarrow (x^i, q^\alpha) \ ,$$

it is clear from (2.46) or (2.39) and (2.41) that any $v \in TC(K)$ for $N > 1$ is projectable; that is

$$\pi_* v = v^i \partial_i + v^\alpha \partial_\alpha \ , \qquad (2.67)$$

and

$$(2.68)\qquad \overset{-1}{\pi}_*(v^i\,\partial_i + v^\alpha\,\partial_\alpha) = v^i\,\partial_i + v^\alpha\,\partial_\alpha + Z_i(v^\alpha - y^\alpha_j v^j)\partial^i_\alpha\ .$$

Further, if $v_a \varepsilon T(G)$, $a = 1,\ldots,r$, forms an r-dimensional Lie algebra, $[v_a,v_b] = C^f_{ab}\,v_f$, then $\overset{-1}{\pi}_*$ applied to this Lie algebra yields a Lie algebra of r dimensions with the same constants of structure that is contained in $TC(K)$. Since the converse of this follows trivially, we see that, for $N > 1$, $TC(K)$ is the extension of $T(G)$ that is generated by $\overset{-1}{\pi}_*$. This is, however, just the statement that a contact structure K with $N > 1$ is the *first group extension* [10] of graph space G with respect to all regular maps of B_n to G ; that is, K with $N > 1$ is a *first group extension structure*

$$(2.69)\qquad E = \{K,\ R(B_n),\ \bar{C},\ TC(K) = \overset{-1}{\pi}_* T(G)\}$$

of graph space G .

THEOREM 2.4. *A contact structure with* $N > 1$ *is the first group extension structure of graph space.*

The importance of this theorem is that it brings together the differential geometric approach with the group extension approach that has been so successful under the hands of Ovsjannikov [10] and Bluman and Cole [11]. It is of interest to note in this context that the equations

$$\Phi^*(v \rfloor C^\alpha) = -\,\Phi^* F^\alpha = 0\ ,$$

where v is an isovector generated by C^α , *and* the equations of evolution, are exactly those equations that give the classic similarity solutions with which the group extension methods deal so successfully (see Section 3.4).

The Exceptional Case, N = 1

We now take up the exceptional case, namely $N = 1$. In this case, Theorem 2.2 gives the orbital equations

$$\text{(2.70)} \qquad d\overset{*}{x}{}^i/ds = \partial\xi(\overset{*}{x},\overset{*}{q},\overset{*}{y})/\partial\overset{*}{y}_i ,$$

$$\text{(2.71)} \qquad d\overset{*}{q}/ds = f(\overset{*}{x},\overset{*}{q}) + (\overset{*}{y}_i\partial/\partial\overset{*}{y}_i - 1)\xi(\overset{*}{x},\overset{*}{q},\overset{*}{y}) ,$$

$$\text{(2.72)} \qquad d\overset{*}{y}_i/ds = \overset{*}{Z}_i(f(\overset{*}{x},\overset{*}{q}) - \xi(\overset{*}{x},\overset{*}{q},\overset{*}{y})) .$$

It is clear, however, from (2.42) that v can be projectable onto G only if $\partial^i_1 v^j = 0$, in which case we must have

$$\text{(2.73)} \qquad v^i = \partial^i_1(f^j(x,q)y_j) = f^i(x,q) ,$$

for some $f^j(x,q)$, and ξ is thus a homogeneous function of degree one in the y's. This then implies, by (2.43), that $v^1 = f^1(\overset{*}{x},\overset{*}{q})$, and hence $\xi(x,q,y)$ being homogeneous of degree one in the y's is both necessary and sufficient for v to be projectable onto G.

<u>THEOREM 2.5</u>. *If* $N = 1$, v *is projectable onto* G *if and only if the function* $\xi(\overset{*}{x},\overset{*}{q},\overset{*}{y})$, *that enters into the defining equations* (2.42) *and* (2.43) *of* v^i *and* v^1, *is a function that is homogeneous of degree one in the* y's. *The projectable elements of* $TC(K)$ *form a Lie subalgebra of* $TC(K)$, *these elements are given by* $\bar{\pi}^{-1}_*T(G)$, *and* $E = \{K, R(B_n), \bar{C}, \bar{\pi}^{-1}_*T(G)\}$ *is a first group extension structure.*

The importance of this theorem is that it shows that for $N = 1$, *the whole contact structure does not reduce to a first group extension structure.* However, if we again set

$$\text{(2.74)} \qquad \overset{*}{F} = -\overset{*}{v}\rfloor\overset{*}{C} = \xi(\overset{*}{x},\overset{*}{q},\overset{*}{y}) - f(\overset{*}{x},\overset{*}{q}) ,$$

then (2.70)-(2.72) are the characteristic equations of the non-linear system

(2.75) $\overset{*}{F} = 0 \, , \quad \overset{*}{y}_i = \partial q(\overset{*}{x}^j)/\partial \overset{*}{x}^i \, .$

The same reasoning that led to Theorem 2.3 now yields the following result.

THEOREM 2.6. *The results of Theorem* 2.3 *hold for arbitrary* N , *with the exception that, for* $N = 1$, $-\overset{*}{v} \rfloor \overset{*}{C}^\alpha = \overset{*}{F}^\alpha = 0$ *is the nonlinear equation*

(2.76) $\xi(\overset{*}{x}^j, \overset{*}{q}^\alpha, \overset{*}{y}^\alpha_j) \;=\; f(\overset{*}{x}^j, \overset{*}{q}^\alpha)$

rather than the quasilinear system

$$f^i(\overset{*}{x}^j, \overset{*}{q}^\beta)\overset{*}{y}^\alpha_i \;=\; f^\alpha(\overset{*}{x}^j, \overset{*}{q}^\beta) \, .$$

There is thus a collection

(2.77) $R(v, B_n) \;=\; \{\Phi \varepsilon R(K) \,|\, \Phi^* F^\alpha \;=\; -\, \Phi^*(v \rfloor C^\alpha) \;=\; 0\}$

of regular maps associated with any $v \varepsilon TC(K)$ *such that* $\Phi_v(s) = \exp(s,v) \circ \Phi$ *maps* $R(v, B_n)$ *onto itself and* $R(v, B_n)$ *is the maximal proper subset of* $R(B_n)$ *with this property.*

2.5 ORBITAL SCALE INVARIANTS AND ISOVECTORS OF THE EXTENDED IDEAL $\bar{I}(C^\alpha, v \rfloor C^\alpha)$

The association of the collection

(2.78) $R(v, B_n) = \{\Phi : B_n \to K \,|\, \Phi^* \mu \neq 0, \; \Phi^* C^\alpha = 0, \; \Phi^*(v \rfloor C^\alpha) = 0\}$

with an element v of $TC(K)$ is not a happenstance of the orbital equations of v alone. We know from Theorem 2.2 that $v \varepsilon TC(K)$ if and only if

(2.79) $\pounds_v C^\alpha \;=\; \partial_\beta(v \rfloor C^\alpha) C^\beta \, .$

If $h(x,q,y)$ is any scalar valued function defined on K ,

$$(2.80)\qquad \pounds_{hv}C^{\alpha} = h\pounds_{v}C^{\alpha} + dh \wedge v\rfloor C^{\alpha} = (v\rfloor C^{\alpha})dh \bmod \bar{C} .$$

Thus,

$$(2.81)\qquad \Phi^{*}(\pounds_{hv}C^{\alpha}) = \Phi^{*}(v\rfloor C^{\alpha})d\Phi^{*}h$$

and Φ^{*} annihilates $\pounds_{hv}C^{\alpha}$ for arbitrary $h(x,q,y)$ if and only if $\Phi^{*}(v\rfloor C^{\alpha}) = 0$; that is, if and only if $\Phi \varepsilon R(v,B_n)$. We thus see that $R(v,B_n)$ consists of all regular sections of K for which

$$(2.82)\qquad \Phi^{*}(\pounds_{hv}C^{\alpha}) = 0$$

for all $h(x^j,q^{\alpha},y^{\alpha}_j)$. This simply says that $R(v,B_n)$ constitutes the maximal subcollection of regular maps that annihilates the deformation of C^{α} for all scale mappings

$$(2.83)\qquad S : v \rightarrow hv$$

of the isovector field v . Further, if $\eta \varepsilon \bar{C}$, then $\eta = A_{\alpha} \wedge C^{\alpha} + B_{\alpha} \wedge dC^{\alpha}$ and

$$\begin{aligned}(2.84)\qquad \pounds_{hv}\eta &= h\pounds_{v}\eta + dh \wedge (v\rfloor\eta) = dh \wedge (v\rfloor\eta) \bmod \bar{C}\\ &= dh \wedge (v\rfloor A_{\alpha}) \wedge C^{\alpha} + (-1)^{a}dh \wedge A_{\alpha}(v\rfloor C^{\alpha})\\ &\quad + dh \wedge (v\rfloor B_{\alpha}) \wedge dC^{\alpha} + (-1)^{a-1}dh \wedge B_{\alpha} \wedge (v\rfloor dC^{\alpha})\\ &= (-1)^{a}dh \wedge \{A_{\alpha}(v\rfloor C^{\alpha}) - B_{\alpha} \wedge (v\rfloor dC^{\alpha})\} \bmod \bar{C} .\end{aligned}$$

However, if $\Phi^{*}(v\rfloor C^{\alpha}) = 0$, then $\Phi^{*}d(v\rfloor C^{\alpha}) = d\Phi^{*}(v\rfloor C^{\alpha}) = 0$, and hence $v\rfloor dC^{\alpha} = \pounds_{v}C^{\alpha} - d(v\rfloor C^{\alpha}) = -d(v\rfloor C^{\alpha}) \bmod \bar{C}$ implies

$$(2.85)\qquad \Phi^{*}(\pounds_{hv}\bar{C}) = 0 \qquad \forall\, h : K \rightarrow R$$

if and only if $\Phi \varepsilon R(v,B_n)$. In other words, if $v \varepsilon TC(K)$, then $\pounds_{v}\bar{C} \subset \bar{C}$, $\pounds_{hv}\bar{C} \not\subset \bar{C}$, but $\Phi^{*}\pounds_{hv}\bar{C} = 0$ for all $\Phi \varepsilon R(v,B_n)$.

Orbital Scale Invariant Sections of K

THEOREM 2.7. *The collection* $R(v,B_n)$ *is the maximal subcollection of* $R(B_n)$ *that annihilates the deformations of all elements of the closed ideal* $\bar{C}$ *that are generated by all orbital scale mappings*

(2.86) $$S : v \longrightarrow hv$$

of the isovector v .

It is thus natural to say that the elements of $R(v,B_n)$ are *orbital scale invariant* sections of K associated with the isovector v .

Again, with $\eta = A_\alpha \wedge C^\alpha + B_\alpha \wedge dC^\alpha$,

$$\begin{aligned} v \rfloor \eta &= (v \rfloor A_\alpha) \wedge C^\alpha + (-1)^a A_\alpha \wedge (v \rfloor C^\alpha) \\ &\quad + (v \rfloor B_\alpha) \wedge dC^\alpha + (-1)^{a-1} B_\alpha \wedge (v \rfloor dC^\alpha) \\ &= (-1)^a \{A_\alpha v \rfloor C^\alpha - B_\alpha \wedge (v \rfloor dC^\alpha)\} \bmod \bar{C} . \end{aligned}$$

However $\pounds_v C^\alpha = v \rfloor dC^\alpha + d(v \rfloor C^\alpha) = 0 \bmod \bar{C}$, and hence

(2.87) $$v \rfloor \eta = (-1)^a \{A_\alpha (v \rfloor C^\alpha) + B_\alpha \wedge d(v \rfloor C^\alpha)\} \bmod \bar{C} .$$

Thus, $v \in TC(K)$ is not a *characteristic vector field of the closed ideal* $\bar{C}$ since $\Phi^*(v \rfloor \eta)$ does not vanish for every $\Phi \in R(B_n)$. However if $\Phi \in R(v,B_n)$, then $\Phi^*(v \rfloor C^\alpha) = 0$ and $\Phi^* d(v \rfloor C^\alpha) = d\Phi^*(v \rfloor C^\alpha) = 0$.

THEOREM 2.8. $R(v,B_n)$ *is the maximal subcollection of* $R(B_n)$ *for which* v *is a characteristic vector field of the closed contact ideal* $\bar{C}$.

Isovectors of the Extended Contact Ideal $\bar{I}(C^\alpha, v \rfloor C^\alpha)$

Let $v \in TC(K)$, then (2.69) yields

$$(2.88) \qquad \pounds_v(v\rfloor C^\alpha) = [v,v]\rfloor C^\alpha + v\rfloor \pounds_v C^\alpha = v\rfloor \pounds_v C^\alpha$$
$$= v\rfloor \{\partial_\beta(v\rfloor C^\alpha)C^\beta\} = \partial_\beta(v\rfloor C^\alpha)v\rfloor C^\beta$$

and

$$(2.89) \qquad \pounds_v d(v\rfloor C^\alpha) = d\pounds_v(v\rfloor C^\alpha) = d\{\partial_\beta(v\rfloor C^\alpha)\}v\rfloor C^\beta$$
$$+ \partial_\beta(v\rfloor C^\alpha)d(v\rfloor C^\beta) .$$

Thus, if $\eta \in \bar{I}(C^\alpha, v\rfloor C^\alpha)$ then $\pounds_v \eta \in \bar{I}(C^\alpha, v\rfloor C^\alpha)$.

<u>LEMMA 2.3</u>. *If* $v \in TC(K)$ *, then* v *is an isovector of the extended closed ideal* $\bar{I}(C^\alpha, v\rfloor C^\alpha)$.

If $v \in TC(K)$, then $v\rfloor C^\alpha \in \bar{I}(C^\alpha, v\rfloor C^\alpha)$, $v\rfloor dC^\alpha = \pounds_v C^\alpha - d(v\rfloor C^\alpha)$ $\in \bar{I}(C^\alpha, v\rfloor C^\alpha)$, $v\rfloor(v\rfloor C^\alpha) = 0 \in \bar{I}(C^\alpha, v\rfloor C^\alpha)$, and $v\rfloor d(v\rfloor C^\alpha)$ $= v\rfloor\{\pounds_v C^\alpha - v\rfloor dC^\alpha\} = v\rfloor \pounds_v C^\alpha = v\rfloor\{\partial_\beta(v\rfloor C^\alpha)C^\beta\} = \partial_\beta(v\rfloor C^\alpha)v\rfloor C^\beta$. Thus, if $\eta \in \bar{I}(C^\alpha, v\rfloor C^\alpha)$ then $v\rfloor \eta \in \bar{I}(C^\alpha, v\rfloor C^\alpha)$.

<u>LEMMA 2.4</u>. *If* $v \in TC(K)$ *, then* v *is a characteristic vector field of* $\bar{I}(C^\alpha, v\rfloor C^\alpha)$ *and any* $\Phi \in R(v, B_n)$ *yields a* Φ^* *that annihilates* $v\rfloor\eta$ *for any* $\eta \in \bar{I}(C^\alpha, v\rfloor C^\alpha)$.

This last result shows that although $v \in TC(K)$ is an isovector of $\bar{I}(C^\alpha, v\rfloor C^\alpha)$, it is not very useful since it is also a characteristic vector of $\bar{I}(C^\alpha, v\rfloor C^\alpha)$. The reason for this is that $v \in TC(K)$ requires $\pounds_v C^\alpha = 0 \bmod \bar{I}(C^\alpha)$ rather than $\pounds_v C^\alpha = 0 \bmod \bar{I}(C^\alpha, v\rfloor C^\alpha)$. Transition from the ideal $\bar{I}(C^\alpha)$ to the ideal $\bar{I}(C^\alpha, v\rfloor C^\alpha)$ thus provides us with the additional scope of allowing $\pounds_v C^\alpha = 0 \bmod \bar{I}(C^\alpha, v\rfloor C^\alpha)$. If $\pounds_v C^\alpha = 0 \bmod \bar{I}(C^\alpha, v\rfloor C^\alpha)$, then

$$(2.90) \qquad \pounds_v C^\alpha = A^\alpha_\beta C^\beta + B^\alpha_\beta\, v\rfloor C^\beta + H^\alpha_\beta d(v\rfloor C^\beta) ,$$

which implies that $\pounds_v dC^\alpha = d\pounds_v C^\alpha = 0 \bmod \bar{I}(C^\alpha, v\rfloor C^\alpha)$. Further,

$$\pounds_v(v\rfloor C^\alpha) = [v,v]\rfloor C^\alpha + v\rfloor \pounds_v C^\alpha = v\rfloor \pounds_v C^\alpha ,$$

so that satisfaction of (2.90) yields

$$\text{(2.91)} \qquad \pounds_v(v\rfloor C^\alpha) = A^\alpha_\beta v\rfloor C^\beta + (v\rfloor B^\alpha_\beta)v\rfloor C^\beta + H^\alpha_\beta v\rfloor d(v\rfloor C^\beta) .$$

However, $v\rfloor d(v\rfloor C^\beta) + d[v\rfloor v\rfloor C^\beta] = v\rfloor d(v\rfloor C^\beta) = \pounds_v(v\rfloor C^\beta)$, and hence (2.91) becomes

$$\text{(2.92)} \qquad \pounds_v(v\rfloor C^\alpha) = (A^\alpha_\beta + v\rfloor B^\alpha_\beta)(v\rfloor C^\beta) + H^\alpha_\beta \pounds_v(v\rfloor C^\beta) .$$

Thus, $\pounds_v(v\rfloor C^\alpha) \in \bar{I}(C^\alpha, v\rfloor C^\alpha)$ provided the matrix $((H^\alpha_\beta))$ has all of its eigenvalues different from one. Under these conditions, (2.92) implies that $\pounds_v d(v\rfloor C^\alpha) \in \bar{I}(C^\alpha, v\rfloor C^\alpha)$. However, $\pounds_v C^\alpha = v\rfloor dC^\alpha + d(v\rfloor C^\alpha)$ so that $d(v\rfloor C^\alpha) = - v\rfloor dC^\alpha + \pounds_v C^\alpha$ and

$$\text{(2.93)} \qquad \bar{I}(C^\alpha, v\rfloor C^\alpha) = I(C^\alpha, dC^\alpha, v\rfloor C^\alpha, v\rfloor dC^\alpha)$$

if (2.90) is satisfied. Under these conditions (2.90) is also equivalent to

$$\pounds_v C^\alpha = A^\alpha_\beta C^\beta + B^\alpha_\beta v\rfloor C^\beta - H^\alpha_\beta v\rfloor dC^\beta + H^\alpha_\beta \pounds_v C^\beta ;$$

that is

$$(\delta^\alpha_\beta - H^\alpha_\beta)\pounds_v C^\beta = A^\alpha_\beta C^\beta + B^\alpha_\beta v\rfloor C^\beta - H^\alpha_\beta(v\rfloor dC^\beta) .$$

Thus, if none of the eigenvalues of $((H^\alpha_\beta))$ is equal to unity, which we can always arrange to be the case, we see that

$$\pounds_v C^\alpha = A^\alpha_\beta C^\beta + B^\alpha_\beta v\rfloor C^\beta + R^\alpha_\beta v\rfloor dC^\beta ,$$

for some matrices of functions $((A^\alpha_\beta))$, $((R^\alpha_\beta))$ and some matrix of 1-forms $((B^\alpha_\beta))$, is necessary and sufficient for v to be an isovector of $\bar{I}(C^\alpha, v\rfloor C^\alpha)$.

Explicit Characterization

THEOREM 2.9. *Necessary and sufficient conditions for* v *to be an isovector of the extended ideal* $\bar{I}(C^\alpha, v\rfloor C^\alpha)$ *are satisfaction of*

$$(2.94)\qquad \pounds_v C^\alpha = A^\alpha_\beta C^\beta + R^\alpha_\beta v \rfloor dC^\beta + B^\alpha_\beta v \rfloor C^\beta$$

for some matrices $((A^\alpha_\beta))$, $((R^\alpha_\beta))$ *of scalar-valued functions such that* $\det(\delta^\alpha_\beta - R^\alpha_\beta) \neq 0$ *and some matrix* $((B^\alpha_\beta))$ *of 1-forms, in which case*

$$(2.95)\qquad \bar{I}(C^\alpha, v\rfloor C^\alpha) = I(C^\alpha, dC^\alpha, v\rfloor C^\alpha, v\rfloor dC^\alpha) .$$

We now proceed exactly as in the case of the proof of Theorem 2.2. If we define the matrix B^α_β of 1-forms by

$$(2.96)\qquad B^\alpha_\beta = B^\alpha_{\beta i}dx^i + B^\alpha_{\beta\gamma}dq^\gamma + B^{\alpha i}_{\beta\gamma}dy^\gamma_i ,$$

(2.94) becomes

$$(2.97)\qquad dv^\alpha - y^\alpha_j dv^j - v^\alpha_i dx^i = A^\alpha_\beta(dq^\beta - y^\beta_i dx^i)$$

$$+ R^\alpha_\beta(v^i dy^\beta_i - v^\beta_i dx^i)$$

$$+ (v\rfloor C^\mu)(B^\alpha_{\mu i}dx^i + B^\alpha_{\mu\beta}dq^\beta + B^{\alpha i}_{\mu\beta}dy^\beta_i) .$$

Since $\{dx^i, dq^\beta, dy^\beta_i\}$ is a basis for $\Lambda^1(K)$, (2.97) will be satisfied if and only if

$$(2.98)\qquad \partial_i v^\alpha - y^\alpha_j \partial_i v^j - v^\alpha_i = -A^\alpha_\beta y^\beta_i - R^\alpha_\beta v^\beta_i + (v\rfloor C^\mu)B^\alpha_{\mu i} ,$$

$$(2.99)\qquad \partial_\beta v^\alpha - y^\alpha_j \partial_\beta v^j = A^\alpha_\beta + (v\rfloor C^\mu)B^\alpha_{\mu\beta} ,$$

$$(2.100)\qquad \partial^i_\beta v^\alpha - y^\alpha_j \partial^i_\beta v^j = R^\alpha_\beta v^i + (v\rfloor C^\mu)B^{\alpha i}_{\mu\beta} .$$

If we solve (2.99) for A^α_β, we obtain

$$(2.101)\qquad A^\alpha_\beta = \partial_\beta(v\rfloor C^\alpha) - (v\rfloor C^\mu)B^\alpha_{\mu\beta} ,$$

in which case (2.98) yields

$$(2.102)\qquad (\delta^\alpha_\beta - R^\alpha_\beta)v^\beta_i = Z_i(v\rfloor C^\alpha) - (v\rfloor C^\mu)(B^\alpha_{\mu i} + y^\beta_i B^\alpha_{\mu\beta}) .$$

An inspection of (2.100) and (2.102) shows that no further eliminations can be made, and we obtain the following result.

THEOREM 2.10. *A vector field* $v \in T(K)$ *is an isovector of the extended ideal* $\bar{I}(C^\alpha, v\rfloor C^\alpha)$ *if and only if* v_i^β *are given by*

$$(2.103) \qquad (\delta^\alpha_\beta - R^\alpha_\beta) v^\beta_i = Z_i(v\rfloor C^\alpha) - (v\rfloor C^\mu)(B^\alpha_{\mu i} + y^\beta_i B^\alpha_{\mu\beta})$$

for some choice of $(R^\alpha_\beta, B^\alpha_{\mu i}, B^\alpha_{\mu\beta})$ *such that*

$$(2.104) \qquad \det(\delta^\alpha_\beta - R^\alpha_\beta) \neq 0$$

and

$$(2.105) \qquad \partial^i_\beta v^\alpha - y^\alpha_j \partial^i_\beta v^j = R^\alpha_\beta v^i + (v\rfloor C^\mu) B^{\alpha i}_{\mu\beta}$$

can be satisfied for some choice of $(B^{\alpha i}_{\mu\beta}, R^\alpha_\beta)$ *satisfying* (2.104).

It is of interest to note that the choice $R^\alpha_\beta = 0$, $B^\alpha_{\mu i} = 0$, $B^\alpha_{\mu\beta} = 0$, $B^{\alpha i}_{\mu\beta} = 0$ gives $v \in TC(K)$, as indeed it should. Further

$$(2.106) \qquad v^i = f^i(x,q), \quad v^\alpha = f^\alpha(x,q)$$

with $R^\alpha_\beta = 0$, $B^{\alpha i}_{\mu\beta} = 0$ satisfy (2.104) and (2.105), in which case

$$(2.107) \qquad v^\beta_i = Z_i(v\rfloor C^\alpha) - (v\rfloor C^\mu)(B^\alpha_{\mu i} + y^\beta_i B^\alpha_{\mu\beta})$$

where $B^\alpha_{\mu i}$ and $B^\alpha_{\mu\beta}$ may be chosen arbitrarily. In this case, v is projectable, but $\pi_*^{-1}(v^i\partial_i + v^\alpha\partial_\alpha) = v^i\partial_i + v^\alpha\partial_\alpha + v^\beta_i\partial^i_\beta$ is now generated by (2.107). This is in sharp contrast to the case where $v \in TC(K)$ in which case π_*^{-1} is generated by $Z_i(v\rfloor C^\alpha)$ only.

If we write out (2.105) fully, we obtain

$$(2.108) \qquad \partial^i_\beta v^\alpha = B^{\alpha i}_{\mu\beta} v^\mu + R^\alpha_\beta v^i + y^\alpha_j \partial^i_\beta v^j - B^{\alpha i}_{\mu\beta} y^\mu_j v^j .$$

Thus, if we freeze the x's and the q's for the moment and consider

$$(2.109) \qquad \underset{\sim}{V} = [v^1,\ldots,v^N]^T \in \Lambda^o_{N,1}(K) \ ,$$

then (2.108) can be written as

$$(2.110) \qquad d\underset{\sim}{V} = \underset{\sim}{\Gamma}\,\underset{\sim}{V} + \underset{\sim}{\Sigma}$$

with

$$(2.111) \qquad \underset{\sim}{\Gamma} = ((B^{\alpha i}_{\mu\beta} dy^{\beta}_{i})) \ , \in \Lambda^1_{N,N}(K)$$

$$(2.112) \qquad \underset{\sim}{\Sigma} = \{(R^{\alpha}_{\beta} v^i + y^{\alpha}_{j} \partial^i_{\beta} v^j - B^{\alpha i}_{\mu\beta} y^{\mu}_{j} v^j) dy^{\beta}_{i}\} \in \Lambda^1_{N,1}(K) \ .$$

We can thus view (2.110) as the first part of a differential system of degree zero with connection $\underset{\sim}{\Gamma}$ and torsion $\underset{\sim}{\Sigma}$. This system is then solvable if and only if

$$(2.113) \qquad \begin{aligned} d\underset{\sim}{\Sigma} &= \underset{\sim}{\Gamma} \wedge \underset{\sim}{\Sigma} - \underset{\sim}{\Theta}\underset{\sim}{V} \\ d\underset{\sim}{\Gamma} &= \underset{\sim}{\Gamma} \wedge \underset{\sim}{\Gamma} + \underset{\sim}{\Theta} \ , \quad d\underset{\sim}{\Theta} = \underset{\sim}{\Gamma} \wedge \underset{\sim}{\Theta} - \underset{\sim}{\Theta} \wedge \underset{\sim}{\Gamma} \end{aligned}$$

can be satisfied, in which case the general solution is given by (1.20)-(1.22). These solutions will yield $\underset{\sim}{V}$ and $\underset{\sim}{\Sigma}$, and hence v^i and v^{α}, provided (2.113) can be satisfied for some choice of $(R^{\alpha}_{\beta}, B^{\alpha i}_{\mu\beta})$. We shall not pursue this matter further for general N since the results are not overly germane to our considerations in the following parts of this study.

The Exceptional Case N=1

The case $N = 1$ is of importance, however. In this event, (2.105) becomes

$$(2.114) \qquad \partial^i v - B^i v = R v^i + y_j(\partial^i v^j - B^i v^j) \ .$$

If, for the moment, we freeze the x's and q, multiplication of (2.114) by dy_i yields

(2.115) $$dv - \Gamma v = \eta + y_j(dv^j - \Gamma v^j)$$

where

(2.116) $$\Gamma = B^i dy_i \ , \qquad \eta = Rv^i dy_i \ .$$

Since we may always write any Γ as

(2.117) $$\Gamma = d\gamma + \theta \qquad (B^i = \partial^i\gamma + \theta^i)$$

where $\theta = Hd\Gamma$ is antiexact,

(2.118) $$\theta^i|_{y_j=0} = 0 \ , \qquad y_i\theta^i = 0 \quad \forall\,(y_i) \ ,$$

(2.115) becomes

(2.119) $$dv - (d\gamma+\theta)v = \eta + y_j(dv^j - (d\gamma+\theta)v^j) \ .$$

As a first step, we consider the equations

(2.120) $$dv^j - (d\gamma+\theta)v^j = h^j \ ,$$

so that (2.119) becomes

(2.121) $$dv - (d\gamma+\theta)v = \eta + y_j h^j \ .$$

Now, (2.120) is equivalent to

(2.122) $$d(e^{-\gamma}v^j) = \theta(e^{-\gamma}v^j) + H^j$$

with

(2.123) $$h^j = e^{\gamma}H^j \ .$$

Since the left-hand side of (2.122) is exact, the right-hand side must likewise be exact if (2.121) is to have solutions. We may thus set

(2.124) $$H^j = d\xi^j - \theta(e^{-\gamma}v^j) \ ,$$

where the ξ^j are scalar-valued functions. In this event, (2.122) becomes $d(e^{-\gamma}v^j - \xi^j) = 0$, and an integration yields

$$(2.125) \qquad v^j = e^{\gamma}\xi^j ,$$

where the constants of integration have been absorbed into the ξ^j since these functions are still arbitrary. A combination of (2.123) and (2.124) now gives

$$(2.126) \qquad h^j = e^{\gamma}\{d\xi^j - \theta\xi^j\} .$$

It thus remains only to solve (2.121).

A substitution of (2.116), (2.125) and (2.126) into (2.121) yields the exterior differential equation

$$dv - (d\gamma+\theta)v = Re^{\gamma}\xi^i dy_i + y_i e^{\gamma}(d\xi^i - \theta\xi^i) ,$$

which is equivalent to

$$(2.127) \qquad d(e^{-\gamma}v - y_i\xi^i) = \theta(e^{-\gamma}v - y_i\xi^i) + (R-1)\xi^i dy_i .$$

Since the left-hand side of (2.127) is exact, the right-hand side must likewise be exact if (2.127) is to have a solution. We may accordingly set

$$(2.128) \qquad d\psi = \theta(e^{-\gamma}v - y_i\xi^i) + (R-1)\xi^i dy_i ,$$

in which case (2.127) becomes $d(e^{-\gamma}v - y_i\xi^i - \psi) = 0$. An integration thus yields

$$(2.129) \qquad v = e^{\gamma}(y_i\xi^i + \psi) ,$$

where the $f(x,q)$ that arises from the integration is absorbed in the arbitrary function ψ. We now substitute (2.129) into (2.128) and obtain

$$(2.130) \qquad d\psi = \theta\psi + (R-1)\xi^i dy_i .$$

This then determines the functions ξ^i by

$$(2.131) \qquad (R-1)\xi^i = \partial^i\psi - \psi\theta^i$$

since $R \neq 1$.

<u>THEOREM 2.11</u>. *For the case* $N = 1$, v *is an isovector of* $\bar{I}(C, v\rfloor C)$ *if and only if*

$$(2.132) \qquad \pounds_v C = AC + Rv\rfloor dC + (B_i dx^i + Bdq + B^i dy_i)(v\rfloor C)$$

with $R \neq 1$. *The most general* v *with this property is given by*

$$(2.133) \qquad (1-R)v_i = \{Z_i - B_i - y_i B\}(v\rfloor C) ,$$

$$(2.134) \qquad (R-1)v^i = e^\gamma(\partial^i\psi - \psi\theta^i) ,$$

$$(2.135) \qquad v^q = e^\gamma\{\psi + \frac{1}{R-1} y_i\partial^i\psi - \frac{1}{R-1}\psi y_i\theta^i\} ,$$

where

$$(2.136) \qquad B^i = \partial^i\gamma + \theta^i$$

$$(2.137) \qquad A = \partial_q(v\rfloor C) - (v\rfloor C)B ,$$

and

$$(2.138) \qquad v\rfloor C = e^\gamma\psi .$$

There is thus an isovector of $\bar{I}(C, v\rfloor C)$ *for each choice of the* $2n+4$ *functions* $R(x^j, q, y_j)$, $B_i(x^j, q, y_j)$, $B(x^j, q, y_j)$, $\gamma(x^j, q, y_j)$, $\theta^i(x^j, q, y_j)$, $\psi(x^j, q, y_j)$ *such that* $R(x^j, q, y_j) \neq 1$, *and* Φ^* *annihilates* $v\rfloor C$ *if and only if*

$$(2.139) \qquad \Phi^*\psi = 0 .$$

Accordingly, since $\bar{I}(C, v\rfloor C) = \bar{I}(C, e^\gamma\psi)$, *the set of vectors generated by* (2.133)-(2.135) *for all choices of the* $2(n+1)$

functions $R(x^j,q,y_j) \neq 1$, $B(x^j,q,y_j)$, $B_i(x^j,q,y_j)$, $\theta^i(x^j,q,y_j)$ *are simultaneous isovectors of the fixed ideal* $\bar{I}(C,e^\gamma\psi)$.

If v is an isovector of $\bar{I}\{C^\alpha, v\rfloor C^\alpha\}$, then (2.94) gives $\pounds_v C^\alpha = A^\alpha_\beta C^\alpha + R^\alpha_\beta v\rfloor dC^\beta + B^\alpha_\beta v\rfloor C^\beta$ and hence $\pounds_v \bar{C}$ does not belong to $\bar{C}$ if $R^\alpha_\beta \neq 0$ or $B^\alpha_\beta \neq 0$. In this event, v *is not an infinitesimal generator of a contact transformation*, as follows immediately from Theorems 2.10 and 2.11. Such vector fields are thus candidates for possible infinitesimal generators of Bäcklund transformation, and will, in fact, be instrumental in the next chapter.

Proper Isovectors of $\bar{I}\{C^\alpha, v\rfloor C^\alpha\}$

There is a very important observation that must be made at this point. If we go back to Theorem 2.10 and solve (2.103) and (2.105) for the terms involving the B's, an elementary rearrangement gives the requirements

$$(2.140) \qquad (v\rfloor C^\mu)(B^\alpha_{\mu i} + y^\beta_i B^\alpha_{\mu\beta}) = -(\delta^\alpha_\beta - R^\alpha_\beta)v^\beta_i + Z_i(v\rfloor C^\alpha) ,$$

$$(2.141) \qquad (v\rfloor C^\mu)B^{\alpha i}_{\mu\beta} = \partial^i_\beta(v\rfloor C^\alpha) + (\delta^\alpha_\beta - R^\alpha_\beta)v^i .$$

If v is chosen so that $v\rfloor C^\alpha = 0$, that is, $v^\alpha = y^\alpha_i v^i$, then (2.140) and (2.141) imply $v^\alpha = v^i = v^\alpha_i = 0$ if $\det(\delta^\alpha_\beta - R^\alpha_\beta) \neq 0$. If $(\delta^\alpha_\beta - R^\alpha_\beta) \neq 0$, $\det(\delta^\alpha_\beta - R^\alpha_\beta) = 0$, then $v\rfloor C^\alpha = 0$ and (2.140), (2.141) imply $v^i = v^\alpha = 0$ and v^β_i belongs to the kernel of $(\delta^\alpha_\beta - R^\alpha_\beta)$ for each value of i . Further, and of more importance, if we make any choice of the quantities $(v^i, v^\alpha, v^\alpha_i)$ as functions of the arguments $(x^i, q^\alpha, y^\alpha_i)$ such that $v\rfloor C^\alpha \neq 0$, then (2.140) and (2.141) can be solved for a collection of B's so as to obtain identical satisfaction of the conditions of Theorem 2.10. In summary, we see that *the isovectors of the extended ideal* $\bar{I}\{C^\alpha, v\rfloor C^\alpha\}$ *consist of all elements of* $T(K)$ *such that* $v\rfloor C^\alpha \neq 0$.

The reason for this lack of determinism can be traced directly to the fact that $\{v \rfloor C^\alpha\}$ consists of forms of degree zero. This, in turn, allows $B^\alpha_\beta(v \rfloor C^\beta)$ to be 1-forms for any collection of 1-forms B^α_β, and hence we may always choose the coefficients of the 1-forms B^α_β so as to secure satisfaction of

$$\pounds_c C^\alpha = A^\alpha_\beta C^\beta + B^\alpha_\beta(v \rfloor C^\beta) + R^\alpha_\beta v \rfloor dC^\beta$$

for any choice of v such that $v \rfloor C^\alpha \neq 0$. This is, of course, well known; a closed ideal that contains nonzero zero forms as generators coincides with $\Lambda(K)$. It is thus clear that we must cut down the ideal $\bar{I}\{C^\alpha, v \rfloor C^\alpha\}$ if we are to obtain definitive results. The obvious thing to do is to continue to require $\pounds_v(v \rfloor C^\alpha)$ to belong to $\bar{I}\{C^\alpha, v \rfloor C^\alpha\}$, but to eliminate the possibility of $\pounds_v C^\alpha$ depending on the generators $v \rfloor C^\alpha$.

Definition 2.3. An isovector of the ideal $\bar{I}\{C^\alpha, v \rfloor C^\alpha\}$ is said to be *proper* if and only if

$$\pounds_v C^\alpha = A^\alpha_\beta C^\beta + R^\alpha_\beta v \rfloor dC^\beta . \tag{2.142}$$

The collection of all proper isovectors of $\bar{I}\{C^\alpha, v \rfloor C^\alpha\}$ is denoted by $PC(K)$.

It is clear from this definition that proper isovectors are obtained from the previous analysis by setting all of the B's equal to zero. A combination of this observation with Theorem 2.10 and (2.140), (2.141) gives the following result.

THEOREM 2.12. *A vector field* $v \in T(K)$ *is a proper isovector field of the extended ideal* $\bar{I}\{C^\alpha, v \rfloor C^\alpha\}$ *if and only if the* v^β_i's *are given by*

$$(\delta^\alpha_\beta - R^\alpha_\beta)v^\beta_i = Z_i(v \rfloor C^\alpha) \tag{2.143}$$

for some choice of the functions R^α_β *such that*

$$\det(\delta^\alpha_\beta - R^\alpha_\beta) \neq 0 \tag{2.144}$$

and

(2.145) $$(R^\alpha_\beta - \delta^\alpha_\beta) v^i = \partial^i_\beta (v \rfloor C^\alpha)$$

are satisfied.

Theorem 2.11 gives the following immediate results on setting all of the B's equal to zero.

Theorem 2.13. $v \in T(K)$ *is a proper isovector of* $\bar{I}\{C^\alpha, v \rfloor C^\alpha\}$ *with* $N = 1$ *if and only if*

(2.146) $$\pounds_v C = AC + R\, v \rfloor dC$$

with $R \neq 1$. *The most general* v *with this property is given by*

(2.147) $$(1-R) v_i = Z_i (v \rfloor C) ,$$

(2.148) $$(R-1) v^i = \partial^i \Psi ,$$

(2.149) $$v^q = \Psi + \frac{1}{R-1} y_i \partial^i \Psi .$$

For $N > 1$, it is a trivial matter to see that (2.145) has solutions if and only if there exist functions ξ^α such that

(2.150) $$(R^\alpha_\beta - \delta^\alpha_\beta) v^i = \partial^i_\beta \xi^\alpha ,$$

(2.151) $$v^\alpha = \xi^\alpha + y^\alpha_i v^i + f^\alpha(x^j, q^\beta) .$$

Thus, since $R^\alpha_\beta - \delta^\alpha_\beta = G^\alpha_\beta$ must be nonsingular, (2.150) gives the conditions

(2.152) $$\delta^\alpha_\rho v^i = \overset{-1}{G}{}^\beta_\rho \partial^i_\beta \xi^\alpha , \quad \partial^i_\rho (G^\alpha_\beta v^i) = \partial^i_\beta (G^\alpha_\rho v^i) .$$

Although a general solution of this system of conditions is not readily available, it is easily seen that a particular solution obtains from the ansatz $\xi^\alpha = f^k(x^j, q^\beta)(R^\alpha_\gamma(x^j, q^\beta) - \delta^\alpha_\gamma) y^\gamma_k$.

THEOREM 2.14. $v \in T(K)$ *is a proper isovector of* $\bar{I}\{C^\alpha, v \rfloor C^\alpha\}$ *with* $N > 1$ *if and only if there exist* N *functions* ξ^α *such that*

$$(R^\alpha_\beta - \delta^\alpha_\beta) v^i = \partial^i_\beta \xi^\alpha , \tag{2.153}$$

$$v^\alpha = \xi^\alpha + y^\alpha_i v^i + f^\alpha(v^j, q^\beta) , \tag{2.154}$$

in which case the v^α_i*'s are given by*

$$(\delta^\alpha_\beta - R^\alpha_\beta) v^\beta_i = Z_i (v \rfloor C^\alpha) \tag{2.155}$$

with

$$\det(R^\alpha_\beta - \delta^\alpha_\beta) \neq 0 . \tag{2.156}$$

All of these conditions are satisfied by

$$v^i = v^i(x^j, q^\beta) , \tag{2.157}$$

$$v^\alpha = f^i(x^j, q^\beta) R^\alpha_\gamma(x^j, q^\beta) y^\gamma_i + f^\alpha(x^j, q^\beta) \tag{2.158}$$

for any choise of the N^2 *functions* $R^\alpha_\beta(x^j, q^\beta)$ *such that* (2.156) *is satisfied. In particular,* TC(K) *is a proper subset of this collection of elements of* PC(K) *that is obtained by the restriction* $R^\alpha_\beta = 0$.

CHAPTER 3

BALANCE STRUCTURES

At this point, we have at our disposal a general contact structure

$$(3.1) \qquad \mathcal{K} = \{K, R(B_n), \bar{C}, TC(K)\} .$$

It is composed of an (n+N+nN)-dimensional Euclidean space K with coordinate functions $(x^i, q^\alpha, y_i^\alpha)$, the collection $R(B_n)$ of all maps $\Phi: B_n \rightarrow K$ such that

$$(3.2) \qquad \Phi^*\mu \neq 0 , \quad \Phi^*(dq^\alpha - y_i^\alpha \, dx^i) = \Phi^* C^\alpha = 0 , \quad \alpha = 1,\dots,N ,$$

the closed ideal $\bar{C}$ of $\Lambda(K)$ that is generated by the N contac 1-forms $\{C^\alpha\}$, and the collection $TC(K)$ of all isovector fields of the ideal $\bar{C}$. Thus, any map Φ belonging to this contact structure assigns the variables y_i^α the values of $\partial\Phi^\alpha(x^j)/\partial x^i$ when q^α is assigned the values $\Phi^\alpha(x^j)$. All is thus in readiness for the study of systems of second order partial differential equations, by viewing such systems as further constraints that select preferred collections of maps from the collection $R(B_n)$ of regular maps; that is, solutions of the system of second order partial differential equations. There is, however, a very important point that must be noted at this juncture. We have assumed throughout that we have n independent variables (x^i) . In order that this may continue to be true, we are forced to take the exterior forms that represent the constraints associated with the systems of second order partial differential equations to be forms of degree n .

3.1 BALANCE FORMS AND THEIR NATURAL EQUIVALENCE CLASSES

Any system of N exterior forms $E_\alpha \varepsilon \Lambda^n(K)$, $\alpha = 1,\ldots,N$, is referred to as a system of *balance forms* on K . If Φ is any element of the collection $R(B_n)$ of regular maps that belongs to the contact structure K , then Φ is said to *solve* the balance system $\{E_\alpha\}$ if and only if

$$(3.3) \qquad \Phi^* E_\alpha = 0 , \quad \alpha = 1,\ldots,N ;$$

that is, Φ solves the balance system E_α if and only if Φ is a regular section of the contact structure K that annihilates the balance forms. For example, suppose that

$$(3.4) \qquad E_\alpha = W_\alpha(x^j,q^\beta,y^\beta_j)\mu - dW^i_\alpha(x^j,q^\beta,y^\beta_j) \wedge \mu_i ,$$

where the $N+nN$ functions (W_α, W^i_α) are given functions of the $n+N+nN$ arguments $(x^j,q^\alpha,y^\alpha_i)$, and $\{\mu_i , i=1,\ldots,n\}$ is the basis for $\Lambda^{n-1}(E_n)$ given by (1.3). If we restrict $\Phi \varepsilon R(B_n)$ so that $x^i = \phi^i(\tau^j) = \tau^i$, which is always possible because $\Phi^*\mu \neq 0$, then $q^\alpha = \Phi^\alpha(x^j)$, $y^\alpha_i = \partial\Phi^\alpha(x^j)/\partial x^i$, and (3.4) yields

$$(3.5) \qquad \Phi^* E_\alpha = \{W_\alpha(x^j,\Phi^\beta,\partial_j\Phi^\beta) - D_i W^i_\alpha(x^j,\Phi^\beta,\partial_j\Phi^\beta)\}\mu$$

when (1.4) is used. In this case, $\Phi \varepsilon R(B_n)$ solves the balance system (3.4) if and only if the functions $\Phi^\beta(x^j)$ satisfy the system of second order partial differential equations

$$(3.6) \qquad D_i W^i_\alpha(x^j,\Phi^\beta,\partial_j\Phi^\beta) = W_\alpha(x^j,\Phi^\beta,\partial_j\Phi^\beta) .$$

As is clear from this example, balance forms provide access to a wide class of problems; in the case of (3.4), to all systems of laws of balance. The reason for this designation in the case of (3.4) is that (3.3) and (3.4) yield the balance laws

$$(3.7)\qquad \int_{B_n} (\Phi^* W_\alpha)\mu = \int_{\partial B_n} (\Phi^* W_\alpha^i)\mu_i$$

since $\Phi^*\mu = \mu$ and $\Phi^*\mu_i = \mu_i$ for the maps considered.

Let E_α , $\alpha = 1,\ldots,N$ be N given elements of $\Lambda^n(K)$ and define the *solution set* $S(E_\alpha, B_n)$ of $R(B_n)$ by

$$(3.8)\qquad S(E_\alpha, B_n) = \{\Phi \in R(B_n) \mid \Phi^* E_\alpha = 0 \ , \quad \alpha = 1,\ldots,N\} \ .$$

The collection $S(E_\alpha, B_n)$ is thus the collection of all maps that solve the balance system E_α since $\Phi \in S(E_\alpha, B_n)$ if and only if

$$(3.9)\qquad \Phi^*\mu \neq 0 \ , \quad \Phi^* C^\alpha = 0 \ , \quad \Phi^* E_\alpha = 0 \ , \quad \alpha = 1,\ldots,N \ .$$

A significant computational simplification results if we introduce matrix notation. We consider balance forms to be defined by the entries of the row matrix of n-forms $\underset{\sim}{E} \in \Lambda^n_{1,N}(K)$. Similarly, we represent the contact forms C^α as the entries of the column matrix of 1-forms $\underset{\sim}{C} \in \Lambda^1_{N,1}(K)$ with

$$(3.10)\qquad \underset{\sim}{C} = d\underset{\sim}{q} - \underset{\sim}{y}_i \, dx^i \ .$$

The closed ideal $\bar{C}$ can then be written as $\bar{C} = \bar{I}\{\underset{\sim}{C}\}$.

A Variational Formulation via Horizontal Isovectors

The conditions that Φ solve a given balance system can be put into another form whereby a direct contact with the calculus of variations is obtained. Let $\underset{\sim}{E} \in \Lambda^n_{1,N}(K)$ be a given balance system. We then construct an (n+1)-form F by the prescription

$$(3.11)\qquad F = \underset{\sim}{E} \wedge \underset{\sim}{C} \ ,$$

so that F belongs to the closed contact ideal $\bar{C}$. The (n+1)-form F was the basis for the study of balanced fields given in [7]. If v is any vector field on K , then (3.11) gives

(3.12) $$v \rfloor F = (v \rfloor \underset{\sim}{E}) \wedge \underset{\sim}{C} + (-1)^n \underset{\sim}{E}\,(v \rfloor \underset{\sim}{C}) ,$$

so that $v \rfloor F$ belongs to the extended ideal $\bar{I}\{\underset{\sim}{C},\ v \rfloor \underset{\sim}{C}\}$. Now, if $\Phi \varepsilon R(B_n)$, (3.12) shows that

(3.13) $$\Phi^*(v \rfloor F) = (-1)^n \Phi^*(\underset{\sim}{E})\ \Phi^*(v \rfloor \underset{\sim}{C}) .$$

Thus, if $\Phi \varepsilon S(E_\alpha, B_n)$ then $\Phi^*(v \rfloor F) = 0$. Now, consider the subspace

(3.14) $$TH(K) = \{v \varepsilon TC(K) \mid v^i = 0\}$$

of $TC(K)$ consisting of *horizontal* isovector fields. It is clear from Theorem 2.2 that any horizontal vector field u has the form

(3.15) $$u = f^\alpha(x^j, q^\beta)\partial_\alpha + Z_i[f^\alpha(x^j, q^\beta)]\partial^i_\alpha ,$$

so that

(3.16) $$u \rfloor C^\alpha = f^\alpha(x^j, q^\beta)$$

and

(3.17) $$\Phi^*(u \rfloor C^\alpha) = f^\alpha(x^j, \phi^\beta(x^j)) \stackrel{\text{def}}{=} \delta\Phi^\alpha(x^j) .$$

It then follows from the arbitrary nature of the functions $f^\alpha(x^j, q^\beta)$, that the functions $\delta\Phi^\alpha(x^j)$ may take on arbitrary values throughout B_n . However, we may also write (3.17) in the equivalent form

(3.18) $$\delta\Phi^\alpha = \Phi^*(u \rfloor dq^\alpha) , \quad \delta\underset{\sim}{\Phi} = \Phi^*(u \rfloor d\underset{\sim}{q}) ,$$

so that we define $\delta\partial_i\Phi^\alpha$ by

(3.19) $$\delta\partial_i\Phi^\alpha = \Phi^*(u \rfloor dy^\alpha_i) , \quad \delta\partial_i\underset{\sim}{\Phi} = \Phi^*(u \rfloor d\underset{\sim}{y}_i) .$$

It now follows from (3.15) that $u \rfloor dy^\alpha_i = Z_i f^\alpha(x^j, q^\beta)$, so that (3.19) gives

$$\delta \partial_i \Phi^\alpha = \Phi^*(Z_i f^\alpha) = D_i \delta \Phi^\alpha . \tag{3.20}$$

The variation process δ that is defined by (3.18) and (3.19) thus has the classic property that it commutes with x-wise differentiation. If we now replace the vector field v in (3.13) by a horizontal vector field u and use (3.17), we obtain

$$\Phi^*(u \rfloor F) = (-1)^n \Phi^*(\underset{\sim}{E}) \, \delta \underset{\sim}{\Phi} , \tag{3.21}$$

and an integration over B_n yields

$$\int_{B_n} \Phi^*(u \rfloor F) = (-1)^n \int_{B_n} \Phi^*(\underset{\sim}{E}) \, \delta \underset{\sim}{\Phi} . \tag{3.22}$$

The right-hand side of (3.22) is of exactly the form for use of the fundamental lemma of the calculus of variations.

<u>LEMMA 3.1</u>. *A necessary and sufficient condition that* $\Phi \in R(B_n)$ *shall belong to the solution set* $S(\underset{\sim}{E}, B_n)$ *is that*

$$\int_{B_n} \Phi^*(u \rfloor F) = 0 \tag{3.23}$$

for all $u \in TH(B_n)$, *where*

$$F = \underset{\sim}{E} \wedge \underset{\sim}{C} . \tag{3.24}$$

Conversely, if $\Phi \in S(\underset{\sim}{E}, B_n)$, *then*

$$\Phi^*(v \rfloor F) = 0 \tag{3.25}$$

for every $v \in T(K)$.

Examples

A few examples may make the situation somewhat clearer for the reader. If we put $q^1 = \rho$, $q^2 = u$ and $c^2 = dP(q^1)/dq^1$, so that $N = 2$, and $x = x^1$, $t = x^2$, so that $n = 2$, then

$$E_1 = dq^1 \wedge dx^1 - q^1 dq^2 \wedge dx^2 - q^2 dq^1 \wedge dx^2$$

$$E_2 = q^1 dq^2 \wedge dx^1 - q^1 q^2 dq^2 \wedge dx^2 - c^2 dq^1 \wedge dx^2$$

yields the well-known equations

$$\frac{\partial \rho}{\partial t} + \rho \frac{\partial u}{\partial x} + u \frac{\partial \rho}{\partial x} = 0 , \quad \rho \frac{\partial u}{\partial t} + \rho u \frac{\partial u}{\partial x} + c^2 \frac{\partial \rho}{\partial x} = 0$$

for any map $\Phi : (q^1, q^2) \longrightarrow (\rho(x,t), u(x,t))$. In this case there are no contact forms because the original equations are of first order. If we put $x^1 = x$, $x^2 = y$, $x^3 = t$, so that $n = 3$, and $N = 1$ with $q^1 = P$, then

$$C^1 = dq - y_1 dx^1 - y_2 dx^2 - y_3 dx^3$$

$$E_1 = \{dy_1 \wedge dx^3 - dq \wedge dx^2 + (ax^1 y_1 + bx^2 y_2 + cx^1 x^2 q) dx^2 \wedge dx^3\} \wedge dx^1$$

yields

$$\frac{\partial^2 P}{\partial x \partial y} + ax \frac{\partial P}{\partial x} + by \frac{\partial P}{\partial y} + cxyP + \frac{\partial P}{\partial t} = 0$$

for any map $\Phi : (q) \longrightarrow (P(x,y,t))$. Finally, with $N = 2$, $U = q^1$, $V = q^2$; $n = 2$, $x^1 = r$, $x^2 = t$,

$$C^1 = dq^1 - y_1^1 dx^1 - y_2^1 dx^2 , \quad C^2 = dq^2 - y_1^2 dx^1 - y_2^2 dx^2$$

$$E_1 = dy_1^1 \wedge dx^2 + dy_2^1 \wedge dx^1 - [e^{-2q^1}((y_2^2)^2 - (y_1^2)^2) - (x^1)^{-1} y_1^1] dx^1 \wedge dx^2$$

$$E_2 = dy_1^2 \wedge dx^2 + dy_2^2 \wedge dx^1 - [2(y_1^1 y_1^2 - y_2^1 y_2^2) - (x^1)^{-1} y_1^2] dx^1 \wedge dx^2$$

yields

$$\frac{\partial^2 U}{\partial r^2} + \frac{1}{r}\frac{\partial U}{\partial r} - \frac{\partial^2 U}{\partial t^2} = e^{-2U}\left\{\left(\frac{\partial V}{\partial t}\right)^2 - \left(\frac{\partial V}{\partial r}\right)^2\right\}$$

$$\frac{\partial^2 V}{\partial r^2} + \frac{1}{r}\frac{\partial V}{\partial r} - \frac{\partial^2 V}{\partial t^2} = 2\left(\frac{\partial V}{\partial r}\frac{\partial U}{\partial r} - \frac{\partial V}{\partial t}\frac{\partial U}{\partial t}\right)$$

for any $\Phi : (q^1, q^2) \longrightarrow (U(r,t), V(r,t))$.

Balance Equivalence and Fibration

Since a system of balance forms is defined by an element of $\Lambda^n_{1,N}(K)$, it is natural to consider $\Lambda^n_{1,N}(K)$ to constitute the space of all balance forms on K. Let $\underset{\sim}{P}$ be an element of $\Lambda^n_{1,N}(K) \cap \bar{C}$; that is, the N entries that comprise the components of the row matrix $\underset{\sim}{P}$ are n-forms that belong to the closed ideal $\bar{C}$. If $\underset{\sim}{R} \varepsilon \Lambda^o_{N,N}(K)$ is any nonsingular matrix valued function on K and $\underset{\sim}{E}_1 \varepsilon \Lambda^n_{1,N}(K)$, then

$$\underset{\sim}{E}_2 = \underset{\sim}{E}_1 \underset{\sim}{R} + \underset{\sim}{P} \tag{3.26}$$

also belongs to $\Lambda^n_{1,N}(K)$. If $\Phi \varepsilon R(K)$, $\Phi^* \underset{\sim}{P} = \underset{\sim}{0}$, and we have

$$\Phi^* \underset{\sim}{E}_2 = \Phi^*(\underset{\sim}{E}_1)\, \Phi^*(\underset{\sim}{R}) . \tag{3.27}$$

Thus, $\Phi^*(\underset{\sim}{E}_2) = \underset{\sim}{0}$ if and only if $\Phi^*(\underset{\sim}{E}_1) = \underset{\sim}{0}$ because $\Phi^*(\underset{\sim}{R})$ is nonsingular on B_n due to the fact that $\underset{\sim}{R}$ is nonsingular throughout K. This shows that $\Phi \varepsilon S(\underset{\sim}{E}_1, B_n)$ if and only if $\Phi \varepsilon S(\underset{\sim}{E}_2, B_n)$; that is, the solution sets of $\underset{\sim}{E}_1$ and $\underset{\sim}{E}_2$ coincide. The relation defined by (3.26) is clearly symmetric, reflexive, and transitive so that it is an equivalence relation on $\Lambda^n_{1,N}(K)$.

Definition 3.1. Two balance forms $\underset{\sim}{E}_1$ and $\underset{\sim}{E}_2$ are said to be *balance equivalent* if and only if there exists a nonsingular $\underset{\sim}{R} \varepsilon \Lambda^o_{N,N}(K)$ and an element $\underset{\sim}{P}$ of $\Lambda^n_{1,N}(K) \cap \bar{C}$ such that

$$\underset{\sim}{E}_2 = \underset{\sim}{E}_1 \underset{\sim}{R} + \underset{\sim}{P} , \tag{3.28}$$

in which case we write

$$\underset{\sim}{E}_2 \simeq \underset{\sim}{E}_1 . \tag{3.29}$$

A fibering of $\Lambda^n_{1,N}(K)$ by the equivalence relation $\simeq$ gives the following immediate result.

LEMMA 3.2. *The balance equivalence relation* $\simeq$ *partitions* $\Lambda^n_{1,N}(K)$ *into equivalence classes*

(3.30) $$\mathcal{B} = \Lambda^n_{1,N}(K)/(\simeq)$$

so that $\Lambda^n_{1,N}(K)$ *becomes a fiber space with base space* $\mathcal{B}$ *and fibers* $F(\ )$ *, where the fiber through* $\underset{\sim}{E}$ *is given by*

(3.31) $$F(\underset{\sim}{E}) = \{\underset{\sim}{E}_1 \varepsilon \Lambda^n_{1,N}(K) \mid \underset{\sim}{E}_1 = \underset{\sim}{E}\,\underset{\sim}{R} + \underset{\sim}{P}\ ,\ \ \forall\ \underset{\sim}{P} \varepsilon \bar{C} \cap \Lambda^n_{1,N}(K)\ ,$$
$$\forall\ \underset{\sim}{R} \varepsilon \Lambda^o_{N,N}(K) \quad \text{such that} \quad \det(\underset{\sim}{R}) \neq 0\}\ ,$$

and

(3.32) $$S(\underset{\sim}{E}, B_n) = S(F(\underset{\sim}{E}), B_n)\ .$$

It is clear from this result that significant simplification can be achieved in the study of a given balance system $\underset{\sim}{E}$ by selection of a "simpler" element of $F(\underset{\sim}{E})$ to replace $\underset{\sim}{E}$. For example, with $N = 1$, $n = 2$, the balance 2-form

(3.33) $$E = (dy_1 - y_2 dx^1) \wedge dx^2$$

gives the 1-dimensional heat equation:

$$\Phi^* E = (\partial_1 \partial_1 \Phi - \partial_2 \Phi)\ dx^1 \wedge dx^2\ .$$

Since $C = dq - y_1\ dx^1 - y_2\ dx^2$ in this case, (3.33) can equally well be written as

(3.34) $$E = dy_1 \wedge dx^2 + dq \wedge dx^1 + dx^1 \wedge C$$
$$\simeq dy_1 \wedge dx^2 + dq \wedge dx^1 = d(y_1 dx^2 + q\ dx^1)\ .$$

We may thus replace (3.33) by the exact 2-form $d(y_1 dx^2 + q\ dx^1)$ without any change in the solution set.

Let us, for the moment, restrict attention to balance systems of the form

(3.35) $$\underset{\sim}{E} = \underset{\sim}{W}\,\mu - d\underset{\sim}{W}^i \wedge \mu_i\ ,$$

where $\underset{\sim}{W}$ and $\underset{\sim}{W}^i$ belong to $\Lambda^0_{1,N}(K)$, $i=1,\ldots,n$. These are the balance forms studied in [7]. Clearly, (3.35) can be reconstructed from the single 1-form [7]

$$(3.36) \qquad W = \underset{\sim}{W}\, d\underset{\sim}{q} + \underset{\sim}{W}^i\, d\underset{\sim}{y}_i \ .$$

Further, the requirement that the simple (n+1)-form $W \wedge \mu$ be closed is both necessary and sufficient in order that there exist a function L on K such that [7]

$$(3.37) \qquad \underset{\sim}{W} = \partial L/\partial \underset{\sim}{q} \ , \qquad \underset{\sim}{W}^i = \partial L/\partial \underset{\sim}{y}_i \ ,$$

in which case

$$(3.38) \qquad \Phi^*(u \rfloor F) = \left\{ \Phi^*\left(\frac{\partial L}{\partial q}\right) - D_i \Phi^*\left(\frac{\partial L}{\partial \underset{\sim}{y}_i}\right) \right\} \delta \underset{\sim}{\Phi}\, \mu \ .$$

Thus,

$$(3.39) \qquad d(W \wedge \mu) = 0$$

is both necessary and sufficient in order that the balance forms (3.35) shall represent the Euler-Lagrange equations that derive from a variational principle

$$\delta \int_{B_n} \Phi^*(L\mu) = 0 \ .$$

This shows that the fiber $F(\underset{\sim}{W}\mu - d\underset{\sim}{W}^i \wedge \mu_i)$ consists of systems of partial differential equations with the same solutions as the Euler-Lagrange equations whenever $d(\underset{\sim}{W}\, d\underset{\sim}{q} + \underset{\sim}{W}^i\, d\underset{\sim}{y}_i) \wedge \mu = 0$.

Equivalence with a Variational System - The Null Class of Balance Forms

As is well known, the question of whether a given system of partial differential equations is equivalent to a variational statement is a difficult one within the context of the classical calculus of variations and has not been answered in a fully satisfactory manner. From the standpoint of balance systems, it would appear to become trivial. Any element of $F(\underset{\sim}{W}\mu - d\underset{\sim}{W}^i \wedge \mu_i)$ is of the form

$$(3.40) \qquad \underset{\sim}{E}_1 = \underset{\sim}{W}\,\underset{\sim}{R}\,\mu - (d\underset{\sim}{W}^i)\underset{\sim}{R} \wedge \mu_i + \underset{\sim}{P}$$

where $\underset{\sim}{P} \in \bar{C} \cap \Lambda^n_{1,N}(K)$. Now, any $\underset{\sim}{P} \in \bar{C} \cap \Lambda^n_{1,N}(K)$ belongs to the fiber $F(\underset{\sim}{0})$ which is simply the *null class* of balance forms since

$$(3.41) \qquad S(F(\underset{\sim}{0}), B_n) = R(B_n) .$$

Written out in full, we have

$$\underset{\sim}{P} = C^\alpha \wedge \underset{\sim}{S}_\alpha + dC^\alpha \wedge \underset{\sim}{T}_\alpha$$

and $\underset{\sim}{S}_\alpha \in \Lambda^{n-1}_{1,N}(K)$, $\underset{\sim}{T}_\alpha \in \Lambda^{n-2}_{1,N}(K)$, $\alpha = 1, \ldots, N$. However, $\underset{\sim}{E}$ was written in terms of the base elements (μ, μ_i) , so let us restrict attention to those $\underset{\sim}{P}$ that can be written in terms of (μ_i, μ_{ij}) . This means that we write

$$(3.42) \qquad \underset{\sim}{P} = C^\alpha \wedge \underset{\sim}{S}^i_\alpha \mu_i + dC^\alpha \wedge \underset{\sim}{T}^{ij}_\alpha \mu_{ij} ,$$

where $\underset{\sim}{S}^i_\alpha$ and $\underset{\sim}{T}^{ij}_\alpha$ are matrices of 0-forms with

$$(3.43) \qquad \underset{\sim}{T}^{ij}_\alpha = - \underset{\sim}{T}^{ji}_\alpha$$

because μ_{ij} is antisymmetric in (i,j) . When $\underset{\sim}{C} = dq - \underset{\sim}{y}_i\, dx^i$ is substituted into (3.42) and (1.4), (1.6) are used, we obtain

$$(3.44) \qquad \underset{\sim}{P} = (\underset{\sim}{S}^j_\alpha\, dq^\alpha - 2\, \underset{\sim}{T}^{ij}_\alpha dy^\alpha_i) \wedge \mu_j - (y^\alpha_i\, \underset{\sim}{S}^i_\alpha)\, \mu .$$

There is one further property that every balance system of the form (3.35) satisfies, namely, that the coefficient of μ_j is an exact 1-form. In order that $\underset{\sim}{P}$ satisfy this additional requirement, it is sufficient to require that

$$(3.45) \qquad (\underset{\sim}{S}^j_\alpha\, dq^\alpha - 2\, \underset{\sim}{T}^{ij}_\alpha dy_i) \wedge \mu_j = d\underset{\sim}{Q}^j \wedge \mu_j + \underset{\sim}{N}\, \mu .$$

Now, this can be the case if and only if

(3.46) $\quad \underset{\sim}{N} = -\partial_i \underset{\sim}{Q}^i, \quad \underset{\sim}{S}^j_\alpha = \partial_\alpha \underset{\sim}{Q}^j, \quad -2\,\underset{\sim}{T}^{ij}_\alpha = \partial^i_\alpha \underset{\sim}{Q}^j,$

where the skew symmetry in (i,j) of $\underset{\sim}{T}^{ij}_\alpha$ requires that

(3.47) $\quad \partial^i_\alpha \underset{\sim}{Q}^j = -\partial^j_\alpha \underset{\sim}{Q}^i .$

When these results are substituted back into (3.42) we obtain the following lemma.

LEMMA 3.3. *The most general element of* $\bar{C} \cap \Lambda^n_{1,N}(K)$ *of the form* $d\underset{\sim}{M}^i \wedge \mu_i - \underset{\sim}{M}\,\mu$ *is given by*

(3.48) $\quad \underset{\sim}{P} = d\underset{\sim}{Q}^j \wedge \mu_j - (Z_i \underset{\sim}{Q}^i)\mu ,$

for some set $\underset{\sim}{Q}^j \in \Lambda^o_{1,N}(K)$ *such that*

(3.49) $\quad \partial^i_\alpha \underset{\sim}{Q}^j + \partial^j_\alpha \underset{\sim}{Q}^i = 0 .$

When (3.48) is put back into (3.40), we obtain a collection of elements of the fiber $F(\underset{\sim}{W}\,\mu - d\underset{\sim}{W}^i \wedge \mu_i)$ that is given by

(3.50) $\quad \underset{\sim}{E}_1 = (\underset{\sim}{W}\,\underset{\sim}{R} - Z_i \underset{\sim}{Q}^i)\mu - ((d\underset{\sim}{W}^i)\underset{\sim}{R} - d\underset{\sim}{Q}^i) \wedge \mu_i$

for any $\underset{\sim}{Q}^i \in \Lambda^o_{1,N}(K)$ such that (3.49) hold. However, it is easily seen that

$$d(\underset{\sim}{W}^i)\underset{\sim}{R} \wedge \mu_i = d(\underset{\sim}{W}^i \underset{\sim}{R}) \wedge \mu_i - \underset{\sim}{W}^i Z_i \underset{\sim}{R}\,\mu - \underset{\sim}{W}^i (\partial_\beta \underset{\sim}{R}) c^\beta \wedge \mu_i$$

if $\underset{\sim}{R} = \underset{\sim}{R}(x^i, q^\beta)$, and hence (3.50) is balance equivalent to

(3.51) $\quad \underset{\sim}{E}_1 = (\underset{\sim}{W}\,\underset{\sim}{R} - Z_i \underset{\sim}{Q}^i - \underset{\sim}{W}^i Z_i \underset{\sim}{R})\mu - d(\underset{\sim}{W}^i \underset{\sim}{R} - \underset{\sim}{Q}^i) \wedge \mu_i .$

We know, however, that the new system will be Euler-Lagrange equations of a variational system if and only if

(3.52) $\quad d\{(\underset{\sim}{W}\,\underset{\sim}{R} - Z_i \underset{\sim}{Q}^i - \underset{\sim}{W}^i Z_i \underset{\sim}{R})d\underset{\sim}{q} + (\underset{\sim}{W}^i \underset{\sim}{R} - \underset{\sim}{Q}^i)d\underset{\sim}{y}_i\} \wedge \mu = 0 .$

This gives specific criteria for the equivalence of the given $\underset{\sim}{E} = \underset{\sim}{W} \mu - d\underset{\sim}{W}^{i} \wedge \mu_{i}$ with a variational principle. Balance equivalence is thus seen to be a very general construct that contains an answer to the question of equivalence with a variational principle as a special case. It should also be abundantly clear that further progress in the general case will be contingent upon the development of suitable criteria for the selection of a useful and simple canonical element from any fiber $F(\underset{\sim}{E})$ of $\Lambda^{n}_{1,N}(K)$.

3.2 CLOSED IDEALS AND CANONICAL FORMS

As it turns out, the search for a useful canonical element of a fiber $F(\underset{\sim}{E})$ is significantly simplified by use of the structure associated with an appropriately defined closed ideal. We first note that any $\Phi \in R(B_n)$ annihilates the closed ideal $\bar{C}$ that is generated by the contact 1-forms $\underset{\sim}{C}$. If $\underset{\sim}{E} \in \Lambda^{n}_{1,N}(K)$ is given, then Φ solves this system of balance forms if and only if $\Phi \in R(B_n)$ and $\Phi^{*}\underset{\sim}{E} = \underset{\sim}{0}$, in which case $\Phi^{*}d\underset{\sim}{E} = d\Phi^{*}\underset{\sim}{E} = \underset{\sim}{0}$. We thus have the following immediate result.

LEMMA 3.4. *The set* $S(\underset{\sim}{E}, B_n)$ *of all solutions of a given balance system* $\underset{\sim}{E}$ *consists of all maps* $\Phi : B_n \longrightarrow K$ *such that* $\Phi^{*}\mu \neq 0$ *and* Φ^{*} *annihilates the closed ideal*

$$(3.53) \qquad \bar{E} = I\{\underset{\sim}{C}, d\underset{\sim}{C}, \underset{\sim}{E}, d\underset{\sim}{E}\}$$

that is generated by $\underset{\sim}{C}$ *and* $\underset{\sim}{E}$.

The Closed Ideal Generated by $\underset{\sim}{C}$ and $\underset{\sim}{E}$: A Fiber Invariant

The relation between the closed ideal $\bar{E}$ and balance equivalence is as follows.

LEMMA 3.5. *All elements of* $F(\underset{\sim}{E})$ *generate the same closed ideal,*

$$(3.54) \qquad \bar{E} = I\{\underset{\sim}{C}, d\underset{\sim}{C}, \underset{\sim}{E}, d\underset{\sim}{E}\} = I\{\underset{\sim}{C}, d\underset{\sim}{C}, F(\underset{\sim}{E}), dF(\underset{\sim}{E})\}.$$

The closed ideal $\bar{E}$ *is thus a fiber invariant and*

$$(3.55) \qquad F(\underset{\sim}{E}) = \bar{E} \cap \Lambda^{n}_{1,N}(K) .$$

Proof. If $\underset{\sim}{E}_1 \varepsilon F(\underset{\sim}{E})$, $\underset{\sim}{E}_1 = \underset{\sim}{E}\,\underset{\sim}{R} + \underset{\sim}{P} = \underset{\sim}{E}\,\underset{\sim}{R} \bmod \bar{C}$ and $d\underset{\sim}{E}_1 = d\underset{\sim}{E}\,\underset{\sim}{R} + (-1)^n \underset{\sim}{E} \wedge d\underset{\sim}{R} + d\underset{\sim}{P} = d\underset{\sim}{E}\,\underset{\sim}{R} + (-1)^n \underset{\sim}{E} \wedge d\underset{\sim}{R} \bmod \bar{C}$ since $\underset{\sim}{P} \varepsilon \bar{C}$ and $\bar{C}$ is closed. Thus, $\underset{\sim}{E} = \underset{\sim}{E}_1 \underset{\sim}{R}^{-1} \bmod \bar{C}$, $d\underset{\sim}{E} = d\underset{\sim}{E}_1 \underset{\sim}{R}^{-1} + (-1)^{n+1} \underset{\sim}{E}_1 \underset{\sim}{R}^{-1} \wedge d\underset{\sim}{R}\,\underset{\sim}{R}^{-1} \bmod \bar{C}$, so that $\underset{\sim}{E} \wedge \underset{\sim}{A} + d\underset{\sim}{E} \wedge \underset{\sim}{B} = \underset{\sim}{E}_1 \wedge \underset{\sim}{R}^{-1}(\underset{\sim}{A} + (-1)^{n+1} d\underset{\sim}{R} \wedge \underset{\sim}{R}^{-1}\,\underset{\sim}{B}) + d\underset{\sim}{E}_1 \wedge \underset{\sim}{R}^{-1}\,\underset{\sim}{B} \bmod \bar{C}$. Thus every element of $F(\underset{\sim}{E})$ generates the same closed ideal, $\bar{E}$ is a fiber invariant, and (3.55) holds.

Closure of the Ideal $I\{\underset{\sim}{C}, d\underset{\sim}{C}, \underset{\sim}{E}\}$: Differential Systems

First, we consider the case where $d\underset{\sim}{E}$ is contained in the ideal $I\{\underset{\sim}{C}, d\underset{\sim}{C}, \underset{\sim}{E}\}$, that is

$$\bar{E} = \bar{I}\{\underset{\sim}{C}, \underset{\sim}{E}\} = I\{\underset{\sim}{C}, d\underset{\sim}{C}, \underset{\sim}{E}\} ,$$

since this case is of importance in its own right. Clearly, $d\underset{\sim}{E} \varepsilon I\{\underset{\sim}{C}, d\underset{\sim}{C}, \underset{\sim}{E}\}$ if and only if there exist a $\underset{\sim}{\Gamma} \varepsilon \Lambda^{1}_{N,N}(K)$ and a $\underset{\sim}{\Sigma} \varepsilon \bar{C} \cap \Lambda^{n+1}_{1,N}(K)$ such that

$$(3.56) \qquad d\underset{\sim}{E} = (-1)^{n+1} \underset{\sim}{E} \wedge \underset{\sim}{\Gamma} + \underset{\sim}{\Sigma} ,$$

where the factor $(-1)^{n+1}$ has been chosen in conformity with $(1.25)_1$. In order that (3.56) be satisfied, $\underset{\sim}{\Gamma}$ and $\underset{\sim}{\Sigma}$ must be such that the integrability conditions of (3.56) are satisfied, and this is the case only if (3.56) extends to a differential system of degree n and order N (Appendix, Section 5-11).

$$(3.57) \qquad \begin{aligned} d\underset{\sim}{E} &= (-1)^{n+1} \underset{\sim}{E} \wedge \underset{\sim}{\Gamma} + \underset{\sim}{\Sigma} , & d\underset{\sim}{\Sigma} &= (-1)^{n+2} \underset{\sim}{\Sigma} \wedge \underset{\sim}{\Gamma} + \underset{\sim}{E} \wedge \underset{\sim}{\Theta} \\ d\underset{\sim}{\Gamma} &= \underset{\sim}{\Gamma} \wedge \underset{\sim}{\Gamma} + \underset{\sim}{\Theta} , & d\underset{\sim}{\Theta} &= \underset{\sim}{\Gamma} \wedge \underset{\sim}{\Theta} - \underset{\sim}{\Theta} \wedge \underset{\sim}{\Gamma} , \end{aligned}$$

where we have dropped the superimposed $\wedge$'s in (1.25), (1.26).

Because $\underset{\sim}{\Sigma} \varepsilon \bar{C}$ and $\bar{C}$ is a closed ideal, $d\Sigma \varepsilon \bar{C}$, and hence $d\underset{\sim}{\Sigma} = (-1)^{n+2}\underset{\sim}{\Sigma} \wedge \underset{\sim}{\Gamma} + \underset{\sim}{E} \wedge \underset{\sim}{\Theta}$ implies that $\underset{\sim}{E} \wedge \underset{\sim}{\Theta} \varepsilon \bar{C}$. Now, $\underset{\sim}{E} \varepsilon \bar{C}$ gives the trivial balance system, so we may assume that $\underset{\sim}{E}$ does not belong to $\bar{C}$. Thus, (3.56) can not be satisfied unless $\underset{\sim}{\Theta} \varepsilon \bar{C}$. An integration of the system (3.57) by the procedure given by (1.28)-(1.31) yields (see (1.29))

$$\underset{\sim}{E} = (d\underset{\sim}{\psi} + \underset{\sim}{\eta} + (-1)^{n+1}H(d\underset{\sim}{\psi} \wedge \underset{\sim}{\theta}))\underset{\sim}{B}^{-1} ,$$

where $\underset{\sim}{B} = \underset{\sim}{B}_o + H(\underset{\sim}{\Gamma}\ \underset{\sim}{B})$, $\det(\underset{\sim}{B}_o) \neq 0$ and

$$\underset{\sim}{\psi} = H(\underset{\sim}{E}\ \underset{\sim}{B}) , \quad \underset{\sim}{\eta} = H(\underset{\sim}{\Sigma}\ \underset{\sim}{B}) , \quad \underset{\sim}{\theta} = H(\underset{\sim}{B}^{-1}\ \underset{\sim}{\Theta}\ \underset{\sim}{B}) .$$

This shows that $\underset{\sim}{E}$ is balance equivalent to

$$\underset{\sim}{E}_1 = d\underset{\sim}{\psi} + \underset{\sim}{\eta} + (-1)^{n+1}H(d\underset{\sim}{\psi} \wedge \underset{\sim}{\theta}) = d\underset{\sim}{\psi} + H(\underset{\sim}{\Sigma}\ \underset{\sim}{B} + (-1)^{n+1} d\underset{\sim}{\psi} \wedge \underset{\sim}{\theta}) .$$

Now, $\underset{\sim}{\Sigma} \varepsilon \bar{C}$ and $\underset{\sim}{\Theta} \varepsilon \bar{C}$ imply that $\underset{\sim}{\Sigma}\ \underset{\sim}{B} \varepsilon \bar{C}$ and $\underset{\sim}{B}^{-1}\underset{\sim}{\Theta}\ \underset{\sim}{B} \varepsilon \bar{C}$ so that

$$\eta = H(\underset{\sim}{\Sigma}\ \underset{\sim}{B}) , \quad \underset{\sim}{\theta} = H(\underset{\sim}{B}^{-1}\underset{\sim}{\Theta}\ \underset{\sim}{B})$$

are homotopy map images of elements of $\bar{C}$. There is thus the possibility that there exists a $\underset{\sim}{P} \varepsilon \bar{C}$ such that

$$\underset{\sim}{\Sigma}\ \underset{\sim}{B} + (-1)^{n+1} d\underset{\sim}{\psi} \wedge \underset{\sim}{\theta} = d\underset{\sim}{P} ,$$

in which case $\underset{\sim}{E}_1$ becomes

$$\underset{\sim}{E}_1 = d\underset{\sim}{\psi} + Hd\underset{\sim}{P} = d\underset{\sim}{\psi} + \underset{\sim}{P} - dH\underset{\sim}{P}$$

$$= d(\underset{\sim}{\psi} - H\underset{\sim}{P}) + \underset{\sim}{P} \simeq d(\underset{\sim}{\psi} - H\underset{\sim}{P}) .$$

THEOREM 3.1. *The ideal* $I\{\underset{\sim}{C}, d\underset{\sim}{C}, \underset{\sim}{E}\}$ *is closed if and only if* $\underset{\sim}{E}$ *forms a differential system of order* N *and degree* n *such that the torsion,* $\underset{\sim}{\Sigma}$ *, and the curvature,* $\underset{\sim}{\Theta}$ *, belong to the closed ideal* $\bar{C}$,

$$(3.58)\qquad \underset{\sim}{dE} = (-1)^{n+1}\underset{\sim}{E}\wedge\underset{\sim}{\Gamma} + \underset{\sim}{\Sigma}\ ,\quad d\underset{\sim}{\Sigma} = (-1)^{n+2}\underset{\sim}{\Sigma}\wedge\underset{\sim}{\Gamma} + \underset{\sim}{E}\wedge\underset{\sim}{\Theta}\ ,$$

$$d\underset{\sim}{\Gamma} = \underset{\sim}{\Gamma}\wedge\underset{\sim}{\Gamma} + \underset{\sim}{\Theta}\ ,\quad d\underset{\sim}{\Theta} = \underset{\sim}{\Gamma}\wedge\underset{\sim}{\Theta} - \underset{\sim}{\Theta}\wedge\underset{\sim}{\Gamma}\ .$$

If these conditions are met, then $\underset{\sim}{E}$ *is balance equivalent to*

$$(3.59)\qquad \underset{\sim}{E}_1 = d\psi + H(\underset{\sim}{\Sigma}\ \underset{\sim}{B} + (-1)^{n+1} d\underset{\sim}{\psi}\wedge\underset{\sim}{\theta})$$

where $\underset{\sim}{B}\in\Lambda^{o}_{N,N}(K)$ *solves the integral equation*

$$(3.60)\qquad \underset{\sim}{B} = \underset{\sim}{B}_o + H(\underset{\sim}{\Gamma}\ \underset{\sim}{B})$$

and $\underset{\sim}{\theta} = H(\underset{\sim}{B}^{-1}\underset{\sim}{\Theta}\ \underset{\sim}{B})$. $\underset{\sim}{E}$ *is thus balance equivalent to an exact balance form if there exists a* $\underset{\sim}{P}\in\bar{C}$ *such that*

$$(3.61)\qquad \underset{\sim}{\Sigma}\ \underset{\sim}{B} + (-1)^{n+1} d\underset{\sim}{\psi}\wedge\underset{\sim}{\theta} = d\underset{\sim}{P}\ ,$$

in which case

$$(3.62)\qquad \underset{\sim}{E} \simeq d(\underset{\sim}{\psi} - H\underset{\sim}{P})$$

and $\underset{\sim}{\psi}$ *and* $\underset{\sim}{\theta}$ *must be such that* $d\underset{\sim}{\psi}\wedge\underset{\sim}{\theta}\in\underset{\sim}{\bar{C}}$.

As an example, consider the case of a system of laws of balance,

$$E_\alpha = W_\alpha\mu - d(W^i_\alpha\ \mu_i)\ .$$

We then have

$$dE_\alpha = (\partial_\beta W_\alpha)dq^\beta\wedge\mu + (\partial^i_\beta W_\alpha)dy^\beta_i\wedge\mu\ .$$

However, $C^\beta\wedge\mu = (dq^\beta - y^\beta_j\ dx^j)\wedge\mu = dq^\beta\wedge\mu$ and $dC^\beta\wedge\mu_i =$ $- dy^\beta_j\ dx^j\wedge\mu_i = - dy^\beta_i\wedge\mu$, and hence

$$dE_\alpha = (\partial_\beta W_\alpha)C^\beta\wedge\mu - (\partial^i_\beta W_\alpha)dC^\beta\wedge\mu_i = 0 \bmod \bar{C}\ .$$

COROLLARY 3.1. *The ideal* $I\{\underset{\sim}{C},d\underset{\sim}{C},\underset{\sim}{E}\}$ *is closed whenever*

$$E_\alpha = W_\alpha\mu - d(W^i_\alpha\mu_i)\ .$$

Although this is a somewhat trivial example of Theorem 3.1, for $\underset{\sim}{\Gamma} \equiv \underset{\sim}{0}$, it figures heavily in applications in view of the relative universality of systems of laws of balance.

The Curvature Free Connection Induced by Balance Equivalence

It is of interest here to note that the above conditions take a particularly simple form if the connection $\underset{\sim}{\Gamma}$ of the system is curvature free, for in this case $\underset{\sim}{\Gamma} = (d\underset{\sim}{B})\underset{\sim}{B}^{-1}$, $\underset{\sim}{\theta} = \underset{\sim}{0}$ and (3.61) becomes $\underset{\sim}{\Sigma}\, \underset{\sim}{B} = d\underset{\sim}{P}$. As it turns out, curvature free connections on K are associated with balance equivalence fibration in a natural manner.

Let $\underset{\sim}{E}$ be an arbitrary balance form. The balance form

$$(3.63) \qquad \underset{\sim}{E}_1 = \underset{\sim}{E}\,\underset{\sim}{B}^{-1} + \underset{\sim}{P}$$

then belongs to $F(\underset{\sim}{E})$ for every nonsingular matrix $\underset{\sim}{B}$ and every $\underset{\sim}{P} \in \bar{C} \cap \Lambda^n_{1,N}(K)$. Since exterior differentiation of (3.63) yields

$$(3.64) \qquad d\underset{\sim}{E}_1 = (-1)^{n+1}\underset{\sim}{E}_1 \wedge \underset{\sim}{\Gamma} + d\underset{\sim}{E}\,\underset{\sim}{B}^{-1} + d\underset{\sim}{P} + (-1)^n \underset{\sim}{P} \wedge \underset{\sim}{\Gamma} ,$$

with

$$(3.65) \qquad \underset{\sim}{\Gamma} = d\underset{\sim}{B}\,\underset{\sim}{B}^{-1} ,$$

we obtain

$$(3.66) \qquad d(\underset{\sim}{E}_1 - \underset{\sim}{P}) = (-1)^{n+1} (\underset{\sim}{E}_1 - \underset{\sim}{P}) \wedge \underset{\sim}{\Gamma} + d\underset{\sim}{E}\,\underset{\sim}{B}^{-1} .$$

Now, $\underset{\sim}{B}$ is a nonsingular matrix function with domain K , so that $\underset{\sim}{B}$ may be realized as a mapping from K into the general linear matrix group $GL(N,\mathbb{R})$,

$$(3.67) \qquad \underset{\sim}{B}: K \longrightarrow GL(N,\mathbb{R}) ,$$

in which case, (3.65) shows that $\underset{\sim}{\Gamma}$ is the natural connection associated with $\underset{\sim}{B}$. Further, since each $\underset{\sim}{B}$ may be thought of as characterizing a horizontal cross-section of all balance fibers mod $\bar{C}$, that is (3.66) is equivalent to

(3.68) $$d\underset{\sim}{E}_1 = (-1)^{n+1} \underset{\sim}{E}_1 \wedge \underset{\sim}{\Gamma} + d\underset{\sim}{E}\,\underset{\sim}{B}^{-1} \bmod \bar{C} ,$$

$\underset{\sim}{\Gamma}$ is a connection that is naturally associated with the horizontal cross-section of the equivalence relation $\underset{\sim}{E}_1 = \underset{\sim}{E}\,\underset{\sim}{B}^{-1} \bmod \bar{C}$.

THEOREM 3.2. *The fibering of the space* $\Lambda^n_{1,N}(K)$ *of balance forms by the balance equivalence relation induces the connection*

(3.69) $$\underset{\sim}{\Gamma} = d\underset{\sim}{B}\,\underset{\sim}{B}^{-1}$$

on the horizontal cross-section of all fibers that is generated by the matrix group $GL(N,\mathbb{R})$ *through the mapping*

(3.70) $$\underset{\sim}{B}: K \longrightarrow GL(N,\mathbb{R}) .$$

If $F(\underset{\sim}{E})$ *is a generic fiber, all* $\underset{\sim}{E}_1 \in F(\underset{\sim}{E})$ *that satisfy*

(3.71) $$d(\underset{\sim}{E}_1 - \underset{\sim}{P}) = (-1)^{n+1}(\underset{\sim}{E}_1 - \underset{\sim}{P}) \wedge \underset{\sim}{\Gamma} + d\underset{\sim}{E}\,\underset{\sim}{B}^{-1} ,$$

for all $\underset{\sim}{P} \in \bar{C} \cap \Lambda^n_{1,N}(K)$, *belong to the horizontal cross-section characterized by* $\underset{\sim}{B}$ *and the differential system* (3.71) *has the general solution*

(3.72) $$\underset{\sim}{E}_1 = \underset{\sim}{E}\,\underset{\sim}{B}^{-1} + \underset{\sim}{P} \in F(\underset{\sim}{E}) .$$

Canonical Forms for Balance Systems

Any $\underset{\sim}{E} \in \Lambda^n_{1,N}(K)$ may be written as

(3.73) $$\underset{\sim}{E} = d\underset{\sim}{\xi} + \underset{\sim}{\eta} ,$$

where

(3.74) $$\underset{\sim}{\xi} = H\underset{\sim}{E} + d\underset{\sim}{\rho} , \quad \underset{\sim}{\eta} = Hd\underset{\sim}{E}$$

and H is the linear homotopy operator on K . A substitution of (3.73) into (3.72) and (3.71) then shows that any $\underset{\sim}{E}_1 \in F(\underset{\sim}{E})$ is given by

$$(3.75) \qquad \underset{\sim}{E}_1 = (d\underset{\sim}{\xi} + \underset{\sim}{\eta})\, \underset{\sim}{B}^{-1} \bmod \bar{C}$$

and satisfies

$$(3.76) \qquad d\underset{\sim}{E}_1 = (-1)^{n+1}\, \underset{\sim}{E}_1 \wedge \underset{\sim}{\Gamma} + d\underset{\sim}{\eta}\, \underset{\sim}{B}^{-1} \bmod \bar{C}\ , \quad \underset{\sim}{\Gamma} = d\underset{\sim}{B}\, \underset{\sim}{B}^{-1}\ .$$

If we define the torsion of the differential system (3.76) by

$$(3.77) \qquad \underset{\sim}{\Sigma} = d\underset{\sim}{\eta}\, \underset{\sim}{B}^{-1}\ ,$$

we obtain the complete differential system

$$(3.78) \qquad d\underset{\sim}{E}_1 = (-1)^{n+1}\, \underset{\sim}{E}_1 \wedge \underset{\sim}{\Gamma} + \underset{\sim}{\Sigma} \bmod \bar{C}$$

$$(3.79) \qquad d\underset{\sim}{\Sigma} = (-1)^{n+2}\, \underset{\sim}{\Sigma} \wedge \underset{\sim}{\Gamma} \bmod \bar{C}$$

with

$$(3.80) \qquad d\underset{\sim}{\Gamma} = \underset{\sim}{\Gamma} \wedge \underset{\sim}{\Gamma}\ .$$

The general solution of such a differential system is given by

$$(3.81) \quad \underset{\sim}{E}_1 = (d\xi + \eta)\, \underset{\sim}{B}^{-1} \bmod \bar{C}\ , \quad \underset{\sim}{\Sigma} = d\underset{\sim}{\eta}\, \underset{\sim}{B}^{-1} \bmod \bar{C}\ , \quad \underset{\sim}{\Gamma} = d\underset{\sim}{B}\, \underset{\sim}{B}^{-1}\ ,$$

where the matrix $\underset{\sim}{B}$ is now solved for in terms of $\underset{\sim}{\Gamma}$ by

$$(3.82) \qquad \underset{\sim}{B} = \underset{\sim}{B}_o + H(\underset{\sim}{\Gamma}\, \underset{\sim}{B})\ , \quad \det(\underset{\sim}{B}_o) \neq 0\ .$$

These considerations show that the requirement that $\underset{\sim}{\Gamma}$ be a flat connection $(d\underset{\sim}{\Gamma} = \underset{\sim}{\Gamma} \wedge \underset{\sim}{\Gamma})$ is sufficient in order that we may generate all elements of the horizontal cross-section of all fibers that is obtained from the $\underset{\sim}{B}$ determined from $\underset{\sim}{\Gamma}$ by (3.82). In other words, use of all possible flat connections generates all possible elements of all possible fibers once the base space is characterized by characterizing $\underset{\sim}{\xi}$ and $\underset{\sim}{\eta}$.

Careful note must be taken of the fact that $d\underset{\sim}{\xi} + \underset{\sim}{\eta}$ and $d\underset{\sim}{\xi} + \underset{\sim}{\eta} + \underset{\sim}{P}$ will belong to the same fiber for any $\underset{\sim}{P} \in \bar{C} \cap \Lambda^n_{1,N}(K)$. Since H does not carry $\bar{C}$ into itself, any $\underset{\sim}{P} \in \bar{C}$ can be represented as

$$(3.83) \qquad \underset{\sim}{P} = d\underset{\sim}{p} + \underset{\sim}{N}$$

with

$$(3.84) \qquad \underset{\sim}{p} = H\underset{\sim}{P}, \quad \underset{\sim}{N} = Hd\underset{\sim}{P}.$$

In this event,

$$(3.85) \qquad d\underset{\sim}{\xi}_1 + \underset{\sim}{\eta}_1 = d(\underset{\sim}{\xi} + \underset{\sim}{p}) + \underset{\sim}{\eta} + \underset{\sim}{N}$$

are equivalent, and the equality (3.85) between exact and anti-exact elements can hold if and only if

$$(3.86) \qquad \underset{\sim}{\xi}_1 = \underset{\sim}{\xi} + \underset{\sim}{p} + d\underset{\sim}{\rho}, \quad \underset{\sim}{\eta}_1 = \underset{\sim}{\eta} + \underset{\sim}{N}$$

for arbitrary $\underset{\sim}{\rho} \in \Lambda^{n-2}_{1,N}(K)$.

THEOREM 3.3. *Every balance system is balance equivalent to a balance system that has the canonical form*

$$(3.87) \qquad \underset{\sim}{E} = d\underset{\sim}{\xi} + \underset{\sim}{\eta}$$

when $\underset{\sim}{\eta}$ *is antiexact and* $\underset{\sim}{\xi}$ *is antiexact to within addition of an exact element of* $\Lambda^{n-1}_{1,N}(K)$. *The forms* $\underset{\sim}{\xi}$ *and* $\underset{\sim}{\eta}$ *are determined on each* $F(\underset{\sim}{E})$ *up to the equivalence relations*

$$(3.88) \qquad \underset{\sim}{\xi} = \underset{\sim}{\xi}_1 + \underset{\sim}{p} + d\underset{\sim}{\rho}, \quad \underset{\sim}{\eta} = \underset{\sim}{\eta}_1 + \underset{\sim}{N},$$

for all

$$(3.89) \qquad \underset{\sim}{p} = H\underset{\sim}{P}, \quad \underset{\sim}{N} = Hd\underset{\sim}{P}, \quad \underset{\sim}{P} \in \bar{C} \cap \Lambda^n_{1,N}(K),$$

and all elements of $F(\underset{\sim}{E})$ *are obtained by solving*

$$(3.90) \qquad d\underset{\sim}{E}_1 = (-1)^n \underset{\sim}{E}_1 \wedge \underset{\sim}{\Gamma} + d\underset{\sim}{\eta}\, \underset{\sim}{B}^{-1} , \quad \underset{\sim}{\Gamma} = d\underset{\sim}{B}\, \underset{\sim}{B}^{-1}$$

for all maps $\underset{\sim}{B}:K \longrightarrow GL(N,\mathbb{R})$.

From now on, we may work directly with the canonical form $d\underset{\sim}{\xi} + \underset{\sim}{\eta}$. This is particularly convenient since any $\Phi \varepsilon S(d\underset{\sim}{\xi} + \underset{\sim}{\eta}, B_n)$ gives $\Phi^*\underset{\sim}{\eta} = -d\Phi^*\underset{\sim}{\xi}$, from which we obtain the integral balance laws

$$(3.91) \qquad \int_{B_n} \Phi^*\underset{\sim}{\eta} = -\int_{\partial B_n} \Phi^*\underset{\sim}{\xi} .$$

3.3 ISOVECTORS OF $\bar{E}$ AND THE GENERATION OF SOLUTIONS

We may now proceed, exactly as with the case of the closed ideal $\bar{C}$, to construct the isovectors of the closed ideal $\bar{E}$.

Definition 3.2. A vector field $v \varepsilon T(K)$ is an *isovector* of the closed ideal $\bar{E}$ if and only if [8, 9]

$$(3.92) \qquad \pounds_v \bar{E} \subset \bar{E} .$$

The collection of all isovectors of $\bar{E}$ is denoted by $T(\underset{\sim}{E},K)$.

THEOREM 3.4. *A vector field* $v \varepsilon T(K)$ *is an isovector of* $\bar{E}$ *if and only if* v *is an isovector of* $\bar{C}$ *and*

$$(3.93) \qquad \pounds_v \underset{\sim}{E} \subset \bar{E} .$$

Thus,

$$(3.94) \qquad T(\underset{\sim}{E},K) = \{v \varepsilon TC(K) \mid \pounds_v \underset{\sim}{E} \subset \bar{E}\} .$$

Proof. Since $\bar{E} = I\{\underset{\sim}{C}, d\underset{\sim}{C}, \underset{\sim}{E}, d\underset{\sim}{E}\}$ and $n \geq 2$, $v \varepsilon T(\underset{\sim}{E},K)$ only if $\pounds_v \bar{C} \subset \bar{C}$; that is, $v \varepsilon TC(K)$. It is thus clear that (3.93) and $v \varepsilon TC(K)$ are necessary in order that $v \varepsilon T(\underset{\sim}{E},K)$. However, $\pounds_v d\underset{\sim}{E} = d\pounds_v \underset{\sim}{E}$, so that the closure of $\bar{E}$ implies that $\pounds_v d\underset{\sim}{E} \subset \bar{E}$ provided $v \varepsilon TC(K)$ and (3.93) are satisfied. In a like manner, any $\eta \varepsilon \bar{E}$ has the form

$$\eta = \underset{\sim}{E} \wedge \underset{\sim}{A} + d\underset{\sim}{E} \wedge \underset{\sim}{B} + \underset{\sim}{M} \wedge \underset{\sim}{C} + \underset{\sim}{N} \wedge d\underset{\sim}{C}$$

and hence $\mathcal{L}_v \eta \in \bar{E}$ for any v satisfying (3.93) and $v \in TC(K)$, for Lie differentiation acts as a derivation with respect to exterior multiplication. This shows that $v \in TC(K)$ and (3.93) are both necessary and sufficient in order that $v \in T(\underset{\sim}{E},K)$, and the characterization (3.94) follows directly.

Algebraic Structure of $T(\underset{\sim}{E},K)$

THEOREM 3.5. $T(\underset{\sim}{E},K)$ *is a Lie subalgebra of* $TC(K)$.

Proof. Since $T(\underset{\sim}{E},K) = \{v \in TC(K) \mid \mathcal{L}_v \underset{\sim}{E} \subset \bar{E}\}$, by Theorem 3.4, $T(\underset{\sim}{E},K) \subset TC(K)$. If v_1 and v_2 belong to $T(\underset{\sim}{E},K)$, then $a\, v_1 + b\, v_2$ and $[v_1, v_2]$ belong to $T(\underset{\sim}{E},K)$ for all real numbers (a, b) by exactly the same argument as that used in the proof of Theorem 2.1. $T(\underset{\sim}{E},K)$ is thus a closed subspace of $TC(K)$ that is also closed under the Lie multiplication [,], and hence $T(\underset{\sim}{E},K)$ is a Lie subalgebra of $TC(K)$.

For $N > 1$, $TC(K)$ is projectable onto graph space, while for $N = 1$, there is a subalgebra of $TC(K)$ that is projectable onto graph space (vid. Theorems 2.4 and 2.5). Thus, for all N, the projectable elements of $TC(K)$ form a Lie subalgebra of $TC(K)$ and this subalgebra is a proper subalgebra if $N = 1$. The subalgebra of projectable vector fields is, however, the Lie algebra associated with the first group extension of graph space. Theorem 3.5 then gives the following result.

COROLLARY 3.2. *The projectable isovectors of* $\bar{E}$ *form a Lie subalgebra of* $T(\underset{\sim}{E},K)$ *and this subalgebra contains all isovectors of* $\bar{E}$ *that belong the first group extension of graph space. For* $N > 1$, *this subalgebra coincides with* $T(\underset{\sim}{E},K)$.

The importance of this corollary is that it gives immediate access to the large body of results that have been established by the method of group extention of graph space [10, 11].

THEOREM 3.6. *If* v *is an isovector of* $\bar{E}$ *, then* v *is an isovector of the closed ideal generated by* $\underset{\sim}{C}$ *and* $\underset{\sim}{E}_1$ *for any* $\underset{\sim}{E}_1 \in F(\underset{\sim}{E})$ *. The Lie algebra* $T(\underset{\sim}{E},K)$ *is thus a fiber invariant; that is*

$$T(\underset{\sim}{E},K) = T(F(\underset{\sim}{E}),K) . \tag{3.95}$$

Proof. Lemma 3.5 showed that $\bar{E} = \bar{I}\{\underset{\sim}{C}, \underset{\sim}{E}\} = \bar{I}\{\underset{\sim}{C}, F(\underset{\sim}{E})\}$, and hence $\bar{E} = \bar{I}\{\underset{\sim}{C}, \underset{\sim}{E}\} = \bar{I}\{\underset{\sim}{C}, \underset{\sim}{E}_1\} = \bar{E}_1$ for any $\underset{\sim}{E}_1 \in F(\underset{\sim}{E})$. Thus, if v is any isovector of $\bar{E}$, $\pounds_v \bar{E}_1 = \pounds_v \bar{E} \subset \bar{E} = \bar{E}_1$ for any $\underset{\sim}{E}_1 \in F(\underset{\sim}{E})$ and hence $T(\underset{\sim}{E},K) = T(F(\underset{\sim}{E}),K)$.

Examples

The following collection of examples, taken from [8], may prove helpful to the reader. For $n = 2$, $N = 1$ and

$$C = dq - y_1 dx^1 - y_2 dx^2 , \quad \underset{\sim}{E} = d(q\,dx^1 + y_1 dx^2) ,$$

we have the linear diffusion equation. The isovectors are

$$\left.\begin{matrix} v_1 \\ \\ v^2 \\ \\ v^q \end{matrix}\right\} = \left\{\begin{matrix} 1, & 0, & 0, & x^1, & -2x^2, & 2x^1x^2, & 0 \\ \\ 0, & 1, & 0, & 2x^2, & 0, & 2(x^2)^2, & 0 \\ \\ 0, & 0, & q, & 0, & x^1q, & -(x^2+\frac{1}{2}(x^1)^2)q, & g(x^1,x^2), \end{matrix}\right.$$

with $\partial^2 g/\partial(x^1)^2 = \partial g/\partial x^2$. We do not list the v_i^α since they are determined by $v_i^\alpha = Z_i(v^\alpha - y_j^\alpha v^j)$. For $n = 2$, $N = 2$ and

$$E_1 = dq^1 \wedge dx - q^1 dq^2 \wedge dt - q^2 dq^1 \wedge dt ,$$

$$E_2 = q^1 dq^2 \wedge dx - q^1 q^2 dq^2 \wedge dt - c^2 dq^1 \wedge dt ,$$

the isovectors are

$$\left.\begin{matrix} v^1 \\ v^2 \\ v^{q^1} \\ v^{q^2} \end{matrix}\right\} = \left\{\begin{matrix} x^1 , & 1 , & q^2F/q^1 - G \\ x^2 , & 0 , & F/q^1 \\ 0 , & 0 , & 0 \\ 0 , & 1 , & 0 \end{matrix}\right.$$

with $\partial G/\partial q^2 = \partial F/\partial q^1$, $(q^1)^2 \partial G/\partial q^1 = c^2 \ \partial F/\partial q^2$, and $c^2 = dP(q^1)/dq^1$. Again with $n = N = 2$,

$$C^1 = dq^1 - y^1_1 dx^1 - y^1_2 dx^2 , \quad C^2 = dq^2 - y^2_1 dx^1 - y^2_2 dx^2$$

$$E_1 = dy^1_1 \wedge dx^2 + dy^1_2 \wedge dx^1 - \left[e^{-2q^1}((y^2_2)^2 - (y^2_1)^2) - (x^1)^{-1} y^1_1\right] dx^1 \wedge dx^2$$

$$E_2 = dy^2_1 \wedge dx^2 + dy^2_2 \wedge dx^1 - \left[2(y^1_1 y^2_1 - y^1_2 y^2_2) - (x^1)^{-1} y^2_1\right] dx^1 \wedge dx^2 ,$$

the isovectors are

$$\left.\begin{matrix} v^1 \\ v^2 \\ v^{q^1} \\ v^{q^2} \end{matrix}\right\} = \left\{\begin{matrix} 0 , & x^1 , & 0 , & 0 , & 0 \\ 1 , & x^2 , & 0 , & 0 , & 0 \\ 0 , & 0 , & 0 , & 1 , & q^2 \\ 0 , & 0 , & 1 , & q^2 , & \frac{1}{2}((q^2)^2 - e^{2q^1}) . \end{matrix}\right.$$

<u>Imbedding of Solutions in Groups of Solutions</u>

<u>THEOREM 3.7</u>. *Any* $v \in T(\underset{\sim}{E}, K)$ *transports* $\underset{\sim}{E}$ *into a balance equivalent system,*

(3.96) $\qquad \exp(s\pounds_v)\underset{\sim}{E} \subset F(\underset{\sim}{E})$,

and hence

(3.97) $\qquad \exp(s\pounds_v)F(\underset{\sim}{E}) \subset F(\underset{\sim}{E})$.

<u>Proof</u>. If $v \in T(\underset{\sim}{E}, K)$, then $\pounds_v \underset{\sim}{E} \subset \bar{E}$ and hence $\exp(s\pounds_v)\underset{\sim}{E} \subset \bar{E}$.

Thus, $\exp(s\pounds_v)\underset{\sim}{E} = \underset{\sim}{E}\,\underset{\sim}{A}(s) + \underset{\sim}{C}^\alpha \wedge \underset{\sim}{R}_\alpha(s) + d\underset{\sim}{C}^\alpha \wedge N_\alpha(s)$, where $\underset{\sim}{A}(s) \in \Lambda^0_{N,N}(K)$, $\underset{\sim}{R}_\alpha(s) \in \Lambda^{n-1}_{1,N}(K)$, $\underset{\sim}{N}_\alpha \in \Lambda^{n-2}_{1,N}(K)$ are analytic functions of s such that $\underset{\sim}{A}(0) = ((\delta^\alpha_\beta))$, $\underset{\sim}{R}_\alpha(0) = \underset{\sim}{0}$, $\underset{\sim}{N}_\alpha(0) = \underset{\sim}{0}$; that is

(3.98) $$\exp(s\pounds_v)\underset{\sim}{E} = \underset{\sim}{E}\,\underset{\sim}{A}(s) + \underset{\sim}{P}(s) , \quad \underset{\sim}{P}(s) \in \bar{C} .$$

Since it is a trivial matter to show that $\det(\underset{\sim}{A}(s)) \neq 0$, (3.98) gives $\exp(s\pounds_v)\underset{\sim}{E} \in F(\underset{\sim}{E})$. However, Theorem 3.6 established that $T(\underset{\sim}{E},K) = T(F(\underset{\sim}{E}),K)$, and hence $\exp(s\pounds_v)F(\underset{\sim}{E}) \subset F(\underset{\sim}{E})$.

THEOREM 3.8. *Let* $v \in T(\underset{\sim}{E},K)$ *and let* $\exp(v,s)$ *be the one parameter family of maps of* K *to* K *that is defined by transport of* K *along the orbits of* v ,

(3.99) $$\exp(v,s):K \longrightarrow K \mid \overset{*}{x}{}^i = \exp(s\pounds_v)x^i = f^i_v(x^j,q^\beta,y^\beta_j;s) ,$$
$$\overset{*}{q}{}^\alpha = \exp(s\pounds_v)q^\alpha = f^\alpha_v(x^j,q^\beta,y^\beta_j;s) ,$$
$$\overset{*}{y}{}^\alpha_i = \exp(s\pounds_v)y^\alpha_i = f^\alpha_{iv}(x^j,q^\beta,y^\beta_j;s) ,$$

in which case

(3.100) $$(\exp(v,s))^*\overset{*}{\Omega} = \exp(s\pounds_v)\Omega .$$

If Φ *is any regular map of* B_n *into* K , *then, for all* $s \in I(\Phi,v)$,

(3.101) $$\Phi_v(s) = \exp(v,s)\circ\Phi$$

is a 1-*parameter family of regular maps of* B_n *into* K *such that*

(3.102) $$\Phi_v(s)^*\underset{\sim}{E} = \Phi^*(\exp(s\pounds_v)\underset{\sim}{E}) = \Phi^*\{\underset{\sim}{E}\underset{\sim}{A}(s) + \underset{\sim}{P}(s)\} , \quad \underset{\sim}{P}(s) \in \bar{C} .$$

Thus, if $\Phi \in S(\underset{\sim}{E},B_n)$, *then* $\Phi_v(s)$ *is a* 1-*parameter family of elements of* $S(\underset{\sim}{E},B_n)$.

Proof. Theorem 2.3 established that the 1-parameter family of maps $\Phi_v(s)$ consists of regular maps of B_n into K for

$s \in I(\Phi, v)$. Noting that (3.100) and (3.101) yield

$$\Phi_v(s)^*\Omega = \Phi^*\{(\exp(v,s))^*\overset{*}{\Omega}\} = \Phi^*(\exp(s\pounds_v)\Omega) \tag{3.103}$$

(3.102) follows directly.

The result just established is a specific motivation for the geometric approach to partial differential equations and constitutes a direct extension of the underlying ideas introduced first by S. Lie. Some further comment would thus seem to be in order. If we have an isovector field v of $\bar{E}$, we solve the orbital equations

$$d\overset{*}{x}{}^i/dx = v^i , \quad d\overset{*}{q}{}^\alpha/ds = v^\alpha , \quad d\overset{*}{y}{}_i^\alpha/ds = v_i^\alpha , \tag{3.104}$$

subject to the initial data $\overset{*}{x}{}^i(0) = x^i$, $\overset{*}{q}{}^\alpha(0) = q^\alpha$, $\overset{*}{y}{}_i^\alpha(0) = y_i^\alpha$, and obtain the map $\exp(v,s)$ that is given by (3.99). Composition with any regular map

$$\Phi: B_n \longrightarrow K \mid x^i = \phi^i(\tau^j) , \quad q^\alpha = \phi^\alpha(\tau^j) , \quad y_i^\alpha = (\partial\phi^\alpha/\partial\tau^k)(\partial\tau^k/\partial x^i)$$

yields the 1-parameter family of maps $\Phi_v(s)$ given by

$$\overset{*}{x}{}^i = f_v^i(\phi^j(\tau^m), \phi^\alpha(\tau^m), (\partial\phi^\alpha/\partial\tau^k)(\partial\tau^k/\partial x^j); s) , \tag{3.105}$$

$$\overset{*}{q}{}^\alpha = f_v^\alpha(\phi^j(\tau^m), \phi^\beta(\tau^m), (\partial\phi^\beta/\partial\tau^k)(\partial\tau^k/\partial x^j); s) , \tag{3.106}$$

$$\overset{*}{y}{}_i^\alpha = f_{iv}^\alpha(\phi^j(\tau^m), \phi^\beta(\tau^m), (\partial\phi^\beta/\partial\tau^k)(\partial\tau^k/\partial x^j); s) , \tag{3.107}$$

where $y_i^\alpha = (\partial\phi^\alpha/\partial\tau^k)(\partial\tau^k/\partial x^i)$ results from the fact that Φ is regular (i.e., $\Phi^*\underset{\sim}{C} = \underset{\sim}{0}$). The theorem then tells us that if Φ solves the balance system $\underset{\sim}{E}$, that is $\Phi^*\underset{\sim}{E} = \underset{\sim}{0}$, then $\Phi_v(s)$ solves the balance system $\underset{\sim}{E}$. In other words, (3.105)-(3.107) define a solution of $\underset{\sim}{E}$ if $(\phi^i, \phi^\alpha, (\partial\phi^\alpha/\partial\tau^k)(\partial\tau^k/\partial x^i))$ defines a solution of $\underset{\sim}{E}$; *an isovector field* v *of* $\bar{E}$ *provides the data whereby we may imbed any solution* Φ *of the balance system* $\underset{\sim}{E}$ *in a 1-parameter group of solutions* $\Phi_v(s)$ *of* $\underset{\sim}{E}$.

In actual fact, we need only (3.105) and (3.106) in order to define the 1-parameter group of solutions that arises from a given solution Φ. This follows from the fact that any $v \in T(\underset{\sim}{E}, K)$ is also an isovector of the closed ideal $\bar{C}$, so that,

$$(3.108) \qquad \Phi_v(s)^* C^\alpha = \Phi^*(\exp(s \pounds_v) C^\alpha) = \Phi^*(A^\alpha_\beta(s)) \Phi^* C^\beta .$$

Thus, if we define C^α_s by

$$(3.109) \qquad C^\alpha_s = d\overset{*}{q}{}^\alpha - \overset{*}{y}{}^\alpha_i \, d\overset{*}{x}{}^i ,$$

then $\Phi_v(s)^* C^\alpha = \Phi_v(s)^* C^\alpha_s = \Phi^*(A^\alpha_\beta(s)) \Phi^* C^\beta$ by (3.108). We thus see that

$$(3.110) \qquad \Phi_v(s) : \{\tau^j\} \longrightarrow \{\overset{*}{\phi}{}^\alpha_i\} = \{(\partial \overset{*}{\phi}{}^\alpha(\tau^j)/\partial\tau^k)(\partial\tau^k/\partial\overset{*}{x}{}^i)\} .$$

There is a very important point that should be made in this context, however. A direct application of Corollary 2.1, together with (3.107) and (3.110) give the following result.

If the functions f^α_{iv} *that occur in* (3.107) *depend explicitly on the variables* $\phi^\beta(\tau^m)$, *then* (3.107) *and* (3.110) *yield*

$$\frac{\partial\overset{*}{\phi}{}^\alpha}{\partial\tau^k}\frac{\partial\tau^k}{\partial\overset{*}{x}{}^i} = f^\alpha_{iv}(\phi^j(\tau^m), \phi^\beta(\tau^m), (\partial\phi^\beta/\partial\tau^k)(\partial\tau^k/\partial x^j); s)$$

and these equations represent generalized contact transformations [13,14] *for any fixed value of* $s \neq 0$.

Thus, although the orbital equations for the $\overset{*}{y}{}^\alpha_i$ are obtained through $v^\alpha_i = Z_i(v^\alpha - y^\alpha_j v^j)$, and hence their content is contained in the structure of the orbital equations for the $\overset{*}{x}{}^i$ and the $\overset{*}{q}{}^\alpha$, their explicit representation leads directly to contact transformations for fixed $s \neq 0$ whenever $Z_i(v^\alpha - y^\alpha_j v^j)$ is not independent of the arguments q^α.

For example, an isovector field associated with the linear diffusion equation, $\partial\phi/\partial t = \partial^2\phi/\partial x^2$, is given by

$$v^1 = -2x^2 , \quad v^2 = 0 , \quad v = x^1 q ,$$

so that $v_i = Z_i(v - y_j v^j)$ yields

$$v_1 = q + x^1 y_1 , \quad v_2 = 2y_1 + x^1 y_2$$

and the orbital equations

$$\frac{d\overset{*}{x}{}^1}{ds} = -2\overset{*}{x}{}^2 , \quad \frac{d\overset{*}{x}{}^2}{ds} = 0 , \quad \frac{d\overset{*}{q}}{ds} = \overset{*}{x}{}^1\overset{*}{q} ,$$

$$\frac{d\overset{*}{y}_1}{ds} = \overset{*}{q} + \overset{*}{x}{}^1\overset{*}{y}_1 , \quad \frac{d\overset{*}{y}_2}{ds} = 2\overset{*}{y}_1 + \overset{*}{x}{}^1\overset{*}{y}_2 .$$

The general solution of these equations, subject to the initial data $\overset{*}{x}{}^1(0) = x^1$, $\overset{*}{x}{}^2(0) = x^2$, $\overset{*}{q}(0) = q$, $\overset{*}{y}_1(0) = y_1$, $\overset{*}{y}_2(0) = y_2$, is given by

$$\overset{*}{x}{}^1 = x^1 - 2x^2 s , \quad \overset{*}{x}{}^2 = x^2 ,$$

$$\overset{*}{q} = q\,\exp(x^1 s - x^2 s^2) ,$$

$$\overset{*}{y}_1 = (y_1 + qs)\exp(x^1 s - x^2 s^2) ,$$

$$\overset{*}{y}_2 = (y_2 + 2y_1 s + qs^2)\exp(x^1 s - x^2 s^2) .$$

Clearly, these solutions, for $s \neq 0$, satisfy the condition that there is explicit dependence of the $\overset{*}{y}_i$ on q , and we obtain the contact transformations

$$\frac{\partial\overset{*}{\phi}}{\partial\overset{*}{x}{}^1} = \left(\frac{\partial\phi}{\partial x^1} + \phi s\right)\exp(x^1 s - x^2 s^2) ,$$

$$\frac{\partial\overset{*}{\phi}}{\partial\overset{*}{x}{}^2} = \left(\frac{\partial\phi}{\partial x^2} + 2\,\frac{\partial\phi}{\partial x^1}\,s + \phi s^2\right)\exp(x^1 s - x^2 s^2)$$

for any $\phi(x^1,x^2)$ that satisfies $\partial\phi/\partial x^2 = \partial^2\phi/\partial(x^1)^2$. The convenient thing to note here, however, is that the contact transformations can be integrated directly to obtain

$$\overset{*}{x}{}^1 = x^1 - 2x^2 s , \quad \overset{*}{x}{}^2 = x^2$$

$$\phi^* = \phi(x^1,x^2)\exp(x^1 s - x^2 s^2)$$

for any $\phi(x^1,x^2)$ that satisfies $\partial\phi/\partial x^1 = \partial^2\phi/\partial(x^2)^2$. Contact transformations can thus be directly associated with any $\exp(v,s)$ for $v \in T(\underset{\sim}{E},K)$ with $\partial_\beta Z_i(v \rfloor C^\alpha) = \partial_\beta v_i^\alpha \not\equiv 0$, but this need not be done since the integrated form can also be obtained directly, namely (3.105) and (3.106).

Note that $\overset{*}{x}{}^1 = x^1 - 2x^2 s$, $\overset{*}{x}{}^2 = x^2$, $\overset{*}{\phi} = \phi(x^1,x^2)\cdot\exp(x^1 s - x^2 s^2)$ gives

$$\frac{\partial\overset{*}{\phi}}{\partial\overset{*}{x}{}^2} - \frac{\partial^2\overset{*}{\phi}}{\partial(\overset{*}{x}{}^1)^2} = \left(\frac{\partial\phi}{\partial x^2} - \frac{\partial^2\phi}{\partial(x^1)^2}\right)\exp(x^1 s - x^2 s^2)$$

and

$$\frac{\partial(\overset{*}{x}{}^1,\overset{*}{x}{}^2)}{\partial(x^1,x^2)} = 1$$

for all s. In this case, we have $I(\Phi,v) = \mathbb{R}$ for all $\Phi \in R(B_n)$ so that $\overset{*}{\phi}(\overset{*}{x}{}^1,\overset{*}{x}{}^2)$ satisfies the linear diffusion equation whenever $\phi(x^1,x^2)$ does, and this holds for all $s \in \mathbb{R}$. If we solve for (x^1,x^2) in terms of $(\overset{*}{x}{}^1,\overset{*}{x}{}^2)$, then, for all $s \in \mathbb{R}$.

$$\overset{*}{\phi}(\overset{*}{x}{}^1,\overset{*}{x}{}^2;s) = \phi(\overset{*}{x}{}^1 + 2s\overset{*}{x}{}^2,\overset{*}{x}{}^2)\exp(\overset{*}{x}{}^1 s + \overset{*}{x}{}^2 s^2) ,$$

which, of course, can equally well be written as $\phi_1(x^1,x^2;s) = \overset{*}{\phi}(x^1,x^2;s) = \phi(x^1 + 2sx^2,x^2)\exp(x^1 s + x^2 s^2)$. This mapping of the solution $\phi(x^1,x^2)$ onto the 1-parameter family of solutions $\phi_1(x^1,x^2;s)$ is exactly the content of Theorem 3.8. If we take another isovector, say $v^1 = x^1$, $v^2 = 2x^2$, $v = 0$, the solution of the orbital equations

$$d\overset{*}{x}{}^1/d\sigma = \overset{*}{x}{}^1 , \quad d\overset{*}{x}{}^2/d\sigma = 2\overset{*}{x}{}^2 , \quad d\overset{*}{q}/d\sigma = 0$$

is

$$\overset{*}{x}{}^1 = e^\sigma x^1 , \quad \overset{*}{x}{}^2 = e^{2\sigma}x^2 , \quad \overset{*}{q} = q .$$

Accordingly, if we take q to be given by the map that defines the solution just obtained, we have

$$\overset{*}{\phi} = \phi(x^1 + 2sx^2, x^2)\exp(x^1 s + x^2 s^2) \quad \text{because} \quad \overset{*}{q} = q .$$

Now, solve for (x^1, x^2) in terms of $(\overset{*}{x}{}^1, \overset{*}{x}{}^2)$. A direct substitution yields

$$\overset{*}{\phi}(\overset{*}{x}{}^1, \overset{*}{x}{}^2; s, \sigma) = \phi(e^{-\sigma}\overset{*}{x}{}^1 + 2se^{-2\sigma}\overset{*}{x}{}^2, e^{-2\sigma}\overset{*}{x}{}^2)\exp(e^{-\sigma}\overset{*}{x}{}^1 s + e^{-2\sigma}\overset{*}{x}{}^2 s^2)$$

which satisfies $\partial\overset{*}{\phi}/\partial\overset{*}{x}{}^2 - \partial^2\overset{*}{\phi}/\partial(\overset{*}{x}{}^1)^2 = 0$ by Theorem 3.8. We thus obtain the 2-parameter family of solutions

$$\phi_2(x^1, x^2; s, \sigma) = \phi(e^{-\sigma}x^1 + 2se^{-2\sigma}x^2, e^{-2\sigma}x^2)\exp(e^{-\sigma}x^1 s + e^{-2\sigma}x^2 s^2) ,$$

for all $(s,\sigma) \in \mathbb{R} \times \mathbb{R}$, that obtains from any specific solution $\phi(x^1, x^2)$. We note in passing that the orbital equations of this latter isovector admit the invariant variety

$$(\overset{*}{x}{}^1)^2/\overset{*}{x}{}^2 = (x^1)^2/x^2 ,$$

which defines one of the classic similarity variables of the diffusion equation.

As a further example,

$$v^1 = -2\alpha x^2 + 2\beta x^1 x^2 , \quad v^2 = 2\beta(x^2)^2$$

$$v = (\alpha x^1 - \beta x^2 - \frac{1}{2}\beta(x^1)^2)q$$

is an isovector of the linear diffusion equation for any constants α and β , and the solution of the orbital equations for this isovector field is given by

$$\overset{*}{x}{}^1 = (x^1 - 2\alpha x^2 s)/(1 - 2\beta x^2 s)$$

$$\overset{*}{x}{}^2 = x^2/(1 - 2\beta x^2 s)$$

$$\overset{*}{q} = q(1 - 2\beta x^2 s)^{1/2} \exp\left[\frac{\alpha^2 s}{2\beta} - \frac{(x^1 - \frac{\alpha}{\beta})^2 \beta s}{2(1 - 2\beta x^2 s)}\right] .$$

If we solve the first two of these equations for x^1 and x^2 in terms of $\overset{*}{x}{}^1$ and $\overset{*}{x}{}^2$, then

$$\overset{*}{\phi}(\overset{*}{x}{}^1,\overset{*}{x}{}^2) = \frac{\phi\left(\dfrac{\overset{*}{x}{}^1 + 2\alpha\overset{*}{x}{}^2 s}{1 + 2\beta\overset{*}{x}{}^2 s}, \dfrac{\overset{*}{x}{}^2}{1 + 2\beta\overset{*}{x}{}^2 s}\right)}{\sqrt{1 + 2\beta\overset{*}{x}{}^2 s}} \exp\left[\alpha\overset{*}{x}{}^1 s + \alpha^2\overset{*}{x}{}^2 s^2 - \frac{\beta(\overset{*}{x}{}^1 + 2\alpha\overset{*}{x}{}^2 s)^2 s}{2(1 + 2\beta\overset{*}{x}{}^2 s)}\right]$$

is a solution of the linear diffusion equation whenever $\phi(x^1,x^2)$ is a solution of that equation.

A further example with $N > 1$ is provided by hydrodynamics in one spatial dimension with a chemical reaction. Let $\Phi^* q^1 = u$ = velocity, $\Phi^* q^2 = p$ = pressure, $\Phi^* q^3 = v$ = specific volume, and $\Phi^* q^4 = \lambda$ = extent of reaction variable of a single chemical reaction for which both the reactant and the product satisfy the polytropic equation of state with the same polytropic index γ. With $t = x^1$ and $X/v_o = x^2$, a Lagrangian spatial variable scaled with respect to a reference volume v_o, the balance forms are

$$E_1 = dq^3 \wedge dx^2 + dq^1 \wedge dx^1 , \quad E_2 = dq^1 \wedge dx^2 - dq^2 \wedge dx^1$$

$$E_3 = (dq^4 - R(q^2,q^3,q^4)dx^1) \wedge dx^2 ,$$

$$E_4 = (q^3\, dq^2 + \gamma\, q^2\, dq^3 + (1-\gamma)\, dq^4) \wedge dx^2 ,$$

where $R(q^2,q^3,q^4)$ defines the rate law for the reaction process. Since there is no y-occurrence (i.e., the equations are of first order), the isovectors are defined by $\pounds_v E_\alpha = A_\alpha^\beta E_\beta$ with

$$v = v^1 \bar{\partial}_1 + v^2 \bar{\partial}_2 + V^1 \partial_1 + V^2 \partial_2 + V^3 \partial_3 + V^4 \partial_4 .$$

They are given by

$$v^1 = (k_3 - k_5)x^1 + k_5 \ , \quad v^2 = (k_4 - k_1)x^2 + k_6 \ ,$$

$$V^1 = k_1 q^1 + k_2 \ , \quad V^2 = (k_3 - k_4 - k_1)q^2 \ ,$$

$$V^3 = (k_3 - k_4 + k_1)q^3 \ , \quad V^4 = 2(k_3 - k_4)q^4 + k_7$$

where the constants k_1 through k_7 are determined by the rate law that must be of the form

$$R = (q^2)^{\frac{k_3-k_4-k_1}{k_3+k_1+2k_4}} H(\alpha,\beta)$$

with

$$\alpha = q^3(q^2)^{\frac{k_4-k_3-k_1}{k_3-k_4-k_1}} \ ,$$

$$\beta = (2(k_3 - k_4)q^4 + k_7)(q^2)^{\frac{2(k_4-k_3)}{k_3-k_4-k_1}} \ .$$

The mapping of solutions onto solutions is then generated by

$$\overset{*}{x}{}^1 = x^1 e^{(k_3-k_1)s} + \frac{k_5}{k_3-k_1}\left(e^{(k_3-k_1)s} - 1\right)$$

$$\overset{*}{x}{}^2 = x^2 e^{(k_4-k_1)s} + \frac{k_6}{k_4-k_1}\left(e^{(k_4-k_1)s} - 1\right)$$

$$\overset{*}{q}{}^1 = q^1 e^{k_1 s} + \frac{k_2}{k_1}\left(e^{k_1 s} - 1\right) , \quad \overset{*}{q}{}^2 = q^2 e^{(k_3-k_4-k_1)s}$$

$$\overset{*}{q}{}^3 = q^3 e^{(k_3-k_4+k_1)s} \ ,$$

$$\overset{*}{q}{}^4 = q^4 e^{2(k_3-k_4)s} + \frac{k_7}{2(k_3-k_4)}\left(e^{2(k_3-k_4)s} - 1\right) .$$

Particular note should be taken of the fact that the absence of a reaction process (i.e., $H(\alpha,\beta) = 0$) gives a 6 parameter family of isovectors, while $H(\alpha,\beta) \neq 0$ will lead to only a single isovector field in the general case.

Contact Transformations and Generalized Bäcklund Transformations

A specific observation would now seem to be in order. Although the mapping $\exp(v,s)$ has been shown to be a contact transformation for any $v \in T(\underset{\sim}{E},K)$, it also maps solutions of the second order partial differential equations $\{E_\alpha\}$ onto solutions of the same system of second order partial differential equations (see (3.102)). In this regard, $\exp(v,s)$ is very difficult to distinguish from an auto-Bäcklund transformation [14, 15] if we did not already know that $\exp(v,s)$ is a contact transformation. It would thus seem that a more precise definition of what does and what does not constitute a Bäcklund transformation would be most useful.

The original work on Bäcklund transformations took the point of view that Lie's theory of contact transformations was specifically tailored to the study of first order partial differential equations. Accordingly, any two surfaces that could be mapped onto one another by a contact transformation were taken as equivalent; that is, contact transformations were explicitly excluded from consideration. This is shown most explicitly in Forsyth's presentation of Bäcklund transformations [14]. We have also seen the analogue of exactly this in the above results, for Theorem 3.7 has shown that $\exp(s\pounds_v)$, for $v \in T(\underset{\sim}{E},K)$, maps $\underset{\sim}{E}$ into the fiber of balance equivalent systems, $F(\underset{\sim}{E})$. This fact notwithstanding, Theorem 3.8 shows that $\exp(v,s)$, for $v \in T(\underset{\sim}{E},K)$, maps every solution of the given balance system $\underset{\sim}{E}$ onto a solution of that same balance system, and it is this fact, above all others, that seems to be what Bäcklund transformations are all about from the practical point of view. We therefore give the following working definition of generalized Bäcklund transformation.

Definition 3.3. A map $\rho: (x^i, \phi^\alpha(x^i)) \longrightarrow (\overset{*}{x}{}^i, \overset{*}{\phi}{}^\alpha(\overset{*}{x}{}^i))$ is a *(generalized) Bäcklund transformation* if and only if ρ maps all solutions of one balance system onto solutions of another balance system, If the two balance systems are balance equivalent, then ρ is referred to as an *auto-Bäcklund transformation*.

Let us go back and reexamine Theorem 3.8 in this new light. First off, if v is any element of $TC(K)$, then (3.102) holds, namely

$$\Phi_v(s)^* \underset{\sim}{E} = \Phi^*(\exp(s\mathfrak{L}_v)\underset{\sim}{E}) .$$

Thus, for fixed s, $\exp(v,s)$ is a (generalized) Bäcklund transformation that maps all solutions Φ of the balance system $\exp(s\mathfrak{L}_v)\underset{\sim}{E}$ onto solutions $\Phi_v(s)$ of the given balance system $\underset{\sim}{E}$. The only troubles here are (1), we do not know what the balance system $\exp(s\mathfrak{L}_v)\underset{\sim}{E}$ is without summing the indicated operator series and (2), we actually have different balance systems for different values of the parameter s. Accordingly, although we may claim that $\exp(v,s)$ generates a generalized Bäcklund transformation for each $v \in TC(K)$ and each value of s, we are unable to find the $\Phi \in S(\exp(s\mathfrak{L}_v)\underset{\sim}{E}_n, B_n)$ because we do not know what the balance system $\exp(s\mathfrak{L}_v)\underset{\sim}{E}$ actually is. Things become considerabl simpler if we require v to belong to the subalgebra $T(\underset{\sim}{E},K)$, for in that event we do not have to sum the operator series $\exp(s\mathfrak{L}_v)\underset{\sim}{E}$ because we know that $\exp(s\mathfrak{L}_v)\underset{\sim}{E} \subset \bar{E}$. Since any $\Phi \in S(\underset{\sim}{E},B_n)$ annihilates $\bar{E}$, we obtain $\Phi_v(s) \in S(\underset{\sim}{E},B_n)$ directly from $\Phi_v(s)^*\underset{\sim}{E} = \Phi^*(\exp(s\mathfrak{L}_v)\underset{\sim}{E}) = 0$. That is, $\exp(v,s)$ is an auto-Bäcklund transformation. Further, this latter result holds for all values of s for which $\Phi_v(s)^*\mu \neq 0$ and for all $v \in T(\underset{\sim}{E},K)$. In fact, if we exponentiate the Lie algebra $T(\underset{\sim}{E},K)$ we obtain a Lie group of auto-Bäcklund transformations of the balance system $\underset{\sim}{E}$. These results are summarized as follows.

<u>COROLLARY 3.3</u>. *Any* $v \in TC(K)$ *generates a generalized Bäcklund transformation* $\exp(v,s)$, *for fixed* s, *that maps all solutions of the balance system* $\exp(s\mathfrak{L}_v)\underset{\sim}{E}$ *onto solutions of the balance system* $\underset{\sim}{E}$, *and* $\exp(T(\underset{\sim}{E},K))$ *is a Lie group of generalized auto-Bäcklund transformations of the balance system* $\underset{\sim}{E}$.

Granted, these results depend upon our definition of generalized Bäcklund transformations and destroy the classic distinction between Bäcklund transformations and contact transformations. However, it seems, at least to this investigator, that the distinction between the two kinds of transformations is of no great consequence, while insistence on the distinction would force us to relinquish the pleasant circumstance of knowing that Bäcklund transformations map solutions of balance systems onto solutions of balance systems whether or not this useful circumstance happens to arise from a contact transformation in the classic sense. We leave it to the reader, however, to judge the merits of Definition 3.3 and its attendant consequences.

A significant simplification occurs if the isovector field v of $\bar{E}$ is projectable, which is always the case if $N > 1$. In this event, the orbital equations reduce to equations for the orbits on graph space and (3.105), (3.106) become

$$\text{(3.111)} \qquad \overset{*}{x}{}^i = f_v^i(\phi^j(\tau^m), \phi^\beta(\tau^m); s) ,$$

$$\text{(3.112)} \qquad \overset{*}{q}{}^\alpha = f_v^\alpha(\phi^j(\tau^m), \phi^\beta(\tau^m); s) .$$

This is also what obtains for general N if we replace the contact structure K by the first group extension structure E. Thus, only for the case $N = 1$, do we obtain results that could not have been obtained by the methods of group extension.

All of these results are dependent upon the arduous task of finding the isovector fields of a given $\bar{E}$. Theorem 3.3 shows that the search must start with isovector fields v of the closed ideal $\bar{C}$. We must thus demand that

$$\text{(3.113)} \qquad v_i^\alpha = Z_i(v \rfloor C^\alpha) = Z_i(v^\alpha - y_i^\alpha v^i)$$

and $v^i = f^i(x^j, q^\beta)$, $v^\alpha = f^\alpha(x^j, q^\beta)$ for $N > 1$, while $v^i = \partial_1^i \xi(x^j, q^\beta, y_j^\beta)$, $v^1 = f(x^j, q^\beta) + (y_i^1 \partial_1^i - 1)\xi(x^j, q^\beta, y_j^\beta)$ for $N = 1$, by Theorem 2.2. With the v's so structured, we must further satisfy $\pounds_{\underset{\sim}{v}} E \subset \bar{E}$; that is

$$(3.114) \qquad \pounds_{\underset{\sim}{v}} E = \underset{\sim}{E}\,\underset{\sim}{R} + C^{\alpha} \wedge \underset{\sim}{S}_{\alpha} + dC^{\alpha} \wedge \underset{\sim}{T}_{\alpha}$$

for some $\underset{\sim}{R} \varepsilon \Lambda^{o}_{N,N}(K)$, $\underset{\sim}{S}_{\alpha} \varepsilon \Lambda^{n-1}_{1,N}(K)$, $\underset{\sim}{T}_{\alpha} \varepsilon \Lambda^{n-2}_{1,N}(K)$. Since $\pounds_{\underset{\sim}{v}} E$ is a system of linear first order differential equations in the v's whose form is determined by $\underset{\sim}{E}$, substitution of (3.113) leads to a system of linear *second* order partial differential equations for the functions f^i, f^{α}, for $N > 1$, or for f and ξ , for $N = 1$, that serve to determine the isovector field. For example, if $\underset{\sim}{E} \simeq \underset{\sim}{d\xi}$ then $\pounds_v \underset{\sim}{d\xi} = d(v \rfloor \underset{\sim}{d\xi})$. We must thus solve

$$d(v \rfloor \underset{\sim}{d\xi}) = \underset{\sim}{d\xi}\,\underset{\sim}{R} + C^{\alpha} \wedge \underset{\sim}{S}_{\alpha} + dC^{\alpha} \wedge \underset{\sim}{T}_{\alpha}$$

for v subject to $v^{\alpha}_{i} = Z_i(v \rfloor C^{\alpha})$ and the remaining conditions of Theorem 2.2. Altogether, this constitutes something of a task of no small degree since $\dim(\Lambda^n(K)) = \binom{n+N+nN}{n}$. Now, it is quite often the case that particular solutions can be readily spotted. Even a particular solution is useful, however, for it provides for the imbedding of any solution of the balance system in a 1-parameter group of solutions. Thus, even if the general form of the isovectors of $\bar{E}$ can not be found, securing any collection of nontrivial isovectors provides a significant amount of additional information concerning the solution set of the given balance system.

An Inverse Problem

There is an interesting inverse problem that can be associated with isovectors. Suppose that we choose nontrivial functions $v^i = f^i(x^j, q^{\beta})$ and $v^{\alpha} = f^{\alpha}(x^j, q^{\beta})$ in some convenient fashion, then $v^{\alpha}_{i} = Z_i(v^{\alpha} - y^{\alpha}_{j} v^j)$ determines v so that $v \varepsilon TC(K)$. An obvious way to choose these functions is such that the orbital equations can be integrated in closed form. We then ask the inverse question: *find all* nontrivial balance systems $\{E_{\alpha}\}$ such that $\pounds_v E_{\alpha} \subset \bar{I}\{\underset{\sim}{C}, \underset{\sim}{E}\}$. The collection of all such balance systems will have the isovector v in common, and hence they will also have the mapping $\exp(v,s)$ in common. Thus, all such balance systems will share a common mapping of solutions onto solutions and this mapping will be given in closed form if the functions f^i and f^α are so chosen that $d\overset{*}{x}{}^i/ds = f^i(\overset{*}{x}{}^j, \overset{*}{q}{}^\beta)$, $d\overset{*}{q}{}^\alpha/ds = f^\alpha(\overset{*}{x}{}^j, \overset{*}{q}{}^\beta)$ can be integrated in closed form subject to the initial conditions $\overset{*}{x}{}^i(0) = x^i$, $\overset{*}{q}{}^\alpha(0) = q^\alpha$. The inverse problem thus has strong computational advantages and may provide unexpected insights into otherwise complex balance systems. This is a virgin area with much needed work.

For example, let $n = 2$, so that $\mu = dx^1 \wedge dx^2$, $\mu_1 = dx^2$, $\mu_2 = - dx^1$, and take

$$v^1 = f'(x^2) , \quad v^2 = 1 , \quad v = 0 .$$

This gives $v \rfloor C = - f'(x^2)y_1 - y_2$, and the associated similarity variables are defined by solving

$$0 = \Phi^*(v \rfloor C) = - f'(x^2)\partial_1\phi - \partial_2\phi ;$$

that is,

$$\phi = \rho(x^1 - f(x^2)) .$$

Let us look for balance systems of the form $E = d(W^i\mu_i)$ for which $\pounds_v E \subset \bar{E}$. Since $v_i = Z_i(v \rfloor C)$, we have

$$v = f'(x^2)\partial_1 + \partial_2 - f''(x^2)y_1\partial^2 ,$$

and hence we obtain the requirement

$$d\pounds_v(W^i\mu_i) = A\, d(W^i\mu_i) + B \wedge C + R\, dC .$$

Clearly a sufficient condition for satisfaction of this requirement is given by

$$\pounds_v(W^1\, dx^2 - W^2\, dx^1) = P(dq - y_1 dx^1 - y_2 dx^2) + \Psi(q)dq .$$

Resolving this expression on the basis (dx^1, dx^2, dq) and elimination of the quantity R yields the following conditions:

$$v(W^2) = - y_1\Psi(q) , \quad v(W^1) = f''(x^2)W^2 + y_2\Psi(q) .$$

If we set

$$W_1 = f'(x^2)W^2 + \eta ,$$

the above equations become

$$v(W^2) = - y_1\Psi(q) , \quad v(\eta) = (f'(x^2)y_1 + y_2)\Psi(q) .$$

The general solution of these equations is given by

$$W^2 = \alpha - \Psi(q)x^2y_1 , \quad \eta = \beta + \Psi(q)x^2(f'(x^2)y_1 + y_2)$$

where α and β are arbitrary functions of the variables

$$x^1 - f(x^2), \quad q, \quad y_1, \quad f'(x^2)y_1 + y_2 .$$

Thus, W^1 and W^2 are given by

$$W^1 = \Psi(q)x^2y_2 + f'(x^2)\alpha + \beta ,$$

$$W^2 = - \Psi(q)x^2y_1 + \alpha .$$

It is of interest to note that

$$W^1 = f'(x^2)\alpha , \; W^2 = \alpha = h(x^1 - f(x^2), \; f'(x^2)y_1 + y_2)$$

gives the solution $\phi = \rho(x^1 - f(x^2))$ of $(D_1\Phi^*W^1 + D_2\Phi^*W^2)\mu = \Phi^*E = 0$. Further, since the solutions of the orbital equations of v are given by

$$\overset{*}{x}{}^1 = x^1 + f(x^2 + s) - f(x^2) , \quad \overset{*}{x}{}^2 = x^2 + s , \quad \overset{*}{q} = q ,$$

any solution $\phi(x^1,x^2)$ of the balance system $\Phi^*E = 0$ can be imbedded in a 1-parameter family of solutions

$$\overset{*}{\phi}(\overset{*}{x}^1, \overset{*}{x}^2) = \phi(\overset{*}{x}^1 - f(\overset{*}{x}^2) + f(\overset{*}{x}^2 - s), \overset{*}{x}^2 - s) .$$

We note in passing, that if ϕ is a similarity solution, $\phi = \rho(x^1 - f(x^2))$, then $\overset{*}{\phi} = \rho(\overset{*}{x}^1 - f(\overset{*}{x}^2))$. This is the rule rather than the exception, for we show in the next section that the similarity solutions associated with any isovector v of $\bar{E}$ are functionally invariant under the mapping $\exp(v,s)$.

3.4 SIMILARITY SOLUTIONS AND ISOVECTORS OF THE EXTENDED IDEAL

Section 2.5 showed that we could associate with each isovector field v of $\bar{C}$ a collection

$$(3.115)\qquad R(v,B_n) = \{\Phi \varepsilon R(B_n) \mid \Phi^*(v \rfloor \underset{\sim}{C}) = \underset{\sim}{0} , \quad v \varepsilon TC(K)\}$$

with the property that $R(v,B_n)$ is the maximal subcollection of $R(B_n)$ for which v is a characteristic vector field of the contact ideal $\bar{C}$. If we take the intersection of $R(v,B_n)$ with $S(\underset{\sim}{E},B_n)$, and require v to be an isovector of $\bar{E}$, we obtain the *similarity* collection

$$(3.116)\qquad S(v,\underset{\sim}{E},B_n) = \{\Phi \varepsilon R(B_n) \mid \Phi^*\underset{\sim}{E} = \underset{\sim}{0} , \quad \Phi^*(v \rfloor \underset{\sim}{C}) = \underset{\sim}{0} , \quad v \varepsilon T(\underset{\sim}{E},K)\} ,$$

associated with the isovector v . Thus, any $\Phi \varepsilon S(v,\underset{\sim}{E},B_n)$ is a solution of the balance system $\underset{\sim}{E}$ that also satisfies the N further conditions

$$(3.117)\qquad \Phi^*(v \rfloor \underset{\sim}{C}) = \Phi^*(v^\alpha - y^\alpha_i v^i) = 0 ;$$

namely, a system of N first order partial differential equations whose characteristic equations are exactly the orbital equations of the $v \varepsilon TC(K)$. The reason for referring to $S(v,\underset{\sim}{E},B_n)$ as similarity solutions is twofold. First, the elements $S(v,\underset{\sim}{E},B_n)$ coincide with what are classically referred to as similarity solutions of the balance system $\underset{\sim}{E}$ [8, 10, 11]. Second, a combination of Theorems 2.3, 2.6 and 3.7 shows that $\Phi_v(s)$ *maps* $S(v,\underset{\sim}{E},B_n)$ *into itself* and that $S(v,\underset{\sim}{E},B_n)$ is a maximal proper subset of $S(\underset{\sim}{E},K)$ with this property. *Similarity solutions of the balance system* E *associated with the isovector* v *of* $\bar{E}$ *constitute a maximal invariant variety under transport along the orbits of* v . Thus, in particular, if $\{\phi^\alpha(x^j)\}$ is given by an element of $S(v,\underset{\sim}{E},B_n)$, then $\Phi_v(s)$ gives the solution

$$\overset{*}{\phi}{}^\alpha(x^j;s) \quad = \quad \phi^\alpha(x^j) \ .$$

This result is of considerable utility since it provides direct access to certain questions concerning whether a balance system $\underset{\sim}{E}$ admits solutions that are asymptotic to similarity solutions. In order to see this, suppose that $\exp(v,s)$ maps one or more of the x's into infinity as s tends to infinity. The solutions that obtain from $\Phi_v(s)$ for *all* $\Phi \in S(\underset{\sim}{E},B_n)$, having the general form $\overset{*}{\phi}(\overset{*}{x}{}^j;s)$, will remain bounded in the limit as s tends to infinity only if the s-dependence cancels out identically. Now, if $\Phi \in S(v,\underset{\sim}{E},B_n)$, then $\overset{*}{\phi}(\overset{*}{x}{}^j;s) = \phi(\overset{*}{x}{}^j)$, and $S(v,\underset{\sim}{E},B_n)$ is the maximal invariant variety with this property. Accordingly, if we demand global boundedness of the solutions, any $\Phi \in S(E,B_n)$ must admit an element of $S(v,\underset{\sim}{E},B_n)$ as an asymptotic representation. This follows simply by noting that one or more of the x's tending to infinity for fixed s is equivalent to s tending to infinity for fixed x's by the imbedding of all solutions in one-parameter groups of solutions through $\Phi_v(s) = \exp(v,s)\circ\Phi$.

The specific example of linear diffusion in two spatial dimensions provides an excellent example of this situation. In this case we have $n = 3$ with $x^0 = t$, $x^1 = x$, $x^2 = y$, $\mu = dx^0 \wedge dx^1 \wedge dx^2$, the contact form is

$$C \quad = \quad dq - y_0\, dx^0 - y_1\, dx^1 - y_2\, dx^2 \ ,$$

and the balance 3-form is

$$E = d(Rq\,\mu_0 - y_1\,\mu_1 - y_2\,\mu_2)$$

where R = constant. If Φ is any map such that $\Phi^*C = 0$, $\Phi^*E = 0$, then $q = \phi(x^0, x^1, x^2)$ must satisfy $R\partial_0\phi = \partial_1\partial_1\phi + \partial_2\partial_2\phi$. A direct calculation† shows that the isovectors of the closed ideal $\bar{E}$ that satisfy $\partial_q\partial_q v = \partial_q v^i = 0$ are given by

$$\left.\begin{matrix} v^0 \\ v^1 \\ v^2 \\ v \end{matrix}\right\} = \left\{\begin{matrix} 1, & 0, & 0, & 0, & 0, & 2x^0 \\ 0, & 1, & 0, & 0, & x^2, & x^1 \\ 0, & 0, & 1, & 0, & -x^1, & x^2 \\ 0, & 0, & 0, & q, & 0, & 0 \end{matrix}\right. ,$$

and

$$\left.\begin{matrix} v^0 \\ v^1 \\ v^2 \\ v \end{matrix}\right\} = \left\{\begin{matrix} 0 & , & 0 & , & (x^0)^2 & , & 0 \\ x^0 & , & 0 & , & x^0x^1 & , & 0 \\ 0 & , & x^0 & , & x^0x^2 & , & 0 \\ -\frac{R}{2}x^1q & , & -\frac{R}{2}x^2q & , & -(x^0+\frac{R}{4}(x^1)^2+\frac{R}{4}(x^2)^2)q & , & r \end{matrix}\right.$$

with $R\partial_0 r = \partial_1\partial_1 r + \partial_2\partial_2 r$. For purposes of discussion, we confine attention to the isovector that is truly representative of 2-dimensional diffusion, namely $v^0 = (x^0)^2$, $v^1 = x^0x^1$, $v^2 = x^0x^2$, $v = -(x^0 + \frac{R}{4}(x^1)^2 + \frac{R}{4}(x^2)^2)q$. The general solution of the orbital equations of this isovector field is given by

$$\overset{*}{x}{}^0 = x^0(1-sx^0)^{-1}, \quad \overset{*}{x}{}^1 = x^1(1-sx^0)^{-1}, \quad \overset{*}{x}{}^2 = x^2(1-sx^0)^{-1},$$

$$\overset{*}{q} = q(1-sx^0)\exp\left\{-\left(\frac{Rs}{4}\right)\frac{(x^1)^2+(x^2)^2}{1-sx^0}\right\},$$

and leads to the following mappings of solution onto solutions:

---

†These results were obtained by Mr. Roy D. Hegedus in connection with a senior project.

$$\overset{*}{\phi}(\overset{*0}{x},\overset{*1}{x},\overset{*2}{x};s) = \phi\left(\frac{\overset{*0}{x}}{1+s\overset{*0}{x}}, \frac{\overset{*1}{x}}{1+s\overset{*0}{x}}, \frac{\overset{*2}{x}}{1+s\overset{*0}{x}}\right) \frac{\exp\left\{-\left(\frac{Rs}{4}\right)\frac{(\overset{*1}{x})^2+(\overset{*2}{x})^2}{1+s\overset{*0}{x}}\right\}}{1+s\overset{*0}{x}}$$

Similarity solutions associated with this isovector field must satisfy the constraint $\overset{*}{\Phi}(v \rfloor C) = 0$. The general solution of this quasilinear first order partial differential equation can be written in the form

$$W\left(\frac{x^1}{x^0}, \frac{x^2}{x^0}, \phi x^0 \exp\left\{\left(\frac{R}{4}\right)\frac{(x^1)^2+(x^2)^2}{x^0}\right\}\right) = 0 ,$$

and hence, the q-wise regular solutions are given by

$$\phi(x^0,x^1,x^2) = M(x^1/x^0, x^2/x^0)(x^0)^{-1} \exp\left\{-\left(\frac{R}{4}\right)\frac{(x^1)^2+(x^2)^2}{x^0}\right\}$$

for any C^2 function $M(z^1,z^2)$ defined over two-dimensional number space. It is now a simple matter to see that the substitution of any such $\phi(x^0,x^1,x^2)$ into the previous equation that defines $\overset{*}{\phi}(\overset{*0}{x},\overset{*1}{x},\overset{*2}{x};s)$ leads to the invariance relation

$$\overset{*}{\phi}(\overset{*0}{x},\overset{*1}{x},\overset{*2}{x};s) = \phi(\overset{*0}{x},\overset{*1}{x},\overset{*2}{x}) .$$

An examination of these results together with the occurrence of the product $(s\overset{*0}{x})$ provides the basis for examining the asymptotic behavior of the solutions for large positive as well as negative values of x^0 .

Mappings of Similarity Solutions

The structure associated with the ideal $\bar{E}$ is somewhat disappointing for Theorem 3.7 shows that transport of the given balance system $\underset{\sim}{E}$ along the orbits of any isovector field v of $\bar{E}$ is nothing more than transport of $\underset{\sim}{E}$ along the fiber $F(\underset{\sim}{E})$ of balance equivalent systems. Thus, transport by isovectors of $\bar{E}$ can only result in new systems that could already be obtained by balance equivalence. If, however, we are willing to consider only similarity solutions, a whole new vista opens up. Section 2.5 showed that if we replace the ideal $\bar{C}$ by the extended ideal $\bar{I}\{\underset{\sim}{C}, v\rfloor\underset{\sim}{C}\}$, then every isovector of $\bar{C}$ is also an isovector of $\bar{I}\{\underset{\sim}{C}, v\rfloor\underset{\sim}{C}\}$ and is, in fact, a characteristic vector field of $\bar{I}\{\underset{\sim}{C}, v\rfloor\underset{\sim}{C}\}$. Further, the isovector fields of $\bar{I}\{\underset{\sim}{C}, v\rfloor\underset{\sim}{C}\}$ are significantly richer than the isovector fields of $\bar{C}$ alone. However, we have seen that inclusion of the zero forms $v\rfloor\underset{\sim}{C}$ as generators of an ideal renders the problem trivial. In fact, we had to replace $\bar{I}\{\underset{\sim}{C}, v\rfloor\underset{\sim}{C}\}$ by $\bar{I}\{\underset{\sim}{C}, d(v\rfloor\underset{\sim}{C})\}$; that is, consideration had to be restricted to proper isovectors of $\bar{I}\{\underset{\sim}{C}, v\rfloor\underset{\sim}{C}\}$. This suggests the immediate extension of $\bar{E}$ to a larger closed ideal

$$(3.118)\qquad \bar{E}_1(v) = \bar{I}\{\underset{\sim}{C}, d(v\rfloor\underset{\sim}{C}), \underset{\sim}{E}\} .$$

There is an obvious system of regular maps that is associated wish such an ideal, namely

$$(3.119)\qquad S_1(v,\underset{\sim}{E},B_n) = \{\Phi \in R(B_n) \mid \Phi^* \underset{\sim}{E}_1(v) = \underset{\sim}{0}\} .$$

Every $\Phi \in S_1(v,\underset{\sim}{E},B_n)$ thus satisfies

$$(3.120)\qquad \Phi^*\mu \neq 0 , \quad \Phi^*\underset{\sim}{C} = \underset{\sim}{0}$$

and

$$(3.121)\qquad \Phi^*\underset{\sim}{E} = \underset{\sim}{0} , \quad \Phi^*(v\rfloor\underset{\sim}{C}) = \underset{\sim}{0} ;$$

any element of $S_1(v,\underset{\sim}{E},B_n)$ solves the balance system $\underset{\sim}{E}$ subject to the constraints $v\rfloor\underset{\sim}{C}$. Accordingly, if $v \in T(\underset{\sim}{E},K)$ then $S_1(v,\underset{\sim}{E},B_n) = S(v,\underset{\sim}{E},B_n)$ and $S_1(v,\underset{\sim}{E},B_n)$ consists of all similarity solutions associated with the isovector v of $\bar{E}$.

The full impact of these considerations arises through introduction of the notion of extended isovector fields.

Definition 3.4. A vector field $v \in T(K)$ is an *extended isovector field* of the balance system $\underset{\sim}{E}$ if and only if v is an isovector of the closed ideal $\bar{E}_1(v) = \bar{I}\{\underset{\sim}{C},\ d(v \rfloor \underset{\sim}{C}),\ \underset{\sim}{E}\}$.

THEOREM 3.9. *A vector field* $v \in T(K)$ *is an extended isovector field of the balance system* $\underset{\sim}{E}$ *if and only if* v *is a proper isovector of* $\bar{I}\{\underset{\sim}{C},\ v \rfloor \underset{\sim}{C}\}$ *and*

$$\pounds_v \underset{\sim}{E} \subset \bar{E}_1(v) \ . \tag{3.122}$$

Proof. The proof proceeds along exactly the same lines as the proof of Theorem 3.4.

Careful note must be taken of the fact that the equations that define an extended isovector v are given in terms of an ideal that depends on the vector field v for its definition. Two different extended isovectors are thus associated with entirely different ideals and give rise to entirely different solution systems $S_1(v_1, \underset{\sim}{E}, B_n)$ and $S_1(v_2, \underset{\sim}{E}, B_n)$. This is most easily seen by writing out (3.122).

$$\pounds_v \underset{\sim}{E} = \underset{\sim}{E}\,\underset{\sim}{R} + C^\alpha \wedge \underset{\sim}{S}_\alpha + dC^\alpha \wedge \underset{\sim}{T}_\alpha + d(v \rfloor C^\alpha) \wedge \underset{\sim}{Q}_\alpha \ . \tag{3.123}$$

There is thus neither a Lie algebra nor a subspace structure that can be associated with extended isovector fields. Be this as it may, significant further progress can be made.

If v is an extended isovector of the system $\underset{\sim}{E}$ that is not an isovector of $\bar{C}$ or of $\bar{E}$, or both, then (3.123) and (2.94) show that

$$\begin{aligned} \exp(-s\pounds_v)\underset{\sim}{E} &= \underset{\sim}{E}\,\underset{\sim}{A}(s) + C^\alpha \wedge \underset{\sim}{B}_\alpha(s) + dC^\alpha \wedge \underset{\sim}{G}_\alpha(s) \\ &\quad + d(v \rfloor C^\alpha) \wedge \underset{\sim}{N}_\alpha(s) \\ &= d(v \rfloor C^\alpha) \wedge \underset{\sim}{N}_\alpha(s) \mod \bar{E} \ , \end{aligned}$$

where at least one of the quantities $\{\underset{\sim}{N}_\alpha(s),\ \alpha=1,\ldots,N\}$ is different from zero for $s \neq 0$. This establishes the following result.

THEOREM 3.10. *If* v *is an isovector field of* $\bar{E}_1(v)$ *that is not contained in* $T(\underset{\sim}{E},K)$, *then* v *deforms* $\underset{\sim}{E}$ *into*

$$(3.124) \qquad \underset{\sim}{E}(v,s) = \exp(-s\pounds_v)\underset{\sim}{E} = d(v \rfloor C^{\alpha}) \wedge \underset{\sim}{N}_{\alpha}(s) \bmod \bar{E}$$

and $\underset{\sim}{E}(v,s)$ *is not an element of* $F(\underset{\sim}{E})$ *for* $s \neq 0$.

Theorem 3.10 shows that use of an extended isovector field allows us to map $\underset{\sim}{E}$ out of the fiber $F(\underset{\sim}{E})$ of balance equivalent systems by Lie deformation of $\underset{\sim}{E}$ along the orbit of the extended isovector field. This was not possible before. If $\Phi \in S_1(v,\underset{\sim}{E},B_n)$, so that Φ defines a similarity solution of $\underset{\sim}{E}$, we define the 1-parameter family of maps

$$\Phi_v(s) = \exp(v,s)\circ\Phi .$$

It is then immediate that

$$\Phi_v(s)^*\underset{\sim}{E}(v,s) = \Phi^*(\exp(s\pounds_v)\underset{\sim}{E}(v,s)) = \Phi^*\underset{\sim}{E} = \underset{\sim}{0}$$

and

$$\Phi_v(s)^*\underset{\sim}{C} = \Phi^*(\underset{\sim}{A}(s))\Phi^*\underset{\sim}{C} + \Phi^*(\underset{\sim}{R}(s))\Phi^*(v \rfloor d\underset{\sim}{C}) = 0 .$$

THEOREM 3.11. *If* v *is an isovector of* $\bar{E}_1(v)$ *that does not belong to* $T(\underset{\sim}{E},K)$, *then the 1-parameter family of maps*

$$(3.125) \qquad \Phi_v(s) = \exp(v,s)\circ\Phi$$

maps all similarity solutions $S_1(v,\underset{\sim}{E},B_n)$ *of* $\underset{\sim}{E}$ *associated with the vector field* v *onto similarity solutions of the balance inequivalent system* $\underset{\sim}{E}(v,s) = \exp(-s\pounds_v)\underset{\sim}{E}$ *associated with the Lie deformation,* $v(s) = v$, *of* v *and* v(s) *is an isovector of the ideal*

$$\bar{I}\{\underset{\sim}{C},\ d(v \rfloor \underset{\sim}{C}),\ \exp(-s\pounds_v)\underset{\sim}{E}\} .$$

Extended isovector fields provide powerful tools for solving a given balance system. In particular, if we can find an extended isovector field that maps the given $\underset{\sim}{E}$ into a simpler balance system $\underset{\sim}{E}(v,s_1)$, then there is a strong possibility that we can construct all similarity solutions of the simpler system $\underset{\sim}{E}(v,s_1)$. The mapping $\exp(s_1 \pounds_v)$ applied to these similarity solutions gives us similarity solutions of the original system $\underset{\sim}{E}$. The isovectors of $\bar{E}$ can then be used to generate 1-parameter groups of new solutions by transport of the similarity solutions. All in all, we obtain access to a large variety of solutions that is not necessarily available through use of the isovectors of $\bar{E}$ alone. In addition, the functions (v^i, v^α) may now depend explicitly on the arguments (y^α_j), in which case the new solutions will depend not only on the old solutions, but on their first derivatives as well. This, then, gives rise to 1-parameter families of Bäcklund transformations [14, 23] for the relations between the old and the new solutions whenever (v^i, v^α) depend explicitly on the arguments (y^α_j), in sharp contrast with the contact transformations that are naturally associated with the orbital equations

$$\frac{d\overset{*}{y}{}^\alpha_i}{ds} = \overset{*}{Z}_j(\overset{*}{v}{}^\alpha - \overset{*}{y}{}^\alpha_1 \overset{*}{v}{}^j) .$$

The reason that we now obtain Bäcklund transformations is that v is not an isovector of $\bar{C}$ if v does not belong to $T(\underset{\sim}{E},K)$, and hence $\exp(v,s)$ does not define a contact transformation. We defer further discussion of this important point until sections 4.6 and 5.3 where an appropriate group-theoretic context can be given (recall that there is no group structure associated with different isovectors since the ideal $\bar{E}_1(v)$ is v-dependent).

3.5 CONSERVED AND BALANCED QUANTITIES

Quantities that are either conserved or balanced as a consequence of satisfying the governing equations of a balance system are of both theoretical and practical importance. As might be expected, the isovector fields of a given balance ideal $\bar{E} = \bar{I}\{\underset{\sim}{C},\underset{\sim}{E}\}$ provide direct access to the construction of such quantities.

Let $\underset{\sim}{E}$ be a given balance system and let v be a vector field on K. If $\underset{\sim}{B}$ is any nonsingular element of $\Lambda^{0}_{N,N}(K)$ and $\underset{\sim}{\rho} \in \Lambda^{n-1}_{1,N}(K)$, then

$$(3.126) \qquad \underset{\sim}{Q} = v \rfloor \underset{\sim}{E}\, \underset{\sim}{B}^{-1} + \underset{\sim}{\rho}$$

is an element of $\Lambda^{n-1}_{1,N}(K)$. Exterior differentiation of (3.126) yields

$$d\underset{\sim}{Q} = d(v \rfloor \underset{\sim}{E}\, \underset{\sim}{B}^{-1}) + d\underset{\sim}{\rho} ,$$

and hence $d(v \rfloor \underset{\sim}{E}\, \underset{\sim}{B}^{-1}) = \pounds_{v}(\underset{\sim}{E}\, \underset{\sim}{B}^{-1}) - v \rfloor d(\underset{\sim}{E}\, \underset{\sim}{B}^{-1})$ shows that

$$(3.127) \qquad d\underset{\sim}{Q} = \pounds_{v}(\underset{\sim}{E}\, \underset{\sim}{B}^{-1}) - v \rfloor d(\underset{\sim}{E}\, \underset{\sim}{B}^{-1}) + d\underset{\sim}{\rho} .$$

However, Theorem 3.3 shows that any balance system $\underset{\sim}{E}$ is balance equivalent to a balance system of the form $d\underset{\sim}{\xi} + \underset{\sim}{\eta}$; that is

$$(3.128) \qquad \underset{\sim}{E} = (d\underset{\sim}{\xi} + \underset{\sim}{\eta} + \underset{\sim}{P})\underset{\sim}{R}$$

for some nonsingular $\underset{\sim}{R}$ and some $\underset{\sim}{P} \in \bar{C}$. When this is substituted into (3.127) and we choose $\underset{\sim}{B}$ by

$$(3.129) \qquad \underset{\sim}{B} = \underset{\sim}{R} ,$$

we see that

$$(3.130) \qquad d\underset{\sim}{Q} = \pounds_{v}(d\underset{\sim}{\xi} + \underset{\sim}{\eta} + \underset{\sim}{P}) - v \rfloor d(\underset{\sim}{\eta} + \underset{\sim}{P}) + d\underset{\sim}{\rho} .$$

Thus, if we choose $\underset{\sim}{\rho}$ by

$$\underset{\sim}{\rho} = - v \rfloor \underset{\sim}{P} ,$$

so that $-v \rfloor d\underset{\sim}{P} + d\underset{\sim}{\rho} = - v \rfloor d\underset{\sim}{P} - d(v \rfloor \underset{\sim}{P}) = - \pounds_{v}\underset{\sim}{P}$, (3.130) becomes

$$(3.131) \qquad d\underset{\sim}{Q} = \pounds_{v}(d\underset{\sim}{\xi} + \underset{\sim}{\eta}) - v \rfloor d\underset{\sim}{\eta} .$$

LEMMA 3.6. *Let* $\underset{\sim}{E}$ *be a given balance system and let* $d\underset{\sim}{\xi} + \underset{\sim}{\eta}$ *be a canonical form equivalent to* $\underset{\sim}{E}$,

$$(3.132) \qquad \underset{\sim}{E} = (d\underset{\sim}{\xi} + \underset{\sim}{\eta} + \underset{\sim}{P})\underset{\sim}{B} , \quad \underset{\sim}{P} \varepsilon \bar{C} .$$

The matrix

$$(3.133) \qquad \underset{\sim}{Q}(v) = v \rfloor (d\underset{\sim}{\xi} + \underset{\sim}{\eta}) = v \rfloor (\underset{\sim}{E}\,\underset{\sim}{B}^{-1} - \underset{\sim}{P})$$

of (n-1)-*forms satisfies*

$$(3.134) \qquad d\underset{\sim}{Q}(v) = \pounds_v(d\underset{\sim}{\xi} + \underset{\sim}{\eta}) - v \rfloor d\underset{\sim}{\eta} .$$

THEOREM 3.12. *If* v *is an isovector field of the closed ideal generated by* C *and* $\underset{\sim}{E} = (d\underset{\sim}{\xi} + \underset{\sim}{\eta} + \underset{\sim}{P})\underset{\sim}{B}$, *then every solution,* $\Phi \varepsilon S(\underset{\sim}{E}, B_n)$, *of the balance system* $\underset{\sim}{E}$ *satisfies the laws of balance*

$$(3.135) \qquad d\Phi^*\underset{\sim}{Q}(v) = - \Phi^*(v \rfloor d\underset{\sim}{\eta})$$

with

$$(3.136) \qquad \Phi^*\underset{\sim}{Q}(v) = \Phi^*\{v \rfloor (d\underset{\sim}{\xi} + \underset{\sim}{\eta})\} = \Phi^*(v \rfloor \underset{\sim}{E}\,\underset{\sim}{B}^{-1}) ,$$

and $\Phi^*\underset{\sim}{Q}(v)$ *is nontrivial if* $v \rfloor \underset{\sim}{E}$ *does not belong to* $\bar{E}$.

Proof. If $v \varepsilon T(\underset{\sim}{E}, K)$, then $v \varepsilon T(F(\underset{\sim}{E}), K)$ by Theorem 3.6. Thus, since $d\underset{\sim}{\xi} + \underset{\sim}{\eta} \varepsilon F(\underset{\sim}{E})$, $\pounds_v(d\underset{\sim}{\xi} + \underset{\sim}{\eta}) \varepsilon \bar{E}$, and every $\Phi \varepsilon S(\underset{\sim}{E}, B_n)$ is such that $\Phi^*\pounds_v(d\underset{\sim}{\xi} + \underset{\sim}{\eta}) = 0$. This establishes (3.135) as a direct consequence of (3.134). Since $\underset{\sim}{P} \varepsilon \bar{C}$, $\Phi^*\underset{\sim}{P} = \underset{\sim}{0}$ and (3.136) is obtained from (3.133). The matrix of (n-1)-forms $\Phi^*\underset{\sim}{Q}(v)$ is nontrivial provided $\Phi^*\underset{\sim}{Q}(v) \not\equiv \underset{\sim}{0}$. However, $\Phi^*(v \rfloor \underset{\sim}{E}\,\underset{\sim}{B}^{-1}) = \Phi^*(v \rfloor \underset{\sim}{E})\Phi^*(B^{-1})$, and hence $\Phi^*\underset{\sim}{Q}(v) \not\equiv 0$ if and only if $v \rfloor \underset{\sim}{E}$ does not belong to the closed ideal $\bar{E}$.

COROLLARY 3.4. *If* $\underset{\sim}{E} = (d\underset{\sim}{\xi} + \underset{\sim}{P})\underset{\sim}{B}$, $\underset{\sim}{P} \varepsilon \bar{C}$ *then each isovector* v *of* $\bar{E}$ *generates a matrix*

(3.137) $$\underset{\sim}{Q}(v) = v \rfloor d\underset{\sim}{\xi} = v \rfloor (\underset{\sim}{E}\,\underset{\sim}{B}^{-1} - \underset{\sim}{P})$$

such that every $\Phi \in S(\underset{\sim}{E}, B_n)$ *satisfies the conservation laws*

(3.138) $$d\Phi^* \underset{\sim}{Q}(v) = \underset{\sim}{0} .$$

The conservation laws (3.138) *are nontrivial for every* $v \in T(\underset{\sim}{E}, K)$ *such that*

(3.139) $$v \rfloor \underset{\sim}{E} \not\subset \bar{E} .$$

Proof. The result follows upon putting $\underset{\sim}{\eta} = \underset{\sim}{0}$ in Theorem 3.12.

The above corollary shows that all solutions of a balance system $\underset{\sim}{E} \simeq d\underset{\sim}{\xi}$ admit as many linearly independent systems of conservation laws as there are linearly independent isovectors of $\bar{E}$ that are not characteristic vectors of $\underset{\sim}{E}$. If we restrict attention to the subsets of solutions of $\underset{\sim}{E}$ that are either similarity solutions or extended similarity solutions, it would seem reasonable to expect a larger collection of independent systems of conservation laws. This is indeed the case as the following theorem shows.

THEOREM 3.13. *If* v *is an isovector of the extended ideal* $\bar{E}_1(v)$ *of* $\underset{\sim}{E} = (d\underset{\sim}{\xi} + \underset{\sim}{\eta} + \underset{\sim}{P})\underset{\sim}{B}$, $\underset{\sim}{P} \in \bar{C}$, *then every extended similarity solution* $\Phi \in S_1(v, \underset{\sim}{E}, B_n)$ *satisfies the laws of balance*

(3.140) $$d\Phi^* \underset{\sim}{Q}(v) = -\Phi^*(v \rfloor d\underset{\sim}{\eta})$$

with

(3.141) $$\Phi^* \underset{\sim}{Q}(v) = \Phi^*\{v \rfloor (d\underset{\sim}{\xi} + \underset{\sim}{\eta})\} = \Phi^*(v \rfloor \underset{\sim}{E}\,\underset{\sim}{B}^{-1}) .$$

Proof. The result follows directly from Lemma 3.6 since $\pounds_v(d\underset{\sim}{\xi} + \underset{\sim}{\eta}) \in \bar{E}_1(v)$ in this case, and hence $\Phi \in S_1(v, \underset{\sim}{E}, B_n)$ implies $\Phi^* \pounds_v(d\underset{\sim}{\xi} + \underset{\sim}{\eta}) = \underset{\sim}{0}$.

We noted in Corollary 3.1 that any balance system that is itself a system of laws of balance,

$$E_\alpha = W_\alpha \mu - d(W_\alpha^i \mu_i)$$

has the property that

$$dE_\alpha = 0 \bmod \bar{C} .$$

Thus, if we set

(3.142) $$\underset{\sim}{\xi} = - \underset{\sim}{W}^i \mu_i , \quad \underset{\sim}{\eta} = W_\alpha \mu$$

then

$$\underset{\sim}{E} = d\underset{\sim}{\xi} + \underset{\sim}{\eta}$$

with

(3.143) $$d\underset{\sim}{E} = d\underset{\sim}{\eta} = \underset{\sim}{0} \bmod \bar{C} .$$

Now, $d\underset{\sim}{\eta} = \underset{\sim}{0} \bmod \bar{C}$ implies that $v \rfloor d\underset{\sim}{\eta} \in \bar{E}_1(v) = \bar{I}\{\underset{\sim}{C}, v \rfloor \underset{\sim}{C}, \underset{\sim}{E}\}$ by (3.118), and hence any $\Phi \in S_1(v, \underset{\sim}{E}, B_n)$ gives

$$\Phi^*(v \rfloor d\underset{\sim}{\eta}) = \underset{\sim}{0} .$$

When this is substituted into (3.140), we obtain the following result.

THEOREM 3.14. *If* v *is an isovector of the extended ideal* $\bar{E}_1(v)$ *of* $\underset{\sim}{E} = d\underset{\sim}{\xi} + \underset{\sim}{\eta}$ *and* $d\underset{\sim}{E} = d\underset{\sim}{\eta} \in \bar{C}$ *, then every extended similarity solution* $\Phi \in S_1(v, \underset{\sim}{E}, B_n)$ *satisfies the conservation laws*

(3.144) $$d\Phi^* \underset{\sim}{Q}(v) = \underset{\sim}{0}$$

with

(3.145) $$\underset{\sim}{Q}(v) = v \rfloor (d\underset{\sim}{\xi} + \underset{\sim}{\eta}) = v \rfloor \underset{\sim}{E} .$$

Thus, every system of laws of balance,

$$(3.146)\qquad \underset{\sim}{E} = \underset{\sim}{W}\mu - d(\underset{\sim}{W}^i \mu_i)$$

gives a matrix

$$(3.147)\qquad \underset{\sim}{Q}(v) = \underset{\sim}{W}v^i \mu_i - v \rfloor d(\underset{\sim}{W}^i \mu_i)$$

such that every $\Phi \varepsilon S_1(v, \underset{\sim}{W}\mu - d(\underset{\sim}{W}^i \mu_i), B_n)$ *satisfies*

$$(3.148)\qquad d\Phi^* \underset{\sim}{Q}(v) = \underset{\sim}{0}$$

for any isovector v *of the extended ideal* $\bar{E}_1(v)$.

The intrinsic utility of isovectors of the extended ideal $\bar{E}_1(V)$ and the concomitant similarity solutions is thus further in evidence.

There is a significantly larger collection of systems of laws of balance that can be exhibited. For simplificity, we assume from now on that any balance system $\underset{\sim}{E}$ has been replaced by the equivalent system $d\tilde{\xi} + \tilde{\eta}$ belonging to $F(\underset{\sim}{E})$.

Let $(b) = (b^i, b^\alpha, b_i^\alpha)$ be the coordinates of the point (b) in K and let $H_{(b)}$ be the linear homotopy operator with center (b) . Thus, if $\omega(x^i, q^\alpha, y_i^\alpha)$ is a k-form,

$$H_{(b)}(\omega)(x^i, q^\alpha, y_i^\alpha) = \int_0^1 \lambda^{k-1}\, \chi \rfloor \tilde{\omega}(\lambda)\, d\lambda$$

where

$$\chi = (x^i - b^i)\partial_i + (q^\alpha - b^\alpha)\partial_\alpha + (y_i^\alpha - b_i^\alpha)\partial_\alpha^i$$

and

$$\tilde{\omega}(\lambda) = \omega(b^i + \lambda(x^i - b^i),\ b^\alpha + \lambda(q^\alpha - b^\alpha),\ b_i^\alpha + \lambda(y_i^\alpha - b_i^\alpha)) .$$

Starting with the balance system $\underset{\sim}{E} = d\underset{\sim}{\xi} + \underset{\sim}{\eta}$ we can construct the system of (n-1)-forms $h_{(b)}$ by

$$\underset{\sim}{h}_{(b)} = H_{(b)}(\underset{\sim}{E}) = H_{(b)}(d\underset{\sim}{\xi} + \underset{\sim}{\eta}) = \underset{\sim}{\xi} - dH_{(b)}(\underset{\sim}{\xi}) + H_{(b)}(\underset{\sim}{\eta}) .$$

Exterior differentiation then yields

$$dh_{\sim(b)} = dH_{(b)}(\underset{\sim}{E}) .$$

However, $dH_{(b)}(\omega) + H_{(b)}(d\omega) \equiv \omega$, since K is Euclidean and hence star-like with respect to any of its points, and we obtain

$$(3.149) \qquad dh_{\sim(b)} = \underset{\sim}{E} - H_{(b)}(d\underset{\sim}{E}) = \underset{\sim}{E} - H_{(b)}(d\underset{\sim}{\eta}) .$$

<u>THEOREM 3.15</u>. *Let* $H_{(b)}$ *be the linear homotopy operator on* K *with center* (b), *then the system of* $(n-1)$-*forms*

$$(3.150) \quad \underset{\sim}{h}_{(b)} = H_{(b)}(\underset{\sim}{E}) = H_{(b)}(d\underset{\sim}{\xi} + \underset{\sim}{\eta}) = \underset{\sim}{\xi} + H_{(b)}(\underset{\sim}{\eta}) - dH_{(b)}(\underset{\sim}{\xi})$$

satisfies

$$(3.151) \qquad dh_{\sim(b)} = \underset{\sim}{E} - H_{(b)}(d\underset{\sim}{\eta})$$

and hence every element Φ *of* $S(\underset{\sim}{E}, B_n)$ *satisfies the system of laws of balance*

$$(3.152) \qquad d\Phi^* \underset{\sim}{h}_{(b)} = -\Phi^* H_{(b)}(d\underset{\sim}{\eta}) .$$

<u>Proof</u>. Any $\Phi \in S(\underset{\sim}{E}, B_n)$ is a solution of $\underset{\sim}{E}$ and hence $\Phi^* \underset{\sim}{E} = \underset{\sim}{0}$. The result then follows directly from (3.149).

CHAPTER 4

PSEUDOPOTENTIALS AND PROLONGATIONS

If we have a system of partial differential equations of the form

$$D_i W^i_\beta(x^j,\ \phi^\alpha(x^j),\ \partial_k\phi^\alpha(x^j)) = W_\beta(x^j,\ \phi^\alpha(x^j),\ \partial_k\phi^\alpha(x^j)) \ ,$$

then a formal potential-like solution can be defined implicitly in terms of functions $S^{ij}_\beta(x^k) = -S^{ji}_\beta(x^k)$ and $h^i_\beta(x^k)$ by

$$W_\beta(x^j,\ \phi^\alpha(x^j),\ \partial_k\phi^\alpha(x^j)) = \partial_i h^i_\beta(x^k) \ ,$$

$$W^i_\beta(x^j,\ \phi^\alpha(x^j),\ \partial_k\phi^\alpha(x^j)) = 2\,\partial_j S^{ij}_\beta(x^k) + h^i_\beta(x^k) \ ,$$

while any solution of the given system can be used to construct explicit functions S^{ij}_β and h^i_β with the above properties. This elementary observation underlies and provides the motivation for a general but equivalent formulation in terms of differential forms, prolongation structures and pseudopotentials that has received such marked attention in the recent literature [9,15-23].

4.1 THE CLOSURE CONDITIONS OF AN EXTENDED IDEAL

We know that any balance system can be written in the form

$$(4.1)\qquad \underset{\sim}{E} = (d\underset{\sim}{\xi} + \underset{\sim}{\rho})\underset{\sim}{R} \bmod \bar{C} ,$$

where $\underset{\sim}{R}$ is a nonsingular element of $\Lambda^{o}_{N,N}(K)$, and that $I\{\underset{\sim}{C}, d\underset{\sim}{C}, \underset{\sim}{E}\}$ is closed if $\underset{\sim}{R}$ and an appropriate element of $\bar{C}$ can be chosen so that $\underset{\sim}{\rho} = \underset{\sim}{0}$. Let $\underset{\sim}{Q}$ belong to $\Lambda^{n-1}_{1,N}(K)$ and construct the ideal

$$(4.2)\qquad E_2 = I\{\underset{\sim}{C}, d\underset{\sim}{C}, \underset{\sim}{E}, d\underset{\sim}{E}, \underset{\sim}{Q}\} .$$

This ideal is closed if and only if there exists a $\underset{\sim}{\Gamma} \in \Lambda^{1}_{N,N}(K)$, an $\underset{\sim}{A} \in \Lambda^{o}_{N,N}(K)$ and a $\underset{\sim}{P} \in \bar{C}$ such that

$$(4.3)\qquad d\underset{\sim}{Q} = (-1)^n \underset{\sim}{Q} \wedge \underset{\sim}{\Gamma} + \underset{\sim}{E}\,\underset{\sim}{A} + \underset{\sim}{P} ;$$

that is

$$d\underset{\sim}{Q} = (-1)^n \underset{\sim}{Q} \wedge \underset{\sim}{\Gamma} \bmod \bar{E} .$$

Quantities $\underset{\sim}{Q}$ with this property are clearly of interest since they constitute the simplest possible generalization of the notion of a system of conservation laws that was studied in Section 3.5. In fact, if $\Phi \in R(E,B_n)$, then (4.3) implies

$$\int_{\partial B_n} \Phi^* \underset{\sim}{Q} = (-1)^n \int_{B_n} \Phi^*(\underset{\sim}{Q} \wedge \underset{\sim}{\Gamma}) .$$

We thus obtain the necessary and sufficient conditions in order that E_2 be a closed ideal.

A substitution of (4.1) into (4.3) leads directly to

$$(4.4)\qquad d\underset{\sim}{Q} = (-1)^n \underset{\sim}{Q} \wedge \underset{\sim}{\Gamma} + (d\underset{\sim}{\xi} + \underset{\sim}{\rho})\underset{\sim}{R}\underset{\sim}{A} + \underset{\sim}{P} .$$

A sequence of elementary manipulations then gives the conditions

$$(4.5)\qquad d(\underset{\sim}{Q} - \underset{\sim}{\xi}\,\underset{\sim}{R}\underset{\sim}{A}) = (-1)^n(\underset{\sim}{Q} - \underset{\sim}{\xi}\,\underset{\sim}{R}\underset{\sim}{A}) \wedge \underset{\sim}{\Gamma}$$
$$+ (-1)^n \underset{\sim}{\xi} \wedge (d(\underset{\sim}{R}\underset{\sim}{A}) + \underset{\sim}{R}\underset{\sim}{A}\underset{\sim}{\Gamma}) + \underset{\sim}{\rho}\,\underset{\sim}{R}\underset{\sim}{A} + \underset{\sim}{P} .$$

The identifications

$$(4.6)\qquad \utilde{\Omega} = \utilde{Q} - \utilde{\xi}\,\utilde{R}\utilde{A} ,$$

$$(4.7)\qquad \utilde{\Sigma} = (-1)^n \utilde{\xi} \wedge (d(\utilde{R}\utilde{A}) + \utilde{R}\utilde{A}\utilde{\Gamma}) + \utilde{\rho}\,\utilde{R}\utilde{A} + \utilde{P} ,$$

reduces (4.5) to

$$(4.8)\qquad d\utilde{\Omega} = (-1)^n\, \utilde{\Omega} \wedge \utilde{\Gamma} + \utilde{\Sigma} .$$

In order that this system be solvable we must adjoin its integrability conditions, which are

$$(4.9)\qquad d\utilde{\Sigma} = (-1)^{n+1}\, \utilde{\Sigma} \wedge \utilde{\Gamma} + \utilde{\Omega} \wedge \utilde{\Theta}$$

$$(4.10)\qquad d\utilde{\Gamma} = \utilde{\Gamma} \wedge \utilde{\Gamma} + \utilde{\Theta} , \qquad d\utilde{\Theta} = \utilde{\Gamma} \wedge \utilde{\Theta} - \utilde{\Theta} \wedge \utilde{\Gamma} .$$

Now, (4.8) through (4.10) constitute a differential system of the form (1.25), (1.26). Thus, $\utilde{\Gamma}$ is a connection and $\utilde{\Theta}$ is a curvature associated with the closure of the ideal E_2, and the general solution of this differential system is given by (1.28)-(1.31).

For the moment, let us assume that $\utilde{\Gamma}$ is given. Equation (1.28) then gives the matrix $\utilde{B}$ through solution of the integral equation

$$(4.11)\qquad \utilde{B} = \utilde{B}_o + H(\utilde{\Gamma}\,\utilde{B}) , \quad \det(\utilde{B}_o) \neq 0 , \quad \utilde{B}_o = \text{constant matrix}$$

and

$$(4.12)\qquad \utilde{\Gamma} = (d\utilde{B} + \utilde{B}\,\utilde{\theta})\utilde{B}^{-1} , \qquad \utilde{\Theta} = \utilde{B}(d\utilde{\theta} - \utilde{\theta} \wedge \utilde{\theta})\utilde{B}^{-1}$$

by (1.31). Since the matrix $\utilde{A}$ is at our disposal, we choose it in accordance with

$$(4.13)\qquad \utilde{R}\,\utilde{A} = \utilde{B}^{-1} .$$

This simplifies (4.6) and (4.7) so that we obtain

$$(4.14) \qquad \underset{\sim}{\Omega} = \underset{\sim}{Q} - (\underset{\sim}{\xi} + \underset{\sim}{P}_1)\underset{\sim}{B}^{-1} ,$$

$$(4.15) \qquad \underset{\sim}{\Sigma} = (\underset{\sim}{\rho} + (-1)^n \underset{\sim}{\xi} \wedge \underset{\sim}{\theta} + \underset{\sim}{P}_2)\underset{\sim}{B}^{-1}$$

where $\underset{\sim}{P}_1$ and $\underset{\sim}{P}_2$ belong to $\bar{C}$. If we substitute (4.15) into (1.33), we obtain

$$(4.16) \qquad \underset{\sim}{\eta} = (-1)^n H(\underset{\sim}{\xi} \wedge \underset{\sim}{\theta}) + H(\underset{\sim}{\rho} + \underset{\sim}{P}_2) ,$$

and (1.29) and (4.14) yield

$$(4.17) \qquad \underset{\sim}{Q} = \{d\underset{\sim}{\psi} + \underset{\sim}{\xi} + (-1)^n H[(d\underset{\sim}{\psi} + \underset{\sim}{\xi}) \wedge \underset{\sim}{\theta}] + H(\underset{\sim}{\rho} + \underset{\sim}{P}_2) + \underset{\sim}{P}_1\}\underset{\sim}{B}^{-1} .$$

THEOREM 4.1. *The ideal* $I\{\underset{\sim}{C}, d\underset{\sim}{C}, \underset{\sim}{E}, d\underset{\sim}{E}, \underset{\sim}{Q}\}$ *with*

$$(4.18) \qquad \underset{\sim}{E} = (d\underset{\sim}{\xi} + \underset{\sim}{\rho})\underset{\sim}{R} \bmod \bar{C} ,$$

is closed if and only if there exists a connection

$$(4.19) \qquad \underset{\sim}{\Gamma} = (d\underset{\sim}{B} + \underset{\sim}{B}\underset{\sim}{\theta})\underset{\sim}{B}^{-1} , \qquad \det(\underset{\sim}{B}) \neq 0$$

on K *and elements* $\underset{\sim}{P}_1$ *and* $\underset{\sim}{P}_2$ *of* $\bar{C}$ *such that*

$$(4.20) \qquad d\underset{\sim}{\Sigma} = (-1)^{n+1} \underset{\sim}{\Sigma} \wedge \underset{\sim}{\Gamma} + \underset{\sim}{\Omega} \wedge \underset{\sim}{\Theta} ,$$

where

$$(4.21) \qquad \underset{\sim}{\Omega} = \underset{\sim}{Q} - (\underset{\sim}{\xi} + \underset{\sim}{P}_1)\underset{\sim}{B}^{-1} , \quad \underset{\sim}{\Sigma} = (\underset{\sim}{\rho} + (-1)^n \underset{\sim}{\xi} \wedge \theta + \underset{\sim}{P}_2)\underset{\sim}{B}^{-1} ,$$

$$\Theta = \underset{\sim}{B}(d\theta - \theta \wedge \theta)\underset{\sim}{B}^{-1} .$$

If these conditions are met, then $\underset{\sim}{Q}$ *is given by*

$$(4.22) \qquad \underset{\sim}{Q} = \{d\underset{\sim}{\psi} + \underset{\sim}{\xi} + (-1)^n H[(d\underset{\sim}{\psi} + \underset{\sim}{\xi}) \wedge \underset{\sim}{\theta}] + H(\underset{\sim}{\rho} + \underset{\sim}{P}_2) + \underset{\sim}{P}_1\}\underset{\sim}{B}^{-1} .$$

It is clear from the above, and, in particular, from the need to satisfy (4.20), that there may not exist any $\underset{\sim}{Q}$ such that

$I\{\underset{\sim}{C}, d\underset{\sim}{C}, \underset{\sim}{E}, d\underset{\sim}{E}, \underset{\sim}{Q}\}$ is closed. This is not an unexpected situation, for the results of Section 3.5 show that balance systems with $\underset{\sim}{\rho} \neq \underset{\sim}{0}$ usually result in laws of balance rather than laws of conservation, a situation which must carry over into any generalization of the notion of laws of conservation. We therefore restrict our attention to balance systems of the form $\underset{\sim}{E} = (d\underset{\sim}{\xi})\underset{\sim}{R} \bmod \bar{C}$. The general case will be treated in Section 4.4. Now, simply observe that with $\underset{\sim}{\rho} = \underset{\sim}{0}$, $\underset{\sim}{P}_2 = \underset{\sim}{0}$ and $\underset{\sim}{\theta} = \underset{\sim}{0}$, we have $\underset{\sim}{\Sigma} = \underset{\sim}{0}$, $\underset{\sim}{\Theta} = \underset{\sim}{0}$ and the condition (4.20) is identically satisfied.

THEOREM 4.2. *If the balance system* $\underset{\sim}{E}$ *is given by*

$$(4.23)\qquad \underset{\sim}{E} = (d\underset{\sim}{\xi})\underset{\sim}{R} \bmod \bar{C} ,$$

then the ideal $I\{\underset{\sim}{C}, d\underset{\sim}{C}, \underset{\sim}{E}, \underset{\sim}{Q}\}$ *is closed for any* $\underset{\sim}{Q}$ *of the form*

$$(4.24)\qquad \underset{\sim}{Q} = (d\underset{\sim}{\Psi} + \underset{\sim}{\xi})\underset{\sim}{B}^{-1} \bmod \bar{C} .$$

For each such $\underset{\sim}{Q}$ *we have*

$$(4.25)\qquad d\underset{\sim}{Q} = (-1)^n \underset{\sim}{Q} \wedge \underset{\sim}{\Gamma} + \underset{\sim}{E}\,\underset{\sim}{A} \bmod \bar{C}$$

where $\underset{\sim}{A} = \underset{\sim}{R}^{-1}\underset{\sim}{B}^{-1}$ *and* $\underset{\sim}{\Gamma}$ *is the curvature free connection*

$$(4.26)\qquad \underset{\sim}{\Gamma} = d\underset{\sim}{B}\,\underset{\sim}{B}^{-1} .$$

4.2 PROLONGATION

If $\underset{\sim}{E} = (d\underset{\sim}{\xi})\underset{\sim}{R} \bmod \bar{C}$, Theorem 4.2 provides us with infinitely many quantities $\underset{\sim}{Q} \in \Lambda^{n-1}_{1,N}(K)$ such that $I\{\underset{\sim}{C}, d\underset{\sim}{C}, \underset{\sim}{E}, \underset{\sim}{Q}\}$ is closed and for which

$$(4.27)\qquad d\underset{\sim}{Q} = (-1)^n \underset{\sim}{Q} \wedge \underset{\sim}{\Gamma} + \underset{\sim}{E}\,\underset{\sim}{A} \bmod \bar{C} .$$

If Φ is any solution of the balance system $\underset{\sim}{E}$, $\Phi \in S(\underset{\sim}{E}, B_n)$, then (4.27) and (4.24) give

$$(4.28)\qquad d\Phi^*\underset{\sim}{Q} = (-1)^n \Phi^*\underset{\sim}{Q} \wedge \Phi^*\underset{\sim}{\Gamma} ,$$

$$(4.29)\qquad \Phi^*\underset{\sim}{Q} = (d\Phi^*\underset{\sim}{\Psi} + \Phi^*\underset{\sim}{\xi})\, \Phi^*\underset{\sim}{B}^{-1} .$$

There is thus little more that can be done if we retain the assumption that $\underset{\sim}{Q} \in \Lambda^{n-1}_{1,N}(K)$. Noting that the collection of (n-2)-forms $\underset{\sim}{\Psi}$ is at our disposal, we are at liberty to introduce an appropriate collection of additional variables by specifying a structure for the quantities that comprise $\underset{\sim}{\Psi}$. This, however, is exactly the underlying concept associated with prolongation of exterior differential structures [8,9,15,17-23].

Let us replace the (n+N+nN)-dimensional space K by a new space K_2 of $n+N+nN+M$ dimensions and coordinate functions $(x^k, q^\alpha, y^\alpha_i, s_a)$, $a = 1,\ldots,M$. Noting that K has the coordinate functions $(x^k, q^\alpha, y^\alpha_i)$, we may structure K_2 through the requirement $K_2 = K \times E_M$. This makes K_2 into a trivial fiber space with base space K , fiber coordinate functions (s_a) , $a = 1,\ldots,M$, and the canonical projection

$$(4.30) \qquad \pi_2 \colon K_2 \longrightarrow K \mid (x^i, q^\alpha, y^\alpha_i, s_a) \longrightarrow (x^k, q^\alpha, y^\alpha_i) \ .$$

Since $\bar{E}$ is a closed ideal of $\Lambda(K)$, it has a trivial prolongation to a closed ideal $\bar{E}^*$ of $\Lambda(K_2)$ that is defined by

$$(4.31) \qquad \bar{E}^* = \pi_2^* \bar{E} \ .$$

Thus, $\bar{E}^*$ consists only of differential forms on $\Lambda(K_2)$ that are independent of the fiber coordinates and their differentials. Accordingly, we may simply write $\bar{E}$, $\underset{\sim}{C}$ and $\underset{\sim}{E}$ with the understanding that they now are considered to belong to $\Lambda(K_2)$. Since $\underset{\sim}{\xi}$ comes from $\underset{\sim}{E}$, $\underset{\sim}{\xi}^* = \pi_2^* \underset{\sim}{\xi}$, and (4.24) shows that a dependence of $\underset{\sim}{Q}$ on the fiber coordinates (s_a) and their differentials can arise only through the quantities $\underset{\sim}{\Psi}$ and $\underset{\sim}{B}$. Now, $\underset{\sim}{\Psi}$ and $\underset{\sim}{B}$ enter into our considerations in completely different manners, for $\underset{\sim}{\Psi}$ is an arbitrary row matrix of (n-2)-forms while $\underset{\sim}{B}$ serves to determine the curvature free connection $\underset{\sim}{\Gamma} = d\underset{\sim}{B}\,\underset{\sim}{B}^{-1}$. There are thus two distinct choices available for prolongation of the ideal $I\{\underset{\sim}{C}, d\underset{\sim}{C}, \underset{\sim}{E}, \underset{\sim}{Q}\}$.

Definition 4.1. $I\{\underset{\sim}{C}, d\underset{\sim}{C}, \underset{\sim}{E}, \underset{\sim}{Q}\}$ is prolonged to an ideal $I\{\underset{\sim}{C}, d\underset{\sim}{C}, \underset{\sim}{E}, \underset{\sim}{Q}\}^*$ of $\Lambda(K_2)$ by a *prolongation of the first kind* if

$$(4.32) \qquad \bar{E}^* = \pi_2^*\bar{E} , \quad \underset{\sim}{B}^* = \pi_2^*\underset{\sim}{B} , \quad (\underset{\sim}{\Gamma}^* = \pi_2^*\underset{\sim}{\Gamma}) ,$$

$$(4.33) \qquad \underset{\sim}{\Psi} = \underset{\sim}{S}^{ij}\mu_{ij} , \quad \underset{\sim}{S}^{ij} = -\underset{\sim}{S}^{ji}$$

and the $Nn(n-1)/2$ independent quantities that comprise $\{\underset{\sim}{S}^{ij}\}$ are taken to be the fiber coordinates (S_a) , $a=1,\ldots,Nn(n-1)/2=M$.

Definition 4.2. $I\{\underset{\sim}{C}, d\underset{\sim}{C}, \underset{\sim}{E}, \underset{\sim}{Q}\}$ is prolonged to an ideal $I\{\underset{\sim}{C}, d\underset{\sim}{C}, \underset{\sim}{E}, \underset{\sim}{Q}\}^*$ of $\Lambda(K_2)$ by a *prolongation of the second kind* if

$$(4.34) \qquad \bar{E}^* = \pi_2^*\bar{E} ,$$

$$(4.35) \qquad \underset{\sim}{\Psi} = \underset{\sim}{S}^{ij}\mu_{ij} , \quad \underset{\sim}{S}^{ij} = -\underset{\sim}{S}^{ji} ,$$

the $Nn(n-1)/2$ independent quantities that comprise $\{\underset{\sim}{S}^{ij}\}$ are taken to be the fiber coordinates (S_a) , $a=1,\ldots,Nn(n-1)/2$, and the matrix $\underset{\sim}{B}$ is taken as an element of $\Lambda^0_{N,N}(K_2)$.

The following result is now immediate.

THEOREM 4.3. *If* $\underset{\sim}{E} = (d\underset{\sim}{\xi})\underset{\sim}{R} \bmod \bar{C}$, *so that* $I\{\underset{\sim}{C}, d\underset{\sim}{C}, \underset{\sim}{E}\}$ *is a closed ideal of* $\Lambda(K)$, *then a prolongation of* $I\{\underset{\sim}{C}, d\underset{\sim}{C}, \underset{\sim}{E}, \underset{\sim}{Q}\}$ *of either the first or second kind yields a closed ideal of* $\Lambda(K_2)$ *for*

$$(4.36) \qquad \underset{\sim}{Q} = \{d(\underset{\sim}{S}^{ij})\wedge\mu_{ij} + \underset{\sim}{\xi}\}\underset{\sim}{B}^{-1} \bmod \bar{C} ,$$

and $\underset{\sim}{B}\in\Lambda^0_{N,N}(K)$ $(\underset{\sim}{B}\in\Lambda^0_{N,N}(K_2))$ *for a prolongation of the first (of the second) kind.*

Now that we have extended K to K_2 and prolonged $\Lambda(K)$ to $\Lambda(K_2)$, the obvious thing to do is to take advantage of the new variables $(\underset{\sim}{S}^{ij})$. In order to do this, we extend any map

$$\Phi\colon B_n \longrightarrow K \mid (\tau^j) \longrightarrow (\phi^i(\tau^j),\ \phi^\alpha,\ \phi^\alpha_i)$$

to a map

$$\Phi_2\colon B_n \longrightarrow K_2 \mid (\tau^j) \longrightarrow (\phi^i,\ \phi^\alpha,\ \phi^\alpha_i,\ \phi^{ij}_\alpha) \ ;$$

that is, Φ_2 assigns the fiber variables S^{ij}_α the values $\phi^{ij}_\alpha(\tau^k)$.

Definition 4.3. A map $\Phi_2\colon B_n \longrightarrow K_2$ is said to *solve* the prolonged balance system $(\underset{\sim}{E},\underset{\sim}{Q})$ if and only if Φ_2 restricted to K is regular and Φ^*_2 annihilates the closed ideal $I\{\underset{\sim}{C},\ d\underset{\sim}{C},\ \underset{\sim}{E},\ \underset{\sim}{Q}\}^*$ of $\Lambda(K_2)$. The collection of all such maps is denoted by $S_2(\underset{\sim}{E},\ \underset{\sim}{Q},\ B_n)$.

Since $\underset{\sim}{R}$ and $\underset{\sim}{B}$ are nonsingular matrices of functions on K and/or K_2 and $\Phi^*_2\mu \neq 0$ can be used to secure $\Phi^*_2 x^i = \phi^i(\tau^j) = \tau^i$, the following Lemma is immediate.

LEMMA 4.1. *A map* $\Phi_2\colon B_n \longrightarrow K_2$ *solves the prolonged balance system* $(\underset{\sim}{E},\underset{\sim}{Q})$ *with* $\underset{\sim}{E} = (d\underset{\sim}{\xi})\underset{\sim}{R}$ *if and only if* $\Phi^*_2\mu \neq 0$, $\Phi^*_2\underset{\sim}{C} = \underset{\sim}{0}$,

(4.37) $$\Phi^*_2\underset{\sim}{E} = (d\Phi^*_2\underset{\sim}{\xi})\ \Phi^*_2\underset{\sim}{R} = \underset{\sim}{0}\ ,$$

(4.38) $$\Phi^*_2\underset{\sim}{Q} = \{d\Phi^*_2\underset{\sim}{S}^{ij} \wedge \mu_{ij} + \Phi^*_2\underset{\sim}{\xi}\}\Phi^*_2\underset{\sim}{B}^{-1} = \underset{\sim}{0}\ ,$$

and this is the case only if

(4.39) $$d(\Phi^*_2\underset{\sim}{\xi}) = \underset{\sim}{0}\ ,$$

(4.40) $$d\Phi^*_2\underset{\sim}{S}^{ij} \wedge \mu_{ij} = -\ \Phi^*_2\underset{\sim}{\xi}\ .$$

Clearly, (4.39) and (4.40) constitute a complete exterior differential system on the domain B_n . Further, since $\underset{\sim}{\xi} \in \Lambda^{n-1}_{1,N}(K)$, and $\{\mu_i\}$ is a basis for $\Lambda^{n-1}(E_n)$, we may always write

$$\Phi_2^* \underset{\sim}{\xi} = \underset{\sim}{\xi}^i(\tau^j, \phi^\alpha(\tau^j), \partial_k\phi^\alpha(\tau^j))\, \mu_i$$

and

$$\Phi_2^* \underset{\sim}{S}^{ij} = \underset{\sim}{\phi}^{ij}(\tau^k) .$$

An elementary application of (1.6) then shows that (4.39) and (4.40) become

$$(4.41) \qquad d\underset{\sim}{\xi}^i \wedge \mu_i = 0 ,$$

$$(4.42) \qquad d\underset{\sim}{\phi}^{ij} \wedge \mu_{ij} = -\, 2(\partial_j \underset{\sim}{\phi}^{ij})\mu_i = -\, \underset{\sim}{\xi}^i \mu_i ;$$

that is

$$(4.43) \qquad \underset{\sim}{\xi}^i = 2\partial_j \underset{\sim}{\phi}^{ij} .$$

Now, (4.42) shows that every choice of the functions $\underset{\sim}{\phi}^{ij}(\tau^k)$ yields functions $\underset{\sim}{\xi}^i$ that satisfy (4.41) formally. Conversely, if the functions $\phi^\alpha(\tau^k)$ that enter into (4.41) are such that (4.41) is satisfied, the linear homotopy operator H_n on B_n (B_n starlike with respect to the coordinate cover (τ^k)) can be used to obtain

$$(4.44) \qquad \underset{\sim}{\phi}^{ij}(\tau^k)\, \mu_{ij} = d\underset{\sim}{\beta} - H_n(\underset{\sim}{\xi}^i \mu_i) ;$$

that is, every solution of the balance system $\underset{\sim}{E} = (d\underset{\sim}{\xi})\underset{\sim}{R} \bmod \bar{C}$ determines a solution of the prolonged balance system $(\underset{\sim}{E}, \underset{\sim}{Q})$.

<u>THEOREM 4.4</u>. *If* $\underset{\sim}{E} = (d\underset{\sim}{\xi})\underset{\sim}{R} \bmod \bar{C}$, *then the general solution of the balance system* $\underset{\sim}{E}$ *is obtained by prolongation of* $I\{\underset{\sim}{C}, d\underset{\sim}{C}, \underset{\sim}{E}\}$ *to a closed ideal* $I\{\underset{\sim}{C}, d\underset{\sim}{C}, \underset{\sim}{E}, \underset{\sim}{Q}\}^*$ *of* $\Lambda(K_2)$ *and solving the prolonged balance system* $(\underset{\sim}{E}, \underset{\sim}{Q})$. *Every solution of this prolonged balance system can be obtained in closed form by*

$$(4.45) \qquad \Phi_2^* \underset{\sim}{\xi} = -\, d\Phi_2^* \underset{\sim}{S}^{ij} \wedge \mu_{ij} .$$

4.3 EXISTENCE OF PSEUDOPOTENTIALS FOR $\underset{\sim}{E} = (d\underset{\sim}{\xi})\underset{\sim}{R} \bmod \bar{C}$

Lemma 4.1 and Theorem 4.4 show that the process of prolongation of $I\{\underset{\sim}{C}, d\underset{\sim}{C}, \underset{\sim}{E}\}$ to $I\{\underset{\sim}{C}, d\underset{\sim}{C}, \underset{\sim}{E}, \underset{\sim}{Q}\}^* \subset \Lambda(K_2)$ with

$$\underset{\sim}{Q} = \{d\underset{\sim}{S}^{ij} \wedge \mu_{ij} + \underset{\sim}{\xi}\}\underset{\sim}{B}^{-1} \tag{4.46}$$

replaces the problem of solving $\underset{\sim}{E} = (d\underset{\sim}{\xi})\underset{\sim}{R} \bmod \bar{C}$ by the problem of solving

$$d\Phi_2^*\underset{\sim}{\xi} = \underset{\sim}{0} , \tag{4.47}$$

$$d\Phi_2^*(\underset{\sim}{S}^{ij}) \wedge \mu_{ij} = - \Phi_2^*\underset{\sim}{\xi} , \tag{4.48}$$

and that (4.48) constitutes the general solution of (4.47) by allowing $\Phi_2^*\underset{\sim}{S}^{ij}$ to range over all possible $Nn(n-1)/2$-tuples of functions defined on B_n . In essence, $\underset{\sim}{S}^{ij}$ plays the role of a potential for the balance quantities $\underset{\sim}{\xi}$, and is, in fact, directly related to the concept of a pseudopotential that enters so prominently in current writings [9,15-22].

In order to make this quite evident, define $\underset{\sim}{p}^{ij}$ by

$$\underset{\sim}{S}^{ij} = \underset{\sim}{p}^{ij}\underset{\sim}{B} . \tag{4.49}$$

When this is substituted into (4.46) and we recall that $\underset{\sim}{\Gamma} = d\underset{\sim}{B}\,\underset{\sim}{B}^{-1}$, we obtain

$$\underset{\sim}{Q} = (d\underset{\sim}{p}^{ij} + \underset{\sim}{p}^{ij}\,\underset{\sim}{\Gamma}) \wedge \mu_{ij} + \underset{\sim}{\xi}\,B^{-1} \bmod \bar{C} , \tag{4.50}$$

with

$$\underset{\sim}{p}^{ij} = - \underset{\sim}{p}^{ji} . \tag{4.51}$$

For $n=2$, $\mu_{11} = \mu_{22} = 0$, $\mu_{12} = -\mu_{21} = -1$, $\underset{\sim}{p}^{ij}\,\mu_{ij} = -2\underset{\sim}{p}^{12} = 2\underset{\sim}{p}^{21}$, and $M=N$. The case $n=2$, which is most often the case in the current literature, is seen to be singular in certain regards, for (4.50) reduces to

(4.52) $$\underset{\sim}{Q} = d\underset{\sim}{p} + \underset{\sim}{p}\,\underset{\sim}{\Gamma} + \underset{\sim}{\xi}\, B^{-1} \bmod \bar{C}$$

with $\underset{\sim}{p} = 2\underset{\sim}{p}^{21}$. Equations (4.52) are, however, exactly the equations that are used to define $\underset{\sim}{p}$ to be pseudopotentials of the balance system $\underset{\sim}{E}$ [9,15]. If $\underset{\sim}{B}$, and hence $\underset{\sim}{\Gamma}$, is defined over K, then $\underset{\sim}{p}$ is referred to as a pseudopotential of the *first kind.* If $\underset{\sim}{B}$, and hence $\underset{\sim}{\Gamma}$ is defined over K_2, then $\underset{\sim}{p}$ is referred to as a pseudopotential of the *second kind.* In this latter event, the dependence of $\underset{\sim}{B}$ and $\underset{\sim}{\Gamma}$ on the variables $\underset{\sim}{p}$ renders $\underset{\sim}{Q}$ nonlinear in these variables. Thus, $\underset{\sim}{Q}$ is linear or nonlinear in the variables $\underset{\sim}{p}$ if the prolongation is of the first kind or of the second kind, respectively. We adopt this same terminology in the general case with $n \geq 2$.

THEOREM 4.4. *If*

(4.53) $$\underset{\sim}{E} = (d\underset{\sim}{\xi})\underset{\sim}{R} \bmod \bar{C},$$

then the balance system $\underset{\sim}{E}$ *possesses a system of* $M = Nn(n-1)/2$ *pseudopotentials* $\underset{\sim}{p}^{ij} = -\underset{\sim}{p}^{ji}$ *that is defined by*

(4.54) $$\underset{\sim}{Q} = (d\underset{\sim}{p}^{ij} + \underset{\sim}{p}^{ij}\,\underset{\sim}{\Gamma}) \wedge \mu_{ij} + \underset{\sim}{\xi}\,\underset{\sim}{B}^{-1} \bmod \bar{C}$$

(4.55) $$\underset{\sim}{\Gamma} = d\underset{\sim}{B}\,\underset{\sim}{B}^{-1}$$

for each choice of the nonsingular matrix $\underset{\sim}{B}$. *If* $\underset{\sim}{B}$ *is defined on* K, *then* $\underset{\sim}{p}^{ij}$ *are pseudopotentials of the first kind. If* $\underset{\sim}{B}$ *is defined on* K_2 *and* $\overset{-1}{\pi}{}^{*}_{2}\,\underset{\sim}{B} \neq \underset{\sim}{B}$, *then* $\underset{\sim}{p}^{ij}$ *are pseudopotentials of the second kind.*

This theorem shows not only that pseudopotentials of the first and second kinds exist for any $\underset{\sim}{E} = (d\underset{\sim}{\xi})\underset{\sim}{R} \bmod \bar{C}$, but, in fact, there are as many as there are distinct matrices $\underset{\sim}{B}$.

It is of interest to note what happens when $\underset{\sim}{E}$ is given by

(4.56) $$\underset{\sim}{E} = d\underset{\sim}{W}^{i} \wedge \mu_i \bmod \bar{C}.$$

In this event, (4.54) becomes

$$(4.57) \qquad \underset{\sim}{Q} = (d\underset{\sim}{p}^{ij} + \underset{\sim}{p}^{ij}\,\underset{\sim}{\Gamma}) \wedge \mu_{ij} + \underset{\sim}{W}^{i} \wedge \mu_{i}\underset{\sim}{B}^{-1} \bmod \bar{C}$$

$$= [d(\underset{\sim}{p}^{ij}\underset{\sim}{B}) \wedge \mu_{ij} + \underset{\sim}{W}^{i}\mu_{i}]\underset{\sim}{B}^{-1} \bmod \bar{C} ,$$

and $\Phi_2^{*}\underset{\sim}{Q} = 0$ yields

$$(4.58) \qquad (\Phi_2^{*}\underset{\sim}{W}^{i})\mu_{i} = - d\Phi_2^{*}(\underset{\sim}{p}^{ij}\underset{\sim}{B}) \wedge \mu_{ij} .$$

There is an obvious further reduction that can be made, although at the expense of losing the ability to generate a large number of solutions. Simply set

$$(4.59) \qquad \underset{\sim}{p}^{ij} = \underset{\sim}{p}\,\underset{\sim}{J}^{ij}$$

where $\underset{\sim}{J}^{ij} = -\underset{\sim}{J}^{ji} \in \Lambda^{o}_{N,N}(K)$ is fixed. This reduces the number of independent fiber coordinates from $Nn(n-1)/2$ to N and (4.54) gives

$$(4.60) \qquad \underset{\sim}{Q} = (d\underset{\sim}{p}\,\underset{\sim}{J}^{ij} + \underset{\sim}{p}\,d\underset{\sim}{J}^{ij} + \underset{\sim}{p}\,\underset{\sim}{J}^{ij}\,\underset{\sim}{\Gamma}) \wedge \mu_{ij} + \underset{\sim}{\xi}\,\underset{\sim}{B}^{-1} \bmod \bar{C} .$$

If $d\underset{\sim}{J}^{ij} = \underset{\sim}{0}$, we obtain the simplified form

$$(4.61) \qquad \underset{\sim}{Q} = (d\underset{\sim}{p} + \underset{\sim}{p}\underset{\sim}{\Gamma}) \wedge \underset{\sim}{J}^{ij}\mu_{ij} + \underset{\sim}{\xi}\,\underset{\sim}{B}^{-1} \bmod \bar{C} .$$

4.4 REPRESENTATIONS OF POTENTIAL TYPE IN THE GENERAL CASE

A completely general balance system has the structure

$$(4.62) \qquad \underset{\sim}{E} = (d\underset{\sim}{\xi} + \underset{\sim}{\eta})\underset{\sim}{R} \bmod \bar{C} ,$$

and thus falls outside the considerations of the previous two sections. There is, however, an alternative approach that provides much the same results.

Lemma 3.3 has shown that

$$(4.63) \qquad \underset{\sim}{P} = d\underset{\sim}{Q}^{j} \wedge \mu_{j} - (Z_{i}\underset{\sim}{Q}^{i})\mu$$

belongs to $\bar{C}$ for any $\underset{\sim}{Q}^j \varepsilon \Lambda^o_{1,N}(K)$ such that

$$\partial^i_\alpha \underset{\sim}{Q}^j + \partial^j_\alpha \underset{\sim}{Q}^i = \underset{\sim}{0} .$$

Thus, for each $\underset{\sim}{E}$ given by (4.62),

$$(4.64) \qquad \underset{\sim}{E}_1 = (d\underset{\sim}{\xi} + \underset{\sim}{\eta})\underset{\sim}{R}\underset{\sim}{A} + \{d\underset{\sim}{Q}^j \wedge \mu_j - (Z_i \underset{\sim}{Q}^i)\underset{\sim}{\mu}\}\underset{\sim}{B}^{-1}$$

belongs to $F(\underset{\sim}{E})$. If we set $\underset{\sim}{R}\underset{\sim}{A} = \underset{\sim}{B}^{-1}$, a simple rearrangement yields

$$(4.65) \qquad \underset{\sim}{E}_1 = d[(\underset{\sim}{\xi} + \underset{\sim}{Q}^j\mu_j + d\underset{\sim}{S}^{ij} \wedge \mu_{ij})\underset{\sim}{B}^{-1}] + (\underset{\sim}{\eta} - Z_i\underset{\sim}{Q}^i\mu)\underset{\sim}{B}^{-1}$$
$$+ (-1)^{n+1}(\underset{\sim}{\xi} + \underset{\sim}{Q}^j\mu_j + d\underset{\sim}{S}^{ij} \wedge \mu_{ij})\underset{\sim}{B}^{-1} \wedge \underset{\sim}{\Gamma}$$

with

$$(4.66) \qquad \underset{\sim}{\Gamma} = d\underset{\sim}{B}\,\underset{\sim}{B}^{-1} .$$

Thus, if we define $\underset{\sim}{Q}$ by

$$(4.67) \qquad \underset{\sim}{Q} = (d\underset{\sim}{S}^{ij} \wedge \mu_{ij} + \underset{\sim}{Q}^j\mu_j + \underset{\sim}{\xi})\underset{\sim}{B}^{-1} ,$$

(4.65) yields

$$(4.68) \qquad d\underset{\sim}{Q} = (-1)^n\, \underset{\sim}{Q} \wedge \underset{\sim}{\Gamma} + \underset{\sim}{E}_1 - (\underset{\sim}{\eta} - Z_i\underset{\sim}{Q}^i\mu)\underset{\sim}{B}^{-1} .$$

This shows that $d\underset{\sim}{Q}$ would belong to the ideal $I\{\underset{\sim}{C}, d\underset{\sim}{C}, \underset{\sim}{E}, d\underset{\sim}{E}, \underset{\sim}{Q}\}$ provided $\underset{\sim}{\eta} = Z_i\underset{\sim}{Q}^i\mu$, which is unduly restrictive. On the other hand, if we rewrite (4.68) in the equivalent form

$$(4.69) \qquad d\underset{\sim}{Q} = \underset{\sim}{E}_1 - \{\underset{\sim}{\eta} - Z_i\underset{\sim}{Q}^i\mu + (-1)^{n+1}\, \underset{\sim}{Q} \wedge d\underset{\sim}{B}\}\underset{\sim}{B}^{-1} ,$$

then $d\underset{\sim}{Q}$ belongs to the closed ideal $\bar{E}$ provided

$$(4.70) \qquad \underset{\sim}{\eta} = Z_i\underset{\sim}{Q}^i\underset{\sim}{\mu} + (-1)^n\, \underset{\sim}{Q} \wedge d\underset{\sim}{B} .$$

<u>THEOREM 4.6</u>. *If* $\underset{\sim}{E} \simeq d\underset{\sim}{\xi} + \underset{\sim}{\eta}$ *and* $\underset{\sim}{Q}^j$, $j=1,\ldots,n$ *is any*

collection of elements of $\Lambda^0_{1,N}(K)$ *such that*

$$(4.71) \qquad \partial^i_\alpha Q^j + \partial^j_\alpha Q^i = 0 ,$$

then $dQ \in \bar{E}$ *when*

$$(4.72) \qquad Q = (dS^{ij} \wedge \mu_{ij} + Q^j \mu_j + \xi) B^{-1} ,$$

$$(4.73) \qquad \eta = (Z_i Q^i)\mu + (-1)^n Q \wedge dB ,$$

for any nonsingular matrix B *and any* S^{ij} *with* $S^{ij} = -S^{ji}$.

We can now proceed directly as in the last two sections and prolong the ideal $I\{C, dC, E, dE, Q\}$ to a closed ideal on K_2 with the fiber coordinates (S^{ij}, Q^j) . A map $\Phi_2 : B_n \longrightarrow K_2$ is then said to solve the prolonged balance system (E, Q) if Φ_2^* annihilates the prolonged ideal and $\Phi^*\mu \neq 0$. Such solving maps Φ_2 , when applied to (4.72) and (4.73) yield

$$\Phi_2^* \xi = - d\Phi_2^* S^{ij} \wedge \mu_{ij} - \Phi_2^* Q^j \mu_j$$

and

$$\Phi_2^* \eta = \Phi_2^*(Z_i Q^i)\mu = D_i (\Phi^* Q^i)\mu$$

in view of (4.71). This, in turn gives

$$\Phi_2^* E \simeq d\Phi_2^* \xi + \Phi_2^* \eta = - d(\Phi_2^* Q^j) \wedge \mu_j + D_i \Phi^* Q^i \mu = 0 ,$$

Φ_2^* annihilates the prolonged ideal if

$$\Phi_2^* C = 0 , \qquad \Phi_2^* Q = 0 ,$$

and

$$(4.74) \qquad \Phi_2^*(\eta - Z_i Q^i \mu + (-1)^{n+1} Q \wedge dB) = 0 .$$

Thus, the general $E \simeq d\xi + \eta$ can be solved by prolongation with the larger set of fiber coordinates, namely (S^{ij}, Q^j)!!

An alternative approach for $\underset{\sim}{E} \simeq d\xi + \underset{\sim}{\eta}$ with $d\eta \in \bar{\underset{\sim}{C}}$ is available directly from Theorem 3.14 if we extend the prolonged ideal so as to include the generators $v \rfloor \underset{\sim}{C}$ and $v \rfloor d\underset{\sim}{C}$ and simultaneously restrict Φ to belong to the class of extended similarity solutions $S_1(v,\underset{\sim}{E},B_n)$, for in this case

$$\underset{\sim}{Q}(v) = v \rfloor (d\underset{\sim}{\xi} + \underset{\sim}{\eta}) + d(\underset{\sim}{S}^{ij}\mu_{ij})$$

will satisfy

$$d\Phi_2^*\underset{\sim}{Q}(v) = \underset{\sim}{0}$$

for any isovector v of the extended and prolonged ideal $\bar{E}_1(v)^*$. Thus, in particular, *we can obtain similarity pseudopotentials for any* $\underset{\sim}{E} = \underset{\sim}{W}\mu - d(\underset{\sim}{W}^i\mu_i)$!

4.5 ISOVECTORS OF THE PROLONGED IDEAL

From now on, we assume that

$$(4.75) \qquad \underset{\sim}{E} = (d\underset{\sim}{\xi})\underset{\sim}{R} \bmod \bar{C} ,$$

so that the ideal $I\{\underset{\sim}{C}, d\underset{\sim}{C}, \underset{\sim}{E}\}$ is closed. It is obvious from what has been presented before, that the thing to do is to construct the isovectors of the prolonged closed ideal $I\{\underset{\sim}{C}, d\underset{\sim}{C}, \underset{\sim}{E}, \underset{\sim}{Q}\}^*$. This is simplified by the following result.

LEMMA 4.2. *If* $I\{\underset{\sim}{C}, d\underset{\sim}{C}, \underset{\sim}{E}\}$ *is closed, then*

$$(4.76) \qquad I\{\underset{\sim}{C}, d\underset{\sim}{C}, \underset{\sim}{E}, \underset{\sim}{Q}\}^* = I\{\underset{\sim}{C}, d\underset{\sim}{C}, \underset{\sim}{Q}, d\underset{\sim}{Q}\}^* = \bar{I}\{\underset{\sim}{C}, \underset{\sim}{Q}\}^*$$

for any $\underset{\sim}{Q}$ *of the form* (4.54). *The closed ideal characterizing the prolonged balance structure* $(\underset{\sim}{E},\underset{\sim}{Q})$ *is thus generated by* $\underset{\sim}{C}$ *and* $\underset{\sim}{Q}$ *alone.*

Proof. It follows directly from (4.54) and (4.55) that

$$(4.77) \qquad \underset{\sim}{Q} = [d(\underset{\sim}{p}^{ij}\underset{\sim}{B}\mu_{ij}) + \underset{\sim}{\xi}]\underset{\sim}{B}^{-1}$$

and hence

$$\underset{\sim}{E} = (d\underset{\sim}{\xi})\underset{\sim}{R} \bmod \bar{C} = d(\underset{\sim}{Q}\underset{\sim}{B} - d(\underset{\sim}{p}^{ij}\underset{\sim}{B}\mu_{ij}))\underset{\sim}{R} \bmod \bar{C}$$

$$= d(\underset{\sim}{Q}\underset{\sim}{B})\underset{\sim}{R} \bmod \bar{C} = \underset{\sim}{0} \bmod \bar{I}\{\underset{\sim}{C}, \underset{\sim}{Q}\}^* .$$

Although we now need only work with the relatively simple closed ideal $\bar{I}\{\underset{\sim}{C}, \underset{\sim}{Q}\}^*$ of $\Lambda(K_2)$, it turns out that the isovector structure is trivial without further extension and prolongation. Consider, therefore a base space $K_3 = K_2 \times E_{Nn^2(n-1)/2}$ with coordinates $(x^i, q^\alpha, y_i^\alpha, p_\alpha^{ij}, h_{\alpha i}^{km})$, where $Nn^2(n-1)/2$ additional variables $h_{\alpha i}^{km}$ enter through the new system of contact forms

$$J_\alpha^{k\ell} = dp_\alpha^{k\ell} - h_{\alpha i}^{k\ell}\, dx^i \tag{4.78}$$

that are naturally associated with the fiber variables $(p_\alpha^{k\ell})$. Our considerations can now be based on the closed ideal $\bar{I}\{\underset{\sim}{C}, \underset{\sim}{J}^{k\ell}, \underset{\sim}{Q}\}$ of $\Lambda(K_3)$. Construction of the isovectors of this ideal is most simply done in a sequential manner by first constructing the isovectors of $\bar{I}\{\underset{\sim}{C}, \underset{\sim}{J}^{k\ell}\}$; namely, vector fields

$$v = v^i\partial_i + v^\alpha\partial_\alpha + v_i^\alpha\partial_\alpha^i + v_\alpha^{k\ell}\partial_{k\ell}^\alpha + v_{\alpha i}^{k\ell}\partial_{k\ell}^{\alpha i} \tag{4.79}$$

of $T(K_3)$, such that $\pounds_v \bar{I}\{\underset{\sim}{C}, \underset{\sim}{J}^{k\ell}\} \subset \bar{I}\{\underset{\sim}{C}, \underset{\sim}{J}^{k\ell}\}$, where

$$\partial_{k\ell}^\alpha \equiv \partial/\partial p_\alpha^{k\ell}, \quad \partial_{k\ell}^{\alpha i} = \partial/\partial h_{\alpha i}^{k\ell} .$$

The proofs of the following results exactly parallel those of Lemma 2.2 and Theorem 2.1.

LEMMA 4.3. A $v \in T(K_3)$ *is an isovector of* $\bar{I}\{\underset{\sim}{C}, \underset{\sim}{J}^{k\ell}\}$ *if and only if*

$$\pounds_v C^\alpha = A_\beta^\alpha C^\beta + B_{k\ell}^{\alpha\beta} J_\beta^{k\ell} , \tag{4.80}$$

(4.81) $$\pounds_v J^{k\ell}_\alpha = R^{k\ell}_{\alpha\beta} C^\beta + T^{\beta k\ell}_{\alpha rs} J^{rs}_\beta$$

hold throughout K_3 *for some collection of functions* $(A^\alpha_\beta, B^{\alpha\beta}_{k\ell}, R^{k\ell}_{\alpha\beta}, T^{\beta k\ell}_{\alpha rs})$.

THEOREM 4.7. $TC(K_3)$ *is a Lie subalgebra of* $T(K_3)$, *where* $TC(K_3)$ *is the collection of all isovectors of the prolonged contact ideal* $\bar{I}\{\underset{\sim}{C}, \underset{\sim}{J}^{k\ell}\}$.

It is now a straightforward, but tedious matter to obtain the following results; simply follow the same steps as used to establish Theorem 2.2.

THEOREM 4.8. *A vector field* v *belongs to* $TC(K_3)$, *for* $N > 1$ *if and only if*

(4.82) $$v^\alpha_i = Z^*_i(v \rfloor C^\alpha) ,$$

(4.83) $$v^{k\ell}_{\alpha i} = Z^*_i(v \rfloor J^{k\ell}_\alpha) ,$$

(4.84) $$\pounds_v C^\alpha = \partial_\beta(v \rfloor C^\alpha) C^\beta + \partial^\beta_{mn}(v \rfloor C^\alpha) J^{mn}_\beta ,$$

(4.85) $$\pounds_v J^{k\ell}_\alpha = \partial_\beta(v \rfloor J^{k\ell}_\alpha) C^\beta + \partial^\beta_{mn}(v \rfloor J^{k\ell}_\alpha) J^{mn}_\beta ,$$

where Z^*_i *is the linear differential operator*

(4.86) $$Z^*_i \equiv \partial_i + y^\beta_i \partial_\beta + h^{mn}_{\beta i} \partial^\beta_{mn}$$

and

(4.87) $$v^i = f^i(x^j, q^\beta, p^{k\ell}_\beta) , \quad v^\alpha = f^\alpha(x^j, q^\beta, p^{k\ell}_\beta) ,$$
$$v^{k\ell}_\alpha = f^{k\ell}_\alpha(x^j, q^\beta, p^{mn}_\beta) .$$

COROLLARY 4.1. *A* $v \in TC(K_3)$ *leaves the contact forms* $J^{k\ell}_\alpha$ *invariant if and only if*

(4.88) $$v^i = g^i(x^j), \quad v^{k\ell}_\alpha = g^{k\ell}_\alpha(x^j), \quad v^\alpha = f^\alpha(x^j, q^\beta, p^{k\ell}_\beta)$$

(4.89) $$v^{k\ell}_{\alpha i} = \partial_i g^{k\ell}_\alpha - h^{k\ell}_{\alpha j}\partial_i g^j$$

(4.90) $$v^\alpha_i = Z^*_i v^\alpha - y^\alpha_j \partial_i g^j \ .$$

Proof. $£_v J^{k\ell}_\alpha = 0$ if and only if $\partial_\beta(v \rfloor J^{k\ell}_\alpha) = 0$, $\partial^\beta_{mn}(v \rfloor J^{k\ell}_\alpha) = 0$, by (4.85). Since $v \rfloor J^{k\ell}_\alpha = v^{k\ell}_\alpha - h^{k\ell}_{\alpha j} v^j$ (4.87) shows that these conditions can be satisfied if and only if (4.88) hold, in which case (4.89) and (4.90) follow directly from (4.82) and (4.83).

Definition 4.4. A $v \in T(K_3)$ is an isovector of the closed ideal $\bar{I}\{\underset{\sim}{C}, \underset{\sim}{J}^{k\ell}, \underset{\sim}{Q}\}$ if and only if

(4.91) $$£_v \bar{I}\{\underset{\sim}{C}, \underset{\sim}{J}^{k\ell}, \underset{\sim}{Q}\} \subset \bar{I}\{\underset{\sim}{C}, \underset{\sim}{J}^{k\ell}, \underset{\sim}{Q}\} \ .$$

The collection of all isovectors of $\bar{I}\{\underset{\sim}{C}, \underset{\sim}{J}^{k\ell}, \underset{\sim}{Q}\}$ is denoted by $T(\underset{\sim}{Q}, K_3)$.

THEOREM 4.9. *A* $v \in T(K_3)$ *is an isovector of* $\bar{I}\{\underset{\sim}{C}, \underset{\sim}{J}^{k\ell}, \underset{\sim}{Q}\}$ *if and only if* $v \in TC(K_3)$ *and*

(4.92) $$£_v \underset{\sim}{Q} = \underset{\sim}{Q}\underset{\sim}{A} \bmod \bar{I}\{\underset{\sim}{C}, \underset{\sim}{J}^{k\ell}\}$$

for some $\underset{\sim}{A} \in \Lambda^0_{N,N}(K_3)$.

Proof. Since $\bar{I}\{\underset{\sim}{C}, \underset{\sim}{J}^{k\ell}, \underset{\sim}{Q}\}$ is closed and $d£_v \equiv £_v d$, $v \in T(\underset{\sim}{Q}, K_3)$ if and only if $£_v \underset{\sim}{C} = \underset{\sim}{0} \bmod \bar{I}\{\underset{\sim}{C}, \underset{\sim}{J}^{k\ell}, \underset{\sim}{Q}\}$, $£_v \underset{\sim}{J}^{k\ell} = \underset{\sim}{0} \bmod \bar{I}\{\underset{\sim}{C}, \underset{\sim}{J}^{k\ell}, \underset{\sim}{Q}\}$, and $£_v \underset{\sim}{Q} = \underset{\sim}{0} \bmod \bar{I}\{\underset{\sim}{C}, \underset{\sim}{J}^{k\ell}, \underset{\sim}{Q}\}$. However, $\underset{\sim}{C}$ and $\underset{\sim}{J}^{k\ell}$ are of degree 1 while $\underset{\sim}{Q}$ is of degree $n-1$. Thus, $£_v \underset{\sim}{C} = \underset{\sim}{0} \bmod \bar{I}\{\underset{\sim}{C}, \underset{\sim}{J}^{k\ell}\}$, $£_v \underset{\sim}{J}^{k\ell} = \underset{\sim}{0} \bmod \bar{I}\{\underset{\sim}{C}, \underset{\sim}{J}^{k\ell}\}$, so that $v \in TC(K_3)$ and (4.92) must also hold.

The following theorem is an immediate consequence of Lemma 4.3 and Theorem 4.9.

THEOREM 4.10. $T(\underset{\sim}{Q}, K_3)$ *is a Lie subalgebra of* $TC(K_3)$.

Particular note should be taken of the contrast between $T(\underset{\sim}{E}, K)$

and $T(\underset{\sim}{Q},K_3)$, for the former requires $v \in TC(K)$ and

$$\pounds_{\underset{\sim}{v}} \underset{\sim}{E} = \underset{\sim}{E}\underset{\sim}{A} \bmod \bar{C} ,$$

where $\underset{\sim}{E} \in \Lambda^n_{1,N}(K)$, while the latter requires $v \in TC(K_3)$ and

$$\pounds_{\underset{\sim}{v}} \underset{\sim}{Q} = \underset{\sim}{Q}\underset{\sim}{A} \bmod \bar{I}\{\underset{\sim}{C}, \underset{\sim}{J}^{k\ell}\} ,$$

where $\underset{\sim}{Q} \in \Lambda^{n-1}_{1,N}(K_3)$; that is, a reduction by one of the degree of the transport equation is obtained together with an increase in the dimension of the underlying space and the number of contact forms that generates the contact ideal. Each of these changes result in an increase in the freedom with which the isovectors are chosen, and this in turn yields a corresponding increase in the scope of problems that can be successfully handled. The situation is further compounded by the fact that the matrix $\underset{\sim}{B}$ that occurs in $\underset{\sim}{Q}$ is still at our disposal and may thus be judiciously chosen to aid the quest for isovectors.

4.6 ORBITS, SOLUTIONS AND BÄCKLUND TRANSFORMATIONS

We now come to the problem of constructing solutions of the prolonged balance system. We have seen that the prolonged balance system is characterized by the closed ideal $\bar{I}\{\underset{\sim}{C}, \underset{\sim}{J}^{k\ell}, \underset{\sim}{Q}\}$ of $\Lambda(K_3)$ whenever $\underset{\sim}{E} = (d\underset{\sim}{\xi})R \bmod \bar{C}$.

<u>Definition 4.5</u>. A map $\Phi_3 \colon B_n \longrightarrow K_3$ such that $\Phi_3^*\mu \neq 0$ is said to *solve* the prolonged balance system characterized by $\bar{I}\{\underset{\sim}{C}, \underset{\sim}{J}^{k\ell}, \underset{\sim}{Q}\}$ if and only if Φ_3^* annihilates $\bar{I}\{\underset{\sim}{C}, \underset{\sim}{J}^{k\ell}, \underset{\sim}{Q}\}$. The set of all solutions is denoted by $S_3(\underset{\sim}{Q}, B_n)$.

Thus, if $\Phi_3 \in S_3(\underset{\sim}{Q},B_n)$ and $\Phi_3(x^i) = \tau^i$, then $\Phi_3^*(y^\alpha_i) = \partial(\Phi_3^* q^\alpha)/\partial\tau^i$, $\Phi_3^*(h^{k\ell}_{\alpha i}) = \partial\Phi_3^*(p^{k\ell}_\alpha)/\partial\tau^i$, and

$$(4.93) \qquad \Phi_3^*\underset{\sim}{\xi} = -\, d\Phi_3^*(\underset{\sim}{p}^{ij}\underset{\sim}{B}\mu_{ij}) ,$$

where (4.93) follows from (4.77) and $\Phi_3^*\underset{\sim}{Q} = \underset{\sim}{0}$. The system of relations (4.93) implies, by exterior differentiation, that

$$(4.94)\qquad d\Phi_3^* \underset{\sim}{\xi} = \underset{\sim}{0} ,$$

so that $\Phi_3 \in S_3(Q, B_n)$ indeed solves the balance system $\underset{\sim}{E} = (d\underset{\sim}{\xi})\underset{\sim}{R} \bmod \bar{C}$.

The usefulness of isovectors for the prolonged balance structure is the same as that established in Chapter 3 for isovectors of the balance structure, namely, the generation of new solutions from old ones by transport along the orbits of isovector fields.

THEOREM 4.11. *Let* v *be an isovector of* $\bar{I}\{\underset{\sim}{C}, \underset{\sim}{J}^{k\ell}, \underset{\sim}{Q}\}$ *and let* $\exp(v,s)$ *be the one parameter family of maps of* K_3 *to* K_3 *that is defined by transport of* K_3 *along the orbits of* v ,

$$(4.95)\qquad \exp(v,s) : K_3 \longrightarrow K_3 \mid \overset{*}{x}{}^i = \exp(s\pounds_v)x^i ,$$

$$\overset{*}{q}{}^\alpha = \exp(s\pounds_v)q^\alpha , \qquad \overset{*}{y}{}^\alpha_i = \exp(s\pounds_v)y^\alpha_i ,$$

$$\overset{*}{p}{}^{k\ell}_\alpha = \exp(s\pounds_v)p^{k\ell}_\alpha , \qquad \overset{*}{h}{}^{k\ell}_{\alpha i} = \exp(s\pounds_v)h^{k\ell}_{\alpha i} ,$$

in which case

$$(4.96)\qquad (\exp(v,s))^*(\overset{*}{\Omega}) = \exp(s\pounds_v)(\Omega) .$$

If Φ_3 *is any regular map of* B_n *into* K_3 , *then*

$$(4.97)\qquad \Phi_{3v}(s) = \exp(v,s)\circ\Phi_3$$

is a 1*-parameter family of regular maps of* B_n *into* K_3 . *If* $\Phi_3 \in S_3(\underset{\sim}{Q}, B_n)$, *then* $\Phi_{3v}(s)$ *is a* 1*-parameter family of elements of* $S_3(\underset{\sim}{Q}, B_n)$; *that is, composition with* $\exp(v,s)$ *generates a map of* $S_3(\underset{\sim}{Q}, B_n)$ *to* $S_3(\underset{\sim}{Q}, B_n)$ *for any value of* s . *Further, those maps have the superposition property*

$$(4.98)\qquad \Phi_{3v}(s_1+s_2) = \exp(v,s_2)\circ\Phi_{3v}(s_1) = \exp(v,s_1)\circ\Phi_{3v}(s_2) .$$

Proof. Since the map $\exp(v,s)$ constitutes transport along the orbits of v from the initial data $(x^i, q^\alpha, y^\alpha_i, p^{k\ell}_\alpha, h^{k\ell}_{\alpha i})$,

$$(4.99)\qquad {}^\backprime\bar{I}\{\underset{\sim}{C}, \underset{\sim}{J}^{k\ell}, \underset{\sim}{Q}\} = \exp(-s\pounds_v)\bar{I}\{\underset{\sim}{C}, \underset{\sim}{J}^{k\ell}, \underset{\sim}{Q}\}$$

and hence

$$\Phi_{3v}(s)^{*}{}^\backprime\bar{I}\{\underset{\sim}{C}, \underset{\sim}{J}^{k\ell}, \underset{\sim}{Q}\} = \Phi_3^*\bar{I}\{\underset{\sim}{C}, \underset{\sim}{J}^{k\ell}, \underset{\sim}{Q}\}$$

because $\Phi_{3v}(s)^* = \Phi_3^*\circ(\exp(s,v))^* = \Phi_3^*(\exp(s\pounds_v))$ by (4.96). Thus, if $\Phi_3 \in S_3(\underset{\sim}{Q},B_n)$, then

$$(4.100)\qquad \Phi_{3v}(s)^{*}{}^\backprime\bar{I}\{\underset{\sim}{C}, \underset{\sim}{J}^{k\ell}, \underset{\sim}{Q}\} = 0 .$$

However, since v is an isovector of $\bar{I}\{\underset{\sim}{C}, \underset{\sim}{J}^{k\ell}, \underset{\sim}{Q}\}$, $\pounds_v\bar{I}\{\underset{\sim}{C}, \underset{\sim}{J}^{k\ell}, \underset{\sim}{Q}\} \subset \bar{I}\{\underset{\sim}{C}, \underset{\sim}{J}^{k\ell}, \underset{\sim}{Q}\}$ and hence

$$(4.101)\qquad \exp(-s\pounds_v)\bar{I}\{\underset{\sim}{C}, \underset{\sim}{J}^{k\ell}, \underset{\sim}{Q}\} \subset \bar{I}\{\underset{\sim}{C}, \underset{\sim}{J}^{k\ell}, \underset{\sim}{Q}\} .$$

The theorem then follows since (4.99) through (4.101) give $\Phi_{3v}(s)^*I\{\underset{\sim}{C}, \underset{\sim}{J}^{k\ell}, \underset{\sim}{Q}\} = 0$.

The same observations and comments given after Theorem 4.8 become applicable here with one further very important observation. Noting that v^i, v^α and $v^{k\ell}_\alpha$ are functions of $(x^j, q^\beta, p^{mn}_\beta)$, but not of $(y^\alpha_j, h^{mn}_{\beta j})$, by Theorem 4.8, the orbital equations that serve to determine $({}^\backprime x^i, {}^\backprime q^\alpha, {}^\backprime p^{k\ell}_\alpha)$ are given by

$$(4.102)\qquad \begin{aligned} \frac{d{}^\backprime x^i}{ds} &= v^i({}^\backprime x^j, {}^\backprime q^\beta, {}^\backprime p^{mn}_\beta) , & {}^\backprime x^i(0) &= x^i , \\ \frac{d{}^\backprime q^\alpha}{ds} &= v^\alpha({}^\backprime x^j, {}^\backprime q^\beta, {}^\backprime p^{mn}_\beta) , & {}^\backprime q^\alpha(0) &= q^\alpha , \\ \frac{d{}^\backprime p^{k\ell}_\alpha}{ds} &= v^{k\ell}_\alpha({}^\backprime x^j, {}^\backprime q^\beta, {}^\backprime p^{mn}_\beta) , & {}^\backprime p^{k\ell}_\alpha(0) &= p^{k\ell}_\alpha , \end{aligned}$$

and give rise to the map $\exp(v,s)$. The map $\exp(v,s)$ thus has the explicit structure

$$\begin{aligned} `x^i &= f^i(x^j,\ q^\beta,\ p_\beta^{mn};\ s)\ , \\ `q^\alpha &= f^\alpha(x^j,\ q^\beta,\ p_\beta^{mn};\ s)\ , \qquad (4.103) \\ `p_\alpha^{k\ell} &= f_\alpha^{k\ell}(x^j,\ q^\beta,\ p_\beta^{mn};\ s)\ , \end{aligned}$$

and $\Phi_{3v}^*(s)$ becomes

$$\begin{aligned} \Phi_{3v}^*(s)`x^i &= f^i(\Phi_3^*x^j,\ \Phi_3^*q^\beta,\ \Phi_3^*p_\beta^{mn};\ s)\ , \\ \Phi_{3v}^*(s)`q^\alpha &= f^\alpha(\Phi_3^*x^j,\ \Phi_3^*q^\beta,\ \Phi_3^*p_\beta^{mn};\ s)\ , \qquad (4.104) \\ \Phi_{3v}^*(s)`p^{k\ell} &= f^{k\ell}(\Phi_3^*x^j,\ \Phi_3^*q^\beta,\ \Phi_3^*p_\beta^{mn};\ s)\ , \end{aligned}$$

for any $\Phi_3 \in S_3(\underset{\sim}{Q},B_n)$. Now, $\Phi_3 \in S_3(\underset{\sim}{Q},B_n)$ gives $\Phi_3^*\underset{\sim}{Q} = \underset{\sim}{0}$, and hence

$$(4.105) \qquad d\Phi_3^*(p_\alpha^{ij}B_\beta^\alpha\mu_{ij}) = -\ \Phi_3^*\xi_\beta\ .$$

For simplicity, let us assume that $\Phi_3^*x^i = x^i$, which is always possible because $\Phi_3^*\mu \neq 0$. Then $\Phi_3^*\underset{\sim}{\xi} \in \Lambda^{n-1}(B_n)$, and hence there exist functions $W_\alpha^i(x^j,\ \Phi_3^*q^\beta,\ \Phi_3^*y_j^\beta) = W_\alpha^i(x^j,\ \Phi_3^*q^\beta,\ \partial_j\Phi_3^*q^\beta)$ such that

$$(4.106) \qquad \Phi_3^*\xi_\alpha = W_\alpha^i\mu_i$$

because $\{\mu_i\}$ is a basis for $\Lambda^{n-1}(B_n)$. Thus, (4.105) becomes

$$(4.107) \qquad d\Phi_3^*(p_\alpha^{ij}B_\beta^\alpha\mu_{ij}) = -\ W_\beta^i\mu_i\ ,$$

which has the particular solution

$$(4.108) \qquad \Phi_3^*(p_\alpha^{ij})\Phi_3^*(B_\beta^\alpha)\mu_{ij} = -\ H_n(W_\beta^i\mu_i)$$

with H_n the linear homotopy operator on B_n . When use is made of (1.5), (1.9) and (1.10), we obtain

(4.109) $$-H_n(W^i_\alpha \mu_i) = \frac{1}{2}[h_n(W^i_\alpha)x^j - h_n(W^j_\alpha)x^i]\mu_{ij} ,$$

where

(4.110) $$h_n(W^i_\alpha) = \int_0^1 W^i_\alpha(\lambda x)\lambda^{n-2}d\lambda ,$$

and hence (4.108) finally yields the explicit evaluations

(4.111) $$\Phi_3^*(p^{ij}_\alpha) = \Phi_3^*(\overset{-1}{B}{}^\beta_\alpha)\frac{1}{2}[h_n(W^i_\beta)x^j - h_n(W^j_\beta)x^i] .$$

This shows that the $(\Phi_3^* p^{ij}_\alpha)$'s depend upon the $(\Phi_3^* q^\alpha)$'s and their *first partial derivatives* and hence the same is true of the $(\Phi_{3v}^*(s)\check{\ }x^i)$'s, the $(\Phi_{3v}^*(s)\check{\ }q^\alpha)$'s and the $(\Phi_{3v}^*(s)\check{\ }p^{k\ell}_\alpha)$'s whenever not all of the functions $(v^i, v^\alpha, v^{k\ell}_\alpha)$ are independent of the arguments $(p^{k\ell}_\alpha)$. The resulting relations implied between the derivatives by the orbital equations for the $(\check{\ }y^\alpha_i)$ gives the following result.

<u>THEOREM 4.12</u>. *If* v *is an isovector of the prolonged ideal* $\bar{I}\{\underset{\sim}{C}, \underset{\sim}{J}^{k\ell}, \underset{\sim}{Q}\}$ *and at least one of the functions* $(v^i, v^\alpha, v^{k\ell}_\alpha)$ *depends on the arguments* $(p^{k\ell}_\beta)$ *, then the new solutions* $\Phi_{3v}(s)$ *and their first derivatives are each functions of the old solutions and their first derivatives for each* $s \neq 0$ *; the maps* $\Phi_{3v}(s)$ *that are generated by* $\exp(v,s)\circ\Phi_3$ *, for* $s \neq 0$ *and* $\Phi_3 \in S_3(Q,\underset{\sim}{B}_n)$ *, are generalized Bäcklund transformations.*

The isovectors of the prolonged ideal $\bar{I}\{\underset{\sim}{C}, \underset{\sim}{J}^{k\ell}, \underset{\sim}{Q}\}$ thus provide not only access to Bäcklund transformations, but rather access to 1-parameter groups of Bäcklund transformations. This result would appear to have far-reaching significance since the group property opens a whole new area of inquiry. Further, $T(\underset{\sim}{Q},K_3)$ is a Lie algebra, by Theorem 4.10, and hence the Bäcklund transformations generated in the manner given above, are not just 1-parameter groups, but rather they are an intrinsic part of a full Lie-group structure $\exp(T(\underset{\sim}{Q},B_n))$. We note, in contrast with the results established in Chapter 3, that this new class of Bäcklund transformations does not integrate directly, for the $\Phi^*_{3v}(s)`q^\alpha$ now depend on the $\Phi^* y^\alpha_i$ as well as on the $\Phi^* x^i$ and the $\Phi^* q^\alpha$!

CHAPTER 5

VARIATIONAL SYSTEMS

If the given balance system happens to obtain from a variational principle, we can make use of the extensive body of results that are already known for variational systems. It is the rule, rather than the exception, however, that a given balance system, as it stands, does not form a variational system. We will show in this chapter that there is a natural prolongation of the original system so that the resulting system is a variational system, and, in particular, that its Lagrangian function assumes a very simple form. All of our previous results can then be used in conjunction with the additional structure offered by the existence of a variational principle. Among other things, we will be able to establish the existence of pseudopotentials for any system of laws of balance and to examine significantly more general transformations than just those that result from transport by isovector fields. Existence of complete and weakly complete figures, in the sense of Carathèodory is also obtained together with the accompanying Hamilton-Jacobi theory for balance equations.

5.1 PROLONGATION TO A VARIATIONAL SYSTEM

We confine our attention throughout this Section to balance systems that represent generalized laws of balance; that is, systems of the form

$$(5.1) \qquad E_\alpha = W_\alpha \mu - d(W_\alpha^i \mu_i) , \quad \alpha = 1,\ldots,N ,$$

where the $N + nN$ quantities $\{W_\alpha, W_\alpha^i\}$ are functions that are defined on the space K and are thus functions of the arguments $(x^i, q^\alpha, y_i^\alpha)$. As in the previous analysis, this balance system is supported by the contact forms

$$(5.2) \qquad C^\alpha = dq^\alpha - y_i^\alpha dx^i , \quad \alpha = 1,\ldots,N ,$$

and we have the closed ideals $\bar{C} = I\{\underset{\sim}{C}, d\underset{\sim}{C}\}$ and $\bar{E} = I\{\underset{\sim}{C}, d\underset{\sim}{C}, \underset{\sim}{E}, d\underset{\sim}{E}\}$.

Definition 5.1. A balance system $\underset{\sim}{E}$ is said to form a *variational system* if and only if $\Phi^*\underset{\sim}{E} = \underset{\sim}{0}$, $\Phi^*\underset{\sim}{C} = \underset{\sim}{0}$ are the Euler-Lagrange equations that arise from a variational principle

$$(5.3) \qquad \delta\int_{B_n} \Phi^* L \mu = 0 , \quad \delta(\Phi^* q^\alpha)\Big|_{\partial B_n} = 0 ,$$

for some function L that is defined on K .

It is clear that sufficient conditions for a balance system (5.1) to form a variational system are satisfaction of

$$(5.4) \qquad W_\alpha = \partial_\alpha L , \quad W_\alpha^i = \partial_\alpha^i L ,$$

for in this case, (5.1) and (5.4) yield

$$\Phi^* E_\alpha = \Phi^*(\partial_\alpha L)\mu - d\Phi^*(\partial_\alpha^i L) \wedge \mu_i = \{\Phi^*(\partial_\alpha L) - D_i \Phi^*(\partial_\alpha^i L)\}\mu$$

and the quantities within the curly brackets are exactly the Euler-Lagrange equations that arise from the variational statement (5.3). We have already remarked in Section 3.1 that a necessary and sufficient condition for the existence of a function L such that (5.4) hold is that the (n+1)-form $W \wedge \mu$ be closed, where

$$W = W_\alpha \, dq^\alpha + W_\alpha^i \, dy_i^\alpha .$$

Since this is an essential aspect of the argument, we give an explicit proof.

LEMMA 5.1. *There exists a function* L *on* K *such that*

$$W_\alpha = \partial_\alpha L , \quad W_\alpha^i = \partial_\alpha^i L \tag{5.5}$$

if and only if the (n+1)-*form* $W \wedge \mu$ *is closed, where*

$$W = W_\alpha \, dq^\alpha + W_\alpha^i \, dy_i^\alpha . \tag{5.6}$$

Proof. In order that $W \wedge \mu$ be closed, we must have $d(W \wedge \mu) = 0$. Since K is Euclidean, and hence star-like with respect to any of its points, the Poincaré Lemma holds, and we conclude that there exists an n-form ρ on K such that $W \wedge \mu = d\rho$. However, μ is a simple n-form and hence $W \wedge \mu$ is a simple (n+1)-form. Thus, $W \wedge \mu = d\rho$ can hold if and only if $d\rho$ is a simple (n+1)-form. Since μ is closed, we obtain the existence of a function L such that $W \wedge \mu = d(L \, \mu) = dL \wedge \mu$, from which (5.5) follows. Conversely, if (5.5) holds, then $W = dL - \partial_i L \, dx^i$ and $W \wedge \mu = dL \wedge \mu = d(L \, \mu)$ is closed. A function L with this property is referred to as the *Lagrangian* function of the variational system.

Now, it is quite clear that a general balance system will not be a variational system. On the other hand, if the particular balance system under examination turns out to be a variational system, the analysis given below will continue to hold and will provide a significant amount of additional information.

In view of the results established in Chapter 4, it seems obvious to ask whether there is a prolongation of the space K so that we may obtain a variational system from a given system of the form (5.1). To this end we construct a new space $K_4 = K \times E_{N+nN}$

with coordinate functions $(x^k, q^\alpha, y^\alpha_i, \bar{q}^\alpha, \bar{y}^\alpha_i)$. We take K_4 to be a fiber space with base space K , fibers E_{N+nN} with fiber coordinates $(\bar{q}^\alpha, \bar{y}^\alpha_i)$, and canonical projection

$$(5.7) \qquad \pi_4 : K_4 \longrightarrow K \mid (x^i, q^\alpha, y^\alpha_i, \bar{q}^\alpha, \bar{y}^\alpha_i) \longrightarrow (x^i, q^\alpha, y^\alpha_i) .$$

The balance system $(\underset{\sim}{E}, \underset{\sim}{C})$ is then prolonged to a balance system in K_4 by

$$(5.8) \qquad \pi_4^* \underset{\sim}{E} = \underset{\sim}{E}^* , \quad \pi_4^* \underset{\sim}{C} = \underset{\sim}{C}^* .$$

Thus, $\underset{\sim}{E}^*$ and $\underset{\sim}{C}^*$ become systems of elements of $\Lambda(K_4)$ that are independent of the fiber coordinates $(\bar{q}^\alpha, \bar{y}^\alpha_i)$, so we may continue to denote them by simply $\underset{\sim}{E}$ and $\underset{\sim}{C}$.

The next thing we do is to introduce a system of conjugate contact 1-form by the relations

$$(5.9) \qquad \bar{C}^\alpha = d\bar{q}^\alpha - \bar{y}^\alpha_i \, dx^i , \qquad \alpha = 1, \ldots, N$$

and a system of conjugate balance forms by the relations

$$(5.10) \qquad \bar{E}_\alpha = \bar{W}_\alpha \, \mu - (d\bar{W}^i_\alpha \, \mu_i) , \qquad \alpha = 1, \ldots, N .$$

These give rise to the *prolonged contact ideal*

$$(5.11) \qquad \bar{C}_4 = I\{C^\alpha, \bar{C}^\alpha, dC^\alpha, d\bar{C}^\alpha\}$$

and to the *prolonged balance ideal*

$$(5.12) \qquad \bar{E}_4 = I\{C^\alpha, \bar{C}^\alpha, dC^\alpha, d\bar{C}^\alpha, E_\alpha, \bar{E}_\alpha, dE_\alpha, d\bar{E}_\alpha\}$$

of $\Lambda(K_4)$.

<u>Definition 5.2.</u> A map $\Phi : B_n \longrightarrow K_4$ is said to *solve* the prolonged balance system if and only if $\Phi^*\mu \neq 0$ and Φ^* annihilates the balance ideal $\bar{E}_4$. The collection of all such maps is denoted by $S_4(E_\alpha, \bar{E}_\alpha, B_n) = S_4(\bar{E}_4, B_n)$.

If $\Phi \in S_4(\bar{E}_4, B_n)$, then $\Phi^*\mu \neq 0$, $\Phi^*\underset{\sim}{C} = \underset{\sim}{0}$, $\Phi^*\underset{\sim}{E} = \underset{\sim}{0}$ and hence Φ solves the original balance system $(\underset{\sim}{E}, \underset{\sim}{C})$. An examination of (5.9) shows that satisfaction of $\Phi^*\underset{\sim}{\bar{C}} = 0$ places no restrictions on the solutions of the original balance system. Thus, if $\Phi^*\underset{\sim}{\bar{E}} = \underset{\sim}{0}$ places no restrictions on the solutions of the original balance system, then $S_4(\bar{E}_4, B_n)$ contains all solutions of the original balance system.

We now note the essential aspect of this construction, namely, that we are free to choose the functions $\bar{W}_\alpha$ and $\bar{W}^i_\alpha$ that make up the conjugate balance system $\underset{\sim}{\bar{E}}$. Clearly, we would like to choose them so that satisfaction of $\Phi^*\underset{\sim}{\bar{E}} = \underset{\sim}{0}$ places no restriction on the solutions of the original system. Second, it would certainly be desirable to have the conjugate balance system $\underset{\sim}{\bar{E}}$ as simple as possible, and what is simpler than a linear system. Finally, we would like to make the choice in such a manner that the prolonged balance system $(\underset{\sim}{E}, \underset{\sim}{\bar{E}})$ is a variational system. The following theorem shows that we can actually achieve all three objectives.

THEOREM 5.1. *Any balance system* $\underset{\sim}{E}$, *with*

$$E_\alpha = W_\alpha \mu - d(W^i_\alpha) \wedge \mu_i \tag{5.13}$$

on a space K *can be prolonged to a balance system* $\bar{E}_4$ *on a space* $K_4 = K \times E_{N+nN}$, *with coordinate functions* $(x^i, q^\alpha, y^\alpha_i, \bar{q}^\alpha, \bar{y}^\alpha_i)$ *and additional exterior differential forms*

$$\bar{C}^\alpha = d\bar{q}^\alpha - \bar{y}^\alpha_i \, dx^i , \tag{5.14}$$

$$\bar{E}_\alpha = \bar{W}_\alpha \mu - d(\bar{W}^i_\alpha) \wedge \mu_i , \tag{5.15}$$

in such a fashion that $\bar{E}_4$ *forms a variational system. This prolongation is achieved by the choice*

$$\bar{W}_\alpha = \partial_\alpha (W_\beta \, \bar{q}^\beta + W^j_\beta \, \bar{y}^\beta_j) , \quad \bar{W}^i_\alpha = \partial^i_\alpha (W_\beta \, \bar{q}^\beta + W^j_\beta \, \bar{y}^\beta_j) , \tag{5.16}$$

which leads to the Lagrangian function

$$(5.17) \qquad L = W_\beta \bar{q}^\beta + W^j_\beta \bar{y}^\beta_j ,$$

the conjugate balance system $\bar{\underset{\sim}{E}}$ *is linear in the fiber variables* $(\bar{q}^\alpha, \bar{y}^\alpha_i)$ *, and satisfaction of the equations* $\Phi^*\bar{\underset{\sim}{E}} = 0$ *places no constraints on the solutions of the original balance system,* $\underset{\sim}{E}$ *.*

Proof. That the prolonged balance system $\bar{E}_4$ is a variational system follows directly from (5.13), (5.15), (5.16) and (5.17), since we then have

$$(5.18) \qquad W_\alpha = \bar{\partial}_\alpha L , \quad W^i_\alpha = \bar{\partial}^i_\alpha L , \quad \bar{W}_\alpha = \partial_\alpha L , \quad \bar{W}^i_\alpha = \partial^i_\alpha L .$$

Here, we have used the notation

$$(5.19) \qquad \bar{\partial}_\alpha \equiv \partial/\partial\bar{q}^\alpha , \quad \bar{\partial}^i_\alpha \equiv \partial/\partial\bar{y}^\alpha_i ,$$

which will be adopted from now on. Thus, in particular, the 1-form W for the prolonged variational system is given by

$$(5.20) \qquad W = W_\alpha\, d\bar{q}^\alpha + W^i_\alpha\, d\bar{y}^\alpha_i + \bar{W}_\alpha\, dq^\alpha + \bar{W}^i_\alpha\, dy^\alpha_i$$

$$= dL - \partial_i L\, dx^i .$$

It thus follows that $W \wedge \mu = d(L\,\mu)$, so that $W \wedge \mu$ is indeed closed. When (5.16) are substituted into (5.15), we obtain

$$(5.21) \qquad \bar{E}_\alpha = \partial_\alpha (W_\beta \bar{q}^\beta + W^j_\beta \bar{y}^\beta_j)\mu - d\{\partial^i_\alpha (W_\beta \bar{q}^\beta + W^j_\beta \bar{y}^\beta_j)\} \wedge \mu_i ,$$

and hence the conjugate balance system is linear in the fiber variables $(\bar{q}^\alpha, \bar{y}^\alpha_i)$. It is thus clear that the equations $\Phi^*\bar{\underset{\sim}{E}} = \underset{\sim}{0}$ place no restriction on the solutions of the original balance system $\underset{\sim}{E}$ and the theorem is established.

What makes the theorem work is the pairing that is chosen in the representation of W , namely,

$$W = W_\alpha\, d\bar{q}^\alpha + W^i_\alpha\, d\bar{y}^\alpha_i + \bar{W}_\alpha\, dq^\alpha + \bar{W}^i_\alpha\, dy^\alpha_i ,$$

and the resulting linearity of the Lagrangian function in the fiber variables $(\bar{q}^{\alpha}, \bar{y}^{\alpha}_{i})$. The imbedding of a system of nonlinear differential equations in a variational system by the introduction of an appropriate collection of new ("adjoint") variables that satisfy linear equations is well known in the classic literature of the calculus of variations and its applications [24,25]. What does seem to be new is the recasting of these classic results within the theory of prolongations of exterior differential structures and the marked advantages that are thus afforded by the geometric approach. For example, if the system $\{E_{\alpha}\}$ results in a system of linear inhomogeneous partial differential equations, then the conjugate system $\{\bar{E}_{\alpha}\}$ will result in the homogeneous adjoint system of the original system. With $x = x^1$, $t = x^2$ and $n = 2$, $\mu_1 = dx^2$, $\mu_2 = -dx^1$ the f-Gordon equation is represented by

$$E \;=\; f(q)\mu - d(-y_1\,\mu_1 + y_2\,\mu_2)\;.$$

The conjugate system is then given by

$$\bar{E} \;=\; \bar{q}\;df(q)/dq\;\mu - d(-\bar{y}_1\,\mu_1 + \bar{y}_2\,\mu_2)\;,$$

as follows directly from the Lagrangian function

$$L \;=\; f(q)\bar{q} - y_1\,\bar{y}_1 + y_2\,\bar{y}_2\;,$$

and hence the section of the prolonged balance system by the map $\Phi : B_2 \longrightarrow K_4$ is given by

$$f(\phi) + \partial_1\partial_1\phi - \partial_2\partial_2\phi \;=\; 0\;,$$

$$(df(\phi)/d\phi)\bar{\phi} + \partial_1\partial_1\bar{\phi} - \partial_2\partial_2\bar{\phi} \;=\; 0\;.$$

Although the f-Gordon equations in already variational, the prolongation system is still nontrivial. An example of a nonvariational balance equation is provided by nonlinear diffusion:

$$E = -d(\Psi(q)\mu_2 - y_1\,\mu_1) .$$

Since the Lagrangian is

$$L = \Psi(q)\bar{y}_2 - y_1\,\bar{y}_1 ,$$

the conjugate equation becomes

$$\bar{E} = (d\Psi(q)/dq)\,\bar{y}_2\mu + d(\bar{y}_1\,\mu_1) .$$

In this example, the section of the prolonged balance system by the map $\Phi : B_2 \longrightarrow K_4$ is given by

$$-(d\Psi(\phi)/d\phi)\,\partial_2\phi + \partial_1\partial_1\phi = 0 ,$$

$$(d\Psi(\phi)/d\phi)\,\partial_2\bar{\phi} + \partial_1\partial_1\bar{\phi} = 0 .$$

5.2 ISOVECTORS OF THE PROLONGED CONTACT IDEAL

We now proceed, exactly as in Chapter 2, and construct the collection of all isovectors of the prolonged contact ideal $\bar{C}_4$.

Definition 5.3. A vector field $v \in T(K_4)$ is an *isovector* of the prolonged contact ideal $\bar{C}_4$ if and only if

(5.22) $$\pounds_v \eta = 0 \bmod \bar{C}_4$$

for all $\eta \in \bar{C}_4$; that is $\pounds_v\,\bar{C}_4 \subset \bar{C}_4$. The collection of all isovectors of $\bar{C}_4$ is denoted by $TC(K_4)$, so that

(5.23) $$TC(K_4) = \{v \in T(K_4) \mid \pounds_v \bar{C}_4 \subset \bar{C}_4\} .$$

There is a very simple observation that allows us to carry over all of the results reported in Chapter 2. If we set

(5.24) $$\{C^a\} = \{C^\alpha, \bar{C}^\alpha\}$$

with $1 < a < 2N$, then $\bar{C}_4 = I\{C^a, dC^a\}$. Thus, if we replace Greek indices by Roman indices, then everything given in Chapter 2 applies directly with the exception that the new value of N is twice the old value, and in particular, there are always at least two q^a's, q^1 and $\bar{q}^1$. Theorems 2.1 and 2.2 thus yield the following immediate results.

THEOREM 5.2. *$TC(K_4)$ is a Lie subalgebra of $T(K_4)$.*

THEOREM 5.3. *A vector field*

$$v = v^i \partial_i + v^\alpha \partial_\alpha + v^\alpha_i \partial^i_\alpha + \bar{v}^\alpha \bar{\partial}_\alpha + \bar{v}^\alpha_i \bar{\partial}^i_\alpha \tag{5.25}$$

belongs to $TC(K_4)$ *if and only if*

$$v^\alpha_i = \bar{Z}_i(v \rfloor C^\alpha) , \quad \bar{v}^\alpha_i = \bar{Z}_i(v \rfloor \bar{C}^\alpha) , \tag{5.26}$$

where $\bar{Z}_i$ *is the linear differential operator*

$$\bar{Z}_i \equiv \partial_i + y^\alpha_i \partial_\alpha + \bar{y}^\alpha_i \bar{\partial}_\alpha \tag{5.27}$$

and

$$\begin{aligned} v^i &= f^i(x^j, q^\beta, \bar{q}^\beta) , \quad v^\alpha = f^\alpha(x^j, q^\beta, \bar{q}^\beta) , \\ \bar{v}^\alpha &= \bar{f}^\alpha(x^j, q^\beta, \bar{q}^\beta) . \end{aligned} \tag{5.28}$$

If these conditions are satisfied, then

$$\pounds_v C^\alpha = \partial_\beta(v \rfloor C^\alpha)\, C^\beta + \bar{\partial}_\beta(v \rfloor C^\alpha)\, \bar{C}^\beta , \tag{5.29}$$

$$\pounds_v \bar{C}^\alpha = \partial_\beta(v \rfloor \bar{C}^\alpha)\, C^\beta + \bar{\partial}_\beta(v \rfloor \bar{C}^\alpha)\, \bar{C}^\beta . \tag{5.30}$$

We take particular note that we always have at least two field quantities q^1 and $\bar{q}^1$ for the prolonged problem, and hence the exceptional case $N = 1$ can not arise. Thus, Theorem 2.4 gives the following important result.

THEOREM 5.4. *The prolonged contact structure K_4 is always a first group extension structure of the prolonged graph space G_4 with coordinates $(x^i, q^\alpha, \bar{q}^\alpha)$.*

Again, since N is always greater than one for the prolonged structure based on K_4, Theorem 2.2 carries over in a more universal formulation.

THEOREM 5.5. *The orbital equations of any* $v \in TC(K_4)$ *are the characteristic equations of the system of first order quasi-linear partial differential equations*

$$(5.31) \qquad \overset{*}{F}{}^{\alpha} = -\overset{*}{v} \rfloor \overset{*}{C}{}^{\alpha} = 0 , \qquad \overset{*}{\bar{F}}{}^{\alpha} = -\overset{*}{v} \rfloor \overset{*}{\bar{C}}{}^{\alpha} = 0$$

$$(5.32) \qquad \overset{*}{y}{}^{\alpha}_{i} = \partial \overset{*}{q}{}^{\alpha}(\overset{*}{x}{}^{j})/\partial \overset{*}{x}{}^{i} , \qquad \overset{*}{\bar{y}}{}^{\alpha}_{i} = \partial \overset{*}{\bar{q}}{}^{\alpha}(\overset{*}{x}{}^{j})/\partial \overset{*}{x}{}^{i} ,$$

and generate a 1*-parameter family of maps*

$$(5.33) \qquad \Phi_v(s) = \exp(v,s)\circ\Phi$$

that map $R_4(B_n)$ *into itself. Here,* $R_4(B_n)$ *is defined by*

$$(5.34) \qquad R_4(B_n) = \{\Phi : B_n \longrightarrow K_4 \mid \Phi^*\mu \neq 0 , \ \Phi^*\bar{C}_4 = 0\} ,$$

and constitutes the collection of all regular maps of B_n *into* K_4.

We will not list the remaining theorems that can be carried over directly from Chapter 2, since they should now be evident to the reader. The theorems given above have been included explicitly, for they prove to be instrumental in the analysis to follow and provide significant simplifications in statements of results. We do remark, however, that Corollary 2.1 is also directly applicable, and hence solutions of the orbital equations of any $v \in TC(K_4)$ yields relations between the old and new collections of y's and $\bar{y}$'s that are generalized contact transformations. However, these contact transformations are completely integrable in the sense that they are solved by the relations between the old and new q's and x's that obtain from solution of the orbital equations for $\overset{*}{\underset{\sim}{q}}(s)$, $\overset{*}{\underset{\sim}{\bar{q}}}(s)$ and $x^i(s)$ subject to the initial data

$\overset{*}{\underset{\sim}{q}}(0) = \underset{\sim}{q}$, $\overset{*}{\bar{\underset{\sim}{q}}}(0) = \bar{\underset{\sim}{q}}$, $\overset{*}{x}{}^i(0) = x^i$.

There is one very important observation that must be made at this juncture. Even though we have carried over all of the results established in Chapter 2, the occurrence of the new fiber variables $(\bar{q}^\alpha, \bar{y}^\alpha_i)$ significantly increases the quantity of isovectors as compared with those that are available for the original contact structure on K alone. In fact, we have the following result.

THEOREM 5.6. *The restriction*

$$v^i = f^i(x^j, q^\beta) , \quad v^\alpha = f^\alpha(x^j, q^\beta) , \quad \bar{v}^\alpha = 0 \tag{5.35}$$

imbeds the projectable elements of TC(K) *as a proper linear subspace of* $TC(K_4)$, *and all members of this subspace have the form*

$$v = f^i\partial_i + f^\alpha\partial_\alpha + Z_i(f^\alpha - y^\alpha_j f^j)\partial^i_\alpha - \bar{y}^\alpha_j Z_i f^j \bar{\partial}^i_\alpha . \tag{5.36}$$

Proof. For the restrictions (5.35), we have

$$v \rfloor C^\alpha = f^\alpha - y^\alpha_i f^i , \quad v \rfloor \bar{C}^\alpha = -\bar{y}^\alpha_i f^i \tag{5.37}$$

and $\bar{Z}_i \equiv Z_i$. Application of Theorem 5.3 shows that such vector fields belong to $TC(K_4)$ if and only if

$$v^\alpha_i = Z_i(v \rfloor C^\alpha) = Z_i(f^\alpha - y^\alpha_j f^j) , \tag{5.38}$$

$$\bar{v}^\alpha_i = Z_i(v \rfloor \bar{C}^\alpha) = Z_i(-\bar{y}^\alpha_j f^j) = -\bar{y}^\alpha_j Z_i f^j . \tag{5.39}$$

However, Theorem 2.2 shows that (5.35) and (5.38) are exactly the conditions under which $v \in TC(K)$ if $N > 1$, in which case TC(K) is projectable, and v belongs to the subset of projectable isovectors of $\bar{C}$ if $N = 1$. A substitution of (5.38) and (5.39) into (5.25) leads directly to (5.36) and hence the projectable elements of TC(K) are imbedded in $TC(K_4)$. Clearly, the reason for the imbedding rather than the lifting of the projectable elements is due to the fact that the nonvanishing of $v^i = f^i(x^j, q^\beta)$ causes changes in the $\bar{C}^\alpha$ through the terms $\bar{y}^\alpha_i \, dx^i$ (see (5.39)). The collection of all such isovector fields obviously forms a subspace of $TC(K_4)$. Further, this is a proper subspace since we may allow $\{v^i\}$, $\{v^\alpha\}$ and $\{\bar{v}^\alpha\}$ to depend on the fiber coordinates $(\bar{q}^\alpha)$ in quite an arbitrary manner and still have elements of $TC(K_4)$ provided only that $\{v^\alpha_i\}$ and $\{\bar{v}^\alpha_i\}$ are determined by (5.26).

5.3 ISOVECTORS OF THE PROLONGED BALANCE IDEAL AND GENERATION OF SOLUTIONS

It should now be clear that we can equally well carry over all of the results of Chapter 3 by the simple expedient of introducing a collective Roman index notation with

(5.40) $$\{E_a\} = \{E_\alpha, \bar{E}_\alpha\}$$

and $1 \le a \le 2N$. We shall therefore confine the discussion in this section to certain pertinent remarks concerning the relation between the results for the prolonged structure characterized by the ideal $\bar{E}_4$ and the results for the original balance system that are characterized by the ideal $\bar{E}$.

First, a direct application of Definition 3.1 shows that two prolonged balance systems $(E_{1\alpha}, \bar{E}_{1\alpha})$ and $(E_{2\alpha}, \bar{E}_{2\alpha})$ are *balance equivalent* if and only if there exist four N-by-N matrices of functions $\underset{\sim}{A}, \underset{\sim}{B}, \bar{\underset{\sim}{A}}, \bar{\underset{\sim}{B}}$ on K_4 such that

(5.41) $$\det \begin{pmatrix} \underset{\sim}{A}, & \underset{\sim}{B} \\ \bar{\underset{\sim}{A}}, & \bar{\underset{\sim}{B}} \end{pmatrix} \neq 0 ,$$

(5.42) $$\underset{\sim}{E}_2 = \underset{\sim}{A}\,\underset{\sim}{E}_1 + \underset{\sim}{B}\,\bar{\underset{\sim}{E}}_1 + \underset{\sim}{P}_1 , \quad \bar{\underset{\sim}{E}}_2 = \bar{\underset{\sim}{A}}\,\underset{\sim}{E}_1 + \bar{\underset{\sim}{B}}\,\bar{\underset{\sim}{E}}_1 + \underset{\sim}{P}_2$$

with $\underset{\sim}{P}_1$ and $\underset{\sim}{P}_2$ elements of $\bar{C}_4$. As before, let $F(\underset{\sim}{E}, \bar{\underset{\sim}{E}})$

denote the fiber of prolonged balance systems that are balance equivalent to $(\underset{\sim}{E}, \underset{\sim}{\bar{E}})$. Lemma 3.2 then shows that $S_4(F(\underset{\sim}{E}, \underset{\sim}{\bar{E}}), B_n) = S_4(\underset{\sim}{E}, \underset{\sim}{\bar{E}}, B_n)$, that is, all prolonged balance systems that belong to the same fiber have exactly the same solutions. It also follows directly from Lemma 3.5 that the prolonged balance ideal $\bar{E}_4$ that is formed from E and $\bar{E}$ is the same as that formed from any pair belonging to the fiber $F(E, \bar{E})$. All of the results concerning fiber invariant quantities and concepts can thus be carried over directly.

We now proceed to the more important aspect, namely the isovectors of the ideal $\bar{E}_4$.

<u>Definition 5.4</u>. A vector field $v \in T(K_4)$ is an isovector of $\bar{E}_4$ if and only if

$$\text{(5.43)} \qquad \pounds_v \bar{E}_4 \subset \bar{E}_4 .$$

The collection of all isovectors of $\bar{E}_4$ is denoted by $T(\underset{\sim}{E}, \underset{\sim}{\bar{E}}, K_4)$.

A direct combination of Theorems 3.4 through 3.6 gives the following result.

<u>THEOREM 5.7</u>. $T(\underset{\sim}{E}, \underset{\sim}{\bar{E}}, K_4)$ *is a Lie subalgebra of* $TC(K_4)$ *that is composed of those elements of* $TC(K_4)$ *that satisfy the further restrictions*

$$\text{(5.44)} \qquad \pounds_v E_\alpha \subset \bar{E}_4 , \quad \pounds_v \bar{E}_\alpha \subset \bar{E}_4 .$$

This Lie subalgebra is an invariant of fibration by balance equivalence, so that

$$\text{(5.45)} \qquad T(F(\underset{\sim}{E}, \underset{\sim}{\bar{E}}), K_4) = T(\underset{\sim}{E}, \underset{\sim}{\bar{E}}, K_4) .$$

Again, as with the case of $TC(K_4)$, the collection of $T(\underset{\sim}{E}, \underset{\sim}{\bar{E}}, K_4)$ is much richer than the corresponding collection $T(\underset{\sim}{E}, K)$ because of the presence of the new variables $(\bar{q}^\alpha)$ and the significant increase in the size of $TC(K_4)$ compared to $TC(K)$. It would appear that use of the prolongation structure associated with K_4 provides new and possibly very productive territory for future work. This is particularly true in view of the following result that obtains directly from Theorem 3.8.

THEOREM 5.8. *Let* $v \in T(\underset{\sim}{E}, \underset{\sim}{\bar{E}}, K_4)$, *then the* 1-*parameter family of maps*

$$(5.46) \qquad \exp(v,s): K_4 \longrightarrow K_4 \mid \overset{*}{x}{}^i = \exp(s\pounds_v)x^i , \quad \overset{*}{q}{}^\alpha = \exp(s\pounds_v)q^\alpha ,$$

$$\overset{*}{y}{}^\alpha_i = \exp(s\pounds_v)y^\alpha_i , \quad \overset{*}{\bar{q}}{}^\alpha = \exp(s\pounds_v)\bar{q}^\alpha ,$$

$$\overset{*}{\bar{y}}{}^\alpha_i = \exp(s\pounds_v)\bar{y}^\alpha_i$$

imbeds any $\Phi \in S_4(\underset{\sim}{E}, \underset{\sim}{\bar{E}}, B_n)$ *in a* 1-*parameter group*

$$(5.47) \qquad \Phi_v(s) = \exp(v,s) \circ \Phi$$

of elements of $S_4(\underset{\sim}{E}, \underset{\sim}{\bar{E}}, B_n)$. *If* v *is such that* $\bar{\partial}_\beta(v \rfloor C^\alpha) \not\equiv 0$, *then* $\pi_4 \circ \exp(v,s)$ *is a* 1-*parameter family of Bäcklund transformations of* K *such that*

$$\Phi_v(s)^* E_\alpha = \Phi^*(\exp(s\pounds_v)E_\alpha)$$

for any $\Phi \in R_4(B_n)$, *and any* $\Phi \in S_4(\underset{\sim}{E}, \underset{\sim}{\bar{E}}, B_n)$ *generates a mapping of a solution of* $\exp(s\pounds_v)(E_\alpha)$ *onto a solution of* E_α *where the functions* $\Phi^*(\bar{q}^\alpha)$ *are considered as subsidiary functions that are determined by*

$$\Phi^*\bar{E}_\alpha = 0 , \quad \Phi^*\bar{C}^\alpha = 0 .$$

Proof. The results (5.46) and (5.47) are direct consequences of Theorem 3.8. If $\bar{\partial}_\beta(v \rfloor C^\alpha) \not\equiv 0$, then (5.29) shows that $\pounds_v C^\alpha = \bar{\partial}_\beta(v \rfloor C^\alpha)\bar{C}^\beta \bmod \bar{C}$. Thus $\pounds_v \bar{C}$ is not contained in $\bar{C}$ and hence $\pi_4 \circ \exp(v,s)$ is not a contact transformation of K .

However, $\Phi_v(s) = \exp(v,s) \circ \Phi$ maps $S_4(\underset{\sim}{E}, \underset{\sim}{\bar{E}}, B_n)$ into $S_4(\underset{\sim}{E}, \underset{\sim}{\bar{E}}, B_n)$ and $\Phi_v(s)^* E_\alpha = \Phi^*(\exp(s\pounds_v)E_\alpha)$, $\Phi_v(s)^*\bar{E}_\alpha = \Phi^*(\exp(s\pounds_v)\bar{E}_\alpha)$. The map $\pi_4 \circ \exp(v,s)$ thus maps solutions of $\exp(s\pounds_v)E_\alpha$ onto solutions of E_α whenever the functions $\Phi^*(\bar{q}^\alpha)$ are chosen so that $\Phi^*\bar{E}_\alpha = 0$, $\Phi^*\bar{C}^\alpha = 0$. The latter result follows from the fact that any $v \in T(\underset{\sim}{E}, \underset{\sim}{\bar{E}}, B_n)$ is such that $\pounds_v \bar{E}_4 \subset \bar{E}_4$ and any $\Phi \in S_4(\underset{\sim}{E}, \underset{\sim}{\bar{E}}, K_4)$ annihilates $\bar{E}_4$. We will not comment further since it would appear unnecessary. There is, however, a significant new body of material that obtains directly from the variational structure that is naturally associated with the prolongation of K_4 . It is this aspect that we concentrate on in the remainder of this Chapter.

5.4 FINITE VARIATIONS AND FREE VARIATIONAL PROBLEMS

Classically, Euler-Lagrange equations arise from the problem of rendering an action functional

$$(5.48) \qquad A[\Phi] = \int_{B_n} \Phi^*(L\mu)$$

stationary in value relative to all variations of Φ that vanish on the boundary of the domain B_n . The reason for the constraint that the variations vanish on ∂B_n is the implicit assumption that Φ and the family of varied maps all satisfy the same given data of Dirichlet type. Such an assumption is certainly not desirable in the case of this study, for we have made absolutely no assumptions concerning the form of the boundary conditions that are to be used for any given balance system. As it turns out, it is not only useful, but also very convenient to obtain a free variational formulation; that is, a variational formulation in which we do not have to impose the constraint that the variations vanish on B_n .

We take particular note of the fact that

$$A[\Phi] = \int_{B_n} \Phi^*(L\mu)$$

may be viewed as a mapping from the collection of all regular maps $R_4(B_n)$ into $\mathbb{R}$. Thus, a comparison of $A[\Phi]$ with $A[\Psi]$ demands that both Φ and Ψ belong to $R_4(B_n)$, and that the imbedding of any given $\Phi \varepsilon R_4(B_n)$ in a continuous family of "neighboring" maps requires that the family consist solely of elements of $R_4(B_n)$. Clearly, the transport of any $\Phi \varepsilon R_4(B_n)$ by the orbits of the isovector fields of the closed ideal $\bar{C}_4$ provides the mechanism for such imbeddings. Indeed, the parameter on the orbits provides the "epsilon" for the usual Gateoux differentiation process, while the different possible choices of the isovector fields provide the functions of position that serve for the function increments; that is

$$\Phi \longrightarrow \Phi + \varepsilon v(\Phi) + o(\varepsilon^2)$$

since transport by an isovector field yields a well defined power series expansion in terms of the orbital parameter ε. The reason for this is that transport by isovector fields of $\bar{C}_4$ preserves membership in $R_4(B_n)$ for sufficiently small values of the orbital parameter. In fact, the classic variation process, "δ", can be generated in this manner by suitable restrictions on the choice of the isovector fields.

Fundamental Forms

The first thing to be done is to render the problem precise. Let $R_4(B_n)$ denote the collection of all regular maps of B_n into K_4:

(5.49) $$R_4(B_n) = \{\Phi : B_n \longrightarrow K_4 \mid \Phi^*\mu \neq 0 ,\ \Phi^*\bar{C}_4 = 0\} .$$

Further, define $TH(K_4)$ by

(5.50) $$TH(K_4) = \{u \in TC(K_4) \mid u^i = 0,\ i = 1,\dots,n\} ,$$

so that $TH(K_4)$ is the collection of all horizontal isovector fields of $\bar{C}_4$. The variations of any $\Phi \in R_4(B_n)$ are defined by

(5.51) $$\delta_u\phi^\alpha = \Phi^*(u \rfloor dq^\alpha) , \quad \delta_u\bar{\phi}^\alpha = \Phi^*(u \rfloor d\bar{q}^\alpha) ,$$

(5.52) $$\delta_u\partial_i\phi^\alpha = \Phi^*(u \rfloor dy_i^\alpha) , \quad \delta_u\partial_i\bar{\phi}^\alpha = \Phi^*(u \rfloor d\bar{y}_i^\alpha) ,$$

so that $u_i^\alpha = \bar{Z}_i(u \rfloor C^\alpha)$, $\bar{u}_i^\alpha = \bar{Z}_i(u \rfloor \bar{C}^\alpha)$ yield

(5.53) $$\delta_u\partial_i\phi^\alpha = D_i\delta_u\phi^\alpha , \quad \delta_u\partial_i\bar{\phi}^\alpha = D_i\delta_u\bar{\phi}^\alpha .$$

The action functional $A[\Phi]$ is then stationary in value if and only if $\Phi \in R_4(B_n)$ satisfies

(5.54) $$\delta_u A[\Phi] = \delta_u\int_{B_n}\Phi^*(L\mu) = \int_{B_n}\Phi^*\pounds_u(L\mu) = 0$$

for all $u \in TH(K_4)$ such that $\Phi^*(u \rfloor dq^\alpha)|_{\partial B_n} = 0$, $\Phi^*(u \rfloor d\bar{q}^\alpha)|_{\partial B_n} = 0$.

For the prolonged balance system on K_4 , we have

(5.55) $$L = W_\alpha \bar{q}^\alpha + W_\alpha^i \bar{y}_i^\alpha ,$$

so that $W = W_\alpha\, d\bar{q}^\alpha + W_\alpha^i\, d\bar{y}_i^\alpha + \bar{W}_\alpha\, dq^\alpha + \bar{W}_\alpha^i\, dy_i^\alpha$ gives

(5.56) $$W \wedge \mu = dL \wedge \mu = d(L\mu) .$$

The following Lemmas provide the computational basis for many of the results to follow.

LEMMA 5.2. *The* n-*form*

(5.57) $$J = \bar{C}^\alpha \wedge W_\alpha^i\, \mu_i + C^\alpha \wedge \bar{W}_\alpha^i\, \mu_i = \bar{C}^\alpha \wedge \bar{\partial}_\alpha^i L\mu_i + C^\alpha \wedge \partial_\alpha^i L\mu_i$$

belongs to $\bar{C}_4$, *and hence*

(5.58) $\Phi^* J = 0$,

(5.59) $\Phi^* \pounds_v J = 0$

for all $\Phi \in R_4(B_n)$ *and all* $v \in TC(K_4)$. *Further,* $v \rfloor J$ *is independent of* v_i^α *and* $\bar{v}_i^\alpha$.

Proof. It is clear from the defining equation (5.57) that J belongs to the closed ideal $\bar{C}_4$, and hence (5.58) follows for any $\Phi \in R_4(B_n)$. Since $v \in TC(K_4)$, $\pounds_v J \in \bar{C}_4$ and hence (5.59) holds for all $\Phi \in R_4(B_n)$. Finally, noting that none of the quantities $v \rfloor C^\alpha$, $v \rfloor \bar{C}^\alpha$, $v \rfloor \mu_i$ depend on v_i^α or $\bar{v}_i^\alpha$, it follows directly from (5.57) that $v \rfloor J$ does not depend on v_i^α or $\bar{v}_i^\alpha$.

LEMMA 5.3. *The* (n+1)-*form*

(5.60) $F = W \wedge \mu + dJ = d(L\mu + J)$

has the evaluation

(5.61) $F = \bar{C}^\alpha \wedge E_\alpha + C^\alpha \wedge \bar{E}_\alpha$.

Thus, F *belongs to the closed ideal* $\bar{C}_4$,

(5.62) $v \rfloor F = (v \rfloor \bar{C}^\alpha)\, E_\alpha - \bar{C}^\alpha \wedge (v \rfloor E_\alpha)$

$+ (v \rfloor C^\alpha)\, \bar{E}_\alpha - C^\alpha \wedge (v \rfloor \bar{E}_\alpha)$

belongs to the closed ideal $\bar{E}_4$, *and*

(5.63) $\Phi^*(v \rfloor F) = \Phi^*(v \rfloor \bar{C}^\alpha)\, \Phi^* E_\alpha + \Phi^*(v \rfloor C^\alpha)\, \Phi^* \bar{E}_\alpha$

for all $\Phi \in R_4(B_n)$.

Proof. We already know that $W \wedge \mu = d(L\mu)$, and hence the second equality in (5.60) follows directly from the first. The evaluation of F that is given by (5.61) follows directly from (5.60) through a straightforward, but cumbersome calculation. The results given by (5.62) and (5.63) obtain directly from (5.61).

Finite Variations

LEMMA 5.4. *The action functional* $A[\Phi]$ *has the evaluation*

$$(5.64) \qquad A[\Phi] = \int_{B_n} \Phi^*(L\mu + J)$$

for all $\Phi \in R_4(B_n)$.

Proof. Lemma 5.2 shows that $\Phi^* J = 0$ for any $\Phi \in R_4(B_n)$, and hence

$$A[\Phi] = \int_{B_n} \Phi^*(L\mu) = \int_{B_n} \Phi^*(L\mu + J) .$$

Lemma 5.4 provides the basis whereby we take advantage of the linearity of L in the variables $(\bar{q}^\alpha, \bar{y}_i^\alpha)$. It follows directly from (5.57) and (5.14) that

$$L\mu + J = (L - y_i^\alpha \partial_\alpha^i L - \bar{y}_i^\alpha \bar{\partial}_\alpha^i L)\mu + (\partial_\alpha^i L\, dq^\alpha + \bar{\partial}_\alpha^i L\, d\bar{q}^\alpha) \wedge \mu_i .$$

Thus, since

$$L = W_\alpha \bar{q}^\alpha + W_\alpha^i \bar{y}_i^\alpha ,$$

we obtain

$$L\mu + J = (W_\alpha \bar{q}^\alpha - y_i^\alpha \partial_\alpha^i (W_\beta \bar{q}^\beta + W_\beta^j \bar{y}_j^\beta))\mu$$

$$+ (W_\alpha^i d\bar{q}^\alpha + \partial_\alpha^i (W_\beta \bar{q}^\beta + W_\beta^j \bar{y}_j^\beta) dq^\alpha) \wedge \mu_i .$$

This explicit form also allows us to take advantage of any homogeneity in the dependence of W_α and W_α^i on the variables y_j^β .

THEOREM 5.9. *Let* $\Phi \in R_4(B_n)$ *and* $v \in TC(K_4)$. *The 1-parameter group of maps*

$$(5.65) \qquad \Phi_v(s) = \exp(v,s) \circ \Phi ,$$

defined by (5.46) *and* (5.47), *belongs to* $R_4(B_n)$ *for all values of* s *such that* $\exp(s\pounds_v)\mu \neq 0$, *and yields*

$$(5.66) \qquad A[\Phi_v(s)] = \int_{B_n} \Phi^*\{\exp(s\pounds_v)(L\mu + J)\} .$$

Proof. The 1-parameter family of maps $\Phi_v(s)$ belongs to $R_4(B_n)$ for all values of s such that $\exp(s\pounds_v)\mu \neq 0$. This follows directly from lifting the results of Theorem 2.2 to the prolonged structure K_4 . A straightforward calculation based on Lemma 5.4 then yields

$$A[\Phi_v(s)] = \int_{B_n} \Phi_v(s)^*(L\mu + J) = \int_{B_n} [\Phi^* \circ (\exp(v,s))^*](L\mu + J) .$$

As with the proof of Theorem 2.3, let a superimposed $*$ denote evaluation on the orbit of $v \in TC(K_4)$ that is given by (5.46). We then have $[\Phi^* \circ (\exp(v,s))^*](L\mu + J) = \Phi^*[(\exp(v,s))^*(\overset{**}{L\mu} + \overset{*}{J})]$ $= \Phi^*\{\exp(s\pounds_v)(L\mu + J)\}$, since $(\exp(v,s))^*(\overset{**}{L\mu} + \overset{*}{J}) = \exp(s\pounds_v)$ $(L\mu + J)$. This establishes (5.66).

Theorem 5.9 provides a natural basis for the definition of the total and the infinitesimal variation of the action functional $A[\Phi]$ that results from the transport of any $\Phi \in R_4(B_n)$ along the orbits of any $v \in TC(K_4)$.

Definition 5.5. The *finite variation* of $A[\Phi]$ that is generated by any $v \in TC(K_4)$ and any $\Phi \in R_4(B_n)$ is defined by

$$(5.67) \qquad \Delta_v(s)\, A[\Phi] = A[\Phi_v(s)] - A[\Phi] .$$

Definition 5.6. The *infinitesimal variation* of $A[\Phi]$ that is generated by any $v \in TC(K_4)$ and any $\Phi \in R_4(B_n)$ is defined by

$$(5.68) \qquad \delta_v A[\Phi] = \lim_{s\to 0}(s^{-1}\, \Delta_v(s)\, A[\Phi]) .$$

THEOREM 5.10. *The finite and the infinitesimal variations*

of $A[\Phi]$ *that are generated by any* $v \in TC(K_4)$ *and any* $\Phi \in R_4(B_n)$ *have the evaluations*

$$(5.69) \qquad \Delta_v(s)\, A[\Phi] = \int_{B_n} \Phi^*(\exp(s\pounds_v) - 1)(L\mu + J) ,$$

$$(5.70) \qquad \delta_v A[\Phi] = \int_{B_n} \Phi^* \pounds_v (L\mu + J) .$$

Proof. The theorem follows directly from Theorem 5.9 and Definitions 5.5, 5.6.

The following two Lemmas will be established simultaneously.

LEMMA 5.5. $\Phi \in S_4(E_\alpha, \bar{E}_\alpha, B_n)$ *if and only if*

$$(5.71) \qquad \delta_u A[\Phi] = 0$$

for all $u \in TH(K_4)$ *such that* $\Phi^*(u \rfloor dq^\alpha)|_{\partial B_n} = 0$, $\Phi^*(u \rfloor d\bar{q}^\alpha)|_{\partial B_n} = 0$.

LEMMA 5.6. $\Phi \in S_4(E_\alpha, \bar{E}_\alpha, B_n)$ *if and only if*

$$(5.72) \qquad \delta_u A[\Phi] = \int_{\partial B_n} \Phi^*(u \rfloor J)$$

for all $u \in TH(K_4)$.

Lemma 5.6 eliminates the constraint $\Phi\ (u \rfloor dq^a)|_{\partial B_n} = 0$, $a = 1, \ldots, 2N$.

Proof. A direct substitution from (5.70) of Theorem 5.10 gives us

$$(5.73) \qquad \delta_u A[\Phi] = \int_{B_n} \Phi^* \pounds_u (L\mu + J) .$$

Now, (5.60) shows that

$$\pounds_u (L\mu + J) = u \rfloor d(L\mu + J) + d\{u \rfloor (L\mu + J)\}$$

$$= u \rfloor F + d\{u \rfloor (L\mu + J)\} ,$$

and hence, for any $u \in TH(K_4)$, we have

$$\pounds_u(L\mu + J) = u \rfloor F + d(u \rfloor J) .$$

Further, for any $u \in TH(K_4)$, $u \rfloor \bar{C}^\alpha = u \rfloor d\bar{q}^\alpha$, $u \rfloor C^\alpha = u \rfloor dq^\alpha$, and $u \rfloor J = (u \rfloor d\bar{q}^\alpha) W^i_\alpha \mu_i + (u \rfloor dq^\alpha) \bar{W}^i_\alpha \mu_i$, as follows directly from (5.57). If we substitute these results into (5.73) and make use of (5.63), we obtain

$$(5.74)\quad \delta_u A[\Phi] = \int_{B_n} \{\Phi^*(u \rfloor d\bar{q}^\alpha)\Phi^*(E_\alpha) + \Phi^*(u \rfloor dq^\alpha)\Phi^*(\bar{E}_\alpha)\} + \int_{\partial B_n} \{\Phi^*(u \rfloor d\bar{q}^\alpha)\Phi^*(W^i_\alpha \mu_i) + \Phi^*(u \rfloor dq^\alpha)\Phi^*(\bar{W}^i_\alpha \mu_i)\} .$$

We now use (5.51) to rewrite (5.74) in terms of the standard form to which the fundamental lemma of the calculus of variations may be applied:

$$(5.75)\qquad \delta_u A[\Phi] = \int_{B_n} \{\Phi^*(E_\alpha)\delta_u \bar{\phi}^\alpha + \Phi^*(\bar{E}_\alpha)\delta_u \phi^\alpha\} + \int_{\partial B_n} \{\Phi^*(W^i_\alpha \mu_i)\delta_u \bar{\phi}^\alpha + \Phi^*(\bar{W}^i_\alpha \mu_i)\delta_u \phi^\alpha\} .$$

This result and the fundamental lemma of the calculus of variations establish Lemma 5.5. Lemma 5.6 is likewise easily established, for substitution of (5.75) into (5.72) leads to

$$\int_{B_n} \{\Phi^*(E_\alpha)\delta_u \bar{\phi}^\alpha + \Phi^*(\bar{E}_\alpha)\delta_u \phi^\alpha\} = 0$$

and the fundamental lemma shows that $\Phi \in S_4(E_\alpha, \bar{E}_\alpha, B_n)$, while the converse is obvious.

The relation between isovector fields of $\bar{E}_4$ and finite variations is given by the following theorem.

THEOREM 5.11. *If* $v \in T(\underset{\sim}{E}, \underset{\sim}{\bar{E}}, K_4)$, *then*

$$(5.76)\qquad \Delta_v(s)\, A[\Phi] = \int_{B_n} \Phi^* P(s) + \int_{\partial B_n} \Phi^* Q(s) ,$$

where $P(s) \in \Lambda^n(K_4) \cap \bar{E}_4$ *and*

$$(5.77) \qquad Q(s) = s\{1 + \frac{s\pounds_v}{2!} + \frac{s^2\pounds_v^2}{3!} + \frac{s^3\pounds_v^3}{4!} + \ldots\}(Lv^i \mu_i + v\rfloor J) .$$

Thus, if $\Phi \in S_4(\underset{\sim}{E}, \underset{\sim}{\bar{E}}, B_n)$, *then the finite variation of* $A[\Phi]$ *is evaluated in terms of the surface integral* $\int_{\partial B_n} \Phi^* Q(s)$ *alone.*

<u>Proof</u>. A trivial manipulation gives

$$(\exp(s\pounds_v) - 1)(L\mu + J) = s(1 + \frac{s\pounds_v}{2!} + \frac{s^2\pounds_v^2}{3!} + \ldots)\pounds_v(L\mu + J) ,$$

while $\pounds_v(L\mu + J) = v\rfloor F + d(v\rfloor(L\mu + J))$. However, Lemma 5.3 showed that $v\rfloor F \in \bar{E}_4$, so that $\pounds_v^n(v\rfloor F) \in \bar{E}_4$ because $v \in T(\underset{\sim}{E}, \underset{\sim}{\bar{E}}, K_4)$. The result then follows on noting that d and $\pounds_v$ commute while $v\rfloor(L\mu + J) = Lv^i \mu_i + v\rfloor J$.

Theorem 5.11 is the variational equivalent of Theorem 5.8. In order to see this, we simply note that (5.76) and (5.67) yield

$$A[\Phi_v(s)] = A[\Phi] + \int_{B_n} \Phi^* P(s) + \int_{\partial B_n} \Phi^* Q(s) .$$

Since (5.77) shows that $Q(s)$ does not depend on dy_i^α or $d\bar{y}_i^\alpha$ for $u \in TH(K_4)$, $\delta_u \int_{\partial B_n} \Phi^* Q(s) = 0$ for all $u \in TH(K_4)$ such that $(u\rfloor dq^a)|_{\partial B_n} = 0$, $a = 1,\ldots,2N$. Thus, for $\Phi \in S_4(\underset{\sim}{E}, \underset{\sim}{E}, B_n)$

$$A[\Phi_v(s)] = A[\Phi] + \int_{\partial B_n} \Phi^* Q(s)$$

gives

$$\delta_u A[\Phi_v(s)] = \delta_u A[\Phi]$$

for all $u \in TH(K_4)$ such that $(u\rfloor dq^a)|_{\partial B_n} = 0$, $a = 1,\ldots,2N$.

5.5 STRONG AND WEAK NOETHERIAN VECTOR FIELDS AND THEIR ASSOCIATED CURRENTS

The classic theorem of E. Noether shows that there is a

conserved current for each isovector field of $\bar{C}_4$ that leaves the action functional, $A[\Phi]$, invariant under transport by that isovector field. Such results are thus directly available for the prolonged balance system that is characterized by $\bar{E}_4$. Significant generalizations of this classic result can be obtained, however, as we shall now show.

Definition 5.7. A vector field $z \in TC(K_4)$ is said to be a *Noetherian vector field* of a given $A[\Phi]$ if and only if

$$(5.78) \qquad \pounds_z(L\mu + J) = 0 .$$

The collection of all Noetherian vector fields of a given $A[\Phi] = \int_{B_n} \Phi^* L\mu$ is denoted by $N(L,K_4)$.

Definition 5.8. A vector field $z \in TC(K_4)$ is said to be a *generalized Noetherian vector field of the first kind* of a given $A[\Phi]$ if and only if

$$(5.79) \qquad \pounds_z(L\mu + J) = 0 \bmod \bar{C}_4 .$$

The collection of all generalized Noetherian vector fields of the first kind of a given $A[\Phi]$ is denoted by $N_1(L,K_4)$.

Definition 5.9. A vector field $z \in TC(K_4)$ is said to be a *generalized Noetherian vector field of the second kind* of a given $A[\Phi]$ if and only if

$$(5.80) \qquad \pounds_z(L\mu + J) = d\beta_z \bmod \bar{C}_4 .$$

The collection of all generalized Noetherian vector fields of the second kind of a given $A[\Phi]$ is denoted by $N_2(L,K_4)$.

THEOREM 5.12. $N(L,K_4)$, $N_1(L,K_4)$ *and* $N_2(L,K_4)$ *are Lie subalgebras of the Lie algebra* $TC(K_4)$ *that satisfy the inclusion relations*

$$(5.81) \qquad N(L,K_4) \subset N_1(L,K_4) \subset N_2(L,K_4) \subset TC(K_4) ,$$

and the law of composition for elements of $N_2(L,K_4)$ *is given by*

(5.82) $$\beta_{[z_1,z_2]} = £_{z_1}\beta_{z_2} - £_{z_2}\beta_{z_1} + d\rho \ .$$

Proof. The inclusion relations (5.81) are immediate consequences of Definitions 5.7 through 5.9. Further, (5.78)-(5.80) show that it is sufficient to establish that $N_2(L,K_4)$ forms a Lie algebra. If $£_{z_1}(L\mu + J) = d\beta_{z_1} \bmod \bar{C}_4$ and $£_{z_2}(L\mu + J) = d\beta_{z_2} \bmod \bar{C}_4$, then $£_{(az_1+bz_2)}(L\mu + J) = d(a\beta_{z_1} + b\beta_{z_2}) \bmod \bar{C}_4$, and hence $N_2(L,K_4)$ is a vector subspace of $TC(K_4)$. Since $z \in TC(K_4)$, $£_z\bar{C}_4 \subset \bar{C}_4$ and $£$ commutes with d, we have

$$\begin{aligned} £_{[z_1,z_2]}(L\mu + J) &= (£_{z_1}d\beta_{z_2} - £_{z_2}d\beta_{z_1}) \bmod \bar{C}_4 \\ &= d(£_{z_1}\beta_{z_2} - £_{z_2}\beta_{z_1}) \bmod \bar{C}_4 \\ &= d\beta_{[z_1,z_2]} \bmod \bar{C}_4 \ , \end{aligned}$$

where $\beta_{[z_1,z_2]}$ is given by the law of composition, (5.82).

The inclusion relations given by the last Theorem show that $N_1(L,K_4)$ and $N_2(L,K_4)$ are, in general, nontrivial generalizations of $N(L,K_4)$. What now remains to be shown is that $N(L,K_4)$ constitute the vector fields that yield the results of the classic Noether theorem.

THEOREM 5.13. *If* $z \in N(L,K_4)$, *then the finite variation,* $\Delta_z(s)A[\Phi]$, *vanishes for all* $\Phi \in R_4(B_n)$ *and we have the global invariance*

(5.83) $$A[\Phi_z(s)] = A[\Phi] \qquad \forall\, \Phi \in R_4(B_n) \ .$$

Further, any $\Phi \in S_4(\underset{\sim}{E}, \underset{\sim}{\bar{E}}, B_n)$ *satisfies the identity*

(5.84) $$d\Phi^*\{z \rfloor (L\mu + J)\} = 0 \ .$$

Proof. If $z \in N(L, K_4)$, then $\pounds_z(L\mu + J) = 0$ by Definition 5.6. It then follows directly from Theorem 5.10 (*vid.* (5.69)) that $\Delta_z(s)A[\Phi] = 0$ for all $\Phi \in R_4(B_n)$, and (5.67) shows that we obtain the global invariance (5.83). For any $z \in N(L, K_4)$, (5.78) gives

$$(5.85) \qquad 0 = \pounds_z(L\mu + J) = z \rfloor d(L\mu + J) + d\{z \rfloor (L\mu + J)\}$$

$$= z \rfloor F + d\{z \rfloor (L\mu + J)\} .$$

Thus, since $z \rfloor F$ belongs to $\bar{E}_4$, by Lemma 5.3, allowing Φ^* to act on both sides of (5.85) shows that (5.84) is satisfied by any $\Phi \in S_4(\underset{\sim}{E}, \underset{\sim}{\bar{E}}, B_n)$.

Theorem 5.13 is exactly the content of the classic Noether theorem. In order to see this, it is only necessary to combine (5.18) and (5.57) in order to obtain

$$(5.86) \qquad z \rfloor (L\mu + J) = z^j(L\,\delta^i_j - \bar{y}^\alpha_j\,\bar{\partial}^i_\alpha L - y^\alpha_j\,\partial^i_\alpha L)\mu_i$$

$$+ (\bar{z}^\alpha\,\bar{\partial}^i_\alpha L + z^\alpha\,\partial^i_\alpha L)\mu_i \mod \bar{C}_4 ,$$

in which case (5.84) becomes

$$(5.87) \quad 0 = D_i\Phi\,\{z^j(L\,\delta^i_j - \bar{y}^\alpha_j\,\bar{\partial}^i_\alpha L - y^\alpha_j\,\partial^i_\alpha L) + \bar{z}^\alpha\,\bar{\partial}^i_\alpha L + z^\alpha\,\partial^i_\alpha L\}\mu .$$

The "momentum-energy" complex

$$(5.88) \qquad -H^i_j = L\,\delta^i_j - \bar{y}^\alpha_j\,\bar{\partial}^i_\alpha L - y^\alpha_j\,\partial^i_\alpha L$$

thus emerges in a natural way, and (5.87) is exactly the form of the identity that is yielded by the classic Noether theorem [25, Chapter IV]. Further, the classic Noether theorem requires $A[\Phi_z(s)](b_n) = A[\Phi](b_n)$ for all connected measurable subsets, b_n, of B_n, where $A[\Phi](b_n) = \int_{b_n} \Phi^*(L\mu + J)$. This requirement leads directly to the condition that $\Phi^*\pounds_z(L\mu + J) = 0$ for all

$\Phi \varepsilon R_4(B_n)$.

THEOREM 5.14. *If* $z \varepsilon N_1(L,K_4)$, *then* $A[\Phi_z(s)] = A[\Phi]$ *for all* $\Phi \varepsilon R_4(B_n)$, *and any* $\Phi \varepsilon S_4(\underset{\sim}{E}, \underset{\sim}{\bar{E}}, B_n)$ *satisfies the identity*

(5.89) $$d\Phi^*\{z\rfloor(L\mu + J)\} = 0 .$$

Proof. By definition, $z \varepsilon N_1(L,K_4)$ if and only if

$$\pounds_z(L\mu + J) = 0 \bmod \bar{C}_4 .$$

Thus, since $\pounds_z\bar{C}_4 \subset \bar{C}_4$, because $z \varepsilon TC(K_4)$, there exists an $M_1(s) \varepsilon \bar{C}_4$ such that

$$(\exp(s\pounds_z) - 1)(L\mu + J) = M_1(s) .$$

When this is substituted into (5.69), we see that $\Delta_z(s)A[\Phi] = 0$ for all $\Phi \varepsilon R_4(B_n)$. This leads directly to the global invariance $A[\Phi_z(s)] = A[\Phi]$ by (5.67). The proof of the identity (5.89) follows the same line as the proof of the identity (5.84).

THEOREM 5.15. *If* $z \varepsilon N_2(L,K_3)$, *then*

(5.90) $$A[\Phi_z(s)] = A[\Phi] + s\int_{\partial B_n} \Phi^*\beta(s) , \quad \forall \Phi \varepsilon R_4(B_n)$$

that is, $A[\Phi]$ *is globally invariant under z-transport to within the addition of a surface integral, and any* $\Phi \varepsilon S_4(\underset{\sim}{E}, \underset{\sim}{\bar{E}}, B_n)$ *satisfies the identity*

(5.91) $$d\Phi^*\{z\rfloor(L\mu + J) - \beta_z\} = 0 .$$

Proof. By definition, $z \varepsilon N_2(L,K_4)$ if and only if

$$\pounds_z(L\mu + J) = d\beta_z \bmod \bar{C}_4 .$$

Thus, since $z \varepsilon TC(K_4)$ implies $\pounds_z\bar{C}_4 \subset \bar{C}_4$, we see that

$$(5.92)\quad (\exp(s\pounds_z) - 1)(L\mu + J) = s(1 + \frac{s}{2!}\pounds_z + \ldots)\pounds_z(L\mu + J) \bmod \bar{C}_4$$

$$= s(1 + \frac{s}{2!}\pounds_z + \ldots)d\beta_z \bmod \bar{C}_4$$

$$= s\ d\beta(s) \bmod \bar{C}_4 ,$$

where

$$(5.93)\quad \beta(s) = d\rho + (1 + \frac{s}{2!}\pounds_z + \ldots)\beta_z .$$

A substitution of (5.91) into (5.69) now gives

$$(5.94)\quad \Delta_z(s)A[\Phi] = s\int_{\partial B_n} \Phi^*\beta(s) ,$$

in which case (5.90) follows directly from (5.67) and (5.94). The proof of the identity (5.91) follows the same lines as those given above.

Although Theorems 5.13 and 5.14 appear similar, they are in fact, quite different. In order that $z \in N(L,K_4)$, we must have $\pounds_z(L\mu + J) = 0$, while for $z \in N_1(L,K_4)$, the requirement is $\pounds_z(L\mu + J) = 0 \bmod \bar{C}_4$. If we introduce the collective indices, a, b, with range $2N$ such that $q^a = \{q^\alpha, \bar{q}^{\alpha}\}$, etc. a simple calculation shows that

$$\pounds_z(L\mu) = (z\rfloor dL + L\,\partial_i z^i)\mu + L\,\partial_b z^i\ dq^b \wedge \mu_i .$$

When dq^b is eliminated between this equation and $c^b = dq^b - y^b_i\,dx^i$, we obtain

$$\pounds_z(L\mu) = (z\rfloor dL + L\,Z_i z^i)\mu + c^b \wedge L\,\partial_b z^i \mu_i .$$

Similarly, (5.57) and $\pounds_z c^a = \partial_b(v\rfloor c^a)\,c^b$ yield

$$\pounds_z J = c^b \wedge \{\partial_b(z\rfloor c^a)\partial^i_a L + z\rfloor d(\partial^i_b L)\}\mu_i$$

$$+ c^b \wedge \partial^i_b L\ dz^j \wedge \mu_{ji} .$$

Thus, $z \in N(L,K_4)$ if and only if

$$(5.95) \qquad \pounds_z(L\mu + J) = (z \rfloor dL + L\, Z_i z^i)\mu$$
$$+ C^b \wedge \{L\, \partial_b z^i + \partial_b(z \rfloor C^a)\partial_a^i L + z \rfloor d(\partial_b^i L)\}\mu_i$$
$$+ C^b \wedge \partial_b^i L\, dz^j \wedge \mu_{ji} = 0 ,$$

while $z \in N_1(L,K_4)$ if and only if

$$(5.96) \qquad (z \rfloor dL + L\, Z_i z^i)\mu = 0 \bmod \bar{C}_4 ,$$

since the remaining terms on the right-hand side of (5.96) belong to the closed ideal $\bar{C}_4$.

There is a very important point that should be noted in regard to Theorems 5.13 and 5.14; namely, that their converses are not true. In order to see this, we note that Definitions 5.7 and 5.8 and (5.85) give

$$z \rfloor F + d\{z \rfloor (L\mu + J)\} = 0$$

for $z \in N(L,K_4)$ and

$$z \rfloor F + d\{z \rfloor (L\mu + J)\} = 0 \bmod \bar{C}_4$$

for $z \in N_1(L,K_4)$, while (5.63) gives

$$\Phi^*(z \rfloor F) = \Phi^*(z \rfloor \bar{C}^\alpha)\Phi^* E_\alpha + \Phi^*(z \rfloor C^\alpha)\Phi^* \bar{E}_\alpha$$

for any $\Phi \in R_4(B_n)$. Thus, since any $\Phi \in R_4(B_n)$ annihilates $\bar{C}_4$, the above relations yield

$$d\Phi^*\{z \rfloor (L\mu + J)\} = - \Phi^*(z \rfloor \bar{C}^\alpha)\Phi^* E_\alpha - \Phi^*(z \rfloor C^\alpha)\Phi^* \bar{E}_\alpha .$$

Accordingly, if $\Phi \in R_4(B_n)$ is such that $\Phi^* \bar{E}_\alpha = 0$, $\Phi^*(z \rfloor \bar{C}^\alpha) = 0$, that is Φ is a similarity solution of $\bar{E}_\alpha$ with the similarity conditions $\Phi^*(z \rfloor \bar{C}^\alpha) = 0$, then Φ satisfies

$d\Phi^*\{z \rfloor (L\mu + J)\} = 0$ whether or not Φ solves E_α .

<u>COROLLARY 5.1</u>. *If* z *belongs either to* $N(L,K_4)$ *or to* $N_1(L,K_4)$, *then any* $\Phi \in R_4(B_n)$ *satisfies the identity*

$$(5.97) \qquad d\Phi^*\{z \rfloor (L\mu + J)\} = - \Phi^*(z \rfloor \bar{C}^\alpha)\Phi^* E_\alpha - \Phi^*(z \rfloor C^\alpha)\Phi^* \bar{E}_\alpha \ .$$

The identity $d\Phi^*\{z \rfloor (L\mu + J)\} = 0$ *thus holds if* (1) $\Phi^*(z \rfloor \bar{C}^\alpha) = 0$, $\Phi^*(z \rfloor C^\alpha) = 0$ *or if* (2) $\Phi^*(z \rfloor \bar{C}^\alpha) = 0$, $\Phi^* \bar{E}_\alpha = 0$, *the collection of all* $\Phi \in R_4(B_n)$ *that satisfy the identity* $d\Phi^*\{z \rfloor (L\mu + J)\} = 0$ *is significantly larger than* $S_4(\underset{\sim}{E}, \underset{\sim}{\bar{E}}, B_n)$, *and* $d\Phi^*\{z \rfloor (L\mu + J)\} = 0$ *for* $z \in N(L,K_4)$ *or* $z \in N_1(L,K_4)$ *does not imply* $\Phi \in S_4(\underset{\sim}{E}, \underset{\sim}{\bar{E}}, B_n)$.

A similar result clearly holds for any variational system. One must therefore be very careful not to confuse maps that satisfy the identities $d\Phi^*\{z \rfloor (L\mu + J)\} = 0$ of a variational system with maps that solve the variational system.

<u>Finite Variations of Similarities</u>

The occurrence of the factors $v \rfloor C^\alpha$ and $v \rfloor \bar{C}^\alpha$ in the above results can be exploited in order to obtain a very interesting conclusion. In order to simplify the analysis, introduce the collective indices a, b, etc. with range 2N and $\{q^a\} = \{q^\alpha, \bar{q}^\alpha\}$. In this event, (5.60) and (5.61) can be written in the equivalent form

$$(5.98) \qquad F = C^a \wedge E_a = d(L\mu + J)$$

with $\{E_a\} = \{\bar{E}_\alpha, E_\alpha\}$. Accordingly, we have

$$(5.99) \qquad \pounds_v(L\mu + J) = v \rfloor F + d\{v \rfloor (L\mu + J)\}$$

$$= (v \rfloor C^a)E_a + d\{v \rfloor (L\mu + J)\} \bmod \bar{C}_4$$

for any $v \in TC(K_4)$. Since any $v \in TC(K_4)$ gives $\pounds_v C^a = \partial_b(v \rfloor C^a)C^b$, it follows that

$$(5.100) \qquad \pounds_v(v \rfloor C^a) = (v \rfloor C^b)\, \partial_b(v \rfloor C^a) .$$

Combination of (5.99) and (5.100) thus show that

$$(5.101) \qquad \pounds_v^2(L\mu + J) = (v \rfloor C^b)\{\partial_b(v \rfloor C^a) + \delta_b^a\, \pounds_v\}E_a$$

$$+ d\pounds_v\{v \rfloor (L\mu + J)\} \bmod \bar{C}_4 .$$

The end result of this is that we obtain

$$(\exp(s\pounds_v) - 1)(L\mu + J)$$

$$= (v \rfloor C^a)M_a(s) + s\, d\eta_v(s) \bmod \bar{C}_4$$

with

$$(5.102) \qquad \eta_v(s) = (1 + \frac{1}{2!}\, s\pounds_v + \frac{1}{3!}\, s^2\pounds_v^2 + \ldots)(v \rfloor (L\mu + J)) ,$$

and hence (5.67) and (5.69) yield

$$(5.103) \quad A[\Phi_v(s)] = A[\Phi] + \int_{B_n} \Phi^*(v \rfloor C^a)\Phi^* M_a(s) + s \int_{\partial B_n} \eta_v(s) .$$

The following result is then clear.

THEOREM 5.16. *If* v *is any element of* $TC(K_4)$, *then*

$$(5.104) \qquad A[\Phi_v(s)] = A[\Phi] + s \int_{\partial B_n} \eta_v(s)$$

for every similarity $\Phi \varepsilon R_4(B_n)$ *such that*

$$(5.105) \qquad \Phi^*(v \rfloor C^\alpha) = 0 , \quad \Phi^*(v \rfloor \bar{C}^\alpha) = 0 ,$$

where $\eta_v(s)$ *is given by* (5.102).

Weak Noetherian Vector Fields

There is an obvious extension of the theory that obtains from replacing the closed ideal $\bar{C}_4$ by the closed ideal $\bar{E}_4$.

Definition 5.10. A vector field $v \in TC(K_4)$ is said to be a *weak Noetherian vector field* of $A[\Phi]$ if and only if

(5.106) $$\pounds_v(L\mu + J) = d\rho_v \mod \bar{E}_4$$

The colleciton of all weak Noetherian vector fields is denoted by $N_3(L,K_4)$.

Weak Noetherian vector fields are not of great use, however, in view of the following results.

LEMMA 5.7. *Every* $v \in TC(K_4)$ *is a weak Noetherian vector field for every* L .

Proof. It follows directly from (5.99) that $v \in TC(K_4)$ gives

$$\pounds_v(L\mu + J) = d\{v \rfloor (L\mu + J)\} \mod \bar{E}_4$$

Hence $\pounds_v(L\mu + J) = d\rho_v \mod \bar{E}_4$ with $\rho_v = d\gamma + v\rfloor (L\mu + J)$ for any $v \in TC(K_4)$ and any L .

Variational Generation of Pseudopotentials

The existence of conserved current densities that are associated with the existence of Noetherian vector fields provides exactly that information needed in order to construct new closed ideals and pseudopotentials in the manner delineated in Chapter 4. We confine our remarks here to the establishment of the underlying theorems from which the specific results on prolongations and pseudopotentials can be constructed. In view of the inclusion relations given in Theorem 5.12, it is sufficient to confine attention to the "largest" collection, namely, generalized Noetherian vector fields of the second kind.

THEOREM 5.17. *Let* z_a , $a = 1,\ldots,r$, *belong to* $N_2(L,K_4)$, *be linearly independent with constant coefficients on* K_4 , *and define the* r *forms* $Q(z_a)$ *of degree* $n-1$ *by*

$$(5.107) \qquad Q(z_a) = dS_a + z_a \rfloor (L\mu + J) - \beta_a .$$

The ideal

$$I_5 = I\{C^\alpha, dC^\alpha, \bar{C}^\alpha, d\bar{C}^\alpha, E_\alpha, dE_\alpha, \bar{E}_\alpha, d\bar{E}_\alpha, Q(z_a)\}$$

is a closed ideal of $\Lambda(K_4)$.

Proof. It follows from Definition 5.9 that $z_a \varepsilon N_2(L,K_4)$ if and only if

$$\pounds_{z_a}(L\mu + J) = z_a \rfloor F + d\{z_a \rfloor (L\mu + J)\}$$

$$= d\beta_a \bmod \bar{C}_4 .$$

Since $z_a \rfloor F \varepsilon \bar{E}_4$, by Lemma 5.3, we obtain the result

$$(5.108) \qquad d\{z_a \rfloor (L\mu + J)\} = d\beta_a \bmod \bar{E}_4 .$$

Now, the exterior derivative of $Q(z_a)$ yields

$$dQ(z_a) = d\{z_a \rfloor (L\mu + J)\} - d\beta_a ,$$

and (5.108) shows that

$$dQ(z_a) = 0 \bmod \bar{E}_4 .$$

The ideal I_5 is thus a closed ideal of $\Lambda(K_4)$ and the Theorem is established.

Theorem 5.17 shows that a prolongation of $(K_4, \Lambda(K_4))$ by the introduction of the systems of r new variables $S_a = p_a^{ij} \mu_{ij}$ and the concomitant extension of $R_4(B_n)$ to $R_5(B_n)$ and $S_4(\underset{\sim}{E}, \underset{\sim}{\bar{E}}, B_n)$ to $S_5(\underset{\sim}{E}, \underset{\sim}{\bar{E}}, Q_a, B_n)$ establishes the *existence of* r *systems of pseudopotentials for the extended balance system.* These pesudopotentials obtain from the demand that $\Phi \varepsilon S_5(\underset{\sim}{E}, \underset{\sim}{\bar{E}}, Q_a, B_n)$ annihilate the Q's; that is

$$d\Phi^* p_a^{ij} \mu_{ij} = \Phi^*\{\beta_a - z_a \rfloor (L\mu + J)\} .$$

There is an important point that should be noted here. We saw in Chapter 4 that the closed ideal $I\{\underset{\sim}{C}, d\underset{\sim}{C}, \underset{\sim}{E}, d\underset{\sim}{E}, \underset{\sim}{Q}, d\underset{\sim}{Q}\}$ could be replaced by the closed ideal $\bar{I}\{\underset{\sim}{C}, \underset{\sim}{Q}\}$ (*vid.* Lemma 4.2), and this fact, to a great extent, accounts for the usefulness of pseudopotentials. In the present case, we have, for $z_a \in N_2(L, K_4)$, $a = 1, \ldots, r$,

$$\pounds_{z_a}(L\mu + J) = z_a \rfloor F + d\{z_a \rfloor (L\mu + J)\}$$

$$= d\beta_a \mod \bar{C}_4 ,$$

and hence (5.107) gives

$$dQ_a = (z_a \rfloor \bar{C}^\alpha) E_\alpha + (z_a \rfloor C^\alpha) \bar{E}_\alpha \mod \bar{C}_4 .$$

Thus, *the ideal* $\bar{I}\{C^\alpha, \bar{C}^\alpha, E_\alpha, \bar{E}_\alpha, Q_\alpha\}$ *can be replaced by the ideal* $\bar{I}\{C^\alpha, \bar{C}^\alpha, Q_\alpha\}$ *if and only if the matrix with entries* $((z_a \rfloor \bar{C}^\alpha, z_a \rfloor C^\alpha))$ *has rank* $2N$, (i.e. E_α and $\bar{E}_\alpha$ can be solved for in terms of dQ_a , $a = 1, \ldots, r \geq 2N$.) Satisfaction of this condition clearly places rather strong demands.

5.6 TRANSFORMATION-THEORETIC METHODS

The only transformations that have been considered thus far are those that result from the point transformations generated by transport along the orbits of isovector fields. The extensive literature on general transformations of variational systems suggests that more general transformations of the prolonged balance system $(\underset{\sim}{E}, \underset{\sim}{\bar{E}})$ on K_4 may provide new and useful results for the solution of the original balance system. This should be particularly true in the instance of $\bar{E}_4$, for the Lagrangian function

$$L = W_\alpha \bar{q}^\alpha + W_\alpha^i \bar{y}_i^\alpha$$

is *linear* in the conjugate variables $(\bar{q}^\alpha, \bar{y}_i^\alpha)$.

We recall, from previous subsections that all quantities governing the prolonged balance system $\bar{E}_4$ are determined in terms of F and J. Since we will need to consider different Lagrangian functions, it is useful to introduce the notation

$$(5.109)\qquad J(L) = \bar{C}^{\alpha} \wedge \bar{\partial}^{i}_{\alpha} L \, \mu_i + C^{\alpha} \wedge \partial^{i}_{\alpha} L \, \mu_i ,$$

$$(5.110)\qquad F(L) = d(L\mu + J(L)) = dL \wedge \mu + dJ(L) .$$

E-Transformations

Definition 5.11. An invertible map

$$(5.111)\qquad \rho\colon K_4 \longrightarrow {}^{\backprime}K_4 \mid {}^{\backprime}x^i = x^i , \quad {}^{\backprime}q^{\alpha} = \rho^{\alpha}(x^j, q^{\beta}, \bar{q}^{\beta}, y^{\beta}_i, \bar{y}^{\beta}_i) ,$$

$${}^{\backprime}\bar{q}^{\alpha} = \bar{\rho}^{\alpha} , \quad {}^{\backprime}y^{\alpha}_i = \rho^{\alpha}_i , \quad {}^{\backprime}\bar{y}^{\alpha}_i = \bar{\rho}^{\alpha}_i$$

is said to constitute an E-*transformation* if and only if there exists an ${}^{\backprime}L({}^{\backprime}x^j, {}^{\backprime}q^{\beta}, {}^{\backprime}\bar{q}^{\beta}, {}^{\backprime}y^{\beta}_j, {}^{\backprime}\bar{y}^{\beta}_j)$ for each L such that

$$(5.112)\qquad \overset{-1}{\rho}{}^{*}F(L) = F({}^{\backprime}L) .$$

THEOREM 5.18. *E-transformations form a group under composition.*

Proof. Since $(\rho_2 \circ \rho_1)^{-1*} = \overset{-1}{\rho}{}^{*}_2 \circ \overset{-1}{\rho}{}^{*}_1$, it follows that $(\rho_2 \circ \rho_1)^{-1*}F(L) = \overset{-1}{\rho}{}^{*}_2(\overset{-1}{\rho}{}^{*}_1 F(L)) = \overset{-1}{\rho}{}^{*}_2 F({}^{\backprime}L) = F({}^{\backprime\backprime}L)$, and hence $\rho_2 \circ \rho_1$ is an E-transformation whenever ρ_1 and ρ_2 are E-transformations.

Noting that ${}^{\backprime}x^i = x^i$, the contact forms on ${}^{\backprime}K_4$ are given by

$$(5.113)\qquad {}^{\backprime}C^{\alpha} = d{}^{\backprime}q^{\alpha} - {}^{\backprime}y^{\alpha}_i \, dx^i , \quad {}^{\backprime}\bar{C}^{\alpha} = d{}^{\backprime}\bar{q}^{\alpha} - {}^{\backprime}\bar{y}^{\alpha}_i \, dx^i .$$

We denote the closed contact ideal of ${}^{\backprime}K_4$ by ${}^{\backprime}\bar{C}_4$ and the isovector fields of ${}^{\backprime}\bar{C}_4$ by $TC({}^{\backprime}K_4)$.

LEMMA 5.8. *If the map* $\rho: K_4 \longrightarrow {}^\backprime K_4$ *is such that* $\overset{-1}{\rho}{}^*$ *maps* $\bar{C}_4$ *onto* ${}^\backprime\bar{C}_4$, *then* ${}^\backprime\Phi = \rho\circ\Phi$ *is a regular map of* B_n *into* ${}^\backprime K$ *whenever* $\Phi \in R_4(B_n)$.

Proof. The definition of ${}^\backprime\Phi$ shows that ${}^\backprime\Phi^*({}^\backprime\bar{C}_4) = \Phi^*\circ\rho^*({}^\backprime\bar{C}_4)$. Under the given hypothesis, we have ${}^\backprime\bar{C}_4 = \overset{-1}{\rho}{}^*(\bar{C}_4)$, and hence ${}^\backprime\Phi^*({}^\backprime\bar{C}_4) = \Phi^*(\bar{C}_4)$. Thus, ${}^\backprime\Phi^*$ annihilates ${}^\backprime\bar{C}_4$ whenever Φ^* annihilates $\bar{C}_4$. Since ${}^\backprime x^i = x^i$, ${}^\backprime\mu = \overset{-1}{\rho}{}^*\mu = \mu$, from which it follows that ${}^\backprime\Phi^*({}^\backprime\mu) = \Phi^*\mu$. Thus ${}^\backprime\Phi$ is regular whenever Φ is regular.

THEOREM 5.19. *If* ρ *is an E-transformation such that* $\overset{-1}{\rho}{}^*$ *maps* $\bar{C}_4$ *onto* ${}^\backprime\bar{C}_4$, *then* ρ *maps* $S_4(\underset{\sim}{E}, \underset{\sim}{\bar{E}}, B_n)$ *onto* $S_4({}^\backprime\underset{\sim}{E}, {}^\backprime\underset{\sim}{\bar{E}}, B_n)$.

Proof. Let $\Phi \in S_4(\underset{\sim}{E}, \underset{\sim}{\bar{E}}, B_n)$. Composition with the map ρ yields the map ${}^\backprime\Phi = \rho\circ\Phi$ so that ${}^\backprime\Phi^* = \Phi^*\circ\rho^*$. Let $v \in T(K_4)$, then ${}^\backprime v = \rho_* v \in T({}^\backprime K_4)$. It then follows from the fact that ρ is an invertible E-transformation that ${}^\backprime v \rfloor F({}^\backprime L) = {}^\backprime v \rfloor \overset{-1}{\rho}{}^* F(L) = \overset{-1}{\rho}{}^*(v \rfloor F(L))$, and hence

$$(5.114) \quad {}^\backprime\Phi^*({}^\backprime v \rfloor F({}^\backprime L)) \quad = \quad \Phi^*\circ\rho^*\circ\overset{-1}{\rho}{}^*(v \rfloor F(L)) \quad = \quad \Phi^*(v \rfloor F(L)) \ .$$

Now, Lemma 5.8 shows that ${}^\backprime\Phi^*$ annihilates ${}^\backprime\bar{C}_4$, and hence (5.114) and (5.62) combine to give

$$(5.115) \qquad {}^\backprime\Phi^*({}^\backprime v \rfloor {}^\backprime\bar{C}^\alpha){}^\backprime\Phi^*({}^\backprime E_\alpha) + {}^\backprime\Phi^*({}^\backprime v \rfloor {}^\backprime C^\alpha){}^\backprime\Phi^*({}^\backprime\bar{E}_\alpha)$$

$$= \quad \Phi^*(v \rfloor \bar{C}^\alpha)\Phi^*(E_\alpha) + \Phi^*(v \rfloor C^\alpha)\Phi^*(\bar{E}_\alpha) \ .$$

Since this holds for all $v \in T(K_4)$ an integration over B_n and a judicious selection of the vector fields v allow us to conclude that ρ maps $S_4(\underset{\sim}{E}, \underset{\sim}{\bar{E}}, B_n)$ onto $S_4({}^\backprime\underset{\sim}{E}, {}^\backprime\underset{\sim}{\bar{E}}, B_n)$ by use of the fundamental lemma of the calculus of variations.

This is a very important result, for it shows just what additional conditions must be imposed upon an E-transformation in order that it constitute a mapping of solutions onto solutions.

In substance, this result is similar to, and in fact was motivated by the analysis given by H. Rund [26].

From now on, we shall write equations like (5.112) in the equivalent form $F(L) = F(\grave{}L)$, where it is understood that the equations of transformation are used to write both sides of the equality in terms of the same system of independent arguments.

THEOREM 5.20. *An invertible map* $\rho: K_4 \longrightarrow \grave{}K_4$ *is an E-transformation if and only if there exists a generating* (n-1)-*form* η *such that*

$$(5.116) \qquad L\mu + J(L) = \grave{}L\mu + J(\grave{}L) + d\eta ,$$

where it is understood that ρ *is used to write both sides of* (5.116) *in terms of the same arguments.*

Proof. Since $F(L) = d(L\mu + J(L))$ by (5.60), (5.112) can hold if and only if

$$d(L\mu + J(L)) = d(\grave{}L\mu + J(\grave{}L)) .$$

The result thus follows from the Poincare Lemma since both K_4 and $\grave{}K_4$ are star-like with respect to any of their points.

Special E-Transformations

It is sufficient to our purposes here to restrict attention to special E-transformations.

Definition 5.12. An E-transformation is said to be *special* if the generating (n-1)-form η is given by

$$(5.117) \qquad \eta = N^i \mu_i ,$$

in which case the n functions $\{N^i\}$ are referred to as the *generating functions* of the E-transformation.

Clearly, everything depends on just what the arguments of the generating functions are taken to be, for this choice completely structures the final results as the next theorem shows.

THEOREM 5.21. *An invertible map* $\rho\colon K_4 \longrightarrow {}^{\backprime}K_4$ *is a special E-transformation with generating (n-1)-form* $\eta = -N^i(x^j, q^\alpha, \bar{q}^\alpha, {}^{\backprime}q^\alpha, {}^{\backprime}\bar{q}^\alpha)\mu_i$ *if and only if the following relations are satisfied:*

$$(5.118) \qquad {}^{\backprime}\partial^i_\alpha {}^{\backprime}L = {}^{\backprime}\partial_\alpha N^i , \qquad {}^{\backprime}\bar{\partial}^i_\alpha {}^{\backprime}L = {}^{\backprime}\bar{\partial}_\alpha N^i ,$$

$$(5.119) \qquad \partial^i_\alpha L = -\partial_\alpha N^i , \qquad \bar{\partial}^i_\alpha L = -\bar{\partial}_\alpha N^i ,$$

$$(5.120) \qquad {}^{\backprime}L = L + \bar{y}^\alpha_i \bar{\partial}_\alpha N^i + {}^{\backprime}\bar{y}^\alpha_i {}^{\backprime}\bar{\partial}_\alpha N^i + y^\alpha_i \partial_\alpha N^i + {}^{\backprime}y^\alpha_i {}^{\backprime}\partial_\alpha N^i + \partial_i N^i .$$

Proof. The results given above follow directly from (5.116) of Theorem 5.20 when (5.57) is used to express $J(L)$ and $J({}^{\backprime}L)$ directly in terms of the basis elements μ, $d{}^{\backprime}q^\alpha \wedge \mu_i$, $d{}^{\backprime}\bar{q}^\alpha \wedge \mu_i$, $dq^\alpha \wedge \mu_i$ and $d\bar{q}^\alpha \wedge \mu_i$ of degree n.

There are still two problems that remain to be solved. The first is that of obtaining restrictions of the E-transformations ρ in order that $\overset{-1}{\rho}{}^*$ is a map of $\bar{C}_4$ onto ${}^{\backprime}\bar{C}_4$. This is clearly necessary in order that ρ and $\overset{-1}{\rho}$ map solutions onto solutions by Theorem 5.19. The second problem is that of obtaining the actual equations that define the E-transformations in an explicit fashion. As it turns out, solution of the second problem is relatively straightforward once the first problem has been solved.

Suppose that we can find collections of functions $k^\alpha = k^\alpha(x^j, q^\beta, \bar{q}^\beta, {}^{\backprime}q^\beta, {}^{\backprime}\bar{q}^\beta)$ and $\bar{k}^\alpha = \bar{k}^\alpha(x^j, q^\beta, \bar{q}^\beta, {}^{\backprime}q^\beta, {}^{\backprime}\bar{q}^\beta)$ such that the Jacobian conditions

$$(5.121) \qquad \frac{\partial(k^\alpha, \bar{k}^\alpha)}{\partial({}^{\backprime}q^\beta, {}^{\backprime}\bar{q}^\beta)} \neq 0 , \qquad \frac{\partial(k^\alpha, \bar{k}^\alpha)}{\partial(q^\beta, \bar{q}^\beta)} \neq 0$$

are satisfied. A direct calculation then gives

$$(5.122)\qquad {}^{\backprime}\partial_\beta k^{\alpha}\,{}^{\backprime}c^\beta + {}^{\backprime}\bar\partial_\beta k^{\alpha}\,{}^{\backprime}\bar c^\beta + \partial_\beta k^\alpha\, c^\beta + \bar\partial_\beta k^\alpha\, \bar c^\beta$$

$$= {}^{\backprime}\partial_\beta k^\alpha\, d{}^{\backprime}q^\beta + {}^{\backprime}\bar\partial_\beta k^\alpha\, d{}^{\backprime}\bar q^\beta + \partial_\beta k^\alpha\, dq^\beta + \bar\partial_\beta k^\alpha\, d\bar q^\beta$$

$$- ({}^{\backprime}y_i^\beta\, {}^{\backprime}\partial_\beta k^\alpha + {}^{\backprime}\bar y_i^\beta\, {}^{\backprime}\bar\partial_\beta k^\alpha + y_i^\beta\, \partial_\beta k^\alpha + \bar y_i^\beta\, \bar\partial_\beta k^\alpha)dx^i$$

with a similar expression when $\bar k^\alpha$ is used in place of k^α. Hence, if we posit the relations

$$(5.123)\qquad {}^{\backprime}y_i^\beta\, {}^{\backprime}\partial_\beta k^\alpha + {}^{\backprime}\bar y_i^\beta\, {}^{\backprime}\bar\partial_\beta k^\alpha + y_i^\beta\, \partial_\beta k^\alpha + \bar y_i^\beta\, \bar\partial_\beta k^\alpha + \partial_i k^\alpha = 0\,,$$

$$(5.124)\qquad {}^{\backprime}y_i^\beta\, {}^{\backprime}\partial_\beta \bar k^\alpha + {}^{\backprime}\bar y_i^\beta\, {}^{\backprime}\bar\partial_\beta \bar k^\alpha + y_i^\beta\, \partial_\beta \bar k^\alpha + \bar y_i^\beta\, \bar\partial_\beta \bar k^\alpha + \partial_i \bar k^\alpha = 0$$

between the old the new y's, (5.122) and its companion become

$$(5.125)\qquad {}^{\backprime}\partial_\beta k^{\alpha}\,{}^{\backprime}c^\beta + {}^{\backprime}\bar\partial_\beta k^{\alpha}\,{}^{\backprime}\bar c^\beta + \partial_\beta k^\alpha\, c^\beta + \bar\partial_\beta k^\alpha\, \bar c^\beta = dk^\alpha\,,$$

$$(5.126)\qquad {}^{\backprime}\partial_\beta \bar k^{\alpha}\,{}^{\backprime}c^\beta + {}^{\backprime}\bar\partial_\beta \bar k^{\alpha}\,{}^{\backprime}\bar c^\beta + \partial_\beta \bar k^\alpha\, c^\beta + \bar\partial_\beta \bar k^\alpha\, \bar c^\beta = d\bar k^\alpha\,.$$

In view of the conditions (5.121), it follows that satisfaction of the independent conditions (5.123), (5.124), and

$$(5.127)\qquad k^\alpha(x^i,\underset{\sim}{q},\underset{\sim}{\bar q},{}^{\backprime}\underset{\sim}{q},{}^{\backprime}\underset{\sim}{\bar q}) = a^\alpha\,,\quad \bar k^\alpha(x^i,\underset{\sim}{q},\underset{\sim}{\bar q},{}^{\backprime}\underset{\sim}{q},{}^{\backprime}\underset{\sim}{\bar q}) = \bar a^\alpha$$

with a^α and $\bar a^\alpha$ constants, is sufficient in order to obtain explicit relations between the old and new q's and to guarantee that the E-transformation ρ will be such that $\overset{-1}{\rho}{}^*$ maps $\bar C_4$ onto ${}^{\backprime}\bar C_4$.

<u>THEOREM 5.22</u>. *A special* E-*transformation* ρ *with generating functions* $N^i(x^j,\underset{\sim}{q},\underset{\sim}{\bar q},{}^{\backprime}\underset{\sim}{q},{}^{\backprime}\underset{\sim}{\bar q})$ *is such that* $\overset{-1}{\rho}{}^*$ *maps* $\bar C_4$ *onto* ${}^{\backprime}\bar C_4$ *provided there exist functions* $k^\alpha(x^j,\underset{\sim}{q},\underset{\sim}{\bar q},{}^{\backprime}\underset{\sim}{q},{}^{\backprime}\underset{\sim}{\bar q})$ *and* $\bar k^\alpha(x^j,\underset{\sim}{q},\underset{\sim}{\bar q},{}^{\backprime}\underset{\sim}{q},{}^{\backprime}\underset{\sim}{\bar q})$ *such that*

$$(5.128)\qquad \frac{\partial(k^\alpha,\bar k^\alpha)}{\partial(q^\beta,\bar q^\beta)} \neq 0\,,\qquad \frac{\partial(k^\alpha,\bar k^\alpha)}{\partial({}^{\backprime}q^\beta,{}^{\backprime}\bar q^\beta)} \neq 0\,,$$

(5.129) $\rho_i k^\alpha = 0 , \quad \rho_i \bar{k}^\alpha = 0 ,$

where ρ_i *are linear differential operators that are defined by*

(5.130) $\rho_i \equiv \partial_i + y_i^\beta \partial_\beta + \bar{y}_i^\beta \bar{\partial}_\beta + \grave{y}_i^\beta \grave{\partial}_\beta + \grave{\bar{y}}_i^\beta \grave{\bar{\partial}}_\beta .$

If these conditions are met, then the equations that define the E-*transformation* ρ *are given by* (5.129) *and*

(5.131) $k^\alpha(x^j, \underset{\sim}{q}, \underset{\sim}{\bar{q}}, \underset{\sim}{\grave{q}}, \underset{\sim}{\grave{\bar{q}}}) = a^\alpha , \quad \bar{k}^\alpha(x^j, \underset{\sim}{q}, \underset{\sim}{\bar{q}}, \underset{\sim}{\grave{q}}, \underset{\sim}{\grave{\bar{q}}}) = \bar{a}^\alpha$

for constants a^α *and* $\bar{a}^\alpha$.

We take particular note of the fact that expansion of (5.129) gives exactly the system of equations (5.123) and (5.124). These latter equations are, however, in view of the requirements (5.121), nothing more than Bäcklund transformations of the most convenient kind, namely, those that are linear in the old and in the new first derivatives. In fact, we have the following result.

COROLLARY 5.1. *If* $\rho : K_4 \longrightarrow \grave{K}_4$ *is a special* E-*transformation that satisfies the hypotheses of Theorems* 5.21 *and* 5.22 *and* $\Phi^*(x^i) = x^i$, *then* ρ *is a variational Bäcklund transformation.*

Proof. Let Φ be any element of $R_4(B_n)$ such that $\Phi^*(x^i) = x^i$, and define $\grave{\Phi}$ by $\grave{\Phi} = \rho \circ \Phi$, then $\grave{\Phi} \varepsilon R_4(B_n)$ because ρ maps $\bar{C}_4$ onto $\grave{\bar{C}}_4$. In this event, we have $\grave{\Phi}^* = \Phi^* \circ \rho^*$, and $\Phi^*(y_i^\alpha) = \partial_i \Phi^* q^\alpha$, $\Phi^*(\bar{y}_i^\alpha) = \partial_i \Phi^* \bar{q}^\alpha$, $\grave{\Phi}^*(\grave{y}_i^\alpha) = \partial_i \grave{\Phi}^* \grave{q}^\alpha$, $\grave{\Phi}^*(\grave{\bar{y}}_i^\alpha) = \partial_i \grave{\Phi}^* \grave{\bar{q}}^\alpha$. Thus, since N^i are functions of only the variables $(x^j, q^\beta, \bar{q}^\beta, \grave{q}^\beta, \grave{\bar{q}}^\beta)$, (5.120) gives

$$\grave{\Phi}^*(\grave{L}) - \Phi^*(L) = D_i[N^i(x^j, \Phi^* q^\beta, \Phi^* \bar{q}^\beta, \grave{\Phi}^* \grave{q}^\beta, \grave{\Phi}^* \grave{\bar{q}}^\beta)] .$$

This relation is, however, what H. Rund gives as the definition of a variational Bäcklund transformation [26].

Clearly, satisfaction of all of the conditions imposed by Theorems 5.21 and 5.22 is no simple task. It is thus evident that E-transformations do not provide a panacea for the solution of all balance systems. There is one advantage that should be noted, however. The functional form of ${}^{\backprime}L$ is still at our disposal if we view (5.120) as an equation that is to be solved for the generating functions N^i. Thus, in particular, if we choose L by the prescription

$$(5.132) \qquad {}^{\backprime}L = {}^{\backprime}y_i^{\alpha}\, {}^{\backprime}\partial_{\alpha} u^i({}^{\backprime}q^{\beta}, {}^{\backprime}\bar{q}^{\beta}) + {}^{\backprime}\bar{y}_i^{\alpha}\, {}^{\backprime}\bar{\partial}_{\alpha} u^i({}^{\backprime}q^{\beta}, {}^{\backprime}\bar{q}^{\beta})$$

then the Euler-Lagrange equations associated with this ${}^{\backprime}L$ are satisfied by any functions ${}^{\backprime}\phi^{\alpha}(x^j)$, ${}^{\backprime}\bar{\phi}^{\alpha}(x^j)$, and use of (5.118) shows that

$$(5.133) \qquad {}^{\backprime}L - {}^{\backprime}y_i^{\alpha}\, {}^{\backprime}\partial_{\alpha}^{i}\, {}^{\backprime}L - {}^{\backprime}\bar{y}_i^{\alpha}\, {}^{\backprime}\bar{\partial}_{\alpha}^{i}\, {}^{\backprime}L \equiv 0 .$$

In this event, (5.120) becomes

$$(5.134) \qquad L + y_i^{\alpha}\, \partial_{\alpha} n^i + \bar{y}_i^{\alpha}\, \bar{\partial}_{\alpha} n^i + \partial_i n^i = 0$$

under the substitution

$$(5.135) \qquad N^i = u^i + n^i(x^j, q^{\beta}, \bar{q}^{\beta}) .$$

Now, (5.134) together with the relations (5.119) is simply the Hamilton-Jacobi equation in Lagrangian form ([7], page 245). This is quite encouraging and suggests that future research along the lines presented here would be very useful in actually solving certain specific classes of problems for which Theorems 5.21 and 5.22 may be implemented in a simple fashion. We should also point out that the prolongation of K to K_4 and the resulting variational system provides access to all of the modern results concerning variational systems. In particular, the full theory of symplectic connections and a different formulation of Hamilton-Jacobi theory becomes available in this manner [27].

There are several generalizations of E-transformations that come to mind. One that would seem useful obtains by replacing (5.112) with $\overset{-1}{\rho}{}^{*}F(L) = F({}^{\backprime}L) \bmod {}^{\backprime}\bar{C}_4$, where ${}^{\backprime}\bar{C}_4 = \bar{I}\{{}^{\backprime}C^{\alpha}, {}^{\backprime}\bar{C}^{\alpha}\}$.

In this event, we lose the group property under composition, but gain the less restrictive requirement $L\mu + J(L) = \grave{L}\mu + J(\grave{L}) + d\eta + \grave{P}$ with $\grave{P} \in \Lambda^n(\grave{K}_4) \cap \grave{\bar{C}}_4$ in place of (5.116). This, in turn, may be generalized by allowing $\grave{P}$ to belong to the closed ideal $\bar{I}\{C^\alpha, \bar{C}^\alpha, \grave{C}^\alpha, \grave{\bar{C}}^\alpha\}$ of $\Lambda(K_4 \cup \grave{K}_4)$, which has been given a preliminary analysis in [7].

5.7 DISCRETE TRANSPORT METHODS

The results of this section come from a very simple observation and Theorem 5.9. If $\Phi \in R_4(B_n)$ and $v \in TC(K_4)$, the map $\Phi_v(s) = \exp(v,s)\circ\Phi$ yields

$$A[\Phi_v(s)] = \int_{B_n} \Phi^*\{\exp(s\pounds_v)(L\mu + J)\}$$

for any value of s such that $\exp(s\pounds_v)\mu \neq 0$. Now, Lemma 5.2 shows that $J \in \bar{C}_4$, in which case $\exp(s\pounds_v)J \in \bar{C}_4$, and hence Φ^* annihilates $\exp(s\pounds_v)J$ because $\Phi \in R_4(B_n)$. Accordingly, we obtain

$$A[\Phi_v(s)] = \int_{B_n} \Phi^*\{\exp(s\pounds_v)(L\mu)\}, \tag{5.136}$$

that is

$$\int_{B_n} \Phi_v(s)^*(L\mu) = \int_{B_n} \Phi^*\{\exp(s\pounds_v)(L\mu)\}.$$

If we restrict v to be a horizontal isovector of $\bar{C}_4$, $v \in TH(K_4)$, then $v^i = 0$, $\pounds_v\mu = d(v \rfloor \mu) = d(v^i\mu_i) = 0$, and hence $\exp(s\pounds_v)\mu = \mu \neq 0$ for all values of s. In this case (5.136) becomes

$$A[\Phi_v(s)] = \int_{B_n} \Phi^*\{\exp(s\pounds_v)L\}\mu. \tag{5.137}$$

Define a 1-parameter family of Lagrangians by

$$L(s) = \exp(s\pounds_v)L. \tag{5.138}$$

The relation (5.137) then becomes

$$(5.139) \qquad A[\Phi_v(s)] = \int_{B_n} \Phi^* L(s)\mu \overset{\text{def}}{=} A_s[\Phi] .$$

Now, simply observe that if Φ is such that $A_k[\Phi]$ is rendered stationary in value, then $A[\Phi_v(k)]$ is likewise rendered stationary in value and hence $\Phi_v(k) \varepsilon S_4(\underset{\sim}{E}, \underset{\sim}{\bar{E}}, B_n)$.

THEOREM 5.23. *If* $v \varepsilon TH(K_4)$ *and* $\Phi \varepsilon R_4(B_n)$ *renders*

$$(5.140) \qquad A_k[\Phi] = \int_{B_n} \Phi^* L(k)\mu ,$$

stationary in value, where

$$(5.141) \qquad L(k) = \exp(k\pounds_v)L ,$$

then

$$(5.142) \qquad \Phi_v(k) = \exp(v,k)\circ\Phi$$

belongs to $S_4(\underset{\sim}{E}, \underset{\sim}{\bar{E}}, B_n)$.

Particular note should be taken of the fact that v is not required to be an isovector of $\bar{E}_4$. Thus, we need only find a $v \varepsilon TH(K_4)$ such that we can solve the Euler-Lagrange equations based on the Lagrangian $L(k) = \exp(k\pounds_v)L$ for some appropriate numerical value of the parameter k in order to obtain solutions of the original balance system. This discrete transport method of solution is thus clearly dependent upon our ability to sum the operator series $\exp(s\pounds_v)L$ in order that we may determine $L(s)$, and hence pick a value k for s so that $L(k)$ gives solvable Euler-Lagrange equations.

Let us start with the defining equation

$$(5.143) \qquad L(s)\mu = \exp(s\pounds_v)L\mu .$$

Although there are a number of methods for summing the operator power series on the right-hand side of (5.143), there is one in particular that is both simple and convenient. Define the operator D by

$$D \equiv d/ds ,$$

then (5.143) shows that

$$(5.144) \qquad D^j L(s)\mu = \exp(s£_v)(£_v^j L)\mu .$$

Suppose that we can find a $v \in TH(K_4)$ such that

$$(5.145) \qquad P_r(£_v)L\mu = P ,$$

where $P \in \bar{C}_4$ and $P_r(z)$ is a polynomial in z or degree r. In this event, (5.144) yields

$$(5.146) \qquad P_r(D)L(s)\mu = \exp(s£_v)P_r(£_v)L\mu = \exp(s£_v)P .$$

However, $v \in TH(K_4)$ implies $£_v P \in \bar{C}_4$, because $P \in \bar{C}_4$, and hence $\exp(s£_v)P = Q(s) \in \bar{C}_4$, where $Q(s)$ has the form $M_\alpha(s) \wedge C^\alpha + N_\alpha(s) \wedge dC^\alpha + \bar{M}_\alpha(s) \wedge \bar{C}^\alpha + \bar{N}_\alpha(s) \wedge d\bar{C}^\alpha$. Accordingly, (5.146) yields

$$(5.147) \qquad P_r(D)(L(s)\mu) = Q(s) = M_\alpha(s) \wedge C^\alpha + N_\alpha(s) \wedge dC^\alpha + \bar{M}_\alpha(s) \wedge \bar{C}^\alpha + \bar{N}_\alpha(s) \wedge d\bar{C}^\alpha .$$

Now, since $P_r(D)$ is a polynomial of degree r in D, (5.147) has the general solution

$$(5.148) \quad L(s)\mu = [Y_1(s),\dots,Y_r(s)][A_1,\dots,A_r]^T + \int_o^s G(s-\tau)Q(\tau)d\tau$$

where $Y_1(s),\dots,Y_r(s)$ are r independent solutions of $P_r(D)Y(s) = 0$. Noting that $Q(s)$ has the form given by (5.147), it follows that $\int_o^s G(s-\tau)Q(\tau)d\tau \in \bar{C}_4$, so that we may write

$$(5.149) \qquad L(s)\mu = [Y_1(s),\dots,Y_r(s)][A_1,\dots,A_r]^T + R(s)$$

where $R(s) \in \bar{C}_4$ and $R(s)$ and its first $r-1$ derivatives with respect to s are zero at $s = 0$. It now follows directly from (5.144) that

$$(5.150) \qquad (D^j L(s)\mu)\big|_{s=0} = \pounds_v^j L\mu \ ,$$

and hence (5.149) yields

$$(5.151) \quad L(s)\mu = [Y_1(s),\dots,Y_r(s)]\underset{\sim}{W}^{-1}[L,\pounds_v L,\dots,\pounds_v^{r-1}L]^T\mu + R(s)$$

where $\underset{\sim}{W}$ is the inverse of the Wronskian matrix evaluated at $s = 0$.

THEOREM 5.24. *If* $v \in TH(K_4)$ *is such that*

$$(5.152) \qquad P_r(\pounds_v)L\mu = 0 \bmod \bar{C}_4 \ ,$$

where $P_r(z)$ *is a polynomial in* z *of degree* r, *then*

$$(5.153) \quad L(s)\mu = [Y_1(s),\dots,Y_r(s)]\underset{\sim}{W}^{-1}[L,\pounds_v L,\dots,\pounds_v^{r-1}L]^T\mu \bmod \bar{C}_4 \ ,$$

where $Y_1(s),\dots,Y_r(s)$ *are* r *independent solutions of* $P_r(D)Y(s) = 0$ *and* $\underset{\sim}{W}$ *is the Wronskian matrix of these solutions evaluated at* $s = 0$. *If* Φ *is any element of* $R_4(B_n)$, *then*

$$(5.154) \quad \int_{B_n} \Phi^* L(s)\mu = [Y_1(s),\dots,Y_r(s)]\underset{\sim}{W}^{-1}\int_{B_n} \Phi^*[L,\pounds_v L,\dots,\pounds_v^{r}L]^T\mu$$

$$= A[\Phi_v(s)] \ .$$

It is now a simple matter to combine Theorems 5.23 and 5.24 in order to obtain a specific discrete transport method. First, we find a nontrivial $v \in TH(K_4)$ such that

$$P_r(\pounds_v)L\mu = 0 \bmod \bar{C}_4 \ ,$$

for some polynomial $P_r(z)$. In this event, we obtain $L(s)\mu$ in the form given by (5.153). It then remains only to find a value k of s for which we can solve the Euler-Lagrange equations based upon $L(k)$. All such solutions, $\Phi(k)$, then give solutions $\Phi_v(k) = \exp(v,k)\circ\Phi(k)$ of the original balance system. For example, if we can find a $v \in TH(K_4)$ such that

$$(\pounds_v^2 + 1)L\mu = 0 \bmod \bar{C}_4 ,$$

then

$$L(s)\mu = L\cos(s) + (\pounds_v L)\sin(s) .$$

Thus, for the discrete choices $s = (2p+1)\pi/2$ we need to determine Φ so that we solve the Euler-Lagrange equations based upon the Lagrangian

$$L(\pi(2p+1)/2) = \pounds_v L = (v \rfloor dL) .$$

We must note, however, that $r > 1$ yields nonlinear equations of degree r for the determination of v . This follows from the fact that $v \in TH(K_4)$ implies $\pounds_v L = (v \rfloor dL)\mu$ and hence $\pounds_v^j L = \{(v\rfloor d)^j L\}\mu$. The discrete transport methods are thus not particularly easy or even effective in all cases. Fortunately, however, we do not require the general solution of $P_r(\pounds_v)L\mu = 0 \bmod \bar{C}_4$ since any one nontrivial such vector field is adequate for the method.

Since we can replace $L\mu$ by $L\mu + J \bmod \bar{C}_4$, and $\exp(s\pounds_v)d\beta = d\beta(s)$, with $d\beta(s)$ independent of dy_i^α and $dy_i^{-\alpha}$ if $\beta = \beta^i\mu_i$, $\partial_\gamma^j\beta^i = 0$, $\bar{\partial}_\gamma^j\beta^i = 0$, it is clear that the proof of Theorem 5.24 remains unchanged if (5.152) is replaced by $P_r(\pounds_v)(L\mu + J) = d\beta^i \wedge \mu_i \bmod \bar{C}_4$.

THEOREM 5.25. *If* $v \in TH(K_4)$ *is such that*

(5.155) $$P_r(\pounds_v)(L\mu + J) = d\beta^i \wedge \mu_i \bmod \bar{C}_4 , \quad \partial_\alpha^j\beta^i = \bar{\partial}_\alpha^j\beta^i = 0 ,$$

then

(5.156) $$L(s)\mu = [Y_1(s),\ldots,Y_r(s)]\underset{\sim}{W}^{-1}[L,\pounds_V L,\ldots,\pounds_V^{r-1}L]^T + d\beta^i(s)\wedge\mu_i \mod \bar{C}_4$$

and the term $\int_{B_n}\Phi^* d\beta^i(s)\wedge\mu_i$ *makes no contribution to the prolonged system of balance equations.*

5.8 DIRECT SUM DECOMPOSITIONS OF $T(K_4)$ AND TRANSVERSALITY

There is an additional structure offered by any variational system that we have yet to take advantage of; namely, the structuring of vector fields by means of the requirements of transversality. As it turns out, this added structure provides a collection of very powerful methods for solution of variational problems along the lines laid down by Carathéodory.

We again introduce collective Roman indices at the beginning of the alphabet so that we may write $\{q^a;\ a=1,\ldots,M=2N\} = \{q^\alpha,\ \bar{q}^\alpha;\ \alpha=1,\ldots,N\}$, in which case K_4 is an $(n+M+nM)$-dimensional number space with coordinates (x^i, q^a, y^a_i). Since the results to be established in this and the following sections hold for any variational system, we will relax the requirement $M = 2N$ that obtained in Section 5.1 in order to secure the variational imbedding of systems of laws of balance by prolongation.

The first thing we note is that $T(K_4)$ admits the trivial direct sum decomposition

(5.157) $$T(K_4) = \mathcal{S}\oplus\mathcal{Y},$$

where the subspaces $\mathcal{S}$ and $\mathcal{Y}$ are defined by

(5.158) $$\mathcal{S} = \{v\in T(K_4) \mid v(y^a_i) = 0,\ a=1,\ldots,M,\ i=1,\ldots,n\},$$

(5.159) $$\mathcal{Y} = \{v\in T(K_4) \mid v(x^i) = 0,\ i=1,\ldots,n;\ v(q^a)=0,\ a=1,\ldots,M\}.$$

Thus, $v \in S$ if and only if $v = v^i \partial_i + v^a \partial_a$ and $v \in \mathcal{Y}$ is and only if $v = v^a_i \partial^i_a$.

The Contact Subspace Z and Its Conjugates

Although S admits a trivial direct sum decomposition in terms of the subspaces spanned by $\{\partial_i\}$ and by $\{\partial_a\}$, there is a much more useful decomposition that directly reflects the structure of any given variational system that is defined on K_4 . To this end, we define the *contact subspace*, Z , of S by

$$(5.160) \qquad Z = \{v \in S \mid v \rfloor C^a = 0,\ a = 1,\ldots,M\} ,$$

where C^a are the contact forms of K_4 . The following Lemma then obtains directly.

LEMMA 5.9. *Z is an n-dimensional subspace of S and a basis for Z is given by*

$$(5.161) \qquad Z_i = \partial_i + y^a_i \partial_a , \quad i = 1,\ldots,n .$$

Definition 5.13. A subspace

$$(5.162) \qquad C = \{v \in S \mid v = A^a c_a\}$$

is said to be *conjugate to* Z *relative to* $\{C^a\}$ if and only if the basis vectors $\{c_a\}$ satisfy

$$(5.163) \qquad c_a \rfloor C^b = \delta^b_a .$$

LEMMA 5.10. *A set of basis vectors for C is given by*

$$(5.164) \qquad c_a = c^i_a Z_i + \partial_a , \quad a = 1,\ldots,M$$

for every choice of the nM *quantities* $c^i_a \in \Lambda^0(K_4)$ *, and* $\dim(C) = M$.

Proof. Since $c_a \in S$, we have $c_a = c^i_a \partial_i + c^b_a \partial_b$, and hence

(5.63) yields the conditions

$$\delta_a^b = c_a \rfloor C^b = - y_i^b \, c_a^i + c_a^b .$$

Thus, $c_a^b = \delta_a^b + y_i^b \, c_a^i$, so that we obtain

$$(5.165) \qquad c_a = c_a^i \partial_i + \partial_a + c_a^i \, y_i^b \partial_b = c_a^i Z_i + \partial_a .$$

Linear independence of these M vectors follows trivially for any choice of the functions c_a^i , and hence $\dim(C) = M$.

Lemma 5.10 shows that there are many subspaces of S that are conjugate to Z relative to $\{C^a\}$ and that these subspaces are parameterized by the functions c_a^i . It is thus convenient to write $C(c_a^i)$ for the specific conjugate subspace that is parameterized by $\{c_a^i\}$.

LEMMA 5.11. *The subspace* S *admits the direct sum decomposition*

$$(5.166) \qquad S = Z \oplus C(c_a^i)$$

for any choice of the functions $\{c_a^i\}$. *In particular, if*

$$(5.167) \qquad u = u^i \partial_i + u^a \partial_a$$

then

$$(5.168) \qquad u = U^i Z_i + U^a c_a ,$$

where

$$(5.169) \qquad U^i = (\delta_j^i + y_j^b \, c_b^i) u^j - c_b^i \, u^b ,$$

$$(5.170) \qquad U^a = u^a - y_j^a \, u^j ,$$

and

$$(5.171) \qquad u^i = U^i + c_b^i \, U^b ,$$

$$(5.172) \qquad u^a = (\delta^a_b + y^a_i c^i_b) U^b + y^a_i U^i .$$

Proof. The $n + M$ vectors $\{Z_i, c_a\}$ belong to the $(n+M)$-dimensional subspace S and are linearly independent for every choice of the functions $\{c^i_a\}$, the latter following directly from (5.161) and (5.164). Thus, $\{Z_i, c_a\}$ forms a basis for S and the direct sum decomposition, (5.166) then follows. The relations (5.169) through (5.172) obtain directly from (5.161) and (5.164) upon resolving any $u \in S$ on the equivalent bases $\{\partial_i, \partial_a\}$ and $\{Z_i, c_a\}$.

The Transverse Subspace $T(L)$

The basic problem that now presents itself is how to single out a particular $C(c^i_a)$ from amongst the embarrassing wealth of such conjugate subspaces. After much deliberation, the very useful article by Rund [28] showed that it is the concept of *transversality* in variational structures that provides the key to the problem. In fact, the following analysis was both motivated and guided by the results given by Rund in [28]. The results established in this and the following sections also have a marked degree of similarity with those presented by Dedecker [29], to which the reader is referred for an alternative account.

Let L be a given element of $\Lambda^0(K_4)$. If v is any element of $T(K_4)$ (v need not belong to $TC(K_4)$), the results leading up to Theorem 5.10 lead to a well defined operator δ_v. Thus, in particular, (5.70) yields

$$\begin{aligned} \delta_v A[\Phi] &= \int_{B_n} \Phi^* \pounds_v (L\mu + J) \\ &= \int_{B_n} \Phi^*(v \lrcorner F) + \int_{\partial B_n} \Phi^*(v \lrcorner (L\mu + J)) . \end{aligned}$$

Classically, in the calculus of variations of multiple integrals, a vector field v is said to be transverse if the surface integral,

$$\int_{\partial B_n} \Phi^*(v \rfloor (L\mu + J))$$

makes no contribution to $\delta_v A[\Phi]$. However, a straightforward calculation based on (5.57) shows that

$$(5.173) \quad v \rfloor (L\mu + J) = \{(L\delta^i_j - y^a_j \partial^i_a L)v^j + v^a \partial^i_a L\}\mu_i \bmod \bar{C}_4 .$$

These considerations lead to the following distinct notions of transversality.

<u>Definition 5.14</u>. A vector field $v \in S$ is said to be *pointwise transverse relative to* L if and only if

$$(5.174) \qquad v^a \partial^i_a L = H^i_j v^j$$

where

$$(5.175) \qquad H^i_j = y^a_j \partial^i_a L - \delta^i_j L$$

is the momentum-energy complex associated with L . The set of all pointwise transverse vector fields relative to L is denoted by $T(L)$, so that

$$(5.176) \qquad T(L) = \{v \in S \mid v^a \partial^i_a L = H^i_j v^j\}$$

is the *transverse subspace of* S *relative to* L .

<u>Definition 5.15</u>. A vector field $v \in S$ is said to be *locally transverse relative to* L if and only if

$$(5.177) \qquad v \rfloor (L\mu + J) = 0 \bmod \bar{C}_4 .$$

<u>Definition 5.16</u>. A vector field $v \in S$ is said to be *globally transverse relative to* L if and only if

$$(5.178) \qquad v \rfloor (L\mu + J) = d\beta \bmod \bar{C}_4 .$$

We shall work from now on only with $T(L)$ since (5.177) and (5.178) show that locally transverse vector fields and globally transverse vector fields do not necessarily form subspaces of S. We have included Definitions 5.15 and 5.16 primarily in order to point out that the collection of all vector fields for which $\int_{\partial B_n} \Phi^*(v \rfloor (L\mu + J)) = 0$ for any B_n is significantly larger than $T(L)$.

LEMMA 5.12. *We have*

$$(5.179) \qquad Z_k \rfloor (L\mu + J) = L\mu_k \bmod \bar{C}_4 ,$$

$$(5.180) \qquad c_a \rfloor (L\mu + J) = (Lc^i_a + \partial^i_a L)\mu_i \bmod \bar{C}_4 .$$

Hence, on the open set $L \neq 0$ *of* K_4,

$$(5.181) \qquad T(L) = C(-L^{-1}\partial^i_a L) ,$$

$\dim(T(L)) = M$, *and a basis for* $T(L)$ *is given by*

$$(5.182) \qquad t_a = \partial_a - L^{-1}(\partial^i_a L)Z_i = - L^{-1}(H^b_a \partial_b + (\partial^i_a L)\partial_i)$$

for $a = 1,\dots,M$, *where*

$$(5.183) \qquad H^b_a = y^b_i \, \partial^i_a L - \delta^b_a L .$$

Proof. The relations (5.179)-(5.180) follow directly from substitutions of (5.161) and (5.164) into (5.173). Thus, since $S = Z \oplus C(c^i_a)$, by Lemma 5.11, (5.179)-(7.180) show that $T(L) = C(-L^{-1}\partial^i_a L)$ on the open set $L \neq 0$ of K_4. Thus, $\dim(T(L)) = M$ on $L \neq 0$, by Lemma 5.10, and (5.182) follow from (5.184) for $c^i_a = - L^{-1}\partial^i_a L$.

The Decomposition $Z \oplus T(L)$

We now have all of the preliminary results that are needed in order to establish the desired direct sum decomposition of S

and hence of $T(K_4)$.

THEOREM 5.26. *The subspace* S *is the direct sum of the contact subspace,* Z *, and the transversality subspace* $T(L)$ *on the open set* $L \neq 0$ *of* K_4 *. Thus, any* $u = u^i\partial_i + u^a\partial_a$ *in* S *can be written uniquely in terms of the sum of a contact part,*

$$(5.184) \qquad u_C = U^i Z_i ,$$

and a transverse part,

$$(5.185) \qquad u_T = U^a t_a ,$$

on $L \neq 0$ *, where*

$$(5.186) \qquad L\,U^i = -H^i_j u^j + (\partial^i_b L) u^b ,$$

$$(5.187) \qquad U^a = u^a - y^a_i u^i ,$$

and

$$(5.188) \qquad L\,u^i = L\,U^i - (\partial^i_b L) U^b ,$$

$$(5.189) \qquad L\,u^a = L\,y^a_i U^i - H^a_b U^b .$$

In addition, the basis vectors

$$(5.190) \qquad Z_i = \partial_i + y^a_i \partial_a , \qquad i = 1,\ldots,n$$

$$(5.191) \qquad t_a = \partial_a - L^{-1}(\partial^i_a L) Z_i , \qquad z = 1,\ldots,M$$

$$= -L^{-1}(H^b_a \partial_b + (\partial^i_a L)\partial_i)$$

of S *on* $L \neq 0$ *satisfy*

$$(5.192) \qquad Z_i \lrcorner C^a = 0 , \qquad t_a \lrcorner C^b = \delta^b_a .$$

Classical multiple integral variational problems, dealing as they do with variational structures for which $A[\Phi]$ is bounded from below with respect to all regular C^2 maps Φ (i.e., problems of minimization rather than stationarization) possess the pleasant circumstance that one may always choose a Lagrangian function L so that it is globally strictly positive on K_4. For problems of stationarity, there is no guarantee that we may choose L so that $L > 0$ on K_4; in fact, it is clear from the results established in Section 5.1 that the Lagrangians that imbed a given system of laws of balance in a variational structure by prolongation will be such that $L = 0$ will have a nonvacuous solution set in K_4. Particular care must thus be exercised when dealing with problems that involve points in K_4 that are near the *singular locus* $L = 0$. The results established in Theorem 5.26 are of considerable use, even when the problem has a nonvacuous singular locus. A case in point is the simplification that results in the characterization of $TC(K)$. We saw in Section 5.2 that $TC(K_4)$ is characterized by the requirements

$$\text{(5.193)} \qquad \pounds_v C^a = A^a_b\, C^b .$$

If we allow $\pounds_v$ to act on the system of equations (5.192) and make use of (5.191), we obtain

$$\text{(5.194)} \qquad 0 = \pounds_v(Z_i \rfloor C^a) = [v,Z_i] \rfloor C^a + Z_i \rfloor A^a_b C^b ,$$

$$\text{(5.195)} \qquad 0 = \pounds_v(t_a \rfloor C^b) = [v,t_a] \rfloor C^b + t_a \rfloor A^b_e C^e .$$

Thus, making use of (5.192) again, we have

$$\text{(5.196)} \qquad [v,Z_i] \rfloor C^a = 0 ,$$

$$\text{(5.197)} \qquad [v,t_a] \rfloor C^b = - A^b_a .$$

It is now a trivial matter to use (5.190) and (5.191) to obtain

(5.198) $$v^a_i = Z_i(v \rfloor c^a)$$

from (5.196) and

(5.199) $$A^b_a = \partial_a(v \rfloor c^b)$$

from (5.197). These are, however, exactly the results that we obtained previously by a more laborious computation. Particular note should be taken of the fact that the factors L^{-1} drop out of the resulting equations so that the results actually hold on the singular locus $L = 0$ as well as at all other points of K_4.

5.9 TRANSVERSE 1-FORMS AND THE TRANSVERSALITY IDEAL

Theorem 5.26 shows that $\{Z_i, t_a\}$ is a basis for S at all points of K_4 for which $L \neq 0$, and that

(5.200) $$Z_i \rfloor c^a = 0, \quad t_a \rfloor c^b = \delta^b_a .$$

The relations (5.200) may obviously be used in order to complete $\{c^a, a = 1, \ldots, M\}$ to a basis for the subspace $\Lambda^1(G_4)$ of $\Lambda^1(K_4)$, i.e. the subspace of all 1-forms $\omega_i dx^i + \omega_a dq^a$ that is the natural dual of S.

Direct Sum Decomposition of the Dual Space $\Lambda^1(G_4)$

LEMMA 5.13. *A set of 1-forms* $\{T^j, j = 1, \ldots, n\}$ *satisfies the conditions*

(5.201) $$Z_i \rfloor T^j = \delta^j_i , \quad t_a \rfloor T^i = 0$$

at all points of K_4 *such that* $L \neq 0$ *if and only if*

(5.202) $$L\, T^j = - H^j_i dx^i + (\partial^j_a L) dq^a = L\, dx^j + (\partial^j_a L) c^a .$$

Proof. Any $T^j \in \Lambda^1(G_4)$ has the form $T^j = T^j_m dx^m + T^j_a dq^a$. Restricting consideration to those points of K_4 for which

$L \neq 0$, (5.190), (5.191) and (5.202) give

$$\delta^j_i = Z_i \rfloor T^j = T^j_i + y^b_i T^j_b ,$$

$$0 = L\, t_a \rfloor T^j = L\, T^j_a - y^b_k (\partial^k_a L) T^j_b - (\partial^i_a L) T^j_i .$$

These equations can be readily solved, and we obtain

$$L\, T^j_a = \partial^j_a L , \quad L\, T^j_i = - H^j_i .$$

The result then follows directly from $LT^j = LT^j_i dx^i + LT^j_a dq^a$, and the fact that $dq^a = C^a + y^a_i dx^i$.

The resulting daulity between $\{Z_i , t_a\}$ and $\{C^a , T^j\}$ motivates the following definition.

Definition 5.17. The 1-forms

$$T^j = - L^{-1}\{H^j_i dx^i - (\partial^j_a L) dq^a\} \tag{5.203}$$

are referred to as *transversality forms* and the closed ideal

$$\bar{T}_4 = I\{T^j , dT^j\} \tag{5.204}$$

is referred to as the *transversality ideal.*

LEMMA 5.14. *At each point of* $L \neq 0$ *we have*

$$T^1 \wedge T^2 \wedge \ldots \wedge T^n = \mu \bmod C_4 , \tag{5.205}$$

$$C^1 \wedge \ldots \wedge C^M \wedge T^1 \wedge \ldots \wedge T^n = dq^1 \wedge \ldots \wedge dq^M \wedge \mu , \tag{5.206}$$

and hence $\{C^a , a = 1,\ldots,M;\ T^j , j = 1,\ldots,n\}$ *is a basis for* $\Lambda^1(G_4)$ *at all points of* K_4 *for which* $L \neq 0$.

Proof. The second of (5.202) shows that $T^j = dx^j \bmod \bar{C}_4$ at all points of K_4 for which $L \neq 0$, so that we obtain (5.205). (5.206) then follows directly from (5.205) since exterior multiplication of any element of the ideal $C_4 = I\{C^a\}$ by $C^1 \wedge \ldots \wedge C^M$ gives zero.

These results provide the basis for the following theorem, that is dual to Theorem 5.26.

THEOREM 5.27. *$\Lambda^1(K_4)$ admits the direct sum decomposition*

$$\Lambda^1(K_4) = \Lambda^1(K_4) \cap \bar{C}_4 + \Lambda^1(K_4) \cap \bar{T}_4 + \Lambda^1(K_4) \cap I\{dy^a_i\}$$

on $L \neq 0$. *Thus, any*

(5.207) $$\Omega = \omega_i dx^i + \omega_a dq^a$$

in $\bar{I}\{C^a, T^j\} \cap \Lambda^1(K_4)$ *can be written uniquely on* $L \neq 0$ *as*

(5.208) $$\Omega = \Omega_i T^i + \Omega_a C^a$$

with

(5.209) $$\Omega_i = \omega_i + y^b_i \omega_b ,$$

(5.210) $$L \Omega_a = - H^b_a \omega_b - (\partial^j_a L) \omega_j ,$$

(5.211) $$L \omega_i = - L y^a_i \Omega_a - H^j_i \Omega_j ,$$

(5.212) $$L \omega_a = L \Omega_a + (\partial^j_a L) \Omega_j .$$

In addition, the bases $\{C^a, T^j\}$ *and* $\{Z_i, t_a\}$ *satisfy*

(5.213) $$Z_i \rfloor C^a = 0 , \quad t_a \rfloor C^b = \delta^b_a ,$$

(5.214) $$Z_i \rfloor T^j = \delta^j_i , \quad t_a \rfloor T^j = 0$$

at all points of K_4 *for which* $L \neq 0$.

Proof. The direct sum decomposition (5.207) of $\Lambda^1(K_4)$ on $L \neq 0$ follows directly from Lemma 5.14. The relations (5.207) through (5.212) then follow by resolving the basis $\{T^j, C^a\}$ on $\{dx^i, dq^a\}$ and vice versa.

An immediate application of these results yield a simplification in the resolution of any $v \in T(K_4)$ on the basis $\{Z_i, t_a, \partial_a^i\}$.

<u>COROLLARY 5.2</u>. *Any* $v \in T(K_4)$ *has the resolution*

(5.215) $$v = v^i Z_i + v^a t_a + v_i^a \partial_a^i$$

on $L \neq 0$, *where*

(5.216) $$v^i = v \lrcorner T^i , \quad v^a = v \lrcorner C^a ,$$

and any $\Omega \in \Lambda^1(K_4)$ *has the resolution*

(5.217) $$\Omega = \Omega_i T^i + \Omega_a C^a + \omega_a^i dy_i^a$$

on $L \neq 0$, *where*

(5.218) $$\Omega_i = \Omega(Z_i) = Z_i \lrcorner \Omega , \quad \Omega_a = \Omega(t_a) = t_a \lrcorner \Omega .$$

<u>Proof</u>. If we take the inner multiple of v with $\{T^i, C^a\}$ and use (5.213)-(5.215), we obtain (5.216). Similarly, allowing Ω to act on the elements of the basis $\{Z_i, t_a\}$ together with (5.213)-(5.215) yields (5.217).

The Null Class of n-Forms

The fairly simple results established thus far have a marked applicability for multiple integral problems in the calculus of variation. If Φ is any regular map of B_n into K_4, then Φ^* annihilates the closed contact ideal $\bar{C}_4$. Thus, since any action functional has the form

$$A[\Phi] = \int_{B_n} \Phi^*(L\mu)$$

we need only consider equalities mod $\bar{C}_4$ in any evaluation of an action integral. We therefore introduce the simplifying notation

$$R \overset{c}{=} S \iff R = S \bmod \bar{C}_4 .$$

We have already noted that $T^1 \wedge \ldots \wedge T^n \overset{c}{=} \mu$ on $L \neq 0$, and hence

(5.219) $$L\mu \overset{c}{=} L\, T^1 \wedge \ldots \wedge T^n$$

is an element of $\Lambda^n(K_4)$. The n-form

(5.220) $$\Omega = L\, T^1 \wedge \ldots \wedge T^n$$

$$= L\, (dx^1 + L^{-1}(\partial^1_{a_1} L) C^{a_1}) \wedge \ldots \wedge (dx^n + L^{-1}(\partial^n_{a_n} L) C^{a_n})$$

is, however, exactly what Carathéodory takes as the fundamental n-form in his monumental memoir [30]. It is of interest to note that $\Omega = L\mu + J + \ldots$.

If $\rho^i \in \Lambda^0(G_4)$, then (5.217) and $T^j \overset{c}{=} dx^j$ yield

$$d\rho^i = (Z_j \rfloor d\rho^i) T^j + (t_a \rfloor d\rho^i) C^a \overset{c}{=} Z_j(\rho^i) T^j \overset{c}{=} Z_j(\rho^i) dx^j .$$

Thus we obtain

$$d(\rho^i \mu_i) \overset{c}{=} Z_j(\rho^i) T^j \wedge \mu_i \overset{c}{=} Z_i(\rho^i)\mu ,$$

so that

$$\int_{\partial B_n} \Phi^*(\rho^i \mu_i) = \int_{B_n} \Phi^* Z_j(\rho^i) T^j \wedge \mu_i = \int_{B_n} \Phi^* Z_i(\rho^i)\mu ,$$

and the surface integral on the left-hand side is constant over all $\Phi \in R_4(B_n)$ such that $\phi^i|_{\partial B_n}$ and $\phi^a|_{\partial B_n}$ are assigned functions on ∂B_n. This simply says that the Euler-Lagrange equations based upon L and those based upon $L + Z_i(\rho^i)$ agree throughout B_n. This motivates the following classic definition.

<u>Definition 5.18</u>. The *null class*, $N(G_4)$, of n-forms is given by

(5.221) $$N(K_4) = \{\omega \in \Lambda^n(G_4) \,|\, \omega \overset{c}{=} d\eta,\ \eta \in \Lambda^{n-1}(G_4)\} .$$

Since $\{T^i, C^a\}$ is a basis for $\Lambda^1(G_4)$, any $\eta \in \Lambda^{n-1}(G_4) \cap \bar{C}_4$ will yield an $\omega = d\eta \overset{c}{=} 0$ that belongs to $N(G_4)$. The set of all $\omega \overset{c}{=} 0$ is referred to as the *trivial null class.* We shall thus consider two elements of $\Lambda^n(K_4)$ to be *null equivalent* if they differ by an element of the trivial null class.

Everything now hinges on the simple observation that $\{T^i, C^a\}$ is a basis for $\Lambda^1(G_4)$ so that any $\omega \in \Lambda^n(G_4)$ is null equivalent to an element of $\bar{T}_4 \cap \Lambda^n(G_4)$. Thus, without loss of generality, we may write any $\omega \in \Lambda^n(G_4)$ as

$$(5.222) \quad \omega \overset{c}{=} \Theta^j_k \, T^k \wedge \mu_j + \Theta^{ij}_{k\ell} \, T^k \wedge T^\ell \wedge \mu_{ij} + \ldots + \Theta \, T^1 \wedge T^2 \wedge \ldots \wedge T^n$$

where the quantities $\Theta^j_i, \Theta^{k\ell}_{ij}, \ldots, \Theta$ belong to $\Lambda^0(K_4)$. Each term may now be analyzed separately as to inclusion in $N(G_4)$. For example, $\Theta^i_k \, T^k \wedge \mu_i \overset{c}{=} d(\rho^i \mu_i) = d\rho^i \wedge \mu_i$, for $\{\rho^i\} \in \Lambda^0(G_4)$, leads to the requirement

$$\Theta^i_i = Z_i(\rho^i)$$

since $\Theta^i_k \, T^k \wedge \mu_i \overset{c}{=} \Theta^i_k \, dx^k \wedge \mu_i = \Theta^i_i \mu$ and $d\rho^i \wedge \mu_i \overset{c}{=} Z_k(\rho^i) T^k \wedge \mu_i$ $\overset{c}{=} Z_i(\rho^i)\mu$. The often stated result

$$\Theta^i_k = Z_k(\rho^i)$$

is thus seen to be sufficient for membership in $N(G_4)$; it is clearly not necessary. In like manner $\Theta^{ij}_{km} \, T^k \wedge T^m \wedge \mu_{ij} \overset{c}{=}$ $d(\rho^i \, d\rho^j \wedge \mu_{ij})$ for $\{\rho^i\} \subset \Lambda^0(G_4)$, leads to the necessary and sufficient condition

$$\Theta^{km}_{km} - \Theta^{mk}_{km} = Z_k(\rho^m) Z_m(\rho^k) - Z_k(\rho^k) Z_m(\rho^m) .$$

Again, the often stated result,

$$2 \, \Theta^{ij}_{km} = Z_k(\rho^i) Z_m(\rho^j) - Z_k(\rho^j) Z_m(\rho^i)$$

is sufficient but not necessary for membership in $N(G_4)$. Lastly, we note that a necessary and sufficient condition that

$\Theta T^1 \wedge T^2 \wedge \ldots \wedge T^n$ belong to $N(G_4)$ is given by[†]

$$\Delta = \det(Z_i(\rho^j)) , \quad \Theta = f(\rho^1,\ldots,\rho^n)\Delta .$$

This notation is taken from Rund's analysis [28] in order to simplify comparisons. The important thing to note here is that $\Theta = f(\rho^k)\Delta$ is both necessary and sufficient for membership in $N(G_4)$, while the quantities Ψ_r, $0 < r < n$ given by Rund satisfy the sufficient conditions but are not necessary for membership in $N(G_4)$. In this regard, the reader is referred to Dedecker's analysis [29].

Equivalent Integrals

The results just established provide a direct answer to the question of the general form of equivalent integrals in multiple integral variational problems; a question whose answer is fundamental to the field theory of Carathéodory [30].

Definition 5.19. Two integrals

$$\int_{B_n} \Phi^*\alpha \quad \text{and} \quad \int_{B_n} \Phi^*\beta$$

for α and β elements of $\Lambda^n(G_4)$, are said to be *equivalent* if and only if there exists an $\eta \in \Lambda^{n-1}(G_4)$ such that

$$\int_{B_n} \Phi^*(\alpha-\beta) = \int_{\partial B_n} \Phi^*\eta \tag{5.223}$$

for all $\Phi \in R_4(B_n)$.

LEMMA 5.15. *Two integrals* $\int_{B_n} \Phi^*\alpha$ *and* $\int_{B_n} \Phi^*\beta$, $\alpha \in \Lambda^n(G_4)$, $\beta \in \Lambda^n(G_4)$, *are equivalent if and only if there exists an element* ω *of* $N(G_4)$ *such that*

---

† Simply set $\Theta T^1 \wedge \ldots \wedge T^n \stackrel{c}{=} d\{\bar{f}(\rho^k)d\rho^2 \wedge \ldots \wedge \rho^n\}$, $\rho^k \in \Lambda^0(G_4)$.

(5.224) $\alpha \overset{c}{=} \beta + \omega$.

<u>Proof.</u> If (5.224) holds for some $\omega \in N(G_4)$, then $\omega \overset{c}{=} d\eta$ with $\eta \in \Lambda^{n-1}(G_4)$ by Definition 5.18, and hence $\int_{B_n} \Phi^*(\alpha-\beta) = \int_{\partial B_n} \Phi^*\eta$ for all $\Phi \in R_4(B_n)$. Conversely, if $\int_{B_n} \Phi^*(\alpha-\beta) = \int_{\partial B_n} \Phi^*\eta$ for all $\Phi \in R_4(B_n)$, $\eta \in \Lambda^{n-1}(G_4)$ then $\int_{B_n} \Phi^*(\alpha-\beta-d\eta) = 0$ for all $\Phi \in R_4(B_n)$, which implies $\alpha \overset{c}{=} \beta+d\eta \overset{c}{=} \beta+\omega$ with $\omega \in N(G_4)$ since $\alpha \in \Lambda^n(G_4)$ and $\beta \in \Lambda^n(G_4)$.

As a specific example, we observe that $\int_B \Phi^* L\mu$ and $\int_{B_n} \Phi^*(L + \det(Z_i(\rho^j))\mu$ are equivalent integrals.

As a last remark, we note that $C^a = dq^a - y^a_i\, dx^i$ and (5.202) give

$$\begin{Bmatrix} dx^i \\ dq^a \end{Bmatrix} = \begin{pmatrix} \delta^i_j & -L^{-1}\partial^i_b L \\ y^a_j & -L^{-1}H^a_b \end{pmatrix} \begin{Bmatrix} T^j \\ C^b \end{Bmatrix} \tag{5.225}$$

and

$$\begin{Bmatrix} T^i \\ C^a \end{Bmatrix} = \begin{pmatrix} -L^{-1}H^i_j & L^{-1}\partial^i_b L \\ -y^a_j & \delta^a_b \end{pmatrix} \begin{Bmatrix} dx^j \\ dq^b \end{Bmatrix} \tag{5.226}$$

so that the matrices on the right-hand sides of these equations are inverses of each other for $L \neq 0$. Thus, in particular, we have the relations

$$y^a_j\, H^j_k = y^b_k\, H^a_b \; . \tag{5.227}$$

5.10 TRANSFORMATIONS OF THE TRANSVERSALITY 1-FORMS: WEAKLY-COMPLETE AND COMPLETE FIGURES OF CARATHÉODORY

Although the transversality 1-forms $\{T^j\}$ provide a useful basis for many calculations, they are not necessarily an optimal choice for the actual problem of solving the balance equations of the system under study. We therefore consider linear homogeneous transformations of the T^j's (i.e., transformations that preserve the ideal generated by the T^j's) in the search for a more efficacious system of 1-forms.

Transversality Transformations

Let each of the 1-forms π^i belong to $\Lambda^1(G_4)$ and let $\{p^i_j\}$ be n^2 elements of $\Lambda^0(K_4)$ such that

$$(5.228)\qquad p = \det(p^i_j) \neq 0 .$$

Definition 5.20. The quantities $\{\pi^i, p^i_j\}$ determine a *transversality transformation* if and only if

$$(5.229)\qquad p^i_k\, T^k = \pi^i , \qquad i = 1,\ldots,n$$

are satisfied at all points of K_4 for which $L \neq 0$.

LEMMA 5.17. *The quantities* $\{\pi^i, p^i_j\}$ *determine a transversality transformation on* $L \neq 0$ *if and only if*

$$(5.230)\qquad p^i_k = Z_k \rfloor \pi^i , \qquad p = \det(Z_k \rfloor \pi^i) \neq 0 ,$$

$$(5.231)\qquad t_a \rfloor \pi^i = 0 .$$

Proof. By definition 5.20, $\{\pi^i, p^i_j\}$ determine a transversality transformation on $L \neq 0$ if and only if (5.228) and (5.229) are satisfied. However, $Z_i \rfloor T^j = \delta^j_i$, $t_a \rfloor T^i = 0$ by (5.214), and hence pulling down (5.229) by Z_m and by t_a yield the first of (5.230) and (5.231), respectively. The second of (5.230) then follows from the first.

LEMMA 5.18. *The quantities* $\{\pi^i, p^i_j\}$ *determine a transversality transformation on* $L \neq 0$ *if and only if*

$$(5.232) \qquad p^i_k = Z_k \rfloor \pi^i , \qquad p = \det(Z_k \rfloor \pi^i) \neq 0$$

$$(5.233) \qquad L^{-1} p^i_k \partial^k_a L = \pi^i_a ,$$

in which case we have

$$(5.234) \qquad \pi^i = \pi^i_k dx^k + \pi^i_a dq^a ,$$

$$(5.235) \qquad \pi^i_j = - L^{-1} p^i_k H^k_j .$$

and

$$(5.236) \qquad p^i_k = \pi^i_k + y^a_k \pi^i_a .$$

Proof. Lemma 5.17 has established the requirements (5.230) and (5.231). Thus, (5.232) must hold. It thus remains to obtain satisfaction of (5.231) on $L \neq 0$. Noting that $t_a = \partial_a - L^{-1}(\partial^i_a L) Z_i$, by (5.191), and that $\pi^i \in \Lambda^1(G_4)$ if and only if (5.234) holds, it follows directly that (5.231) gives the conditions

$$0 = t_a \rfloor \pi^i = \pi^i_a - L^{-1}(\partial^k_a L) Z_k \rfloor \pi^i .$$

Equations (5.233) thus follow upon solving for π^i_a and using the first of (5.232). The remaining equations, (5.235), then obtain from the direct calculation

$$p^i_k T^k = (Z_k \rfloor \pi^i)\{-LH^k_m dx^m + L^{-1}(\partial^k_a L) dq^a\}$$

$$= \pi^i = \pi^i_m dx^m + \pi^i_a dq^a .$$

It is easily seen, however, that satisfaction of (5.232) and (5.233) imply the satisfaction of (5.235); simply use the definition of H^k_m.

LEMMA 5.19. *If* $\{\pi^i, p^i_j\}$ *determine a transversality transformation on* $L \neq 0$, *then the quantities* P^i_k, *defined on* $L \neq 0$ *by*

$$(5.237) \qquad P^i_k = \partial p/\partial p^k_i ,$$

satisfy

$$(5.238) \qquad P^k_i \, p^i_j = p \, \delta^k_j ,$$

$$(5.239) \qquad p \, Z_k(P^k_m) = P^k_i \, P^j_m \{Z_j(Z_k \lrcorner \pi^i) - Z_k(Z_j \lrcorner \pi^i)\} ,$$

and we have

$$(5.240) \qquad p \, \partial^k_a L = L \, P^k_i \, \pi^i_a .$$

Proof. Since P^i_k is simply the cofactor of p^k_i in p, by (5.237), (5.238) follows directly. Allowing Z_k to act on both sides of (5.238) gives

$$Z_k(P^k_i) \, p^i_j + P^k_i \, Z_k(p^i_j) = Z_j p = P^k_i \, Z_j(p^i_k) .$$

When this is solved for $Z_k(P^k_i)$, by use of (5.238), and we note that $p^i_j = Z_j \lrcorner \pi^i$, we obtain (5.239). The relations (5.240) then follow directly from (5.233) upon using (5.238) to solve for $\partial^k_a L$.

It is of interest to note at this point that (5.235), (5.238) and (5.240) give us the system of relations

$$p \, \partial^k_a L = L \, P^k_i \, \pi^i_a , \quad p \, H^i_j = - L \, P^i_m \, \pi^m_j .$$

These agree exactly with what Rund gives as the equations that define the "canonical coordinates" π^i_a and π^i_j ([28], page 518) if we adjoin the added requirements $p = L$ and $\det(H^i_j) \neq 0$. Thus, this system may be considered as the precursor of the canonical relations given by Rund. One very important conclusion can be drawn at this point in view of the coincidence just noted;

namely, Rund's results can be obtained without the restriction $\det(H^i_j) \neq 0$. Although the restriction $\det(H^i_j) \neq 0$ is thus removable, the remaining restriction, namely $p = L$, is an entirely different matter. It centers heavily in the procedure that will be obtained for solving systems of variational balance equations.

Regularity Condition and the Induced Mapping of K_4

LEMMA 5.20. *If* $\{\pi^i, p^i_j\}$ *determine a transversality transformation on* $L \neq 0$, *then the equations* (5.233) *can be solved for the* y's *on* $L \neq 0$ *so as to obtain relations of the form*

$$(5.241) \qquad y^a_i = Y^a_i(x^j, q^b, \pi^i_j, \pi^i_a)$$

if and only if the nM-by-nM *matrix with entries*

$$(5.242) \qquad Q^{mj}_{ba} = p\{L\, \partial^m_b \partial^j_a L - (\partial^m_b L)(\partial^j_a L) + (\partial^j_b L)(\partial^m_a L)\}$$

$$+ L(\partial^k_a L) p^j_i Z_k \lrcorner (\partial^m_b \pi^i)$$

is nonsingular on $L \neq 0$. *Thus, in this case, in which the variational problem is said to be regular, a transversality transformation induces a mapping on* $L \neq 0$ *in* K_4 *of the form*

$$(x^j, q^a, y^a_j) \longrightarrow (x^j, q^a, \pi^i_j, \pi^i_a) .$$

Proof. The equations (5.233) can be solved for the y's on $L \neq 0$ only if $\det(\bar{Q}^{mi}_{ba}) \neq 0$, where

$$(5.243) \qquad \bar{Q}^{mi}_{ba} = \partial^m_b (L^{-1} p^i_k \partial^k_a L) , \quad p^i_k = Z_k \lrcorner \pi^i .$$

Now,

$$\bar{Q}^{mi}_{ba} = p^i_k L^{-2}\{L\partial^m_b \partial^k_a L - (\partial^m_b L)(\partial^k_a L)\} + L^{-1}(\partial^k_a L)\partial^m_b(p^i_k) ,$$

and

$$\partial_b^m(p_k^i) = \partial_b^m(Z_k \rfloor \pi^i) = \partial_b^m(\pi_k^i + y_k^e \pi_e^i)$$

$$= \delta_k^m \pi_b^i + Z_k \rfloor (\partial_b^m \pi^i)$$

yield

$$\bar{Q}_{ba}^{mi} = p_k^i L^{-2}\{L\partial_b^m \partial_a^k L - (\partial_b^m L)(\partial_a^k L) + (\partial_b^k L)(\partial_a^m L)\}$$

$$+ L^{-1}(\partial_a^k L)(Z_k \rfloor (\partial_b^m \pi^i)) .$$

Multiplication of this result by P_i^j and summing on i then leads to the equivalent matrix Q_{ba}^{mj}. It might seem, on first glance, that (5.233) would yield $y_i^a = Y_i^a(x^j, q^b, \pi_b^j)$ only. However, (5.236) shows that the term $p_k^i = \pi_k^i + y_k^b \pi_b^i$ that occurs in (5.233) induces a dependence on the variables π_j^i.

Transversality Mapping Methods for Solving Variational Problems

All of the preliminary results are now at hand for the construction of a general collection of methods for solving variational problems. We first note that $p_j^i T^j = \pi^i$ implies $\pi^1 \wedge \ldots \wedge \pi^n = p\, T^1 \wedge \ldots \wedge T^n$, and hence we obtain

$$(5.244) \qquad L\mu \overset{c}{=} L\, T^1 \wedge \ldots \wedge T^n = L\, p^{-1}\, \pi^1 \wedge \ldots \wedge \pi^n ,$$

where we have again used the convention that $\overset{c}{=}$ means equality modulo the closed contact ideal $\bar{C}_4$. With all of the freedom at hand to select the π's, the obvious and most direct simplification would obtain if we can achieve

$$(5.245) \qquad L\,\mu \overset{c}{=} d\eta , \quad \eta \in \Lambda^{n-1}(G_4) ,$$

that is, $L\,\mu$ becomes an element of the null class. This possibility might seem surprising on first reading for we would then have $\Phi^* E_a = 0$ for every $\Phi \in R_4(B_n)$. However, Lemma 5.20 shows that a transversality transformation of a regular variational problem induces the mapping $y_i^a = Y_i^a(x^j, q^b, \pi_k^j, \pi_b^j)$, and hence the Φ's that belong to $S_4(B_n)$ are those that satisfy the *first order constraints*

$$(5.246) \qquad \Phi^* C^a = \Phi^*(dq^a - Y_i^a \, dx^i) = 0 .$$

With this overview in mind, we turn to the specifics of implementing the requirement (5.245). In view of (5.244), the condition (5.245) takes the form

$$(5.247) \qquad L \, p^{-1} \, \pi^1 \wedge \ldots \wedge \pi^n \stackrel{c}{=} d\eta \, , \qquad \eta \in \Lambda^{n-1}(G_4) \, .$$

Since the left-hand side of (5.247) is a simple n-form, the right-hand side must likewise be a simple n-form. This can clearly be achieved if we have

$$(5.248) \qquad \eta = f(s^1,\ldots,s^n) \, ds^2 \wedge \ldots \wedge ds^n$$

where the n functions $\{s^i\}$ constitute a functionally independent system of elements of $\Lambda^0(G_4)$. Clearly, the restriction to $\Lambda^0(G_4)$ is both necessary and sufficient in order that $\eta \in \Lambda^{n-1}(G_4)$. We now combine (5.247) and (5.248) to obtain

$$(5.249) \quad L \, p^{-1} \, \pi^1 \wedge \ldots \wedge \pi^n \stackrel{c}{=} \frac{\partial f(s^1,\ldots,s^n)}{\partial s^1} \, ds^1 \wedge ds^2 \wedge \ldots \wedge ds^n \, .$$

The desired purpose is thus achieved by setting

$$(5.250) \qquad L \, p^{-1} = \partial f(s^1,\ldots,s^n)/\partial s^1 \, ,$$

$$(5.251) \qquad \pi^i \stackrel{c}{=} ds^i \, ,$$

provided these relations are consistent with our previous results. We note, however, that $p_k^i = Z_k \rfloor \pi^i$ and $Z_k \rfloor C^a = 0$ show that (5.251) imply

$$(5.252) \qquad p_k^i = Z_k(s^i) \, ,$$

and hence

$$(5.253)\qquad p = \det(Z_k(s^i)) = \Delta ,$$

where Δ is known to be an element of $N(G_4)$ from the results given in the previous section. When this result is substituted back into (5.250), we have the requirement

$$(5.254)\qquad L = \Delta\, \partial f(s^1,\dots,s^n)/\partial s^1 .$$

We must thus restrict the function f to be such that

$$(5.255)\qquad \partial f/\partial s^1 \neq 0 \quad \text{on} \quad L \neq 0 ,$$

and the functions $\{s^i\}$ to satisfy

$$(5.256)\qquad ds^1 \wedge \dots \wedge ds^n \neq 0 , \quad \Delta = \det(Z_k(s^i)) \neq 0 .$$

It is interesting to note in this context that (5.254) provides an elementary means of accommodating the possibility $L = 0$ on some proper subset of K_4 by an appropriate choice of the function $f(s^1,\dots,s^n)$. The added freedom afforded by inclusion of the function $f(s^1,\dots,s^n)$ is thus fundamental in problems for which L is not bounded from above and not bounded from below as a function on K_4 (classic nonextremal problems).

What now remains is an implementation of the conditions implied by the choice $\pi^i \overset{c}{=} ds^i$. Clearly, these conditions are satisfied if and only if

$$(5.257)\qquad \pi^i = ds^i + A^i_a\, C^a$$

for some choice of the functions $A^i_a \in \Lambda^0(K_4)$. We then simply read off the results

$$(5.258)\qquad \pi^i_j = \partial_j s^i - y^b_j\, A^i_b ,$$

$$(5.259)\qquad \pi^i_a = \partial_a s^i + A^i_a ,$$

in which case the fundamental relations (5.240) become

$$(5.260) \qquad \Delta\, \partial^k_a L = L\, P^k_i\, (\partial_a S^i + A^i_a) .$$

It is then a simple matter to see that (5.242) gives

$$(5.261) \qquad Q^{mj}_{ba} = \Delta\Big\{L\, \partial^m_b \partial^j_a L - (\partial^m_b L)(\partial^j_a L) + (\partial^j_b L)(\partial^m_a L) - L\,\Delta^{-1}\, P^j_i\, A^i_b\, \partial^m_a L\Big\}$$

provided we choose the A;s so that $\partial^i_a A^j_b = 0$; that is, $A^j_a \in \Lambda^0(G_4)$.

Weakly Complete Figures: A Hamilton-Jacobi Process

Now, simply observe that if $\det(Q^{mj}_{ba}) \neq 0$ on $L \neq 0$, then we can solve (5.260) for the y's so as to obtain

$$y^a_i = X^a_i(x^j, q^b, \pi^j_k, \pi^j_b) .$$

When (5.258) and (5.259) are used to eliminate the π's, we have

$$(5.262) \qquad y^a_i = X^a_i(x^j, q^b, \partial_k S^j - y^b_k A^j_b, \partial_b S^j + A^j_b) ,$$

which is a system of implicit equations for the determination of the y's unless $A^j_b = 0$, $j = 1,\ldots,n$, $b = 1,\ldots,M$. Since $A^1_a \in \Lambda^0(G_4)$, let us assume that they can be so constrained that (5.262) can be solved for the y's so as to obtain

$$(5.263) \qquad y^a_i = Y^a_i(x^j, q^b, \partial_k S^j, \partial_b S^j, A^j_b) .$$

This is clearly possible since (5.262) gives

$$\partial^e_m(y^a_i - X^a_i) = \delta^a_e\, \delta^m_i - (\partial X^a_i/\partial \pi^j_m)\, A^j_e$$

whose determinant can be made nonzero by appropriate restrictions on the ranges of the A's. However, $S^i \in \Lambda^0(G_4)$, $A^j_b \in \Lambda^0(G_4)$ show that (5.263) defines a map

(5.264) $\mathcal{S}:G_4 \longrightarrow K_4 | x^i = x^i,\ q^a = q^a,\ y_i^a = Y_i^a(x^j, q^b, \partial_k s^j, \partial_b s^j, A_b^j)$

for each choice of the functions $(s^i, A_b^i) \in \Lambda^0(G_4)$. If we use this map to eliminate the y's from $L - \Delta\, \partial f/\partial s^1$ (i.e., we compute $\mathcal{S}^*(L - \Delta\, \partial f/\partial s^1)$, we obtain the *Hamiltonian function*

(5.265) $$H(x^j, q^b, \partial_k s^j, \partial_b s^j, A_b^j) = \mathcal{S}^*(L - \Delta\, \partial f/\partial s^1)$$

on G_4 and the condition (5.254) gives the *generalized Hamilton-Jacobi equation*

(5.266) $$H(x^j, q^b, \partial_k s^j, \partial_b s^j, A_b^j) = 0 .$$

This, however, is only one equation in the $n(1+M)$ unknowns $s^i(x^j, q^b)$, $A_a^i(x^j, q^b)$. The requirements are completed by adjoining the conditions

(5.267) $$0 = \mathcal{S}^* C^a = dq^a - Y_i^a(x^j, q^b, \partial_k s^j, \partial_b s^j, A_b^j) dx^i ,$$

in which case the M 1-forms $\mathcal{S}^* C^a \in \Lambda^1(G_4)$ must form a completely integrable system of 1-forms on G_4.

In order that $\{\mathcal{S}^* C^a\}$ form a completely integrable system of 1-forms on G_4, the system of characteristic vector fields of $\{\mathcal{S}^* C^a\}$ must form a Lie subalgebra of $T(G_4)$ (see Appendix, Section 4-4). Since $v \rfloor \mathcal{S}^* C^a = 0$ has the n solutions

$$v_i = \partial_i + Y_i^a\, \partial_a , \qquad i = 1, \ldots, n$$

and

$$[v_i, v_j] = (Y_i^a\, \partial_a Y_j^b - Y_j^a\, \partial_a Y_i^b + \partial_i Y_j^b - \partial_j Y_i^b)\partial_b ,$$

the characteristic vector fields of $\{\mathcal{S}^* C^a\}$ form a Lie subalgebra of $T(G_4)$ if and only if the S's and the A's are chosen so as to secure satisfaction of the $Mn(n-1)/2$ conditions

$$Y_i^a\, \partial_a Y_j^b + \partial_i Y_j^b = Y_j^a\, \partial_a Y_i^b + \partial_j Y_i^b ,$$

in which case the Lie subalgebra is Abelian. It is of interest to note in this context that v_i is exactly Z_i under the restriction $y^a_i = Y^a_i$ of S and that (5.269) is equivalent to $Z_i(Y^b_j) = Z_j(Y^b_i)$ which are the integrability conditions of $\Phi^* y^a_i = \partial_i \phi^a(x^j) = Y^a_i(x^j, q^b, \partial_k S^j, \partial_b S^j, A^j_b)$.

When conditions (5.269) are met, the Frobenius theorem shows that there exist a nonsingular matrix $((K^a_b))$ of elements of $\Lambda^0(G_4)$ and M independent elements $\{\Psi^a\}$ of $\Lambda^0(G_4)$ such that

$$S^* C^a = K^a_b \, d\Psi^b \,, \qquad \det(\partial_a \Psi^b) \neq 0 \,,$$

in which case $\{S^* C^a\}$ form a torsion-free and curvature-free differential system

$$dS^* C^a = \Gamma^a_b \wedge S^* C^b$$

of degree one and class M with the connection 1-forms

$$\underset{\sim}{\Gamma} = (d\underset{\sim}{K}) \, \underset{\sim}{K}^{-1} \in \Lambda^{M,1}_M(G_4) \,.$$

We further note that (5.257) gives

$$S^* \pi^i = dS^i + A^i_a S^* C^a \,,$$

and hence

$$dS^* \pi^i = (dA^i_a + A^i_b \, \Gamma^b_a) \wedge S^* C^a \,.$$

The $n+M$ 1-forms $\{S^* \pi^i, S^* C^a\}$ thus form a torsion-free and curvature-free differential system of degree one and class $n+M$ on G_4 . However, we saw in the last Section that $\{T^i, C^a\}$ form a basis for $\Lambda^1(G_4)$, and hence $\{\pi^i, C^a\}$ also form a basis for $\Lambda^1(G_4)$ since the π's obtain from the T's by a nonsingular linear homogeneous transformation. Since S^* restricted to $\Lambda^1(G_4)$ is regular,

$$S^*(\pi^1 \wedge \ldots \wedge \pi^n \wedge C^1 \wedge \ldots \wedge C^M) = dS^1 \wedge \ldots \wedge dS^n \wedge d\Psi^1 \wedge \ldots \wedge d\Psi^M \neq 0$$

and the $n+M$ functions (s^i, Ψ^a) are independent on G_4. There thus exists a coordinate map

$$C\Phi : G_4 \longrightarrow G_4 \mid {}^{\backprime}x^i = s^i(x^j, q^b), \; {}^{\backprime}q^a = \Psi^a(x^j, q^b)$$

such that solutions to the variational problem are given by

$${}^{\backprime}q^a = \Psi^a(x^j, q^b) = k^a$$

with the k's constants; that is

$$(C\Phi)^*(C^{-1}\Phi)^*(K^a_b)(C\Phi)^{*\backprime}q^b = K^a_b d\Psi^b = S^*C^a .$$

We have thus established the following basic result.

THEOREM 5.28. *Let* L *be a given Lagrangian function on* K_4 *such that* $\det(Q^{mj}_{ba}) \neq 0$ *or* $L \neq 0$ *and let*

$$S : G_4 \longrightarrow K_4 \mid x^i = x^i, \; q^a = q^a, \; y^a_i = Y^a_i(x^j, q^b, \partial_k s^j, \partial_b s^j, A^j_b)$$

be a map that is implicitly defined in terms of the functions $s^i(x^j, q^b)$, $A^i_a(x^j, q^b)$, *where the functions* Y^a_i *are obtained in the manner defined above. If the functions* $s^i(x^j, q^b)$ *and* $A^i_a(x^j, q^b)$ *satisfy the generalized Hamilton-Jacobi equation*

$$(5.268) \qquad 0 = H(x^j, q^b, \partial_k s^j, \partial_b s^k, A^j_b) \stackrel{\text{def}}{=} S^*(L - \Delta\partial f/\partial s^1)$$

and the $Mn(n-1)/2$ *conditions*

$$(5.269) \qquad Y^a_i \, \partial_a Y^b_j + \partial_i Y^b_j = Y^a_j \, \partial_a Y^b_i + \partial_j Y^b_i ,$$

then there exist a nonsingular matrix $((K^a_b))$ *of elements of* $\Lambda^0(G_4)$ *and* M *elements* $\{\Psi^a\}$ *of* $\Lambda^0(G_4)$ *such that*

$$(5.270) \qquad S^*C^a = K^a_b \, d\Psi^b , \quad \det(\partial_a \Psi^b) \neq 0 ,$$

and the conditions $S^*C^a = 0$ *yield an implicitly defined solution of the stationarization problem* $0 = \delta\int_{B_n} L\,\mu$ *through the relations*

(5.271) $\Psi^a(x^j, q^b) = k^a \; , \quad a = 1,\ldots,M \; .$

In this event, $(S^*\pi^i, S^*c^a)$ *form a torsion-free and curvature free differential system*

$$S^*d\pi^i = (dA^i_a + A^i_b \Gamma^b_a) \wedge S^*c^a \; , \quad S^*dc^a = \Gamma^a_b \wedge S^*c^b$$

of degree 1 *and class* $n+M$ *on* G_4 *that assigns a flat, curvature-free connection on* G_4 *and there exists a regular coordinate map*

$$C\Phi : G_4 \longrightarrow G_4 \mid {}^{\backprime}x^i = s^i(x^j, q^b) \; , \quad {}^{\backprime}q^a = \Psi^a(x^j, q^b)$$

for which solutions of the stationarization problem are given by ${}^{\backprime}q^a = k^a$ *.*

This general case is referred to as the case of the *weakly complete figure* in the sense of Carathéodory, for the particularization that obtains by the restrictions $L \neq 0$, $A^j_b = 0$, $f(S^1,\ldots,S^n) = S^1$ gives the classic case of the *complete figure* in the sense of Carathéodory. The reader is referred to Rund's analysis [28] for the accompanying construction of a full canonical formalism in the classic case. Particular note should be taken of the fact that $A^i_a = 0$ yields $\pi^i = dS^i$, and hence $p^i_j T^j = \pi^i = dS^i$ shows that the transversality 1-forms constitute a completely integrable Pfaffian system in the classic case. Since the general case obtains under the conditions $\pi^i = dS^i + A^i_a c^a$, it is clear that generalization of the classic case can not proceed from the notion of complete integrability of the transversality 1-forms. On the other hand, the general case gives the completely integrable differential system $\{S^*\pi^i, S^*c^a\}$, so that the construct of a completely integrable system is not lost; rather, it is simply expanded to include the elements $\{S^*c^a\}$ that must themselves form a completely integrable system in both the classic and the general cases. In this regard, it is also useful to note that for $n = 1$ (classical mechanics), the additional integrability conditions (5.255) disappear and we may choose $A^1_b = 0$ with no loss of generality. This, however, is just the case of M 1-forms $\{S^*C^a\}$ on an $M+1$ dimensional space G_4 which are always completely integrable. The transition from $n = 1$ to multiple integral variational problems $(n>1)$, with the concomitant imposition of the $Mn(n-1)/2$ conditions (5.269) in addition to the Hamilton-Jacobi equation and the replacement of the single generating function S^1 by the collection (S^i, A^i_a), is definitely nontrivial. In fact, for large n and M, the satisfaction of the $1 + Mn(n-1)/2$ conditions (5.268), (5.269) by the $n(1+M)$ functions (S^i, A^i_a) of the $n+M$ arguments (x^i, q^a) is anything but obvious. Thus, there may exist no complete figures or weakly complete figures for a given problem with n and M large.

It is instructive to look at the problem $S^*C^a = 0$ from the standpoint of Theorem 4-3.6 of the Appendix. The closed ideal of $\Lambda(G_4)$ that contains $\{S^*C^a\}$ is given by

$$\bar{I} = I\{S^*C^a, dS^*C^a\} = S^*\bar{C}.$$

A vector field $v \in T(G_4)$ is thus a characteristic vector field of $\bar{I}$ if and only if $v \rfloor S^*C^a = 0$, $v \rfloor S^*dC^a = W^a_b\, S^*C^b$ are satisfied. Since $v \rfloor S^*C^a = v^a - Y^a_i v^i = 0$ implies $v^a = Y^a_i v^i$, we have

$$- v \rfloor S^*dC^a = v^i(\partial_b Y^a_i)S^*C^b + v^k\Big\{Y^b_k\, \partial_b Y^a_j + \partial_k Y^a_j - Y^b_j\, \partial_b Y^a_k - \partial_j Y^a_k\Big\}dx^j ,$$

and hence $v = v^i\,\partial_i + v^a\,\partial_a$ is a characteristic vector field of $\bar{I}$ if and only if

$$v^a = Y^a_i v^i ,$$

$$v^k\Big\{Y^b_k\, \partial_b Y^a_j + \partial_k Y^a_j - Y^b_j\, \partial_b Y^a_k - \partial_j Y^a_k\Big\} = 0$$

are satisfied throughout G_4 . We note that the quantities within the curly brackets are just the $Mn(n-1)/2$ independent terms that must vanish in order that $S^*C^a = 0$ be completely integrable (see (5.269)). Suppose that some of the expressions given by (5.269) can not be made to vanish for any choice of the functions (S^i, A^i_a) . If this were the case, the n functions $\{v^k\}$ could not be assigned arbitrarily and we would have only an $(n-r)$-dimensional characteristic subspace of $\bar{I}$. Since $\bar{I}$ is closed, the characteristic subspace of $\bar{I}$ forms a Lie subalgebra of $T(G_4)$ and there would exist $n + M - (n-r) = M + r$ functions $g^1(x^j, q^a), \ldots, g^{M+r}(x^j, q^a)$ that are annihilated by the characteristic subspace of $\bar{I}$. In this event, Theorem 4-3.6 shows that $\bar{I}$ and hence S^*C^α would vanish on the $n + M - (M+r) = n - r$ dimensional subsets $g^1 = k^1, \ldots, g^{M+r} = k^{M+r}$. Since the g's are independent, we can solve for $M + r$ of the variables (x^i, q^a) in terms of the remaining $n - r$ variables, but we are unable in general to solve for the q's in terms of the x's. In the ideal case, we would obtain the q's in terms of the x's only for those x's that belong to certain subsets of E_n of dimension $n - r$. What this says is as follows: if we do not demand the complete integrability of $S^*C^a = 0$ then the resulting Hamilton-Jacobi procedure will only yield solutions for which there are relations that must be satisfied amongst the independent variables $x^1, \ldots, x^n$. There is thus no alternative to demanding that the functions (S^i, A^i_a) satisfy the $Mn(n-1)/2$ conditions (5.269) if the Hamilton-Jacobi procedure is to lead to solutions in the usual sense.

There is a further generalization that is suggested by the results given above. If we replace

$$\pi^i = dS^i + A^i_a C^a$$

by

$$\pi^i = M^i_j\, dS^j + A^i_a\, C^a , \qquad \det(M^i_j) \neq 0 ,$$

then $\{S^*\pi^i, S^*C^a\}$ will still form a completely integrable system. The only changes occur in replacing (5.252) and (5.253) by

$$p^i_k = Z_k \lrcorner\, \pi^i = M^i_j\, Z_k(S^j)$$

and

$$p = \det(Z_k \lrcorner\, \pi^i) = M \det(Z_k(S^i)) = M\Delta$$

with $M = \det(M^i_j)$. This introduces an additional n^2 of elements of $\Lambda^o(G_4)$, $((M^i_j))$, so that we have the $1 + Mn(n-1)/2$ conditions (5.268), (5.269) that must be satisfied by choice of the $n + Mn + n^2$ functions (S^i, A^i_a, M^i_j) of the $n + M$ arguments (x^i, q^a). A wider range of problems may thus be handled by these techniques, although a significant amount of additional work is required in this area before a generally applicable procedure is obtained.

Examples

The simplest way of seeing just what is involved is to study explicit examples. One of particular interest is that for which $n = 2$, $N = 1$ and the Lagrangian functions is given by

$$L = y_1 y_2 + F(x^1, x^2, q) ,$$

in which case the Euler-Lagrange equation is the nonlinear wave equation

$$2\, \partial_1 \partial_2 \phi = \partial F(x^1, x^2, \phi)/\partial\phi .$$

In this instance, we have

$$\pi^i = \pi^i_j\, dx^j + \hat{\pi}^i\, dq , \quad p^i_j = \pi^i_j + y_j \hat{\pi}^i$$

and hence

$$((P^i_j)) = \begin{pmatrix} \pi^2_2 + y_2\hat{\pi}^2 & -\pi^1_2 - y_2\hat{\pi}^1 \\ -\pi^2_1 - y_1\hat{\pi}^2 & \pi^1_1 + y_1\hat{\pi}^1 \end{pmatrix} .$$

Thus, $p\,\partial^i L = LP^i_j\,\hat{\pi}^j$ yields

$$py_1 = (\pi^1_1\hat{\pi}^2 - \pi^2_1\hat{\pi}^1)L , \quad py_2 = (\pi^2_2\hat{\pi}^1 - \pi^1_2\hat{\pi}^2)L .$$

We demand satisfaction of the Hamilton-Jacobi equation

$$H = L - p = 0$$

with $f = S^1$ in all instances, from which it follows that

$$y_1 = \pi^1_1\hat{\pi}^2 - \pi^2_1\hat{\pi}^1 , \quad y_2 = \pi^2_2\hat{\pi}^1 - \pi^1_2\hat{\pi}^2$$

and hence

$$p = \det(p^i_j) = \pi^1_1\pi^2_2 - \pi^1_2\pi^2_1 + 2y_1y_2 .$$

The basic quantities, $H = L-p$ and $C = dq - y_i dx^i$ thus become

$$H = F(x^1, x^2, q) - (\pi^1_1\hat{\pi}^2 - \pi^2_1\hat{\pi}^1)(\pi^2_2\hat{\pi}^1 - \pi^1_2\hat{\pi}^2)$$
$$+ \pi^1_2\pi^2_1 - \pi^1_1\pi^2_2 ,$$

$$C = dq - (\pi^1_1\hat{\pi}^2 - \pi^2_1\hat{\pi}^1)dx^1 - (\pi^2_2\hat{\pi}^1 - \pi^1_2\hat{\pi}^2)dx^2 .$$

Underlying these relations is the requirement that the 1-forms π^i satisfy the conditions $\pi^i = dS^i \bmod \bar{C}$, and this is the case if and only if

$$\pi^i_j = \partial_j S^i - y_j A^i , \quad \hat{\pi}^i = \partial S^i + A^i$$

for some functions A^1, A^2 of the variables (x^1, x^2, q). The y_i's have already been determined as functions of the $\hat{\pi}^i$'s and the π^i_j's and hence the first system of relations just given constitutes a system of relations that must be solved for the π^i_j's if the A^i's are not identically zero. The solution is straightforward, so we simply state the results:

$$\Delta_1 = 1 + A^1\hat{\pi}^2 - A^2\hat{\pi}^1 , \quad \Delta_2 = 1 + A^2\hat{\pi}^1 - A^1\hat{\pi}^2$$

$$\pi^1_1 = \Delta_1^{-1}\{(1-A^2\hat{\pi}^1)\partial_1 S^1 + A^1\hat{\pi}^1 \partial_1 S^2\} ,$$

$$\pi^2_1 = \Delta_1^{-1}\{-A^2\hat{\pi}^2 \partial_1 S^1 + (1+A^1\hat{\pi}^2)\partial_1 S^2\} ,$$

$$\pi^1_2 = \Delta_2^{-1}\{(1+A^2\hat{\pi}^1)\partial_2 S^1 - A^1\hat{\pi}^1 \partial_a S^2\} ,$$

$$\pi^2_2 = \Delta_2^{-1}\{A^2\hat{\pi}^2 \partial_2 S^1 + (1-A^1\hat{\pi}^2)\partial_2 S^2\} ,$$

$$y_1 = \Delta_1^{-1}(\hat{\pi}^2 \partial_1 S^1 - \hat{\pi}^1 \partial_1 S^2) , \quad y_2 = \Delta_2^{-1}(\hat{\pi}^1 \partial_2 S^2 - \hat{\pi}^2 \partial_2 S^1) .$$

The problem thus reduces to finding the four functions A^1, A^2, S^1, S^2 of the arguments (x^1, x^2, q) so that $H = 0$ and $\Phi^*C = 0$.

Let us now restrict attention to the classic case of a complete figure; that is $A^1 = A^2 = 0$. In this instance, we have

$$S^*C = dq - (\partial S^2 \partial_1 S^1 - \partial S^1 \partial_1 S^2)dx^1 - (\partial S^1 \partial_2 S^2 - \partial S^2 \partial_2 S^1)dx^2 .$$

If we make the restriction

$$S^2 = k S^1 + f(x^1, x^2) ,$$

we obtain

$$S^*C = dq - (\partial S^1)\{\partial_1(-f)dx^1 + \partial_2 f\ dx^2\}$$

and the quantity within the braces is a total differential if $\partial_1\partial_2 f = 0$. Thus, if we set

$$f(x^1, x^2) = u(x^1) + v(x^2) ,$$

then

$$S^*C = dq - (\partial S^1)d\{v(x^2) - u(x^1)\}$$

and

$$H = F(x^1, x^2, q) + (\partial S^1)^2 \frac{du}{dx^1}\frac{dv}{dx^2} + \frac{du}{dx^1}\partial_2 S^1 - \frac{dv}{dx^2}\partial_1 S^1 .$$

Since we have already made a number of particularizations, the simplest thing to do at this point is just to set

$$S^1 = k\,\eta(q) + C_1 u + C_2 v .$$

This leads to a consistent system provided $F(x^1, x^2, q)$ has the form

$$F(x^1, x^2, q) = \rho(q)\,\frac{du(x^1)}{dx^1}\,\frac{dv(x^2)}{dx^2} ,$$

in which case $H = 0$ gives

$$k\left(\frac{d\eta}{dq}\right)^2 = C_1 - C_2 - \rho(q)$$

and $\Phi^*C = 0$ yields the quadrature relation

$$k\int_a^{\phi(x^1,x^2)} \frac{dq}{\sqrt{C_1 - C_2 - \rho(q)}} = v(x^2) - u(x^1) + C_3 .$$

It is clear that the solutions we have obtained are very special in view of the large number of restrictions that have been imposed in the interests of expediency. However, if we are given the nonlinear wave equation

$$\partial_1\partial_2\phi = \left(\int\rho(\phi)\,d\phi\right)U(x^1)\,V(x^2)$$

then we obtain the solutions

$$k\int_a^{\phi} \frac{dq}{\sqrt{C_1 - \rho(q)}} = \frac{1}{\lambda}\int_b^{x^1} V(t_2)dt_2 - \lambda\int_c^{x^2} U(t_1)dt_1 + C_3$$

for all values of the constants a, b, c, C_1, C_3 and for all $\lambda \neq 0$, $k \neq 0$.

It is of interest to note that similar results obtain without the restriction to the classical case for which $A^1 = A^2 = 0$. This is most easily seen by considering the diametrically opposite case in which we put $S^i = x^i$, for then $\pi^i_j = \partial_j S^i - y_j A^i = \delta^i_j - y_j A^i$, $\hat{\pi}^i = \partial S^i + A^i = A^i$ and hence $\Delta_1 = \Delta_2 = 1$, $y_1 = A^2$, $y_2 = A^1$,

$$H = F(x^1, x^2, q) + A^1 A^2 - 1,$$

$$S^*C = dq - A^2 dx^1 - A^1 dx^2.$$

A necessary and sufficient condition for $S^*C = 0$ to be completely integrable is the existence of a $\Psi(x^1, x^2, q)$ such that

$$A^2 \partial\Psi = -\partial_1\Psi, \quad A^1 \partial\Psi = -\partial_2\Psi,$$

in which case the corresponding solutions are given in implicit form by $\Psi(x^1, x^2, q) =$ constant. When these evaluations of the A's are put into $H = 0$, we obtain the requirement that Ψ satisfy the Hamilton-Jacobi equation

$$(\partial_1\Psi)(\partial_2\Psi) = (\partial\Psi)^2(1 - F(x^1, x^2, q)).$$

If we now put $\Psi = \alpha(q) f(x^1, x^2) + g(x^1, x^2)$ and $F = \rho(q) \dfrac{du(x^1)}{dx^1} \dfrac{dv(x^2)}{dx^2}$, then results similar to those given above are readily obtained.

A one-dimensional nonlinear diffusion process provides a useful example of a system that is not variational to begin with. Prolongation to the space K_4 through introduction of the "adjoint" variable $\bar{q}$ gives the Lagrangian function

$$L = \alpha(q)\, \bar{y}_2 - y_1 \bar{y}_1$$

and the contact forms

$$C = dq - y_1\, dx^1 - y_2\, dx^2 , \quad \bar{C} = d\bar{q} - \bar{y}_1\, dx^1 - \bar{y}_2\, dx^2 .$$

Since

$$((p^i_j)) = \begin{pmatrix} \pi^1_1 + y_1\pi^1 + \bar{y}_1\bar{\pi}^1 & \pi^1_2 + y_2\pi^1 + \bar{y}_2\bar{\pi}^1 \\ \pi^2_1 + y_1\pi^2 + \bar{y}_2\bar{\pi}^2 & \pi^2_2 + y_2\pi^2 + \bar{y}_2\bar{\pi}^2 \end{pmatrix}$$

gives

$$((P^i_j)) = \begin{pmatrix} \pi^2_2 + y_2\pi^2 + \bar{y}_2\bar{\pi}^2 & -(\pi^1_2 + y_2\pi^2 + \bar{y}_2\bar{\pi}^1) \\ -(\pi^2_1 + y_1\pi^2 + \bar{y}_1\bar{\pi}^2) & \pi^1_1 + y_1\pi^1 + \bar{y}_1\bar{\pi}^1 \end{pmatrix} ,$$

the equations

$$p\,\partial^k L = L P^k_i\, \pi^i , \quad p\,\bar{\partial}^k L = L P^k_i\, \bar{\pi}^i$$

together with the constraint $L = p = \det(p^i_j)$, lead to a system of equations that can be solved for the y's and the $\bar{y}$'s. This yields the relations

$$y_1 = \lambda^{-1}(\pi^2_1\,\bar{\pi}^1 - \pi^1_1\,\bar{\pi}^2 + \alpha)$$

$$y_2 = \lambda^{-1}(\pi^2_2\,\bar{\pi}^1 - \pi^1_2\,\bar{\pi}^2) + \lambda^{-2}(\pi^2_1\,\bar{\pi}^1 - \pi^1_1\,\bar{\pi}^2 + \alpha)$$

$$\bar{y}_1 = \lambda^{-1}(\pi^1_1\,\pi^2 - \pi^2_1\,\pi^1)$$

$$\bar{y}_2 = \lambda^{-1}(\pi^1_2\,\pi^2 - \pi^2_2\,\pi^1) + \lambda^{-2}(\pi^2_1\,\pi^i - \pi^1_1\,\pi^2)$$

with

$$\lambda = \pi^1\,\bar{\pi}^2 - \bar{\pi}^1\,\pi^2 \neq 0 .$$

When these relations are used to eliminate the y's and the $\bar{y}$'s from C, $\bar{C}$ and $H = S^*(L-\Delta) = S^*(L-p)$ for $f = s^1$, the Hamilton-Jacobi equation, $H = 0$, and the contact equations, $S^*C = 0$, $S^*\bar{C} = 0$ become

$$0 = \lambda^2 H = \alpha\lambda\,(\pi_2^1\,\pi^2 - \pi_2^2\,\pi^1) + \alpha\,(\pi_1^2\,\pi^1 - \pi_1^1\,\pi^2)$$
$$- (\pi_1^2\,\bar{\pi}^1 - \pi_1^1\,\bar{\pi}^2)(\pi_1^1\,\pi^2 - \pi_1^2\,\pi^1)\ ,$$

$$0 = \lambda^2 S^*C = \lambda^2\,dq + \lambda(\pi_1^2\,\bar{\pi}^1 - \pi_1^1\,\bar{\pi}^2 + \alpha)\,dx^1$$
$$+ \lambda(\pi_2^2\,\bar{\pi}^1 - \pi_2^1\,\bar{\pi}^2)\,dx^2 + (\pi_1^2\,\bar{\pi}^1 - \pi_1^1\,\bar{\pi}^2 + \alpha)\,dx^2\ ,$$

$$0 = \lambda^2 S^*\bar{C} = \lambda^2\,d\bar{q} + \lambda(\pi_1^1\,\pi^2 - \pi_1^2\,\pi^1)\,dx^1$$
$$+ \lambda(\pi_2^1\,\pi^2 - \pi_2^2\,\pi^1)\,dx^2 + (\pi_1^2\,\pi^1 - \pi_1^1\,\pi^2)\,dx^2\ .$$

In the case of a complete figure, these equations are supplimented by the relations

$$\pi^i = \partial s^i\ ,\quad \bar{\pi}^i = \bar{\partial} s^i\ ,\quad \pi_j^i = \partial_j s^i$$

with $s^i = s^i(x^1, x^2, q, \bar{q})$. This is the general problem.

There are several special cases for which the general problem assumed a more tractable form. For separable solutions of the form

$$s^i = B^i(x^1, x^2) + s^i(q, \bar{q})\ ,$$

the governing equations become

$$0 = \lambda^2 H = -\lambda\,\alpha\,\partial_2\partial\rho + \partial_1\bar{\partial}(\rho + x^1\bar{q}\alpha)\cdot\partial_1\partial\rho\ ,$$

$$0 = \lambda^2 S^*C = \lambda^2\,dq + \lambda\,\partial_1\bar{\partial}(\rho + x^1\bar{q}\alpha)\,dx^1$$
$$+ \lambda\,\partial_2\bar{\partial}(\rho + x^1\bar{q}\alpha)\,dx^2 + \partial_1\bar{\partial}(\rho + x^1\bar{q}\alpha)\,dx^2\ ,$$

$$0 = \lambda^2 S^*\bar{C} = \lambda^2 d\bar{q} - \lambda\, \partial_1 \partial\rho\, dx^1 - \lambda\, \partial_2 \partial\rho\, dx^2 + \partial_1 \partial\rho\, dx^2$$

with

$$\lambda = \partial s^1\, \bar{\partial} s^2 - \partial s^2\, \bar{\partial} s^1 , \qquad \rho = B^2\, s^1 - B^1\, s^2 .$$

The last two relations show that $\lambda = \lambda(q, \bar{q})$ and $\rho = \rho(q, \bar{q}, x^1, x^2)$ are independent. If we effect the further restriction

$$\rho = x^1\, f(q, \bar{q}) + x^2\, g(q, \bar{q}) ,$$

we obtain the three equations

$$0 = \lambda^2 H = -\lambda\, \alpha\, \partial g + (\bar{\partial} f + \alpha)\partial f ,$$

$$0 = \lambda^2 S^* C = \lambda^2\, dq + \lambda(\bar{\partial} f + \alpha) dx^1 + (\bar{\partial} f + \alpha + \lambda\, \bar{\partial} g) dx^2 ,$$

$$0 = \lambda^2 S^*\bar{C} = \lambda^2\, d\bar{q} - \lambda\, \partial f\, dx^1 + (\partial f - \lambda\, \partial g) dx^2$$

for the determination of the three functions λ, f, g, as functions of the two variables q and $\bar{q}$. If we write the second and third of these equations in the form

$$0 = S^* C = dq + a(q, \bar{q}) dx^1 + b(q, \bar{q}) dx^2 ,$$

$$0 = S^* \bar{C} = d\bar{q} + \bar{a}(q, \bar{q}) dx^1 + \bar{b}(q, \bar{q}) dx^2 ,$$

then

$$S^* dC = -(\partial a\, dx^1 + \partial b\, dx^2) \wedge S^* C - (\bar{\partial} a\, dx^1 + \bar{\partial} b\, dx^2) \wedge S^* \bar{C}$$
$$- (a\, \partial b - b\, \partial a + \bar{a}\, \bar{\partial} b - \bar{b}\, \bar{\partial} a)\, dx^1 \wedge dx^2 ,$$

$$S^* d\bar{C} = -(\partial \bar{a}\, dx^1 + \partial \bar{b}\, dx^2) \wedge S^* C - (\bar{\partial} \bar{a}\, dx^1 + \bar{\partial} \bar{b}\, dx^2) \wedge S^* \bar{C}$$
$$- (a\, \partial \bar{b} - b\, \partial \bar{a} + \bar{a}\, \bar{\partial} \bar{b} - \bar{b}\, \bar{\partial} \bar{a})\, dx^1 \wedge dx^2 .$$

The Frobenius theorem then shows that complete integrability obtains if and only if

$$0 = a\,\partial b - b\,\partial a + \bar{a}\,\bar{\partial} b - \bar{b}\,\bar{\partial} a\ ,$$

$$0 = a\,\partial \bar{b} - b\,\partial \bar{a} + \bar{a}\,\bar{\partial} \bar{b} - \bar{b}\,\bar{\partial} \bar{a}\ ,$$

in which case

$$S^*C = K^1_1\,d\Psi + K^1_2\,d\bar{\Psi}\ ,\quad S^*\bar{C} = K^2_1\,d\Psi + K^2_2\,d\bar{\Psi}$$

where Ψ and $\bar{\Psi}$ are functions of $(x^1, x^2, q, \bar{q})$ and $\det(K^i_j) \neq 0$. Thus, satisfaction of the three conditions

$$0 = a\,\partial b - b\,\partial a + \bar{a}\,\bar{\partial} b - \bar{b}\,\bar{\partial} a\ ,$$

$$0 = a\,\partial \bar{b} - b\,\partial \bar{a} + \bar{a}\,\bar{\partial} \bar{b} - \bar{b}\,\bar{\partial} \bar{a}\ ,$$

$$0 = -\lambda\,\alpha\,\partial g + (\bar{\partial} f + \alpha)\partial f\ ,$$

with

$$a = \lambda^{-1}(\bar{\partial} f + \alpha)\ ,\qquad b = \lambda^{-2}(\bar{\partial} f + \alpha + \lambda\,\bar{\partial} g)\ ,$$

$$\bar{a} = -\lambda^{-1}\,\partial f\ ,\qquad \bar{b} = \lambda^{-2}(\partial f - \lambda\,\partial g)\ ,$$

will lead to solutions defined implicitly by

$$\Psi(x^1, x^2, q, \bar{q}) = k\ ,\quad \bar{\Psi}(x^1, x^2, q, \bar{q}) = \bar{k}$$

with Ψ and $\bar{\Psi}$ linear in the arguments x^1, x^2 . It is of interest to note that the classic method of the complete figure of Carathéodory leads to methods for solving the 1-dimensional nonlinear diffusion equation even under the stringent restrictions that we have imposed (separable solutions together with $\rho = x^1 f(q, \bar{q}) + x^2 g(q, \bar{q})$. In this regard, it should be noted that the restriction $\rho = x^1 f(q, \bar{q}) + x^2 g(q, \bar{q})$ has the effect of rendering all coefficients in C and $\bar{C}$ independent of x^1 and x^2. This is thus a very severe constraint and leads to somewhat uninteresting solutions. The same procedure can be followed without assuming $\rho = x^1 f + x^2 g$, in which case the solutions are more interesting but require significantly more complex calculations. Some degree of simplification obtains by making use of the linear homotopy operator to construct the matrix $((K^i_j))$ and the functions Ψ and $\bar{\Psi}$ by the methods delineated in Chapter V of the Appendix. In any event, it should be abundantly clear that the major source of computational difficulty stems from the requirements that the resulting 1-forms S^*C and $S^*\bar{C}$ be completely integrable rather than from the Hamilton-Jacobi equation, $H = 0$.

5.11 PROBLEMS WITH DIFFERENTIAL CONSTRAINTS OF ARBITRARY DEGREE

Isovector methods and the process of prolongation provide immediate access to problems involving the stationarization of an action functional subject to exterior differential constraints of arbitrary degree.

For simplicity, let $`E_n$ be the n-dimensional space with the global coordinate cover $(x^1, x^2, \ldots, x^n)$, let E_n be the n-dimensional domain space with the global coordinate cover $(\tau^1, \ldots, \tau^n)$, let G be the $(n+N)$-dimensional graph space with global coordinate cover (x^k, q^α), and let K be the $(n+N+nN)$-dimensional contact space with the global coordinate cover $(x^i, q^\alpha, y^\alpha_i)$. The action functional we wish to consider is

$$(5.272) \qquad A[\Phi] = \int_{B_n} \Phi^* L(x^j, q^\beta, y^\beta_i)\mu ,$$

where L is a given element of $\Lambda^0(K)$, $\mu = dx^1 \wedge dx^2 \wedge \ldots \wedge dx^n$, Φ belongs to the class of regular maps

$$(5.273) \qquad R(B_n) = \{\Phi : B_n \subset E_n \longrightarrow K \mid \Phi^*\mu \neq 0,\ \Phi^* C^\alpha = 0\} ,$$

and the C^α's are the contact forms

$$(5.274) \qquad C^\alpha = dq^\alpha - y_i^\alpha \, dx^i .$$

The closed ideal generated by the C^α's is again denoted by $\bar{C}$.

Let ω be a given element of $\Lambda^k(K)$ with $k \leq n$.

Definition 5.21. A regular map Φ is said to satisfy the *constraint* generated by ω if and only if

$$(5.275) \qquad \Phi^*\omega = 0 .$$

For example, if

$$\omega = w_i \, dx^i + w_\alpha \, dq^\alpha + w_\alpha^j \, dy_j^\alpha ,$$

with $(w_i, w_\alpha, w_\alpha^j)$ elements of $\Lambda^0(K)$, and Φ is realized by

$$x^i = \tau^i , \quad q^\alpha = \phi^\alpha(\tau^j) , \quad y_i^\alpha = \partial_i \phi^\alpha(\tau^j) ,$$

then Φ satisfies the constraint generated by ω if and only if

$$0 = \{\Phi^* w_i + \Phi^* w_\alpha \, \partial_i \phi^\alpha + \Phi^* w_\alpha^j \, \partial_i \partial_j \phi^\alpha\} d\tau^i .$$

Since the functions $(w_i, w_\alpha, w_\alpha^j)$ may depend on the variables (y_j^β) in a quite general way, the only restriction on the constraints of degree one is that they be linear in the second derivatives $\partial_i \partial_j \phi^\alpha(\tau^k)$. A little reflection shows that constraints of degree k are restricted to be of degree k or less in the second derivatives; for example $\omega = dy_1^1 \wedge dy_2^2 \wedge dy_1^3 \wedge dx^6$. We further note that $\Phi^*\omega \in \Lambda^k(E_n)$ for $\omega \in \Lambda^k(K)$, and hence the constraint $\Phi^*\omega = 0$ represents $\binom{n}{k}$ constraint equations if none of the coefficients of ω vanish identically. Thus, for $n = 4$ and

$$\omega = (q^1 \, dx^1 - dq^1 + q^1 \, dq^2) \wedge dx^2 \wedge dx^3 ,$$

the constraint $\Phi^*\omega = 0$ yields the conditions

$$0 = \phi^1 - \partial_1\phi^1 + \phi^1\,\partial_1\phi^2 ,$$

$$0 = \phi^1 - \partial_4\phi^1 + \phi^1\,\partial_4\phi^2 .$$

Let $TH(K)$ be the collection of all horizontal isovector fields of $\bar{C}$,

(5.276) $$TH(K) = \{v \in TC(K) \mid v(x^i) = 0 , \quad i = 1,\ldots,n\} .$$

The first variation of the action functional $A[\Phi]$ that is generated by any $u \in TH(K)$ is given by

(5.277) $$\delta_u A[\Phi] = \int_{B_n} \Phi^* \pounds_u (L\mu) = \int_{B_n} \Phi^* \pounds_u (L\mu + J)$$

with

(5.278) $$J = C^\alpha \wedge \partial^i_\alpha L\, \mu_i .$$

Since a constrained problem requires that $\Phi \in R(K)$ satisfy $\Phi^*\omega = 0$, the available horizontal isovector vector fields that generate the variations have to satisfy the "constraint on variations" condition

(5.279) $$\Phi^*(u \rfloor \omega) = 0 .$$

This is quite distinct from the "variation of constraint" that would be given by

$$\Phi^* \pounds_u \omega = \Phi^*(u \rfloor d\omega) + \Phi^* d(u \rfloor \omega) .$$

When the constraint on variations condition is substituted into this expression, we obtain

$$\Phi^* \pounds_u \omega = \Phi^*(u \rfloor d\omega) .$$

Thus, the varied constraint is satisfied only when $\Phi^*(u \rfloor d\omega)$ also vanishes, which is not assumed here. What this says is that there need be no "neighboring" Φ's that satisfy the constraint if Φ satisfies the constraint. Another way of viewing this situation is to consider the 1-parameter family of horizontal isovector fields $\bar{u} = s\,u$. In this case,

$$\Phi^*\pounds_{\bar{u}}\omega \;=\; s\,\Phi^*\pounds_u\omega + ds\wedge\Phi^*(u\rfloor\omega) \;=\; ds\wedge\Phi^*(u\rfloor\omega) + 0(s) ,$$

and hence $\Phi^*\pounds_{\bar{u}}\omega = 0(s)$. The 1-parameter family of regular elements $\Phi_u(s) = \exp(u,s)\circ\Phi$ thus gives $\Phi_u(s)^*\omega = \Phi^*\omega + 0(s)$ if $\Phi^*(u\rfloor\omega) = 0$; that is $\Phi_u(s)^*\omega$ vanishes only for $s = 0$. The classic assumption that the varied maps, $\Phi_u(s)$, also satisfy the constraint is relaxed by the constraint on variations requirement (5.279).

Definition 5.22. A map $\Phi\in R(B_n)$ is said to satisfy a *constrained variational* problem with constraint form $\omega\in\Lambda^k(K)$, $k\leq n$, and action functional $A[\Phi] = \int_{B_n}\Phi^*L\mu$, $L\in\Lambda^0(K)$, if and only if Φ is such that

$$(5.280)\qquad \Phi^*\omega \;=\; 0$$

and

$$(5.281)\qquad \int_{B_n}\Phi^*\pounds_u(L\mu) \;=\; 0$$

for all $u\in TH(K)$ that satisfy the constraint on variations condition

$$(5.282)\qquad \Phi^*(u\rfloor\omega) \;=\; 0\ .$$

A problem with more than one given constraint form may easily be handled once the constrained variational problem has been fully treated. Such problems are thus left to the reader.

The essential aspect of this, as with any other constrained variational problem, is to remove the restriction $\Phi^*(u\rfloor\omega) = 0$ on the horizontal isovector fields by a Lagrange multiplier technique. The usual technique allows this to occur only when Φ can be imbedded in a "neighborhood" of maps all of which satisfy the constraint. The following theorem shows that a modified Lagrange multiplier method is also available for the general problem posed here, but the proof relies heavily on the ability to generate an appropriate prolongation of the given problem. It is also of interest to note that the prolongation method does not require more than one family of variations at a time, in contrast with the classic method of proof of the Lagrange multiplier method.

THEOREM 5.29. *Let* $N_1 = \binom{n}{n-k}$ *and let* $\{b_\Gamma\}$ *, with* $\Gamma = 1,\ldots,N_1$ *, be a basis for* $\Lambda^{n-k}(E_n)$ *such that*

$$(5.283)\qquad u \rfloor b_\Gamma = 0 \quad u \in TH(K) \ , \quad db_\Gamma = 0 \ , \quad \Gamma = 1,\ldots,N_1 \ .$$

A map $\Phi \in R(B_n)$ *satisfies a constrained variational problem with constraint form* $\omega \in \Lambda^k(K)$ *,* $k \leq n$ *, and action functional* $\int_{B_n} \Phi^*(L\mu)$ *if and only if there exist* N_1 *elements* $\{\lambda^\Gamma\}$ *of* $\Lambda^0(E_n)$ *such that*

$$(5.284)\qquad \Phi^*\omega = 0 \ ,$$

$$(5.285)\qquad \int_{B_n} \Phi^*\{\pounds_u(L\mu) + (u \rfloor \omega) \wedge d(\lambda^\Gamma b_\Gamma)\} = 0$$

for all $u \in TH(K)$.

Proof. We first imbed the space K in a space $\hat{K} = K \times E_{N_1}$ with global coordinate cover $(x^i, q^\alpha, y^\alpha_i, p^\Gamma)$, so that $\hat{K}$ is a trivial fiber space with canonical projection

$$(5.286)\qquad \pi : \hat{K} \longrightarrow K \mid (x^i, q^\alpha, y^\alpha_i, p^\Gamma) \longrightarrow (x^i, q^\alpha, y^\alpha_i) \ .$$

All differential forms on K may then be prolonged to differential forms on $\hat{K}$ by the action of π^* . Any regular map of

E_n to K is extended to a regular map $\hat{\Phi}$ of E_n to $\hat{K}$ by requiring $\hat{\Phi}^*(x^i, q^\alpha, y^\alpha_i, p^\Gamma) = (\Phi^* x^i, \Phi^* q^\alpha, \Phi^* y^\alpha_i, \lambda^\Gamma(\tau^k))$. The collection of all such regular maps is denoted by $R_5(B_n)$. Likewise, any $u \in TH(K)$ is prolonged to a horizontal isovector field $\hat{u}$ of $\pi^* \bar{C}$ by setting

$$\hat{u} = u + f^\Gamma(x^j)\, \partial/\partial p^\Gamma . \tag{5.287}$$

The collection of all such isovector fields is denoted by $TH(\hat{K})$. We now consider all $\hat{\Phi}^* \in R_5(B_n)$ such that $\Phi^* x^i = \tau^i$ and

$$0 = \int_{B_n} \hat{\Phi}^* \{ \pounds_u L\mu + \hat{u} \rfloor (\omega \wedge dp^\Gamma \wedge b_\Gamma) \} \tag{5.288}$$

for all $\hat{u} \in TH(K)$. Since $\hat{u} \rfloor b_\Gamma = 0$,

$$\begin{aligned} \hat{u} \rfloor (\omega \wedge dp^\Gamma \wedge b_\Gamma) &= (\hat{u} \rfloor \omega) \wedge dp^\Gamma \wedge b_\Gamma + (-1)^k \omega \wedge b_\Gamma \hat{u}(p^\Gamma) \\ &= (u \rfloor \omega) \wedge d(p^\Gamma b_\Gamma) + (-1)^k \omega \wedge b_\Gamma f^\Gamma(x^j) , \end{aligned}$$

and $\hat{\Phi}^* \pounds_{\hat{u}}(L\mu) = \Phi^* \pounds_u(L\mu)$,

$$\begin{aligned} \hat{\Phi}^* \{ \hat{u} \rfloor (\omega \wedge dp^\Gamma \wedge b_\Gamma) \} &= \Phi^*(u \rfloor \omega) \wedge d(\lambda^\Gamma(\tau^k) b_\Gamma) \\ &\quad + (-1)^k \Phi^* \omega \wedge b_\Gamma f^\Gamma(\tau^k) . \end{aligned}$$

Thus, the term $\hat{\Phi}^* \{ \hat{u} \rfloor (\omega \wedge dp^\Gamma \wedge b_\Gamma) \}$ vanishes whenever $\Phi^* \omega = 0$, $\Phi^*(u \rfloor \omega) = 0$ and (5.288) reduces to the stationarity requirement $0 = \int_{B_n} \hat{\Phi}^* \pounds_{\hat{u}} L\mu$. The full requirement (5.288) now becomes

$$\begin{aligned} 0 &= \int_{B_n} \Phi^* \{ \pounds_u(L\mu) + (u \rfloor \omega) \wedge d(\lambda^\Gamma(\tau^k) b_\Gamma) \} \\ &\quad + (-1)^k \int_{B_n} \Phi^* \omega \wedge b_\Gamma f^\Gamma(\tau^k) \end{aligned} \tag{5.289}$$

for all $u \in TH(K)$ and for all $f^\Gamma(\tau^k) \in \Lambda^0(E_n)$. In order for this to be the case, (5.289) must hold for $u = 0$ and all

$f^\Gamma(\tau^k) \in \Lambda^0(E_n)$; that is

$$0 = \int_{B_n} \Phi^*\omega \wedge b_\Gamma f^\Gamma(\tau^k) \qquad \forall\, f^\Gamma(\tau^k) \in \Lambda^0(E_n) .$$

The fundamental lemma of the calculus of variations thus gives the requirements

$$(5.290) \qquad \Phi^*\omega \wedge b_\Gamma = 0 , \quad \Gamma = 1,\ldots,N_1 .$$

Thus, since $\Phi^*\omega \in \Lambda^k(E_n)$ and $\{b_\Gamma\}$ is a basis for $\Lambda^{n-k}(E_n)$, (5.290) is satisfied if and only if

$$(5.291) \qquad \Phi^*\omega = 0 .$$

Under this condition, (5.289) becomes

$$(5.292) \qquad 0 = \int_{B_n} \Phi^*\{\pounds_u(L\mu) + (u \rfloor \omega) \wedge d(\lambda^\Gamma b_\Gamma)\}$$

for all $u \in TH(K)$. It thus follows that $0 = \int_{B_n} \Phi^*\pounds_u(L\mu)$ for all $u \in TH(K)$ such that $\Phi^*(u \rfloor \omega) = 0$. Thus, if prolongation variables $\{\lambda^\Gamma(\tau^k)\}$ exist so that $\Phi \in R(E_n)$ satisfies (5.284) and (5.285) for all $u \in TH(K)$, then Φ satisfies the constrained variational problem.

A direct contact obtains with the classic results when the constraint form is "integrable". In the case of only one constraint form $\omega \in \Lambda^k(K)$, this means $\omega = A\, d\beta$ for $A \in \Lambda^0(K)$ and $\beta \in \Lambda^{k-1}(K)$. For several constraint forms of the same degree, we must have $\omega^a = A^a_b\, d\beta^b$, a situation that is considered in the last chapter of the Appendix. It is clear that any regular Φ satisfies $\Phi^*\omega = 0$ with $\omega = A\, d\beta$ if and only if Φ satisfies $\Phi^* d\beta = d\phi^*\beta = d\Phi^*(\beta - d\gamma) = 0$. Thus, setting $\bar{\beta} = \beta - d\gamma$, $\Phi^*\omega = 0$ holds if and only if $\Phi^*\bar{\beta} = 0$.

<u>THEOREM 5.30</u>. *A map* $\Phi \in R(B_n)$ *satisfies a constrained variational problem with constraint form* $\omega = A\, d\beta$, $\beta \in \Lambda^{k-1}(K)$

and action functional $\int_{B_n} \Phi^*(L\mu)$ *if and only if there exist* N_1 *elements* $\{\lambda^\Gamma\}$ *of* $\Lambda^0(E_n)$ *such that*

$$(5.293) \qquad \Phi^*\beta = 0$$

and

$$(5.294) \qquad \int_{B_n} \Phi^*\{\pounds_u(L\mu) + u \rfloor d(\beta \wedge d\lambda^\Gamma \wedge b_\Gamma)\} = 0$$

holds for all $u \in TH(K)$. *This latter set of conditions is equivalent to the requirement that*

$$(5.295) \quad \delta_u \int_{B_n} \Phi^*\{L\mu + \beta \wedge d(\lambda^\Gamma b_\Gamma)\}$$

$$= \int_{B_n} \Phi^* \pounds_u \{L\mu + \beta \wedge d(\lambda^\Gamma b_\Gamma)\} = \int_{\partial B_n} \Phi^*\{u \rfloor [\beta \wedge d(\lambda^\Gamma b_\Gamma)]\}$$

for all $u \in TH(K)$. *The conditions* (5.293) *and* (5.295) *are satisfied if and only if there exists a map* $\hat{\Phi} \in R_5(B_n)$ *of the prolonged problem such that*

$$(5.296) \qquad \delta_{\hat{u}} \int_{B_n} \hat{\Phi}^*\{L\mu + \beta \wedge dp^\Gamma \wedge b_\Gamma\} = \int_{\partial B_n} \Phi^*\{u \rfloor [\beta \wedge dp^\Gamma \wedge b_\Gamma]\}$$

for all $\hat{u} \in TH(K)$. *Thus, if* $\partial^i_\alpha \rfloor \beta = 0$ *and if the variations* $\Phi^* u(q^\alpha)$ *vanish on* ∂B_n , *then* (5.296) *becomes the homogeneous prolonged variational problem*

$$(5.297) \qquad \delta_{\hat{u}} \int_{B_n} \hat{\Phi}^*\{L\mu + \beta \wedge dp^\Gamma \wedge b_\Gamma\} = 0$$

for all $u \in TH(\hat{K})$ *such that* $\Phi^* u(q^\alpha)\Big|_{\partial B_n} = 0$.

Proof. Since $\omega = A\, d\beta$, $\Phi^*\omega = \Phi^* A\, d\Phi^*\beta$ and hence we may replace the constraint ω with the constraint $d\Phi^*\beta = d\Phi^*(\beta + d\gamma) = 0$. We thus have $\Phi^*\beta + d\Phi^*\gamma = d\alpha$, and the choice $d\Phi^*\gamma = d\alpha$ gives us $\Phi^*\beta = 0$. This establishes (5.293). The remainder of the results follow directly from Theorem 5.29 and the definition of the Lie derivative.

For $k = 1$, the general form of the constraint is

(5.298) $$\omega = w_i \, dx^i + w_\alpha \, dq^\alpha + w_\alpha^i \, dy_i^\alpha$$

and $\{b_\Gamma\} = \{\mu_i\}$. Thus,

(5.299) $$(u \rfloor \omega) \wedge d(\lambda^\Gamma b_\Gamma) = (w_\alpha \, u(q^\alpha) + w_\alpha^j \, u(y_j^\alpha))(\partial_i \lambda^i)\mu \, .$$

If we introduce the function λ by $\lambda = \partial_i \lambda^i$ and note that

$$u \rfloor d(C^\alpha \wedge \lambda w_\alpha^i \mu_i) = - u(y_j^\alpha)\lambda w_\alpha^j \mu - u(q^\alpha) d(\lambda w_\alpha^i \mu_i) \mod \bar{C} \, ,$$

then (5.299) becomes

(5.300) $$\begin{aligned}(u \rfloor \omega) \wedge d(\lambda^\Gamma b_\Gamma) &= u(q^\alpha)\{\lambda w_\alpha \mu - d(\lambda w_\alpha^i \mu_i)\} \\ &\quad + u \rfloor d(C^\alpha \wedge \lambda w_\alpha^i \mu_i) \mod \bar{C} \\ &= u(q^\alpha)\{\lambda w_\alpha \mu - d(\lambda w_\alpha^i \mu_i)\} + \pounds_u (C^\alpha \wedge \lambda w_\alpha^i \mu_i) \\ &\quad - d\{u(q^\alpha)\lambda w_\alpha^i \mu_i\} \mod \bar{C} \, .\end{aligned}$$

Thus, (5.285) yields the conditions

(5.301) $$\begin{aligned}&\int_{B_n} \Phi^* \left\{ \frac{\partial L}{\partial q^\alpha} \mu + \lambda w_\alpha \mu - d\left(\frac{\partial L}{\partial y_i^\alpha} + \lambda w_\alpha^i \right) \wedge \mu_i \right\} \Phi^* u(q^\alpha) \\ &= \int_{\partial B_n} \Phi^* \left\{ u(q^\alpha)(\lambda w_\alpha^i + \frac{\partial L}{\partial y_i^\alpha})\mu_i \right\}\end{aligned}$$

for all $u \in TH(K)$. If we set $\Phi^*\{w_i, w_\alpha, w_\alpha^i\} = \{\bar{w}_i, \bar{w}_\alpha, w_\alpha^i\}$, the constraint, $\Phi^*\omega = 0$ yields the n conditions

(5.302) $$\bar{w}_i + \bar{w}_\alpha \, \partial_i \phi^\alpha + \bar{w}_\alpha^j \, \partial_i \partial_j \phi^\alpha = 0 \, ,$$

while (5.301) with the standard boundary conditions

$\Phi^* u(q^\alpha)\Big|_{\partial B_n} = 0$ yields the N Euler equations

$$(5.303) \qquad \Phi^*(\frac{\partial L}{\partial q^\alpha}) - D_i\Phi^*(\frac{\partial L}{\partial y_i^\alpha}) = D_i(\lambda \bar{w}_\alpha^i) - \lambda\bar{w}_\alpha .$$

This example shows that we thus have the n partial differential constraints (5.302), but the N Euler equations (5.303) only contain the single Lagrange multiplier function $\lambda(\tau^k)$. This is the rule rather than the exception in all cases but those for which the constraint form is completely integrable. The disparity between the number of available multipliers and the number of partial differential constraint equations may explain the lack of results along classic lines that have been reported in the literature for such constraint problems.

If ω is a constraint form of degree k , then $\Phi^*\omega = 0$ can depend on the second derivatives $(\partial_i\partial_j\phi^\alpha(x^m))$ up to and including terms of algebraic degree k in these variables. This, in turn, will lead to stationarity conditions that involve up to third order x-wise derivatives of the $\phi(x^m)$'s and to the inability to kill the resulting surface integral by the requirement that $\Phi^*(u \rfloor dq^\alpha) = \delta\phi^\alpha$ vanish on the boundary. There are two ways out of this difficulty when $k > 1$. The first is to consider only boundary conditions that lead to an identical vanishing of the integrand of the resulting surface integral. Such modified boundary conditions could have been anticipated from the fact that there need exist no neighborhood of a given Φ all of whose elements satisfies the same Dirichlet data when Φ satisfies $\Phi^*\omega = 0$; that is, the imposition of the constraint $\Phi^*\omega = 0$ necessitates either Neumann data or data of a mixed type. The second alternative is to restrict consideration to constraint forms that contain only first order terms in the basis elements $\{dy_i^\alpha\}$.

We now restrict attention to constraint forms of degree k that are linear in the basis elements $\{dy_i^\alpha\}$. Let $\eta \in \Lambda^k(K)$

and $\rho^i_\alpha \in \Lambda^{k-1}(K)$ satisfy

$$(5.304) \qquad \partial^j_\beta \rfloor \eta = 0 , \quad \partial^j_\beta \rfloor \rho^i_\alpha = 0$$

and set

$$(5.305) \qquad \omega = \eta + dy^\alpha_i \wedge \rho^i_\alpha .$$

Introducing the notation

$$(5.306) \qquad \Theta = d(\lambda^\Gamma b_\Gamma) , \quad u \rfloor \eta = u^\beta \eta_\beta , \quad u \rfloor \rho^i_\alpha = u^\beta \rho^i_{\alpha\beta} ,$$

we have

$$(5.307) \qquad (u \rfloor \omega) \wedge d(\lambda^\Gamma b_\Gamma) = u^\beta(\eta_\beta - dy^\alpha_i \wedge \rho^i_{\alpha\beta}) \wedge \Theta + u^\alpha_i \rho^i_\alpha \wedge \Theta .$$

In view of the second of (5.304) and $dq^\alpha = y^\alpha_i \, dx^i \bmod \bar{C}$, there exists 0-forms $\{P^i_\alpha\}$ such that

$$(5.308) \qquad \rho^i_\alpha \wedge \Theta = P^i_\alpha \, \mu \bmod \bar{C} .$$

If we now define the n-form j by

$$(5.309) \qquad j = C^\alpha \wedge P^i_\alpha \, \mu_i$$

then $j = 0 \bmod \bar{C}$, $\pounds_u j = 0 \bmod \bar{C}$, and (5.307) becomes

$$(u \rfloor \omega) \wedge d(\lambda^\Gamma b_\Gamma) = \{u^\beta(\eta_\beta - dy^\alpha_i \wedge \rho^i_{\alpha\beta}) + u^\alpha_i \rho^i_\alpha\} \wedge \Theta$$

$$+ u \rfloor dj + d(u \rfloor j) \bmod \bar{C}$$

$$= u^\beta\{(\eta_\beta - dy^\alpha_i \wedge \rho^i_{\alpha\beta}) \wedge \Theta - dP^i_\beta \wedge \mu_i\} + d(u \rfloor j) \bmod \bar{C} .$$

The condition (5.285) is thus equivalent to

$$(5.310) \qquad 0 = \int_{B_n} \Phi^* u^\beta \Phi^* \{\partial_\beta L\mu - d(\partial^i_\beta L) \wedge \mu_i + (\eta_\beta - dy^\alpha_i \wedge \rho^i_{\alpha\beta}) \wedge \Theta - dP^i_\beta \wedge \mu_i\} + \int_{\partial B_n} \Phi^* \{u \rfloor (J+j)\} .$$

Accordingly, since

$$u \rfloor (J+j) = (u \rfloor C^\alpha)(\partial^i_\alpha L + P^i_\alpha)\, \mu_i \mod \bar{C} ,$$

the requirement that $\delta\phi^\alpha = \Phi^*(u \rfloor C^\alpha) = \Phi^*(u \rfloor dq^\alpha)$ vanish on the boundary leads to the stationarization conditions

$$(5.311) \qquad \Phi^* \{\partial_\beta L\, \mu - d(\partial^i_\beta L) \wedge \mu_i + (\eta_\beta - dy^\alpha_i \wedge \rho^i_{\alpha\beta}) \wedge \Theta - dP^i_\beta \wedge \mu_i\} = 0 .$$

It should be noted that the imposition of a constraint not only changes the Euler equations, it also changes the meaning of transverse vector fields: $v \rfloor (J+j) = 0$. If a constraint of degree $k > 1$ is not linear in the basis elements $\{dy^\alpha_i\}$, the surface integral that would appear in (5.310) would also contain terms involving $\Phi^*(u \rfloor dy^\alpha_i) = \partial_i \delta\phi^\alpha$ and the vanishing of $\delta\phi^\alpha$ on the boundary would not kill the surface integral. The Euler equations would obtain in this case only if conditions at the boundary were such that the coefficients of $\Phi^*(u \rfloor dy^\alpha_i)$ were to vanish on the boundary. Additional boundary conditions would thus obtain that compensate for the fact that the Euler equations would now be of third order from the terms involving the constraints. There is thus much additional work to be done in the general case of constraints of degree $k > 1$.

CHAPTER 6

THE METHOD OF EXPLICIT EXTENSION

We have seen in previous Sections that exp(v,s) generates other than contact transformations of K only when the ideal $I\{C^\alpha, dC^\alpha\}$ is enlarged in some fashion. One method is that provided by prolongation by means of pseudopotentials while another obtains by prolongation to a variational system. There is yet another method that is often quite simple and direct. We refer to this method as the method of *explicit extension.*

6.1 THE EXTENDED CONTACT IDEAL AND ITS ISOVECTORS

We start, as with the previous discussions, with a Euclidean space K of n+N+nN dimensions and coordinate functions $(x^i, q^\alpha, y_i^\alpha)$. This space supports the contact forms

(6.1) $$C^\alpha = dq^\alpha - y_i^\alpha \, dx^i \ , \quad \alpha = 1,\dots,N$$

and a given system of balance forms E_α of degree n . These give rise to the closed contact ideal

(6.2) $$\bar{C} = \bar{I}\{C^\alpha\}$$

and the closed balance ideal

(6.3) $$\bar{E} = \bar{I}\{C^{\alpha}, E_{\alpha}\} .$$

Let $\hat{K} = K \times E_{M(1+n)}$ be a Euclidean space that is an extension of K and let $\hat{K}$ have coordinate functions $(x^k, q^{\alpha}, y^{\alpha}_i, p^a, h^a_i)$ with $a = 1,\ldots,M$. The extended space $\hat{K}$ is then a fiber space with K as the base space and $E_{M(1+n)}$ as fibers. We thus have the canonical projection

(6.4) $$\hat{\pi}:\hat{K} \longrightarrow K|(x^i, q^{\alpha}, y^{\alpha}_i, p^a, h^a_i) \longrightarrow (x^i, q^{\alpha}, y^{\alpha}_i) .$$

The balance system (E_{α}, C^{α}) is then prolonged to a balance system of $\hat{K}$ by

(6.5) $$\hat{\pi}^* C^{\alpha} = C^{*\alpha} , \quad \hat{\pi}^* E_{\alpha} = E^*_{\alpha}$$

so that $\underset{\sim}{E}^*$ and $\underset{\sim}{C}^*$ become elements of $\Lambda(\hat{K})$ that are independent of the fiber coordinates (p^a, h^a_i). Thus, we may again denote $\underset{\sim}{E}^*$ and $\underset{\sim}{C}^*$ by simply $\underset{\sim}{E}$ and $\underset{\sim}{C}$.

The second step in the process is to extend the contact ideal $\bar{C}$. This is accomplished by introducing M new contact forms

(6.6) $$C^a = dp^a - h^a_i \, dx^i$$

so that $\bar{C}$ can be replaced by the extended contact ideal

(6.7) $$\hat{C} = \bar{I}\{C^{\alpha}, C^a\}$$

that contains $\bar{C}$ as a closed subideal.

<u>Definition 6.1.</u> A map $\Phi:B_n \longrightarrow \hat{K}$ is said to be *regular* if and only if

(6.8) $$\Phi^*\mu \neq 0 , \quad \Phi^*\hat{C} = 0 .$$

The collection of all regular maps is denoted by $\hat{R}(B_n)$. Thus, any $\Phi \in \hat{R}(B_n)$ has the property that

$$\Phi^*(y_i^\alpha)d\Phi^*(x^i) = d\Phi^*(q^\alpha) , \quad \Phi^*(h_i^a)d\Phi^*(x^i) = d\Phi^*(p^a) .$$

Regular maps thus correlate the y's with the x-wise first partial derivatives of the $\Phi^*(q)$'s and the h's with the x-wise first partial derivatives of the $\Phi^*(p)$'s .

Definition 6.2. A vector field $v \in T(\hat{K})$ is an *isovector* of the extended contact ideal $\hat{C}$ if and only if

(6.9) $$\pounds_v \hat{C} \subset \hat{C} .$$

The collection of all isovectors of $\hat{C}$ is denoted by $TC(\hat{K})$ so that

(6.10) $$TC(\hat{K}) = \{v \in T(\hat{K}) \,|\, \pounds_v \hat{C} \subset \hat{C}\} .$$

The following theorems are obtained by exactly the same arguments as given in Section 5.2.

THEOREM 6.1. $TC(\hat{K})$ *is a Lie subalgebra of* $T(\hat{K})$.

Let $\hat{\partial}_a \equiv \partial/\partial p^a$ and $\hat{\partial}_a^i \equiv \partial/\partial h_i^a$.

THEOREM 6.2. *A vector field*

(6.11) $$v = v^i\partial_i + v^\alpha\partial_\alpha + v_i^\alpha\partial_\alpha^i + \hat{v}^a\hat{\partial}_a + \hat{v}_i^a\hat{\partial}_a^i$$

belongs to $TC(\hat{K})$ *if and only if*

(6.12) $$v_i^a = \hat{Z}_i(v \rfloor C^\alpha) , \quad \hat{v}_i^a = \hat{Z}_i(v \rfloor C^a) ,$$

where $\hat{Z}_i$ *is the linear differential operator*

(6.13) $$\hat{Z}_i = \partial_i + y_i^\alpha\partial_\alpha + h_i^a\hat{\partial}_a ,$$

and

(6.14) $$v^i = f^i(x^j, q^\beta, p^b) , \quad v^\alpha = f^\alpha(x^j, q^\beta, p^b)$$

$$\hat{v}^a = \hat{f}^a(x^j, q^\beta, p^b) .$$

If these conditions are satisfied, then

(6.15) $$\pounds_v C^\alpha = \partial_\beta(v \rfloor C^\alpha) C^\beta + \hat{\partial}_b(v \rfloor C^\alpha) C^b ,$$

(6.16) $$\pounds_v C^a = \partial_\beta(v \rfloor C^a) C^\beta + \hat{\partial}_b(v \rfloor C^a) C^b ,$$

and the extended contact structure

$$\hat{K} = \{K, \hat{R}(B_n), \hat{C}, TC(\hat{K})\}$$

is always a first group extension structure of the extended graph space $\hat{G}$ *with coordinate functions* (x^i, q^α, p^a) .

THEOREM 6.3. *The orbital equations of any* $v \in TC(\hat{K})$ *are the characteristic equations of the system of first order partial differential equations*

(6.17) $$\overset{*}{F}{}^\alpha = -\overset{*}{v} \rfloor \overset{*}{C}{}^\alpha = 0 , \quad \overset{*}{\hat{F}}{}^a = -\overset{*}{v} \rfloor \overset{*}{C}{}^a = 0 .$$

The solutions $\exp(v,s) : \hat{K} \longrightarrow \hat{K}|$

(6.18) $$\overset{*}{x}{}^i = \exp(s\pounds_v)x^i , \quad \overset{*}{q}{}^\alpha = \exp(s\pounds_v)q^\alpha , \quad \overset{*}{y}{}^\alpha_i = \exp(s\pounds_v)y^\alpha_i ,$$

$$\overset{*}{p}{}^a = \exp(s\pounds_v)p^a , \quad \overset{*}{h}{}^a_i = \exp(s\pounds_v)h^a_i$$

of the orbital equations, subject to the initial data $\overset{*}{x}{}^i(0) = x^i$, $\overset{*}{q}{}^\alpha(0) = q^\alpha$, $\overset{*}{y}{}^\alpha_i(0) = y^\alpha_i$, $\overset{*}{p}{}^a(0) = p^a$, $\overset{*}{h}{}^a_i(0) = h^a_i$, *are contact transformations of* $\hat{K}$ *and define a 1-parameter family of maps*

(6.19) $$\Phi_v(s) = \exp(v,s) \circ \Phi$$

that maps $\hat{R}(B_n)$ *into itself.*

The following two theorems constitute the basis upon which the method of specific extension will be shown to rest.

THEOREM 6.4. *The restriction*

(6.20) $$v^i = f^i(x^j, q^\beta) , \quad v^\alpha = f^\alpha(x^j, q^\beta) , \quad \hat{v}^a = 0$$

imbeds the projectable elements of TC(K) *as a proper linear subspace of* $TC(\hat{K})$ *and all members of this subspace have the form*

$$(6.21)\qquad v = f^i\partial_i + f^\alpha\partial_\alpha + Z_i(f^\alpha - y^\alpha_j f^j)\partial^i_\alpha - h^a_j Z_i(f^j)\hat{\partial}^i_a .$$

Proof. The restriction (6.20) yields

$$(6.22)\qquad v\rfloor C^\alpha = f^\alpha - y^\alpha_j f^j , \qquad v\rfloor C^a = - h^a_j f^j ,$$

and hence $\hat{Z}_i(v\rfloor C^\alpha) \equiv Z_i(v\rfloor C^\alpha)$, $\hat{Z}_i(v\rfloor C^a) \equiv Z_i(v\rfloor C^a)$. Application of Theorem 6.2 shows that such vector fields belong to $TC(\hat{K})$ if and only if

$$(6.23)\qquad v^\alpha_i = Z_i(v\rfloor C^\alpha) = Z_i(f^\alpha - y^\alpha_j f^j) ,$$

$$(6.24)\qquad \hat{v}^a_i = Z_i(v\rfloor C^a) = Z_i(-h^a_j f^j) = - h^a_j Z_i f^j .$$

However, Theorem 2.2 shows that (6.20) and (6.23) are exactly the conditions under which $v \in TC(K)$ if $N > 1$, in which case TC(K) is projectable, and v belongs to the subset of TC(K) that are projectable if $N = 1$. A substitution of (6.23) and (6.24) into (6.11) leads directly to (6.21) and hence the projectable elements of TC(K) are imbedded in $TC(\hat{K})$. The collection of all vector fields of the form (6.21) is obviously closed under addition and multiplication by real numbers, so it forms a linear space. That this linear space is a proper linear subspace of $TC(\hat{K})$ is obvious.

THEOREM 6.5. *If* $v \in TC(\hat{K})$ *is such that*

$$\hat{\partial}_b(v\rfloor C^\alpha) \not\equiv 0 ,$$

then v *restricted to* K *is not an infinitesimal generating vector of a contact transformation of* K *and* exp(v,s) *restricted to* K *is not a contact transformation of* K .

Proof. The restriction of any $v \in TC(\hat{K})$ to K gives

$$\hat{\pi}_* v = v^i \partial_i + f^\alpha \partial_\alpha + \hat{Z}_i (f^\alpha - y^\alpha_j f^j) \partial^i_\alpha$$

and (6.15) yields

$$\pounds_{\hat{\pi}_* v} C^\alpha = \partial_\beta (v \rfloor C^\alpha) C^\beta + \hat{\partial}_b (v \rfloor C^\alpha) C^b = \hat{\partial}_b (v \rfloor C^\alpha) C^b \bmod \bar{C} ,$$

since $\pounds_{\hat{\pi}_* v} C^\alpha \equiv \pounds_{\underline{v}} C^\alpha$. Thus, if $\hat{\partial}_b (v \rfloor C^\alpha) \not\equiv 0$ then $\pounds_v C^\alpha$ is not contained in $\bar{C}$ and hence v is not an element of $TC(K)$. Now, $\hat{\pi}_* v$ is an infinitesimal generating vector of a contact transformation of K if and only if $\hat{\pi}_* v \in TC(K)$ (see [13]), so $\hat{\pi}_* v$ is not an infinitesimal generator of a contact transformation of K . It thus follows that the restriction of $\exp(v,s)$ to K is not a contact transformation of K .

We note that $\exp(v,s)$ is a contact transformation of $\hat{K}$ and v is an infinitesimal generating vector of a contact transformation of $\hat{K}$. The purpose of the additional variables $q\{p^a, h^a_i\}$ can now be clarified, for their inclusion allows for the possibility of $\hat{\partial}_b (v \rfloor C^\alpha) \not\equiv 0$. Thus, if $v \in TC(\hat{K})$ is such that $\hat{\partial}_b (v \rfloor C^\alpha) \not\equiv 0$, we obtain a contact transformation of $\hat{K}$ but *not* a contact transformation of just K alone. This then provides the possibility of obtaining a much wider class of transformations of K than just contact transformations, and yet preserve the possibility of obtaining trivial closed form evaluations of $\Phi_v(s)^* E_\alpha$. This turns out to be particularly useful in view of the fact that we are free to choose a collection of balance forms for the fiber variables (p^a, h^a_i) in an efficacious manner.

6.2 EXTENSION OF THE BALANCE IDEAL

The third step in the method of explicit extension is the selection of an explicit set of equations of balance for the fiber variables $\{p^a\}$. There are two obvious criteria for this selection. The first is that the equations of balance for the variables $\{p^a\}$ should not involve the original variables $\{q^\alpha\}$ or their derivatives. The second is that the equations of balance for the variables $\{p^a\}$ should be readily solvable. If these criteria are met, then the equations of balance for the $\{p^a\}$ can be solved directly as an explicit problem that is independent of the problem of solving the original balance system. Now, a linear system is readily solvable, so we shall take the n-forms E_a , $a=1,\ldots,M$, that specify the balance forms to be

$$(6.25)\qquad E_a = d(\alpha^i_{ab}p^b + \beta^{ij}_{ab}h^b_j)\wedge\mu_i + (\gamma_{ab}p^b + \delta^j_{ab}h^b_j)\mu ,$$

where α^i_{ab} , β^{ij}_{ab} , γ_{ab} , δ^j_{ab} are constants. A section of $\hat{K}$ by any $\Phi\in\hat{R}(B_n)$ thus gives

$$(6.26)\qquad \Phi^* E_a = \{D_i(\alpha^i_{ab}\phi^b + \beta^{ij}_{ab}\partial_j\phi^b) + \gamma_{ab}\phi^b + \delta^j_{ab}\partial_j\phi^b\}\mu .$$

Now that we have the M balance forms E_a , we may form the extended balance ideal

$$(6.27)\qquad \hat{E} = \bar{I}\{E_\alpha, E_a, C^\alpha, C^a\} .$$

<u>Definition 6.3</u>. A map $\Phi : B_n \longrightarrow \hat{K}$ is said to *solve* the extended balance system if and only if

$$(6.28)\qquad \Phi^*\mu \neq 0 , \quad \Phi^*\hat{E} = 0 .$$

The collection of all such maps is denoted by $\hat{S}(\hat{E},B_n)$.

We note that any $\Phi\in\hat{S}(\hat{E},B_n)$ is such that $\Phi^*\mu\neq 0$, $\Phi^*\hat{C}=0$, $\Phi^*E_\alpha = 0$, $\Phi^*E_a = 0$, and hence Definition 6.1 shows that

$$(6.29)\qquad \hat{S}(\hat{E},B_n) = \{\Phi\in\hat{R}(B_n) \mid \Phi^*E_\alpha = 0 , \Phi^*E_a = 0\} .$$

Any $\Phi\in\hat{S}(\hat{E},B_n)$ thus yields a solution of the original balance system. However, $\{E_\alpha\}$ and $\{E_a\}$ are independent and uncoupled, so that solving $\{E_a\}$ places no constraints on the solution of

$\{E_\alpha\}$. Thus $\hat{S}(\hat{E},B_n)$ *contains all solutions of the original balance system* $\{E_\alpha\}$.

A direct application of Definition 3.1 with $\{E_\Gamma\} = \{E_\alpha, E_a\}$, $\Gamma = 1,\ldots,N+M$, shows that two extended balance systems $\{E_{1\alpha}, E_{1a}\}$ and $\{E_{2\alpha}, E_{2a}\}$ are *balance equivalent* if and only if

$$(6.30) \qquad E_{2\alpha} = A^\beta_\alpha E_{1\beta} + B^b_\alpha E_{1b} + P_\alpha , \quad E_{2a} = \hat{A}^\alpha_a E_{1\alpha} + \hat{B}^b_a E_{1b} + P_a ,$$

where $\{P_\alpha\}$ and $\{P_a\}$ belong to $\hat{C}$ and

$$\det \begin{pmatrix} \underset{\sim}{A} & \underset{\sim}{B} \\ \underset{\sim}{A} & \underset{\sim}{B} \end{pmatrix} \neq 0 .$$

As before, let $F(\hat{E})$ denote the fiber of extended balance systems that are balance equivalent to $\{E_\alpha, E_a\}$. Lemma 3.2 shows that

$$(6.31) \qquad \hat{S}(F(\hat{E}),B_n) = \hat{S}(\hat{E},B_n) ;$$

that is, all extended balance systems that belong to the same balance equivalent fiber have exactly the same solutions. It also follows from Lemma 3.5 that the ideal $\hat{E}$ that is formed from $\{E_\alpha, E_a\}$ is the same as the ideal formed from any $\{E_{1\alpha}, E_{1a}\} \in F(\hat{E})$. All of the results concerning balance fiber invariants can thus be carried over directly from Chapter 3.

6.3 ISOVECTORS OF THE BALANCE IDEAL AND GENERATION OF SOLUTIONS

The essential tool of the method of explicit extension is the structure associated with the isovectors of the closed balance ideal $\hat{E}$.

Definition 6.4. A vector field $v \in T(\hat{K})$ is an *isovector* of $\hat{E}$ if and only if

$$(6.32) \qquad \pounds_v \hat{E} \subset \hat{E} .$$

The collection of all isovectors of $\hat{E}$ is denoted by $T(\hat{E},\hat{K})$.

A direct combination of Theorems 3.4 through 3.6 gives the following result.

THEOREM 6.6. *$T(\hat{E},\hat{K})$ is a Lie subalgebra of $TC(\hat{K})$ that is composed of those elements of $TC(K)$ that satisfy the further restrictions*

(6.33) $$\pounds_v E_\alpha \subset \hat{E} , \quad \pounds_v E_a \subset \hat{E} .$$

This Lie subalgebra is an invariant of fibration by balance equivalence, so that

(6.34) $$T(F(\hat{E}),\hat{K}) \equiv T(\hat{E},\hat{K}) .$$

Again, as with the case of $T(E,\bar{E},K_4)$, the collection $T(\hat{E},\hat{K})$ is significantly richer than the corresponding collection $T(\underset{\sim}{E},K)$ because of the presence of the new variables $\{p^a\}$ and the significant increase in the size of $TC(\hat{K})$ compared to $TC(K)$. This is particularly true in view of the following result that follows directly from Theorem 3.8.

THEOREM 6.7. *If $v \in T(\hat{E},\hat{K})$, then the map $\exp(v,s):\hat{K} \longrightarrow \hat{K}$ given by (6.18), that solves the orbital equations*

(6.35)
$$\frac{d\overset{*}{x}^i}{ds} = \overset{*}{v}^i , \quad \frac{d\overset{*}{q}^\alpha}{ds} = \overset{*}{v}^\alpha , \quad \frac{d\overset{*}{y}^\alpha_i}{ds} = \overset{*}{v}^\alpha_i ,$$
$$\frac{d\overset{*}{p}^a}{ds} = \overset{*}{v}^a , \quad \frac{d\overset{*}{h}^a_i}{ds} = \overset{*}{\hat{v}}^a_i ,$$

subject to the initial data $\overset{*}{x}^i(0) = x^i$, $\overset{*}{q}^\alpha(0) = q^\alpha$, $\overset{*}{y}^\alpha_i(0) = y^\alpha_i$, $\overset{*}{p}^a(0) = p^a$, $\overset{*}{h}^a_i(0) = h^a_i$, *imbeds any* $\Phi \in \hat{S}(\hat{E},B_n)$ *in a 1-parameter group*

(6.36) $$\Phi_v(s) = \exp(v,s)\circ\Phi$$

of elements of $\hat{S}(\hat{E},B_n)$.

This result is of sufficient importance that a detailed examination of its implications is called for. We first note that (6.36) implies

$$(6.37)\qquad \begin{aligned} \Phi_v(s)^* E_\alpha &= \Phi^*(\exp(s\pounds_v)E_\alpha) \\ \Phi_v(s)^* E_a &= \Phi^*(\exp(s\pounds_v)E_a) \ . \end{aligned}$$

Thus, since any $v \in T(\hat{E},\hat{K})$ is such that $\pounds_v(\hat{E}) \subset \hat{E}$, the operator series $\exp(s\pounds_v)E_\alpha$ and $\exp(s\pounds_v)E_a$ do not have to be summed since we know that $\exp(s\pounds_v)E_\alpha \subset \hat{E}$ and $\exp(s\pounds_v)E_a \subset \hat{E}$. Accordingly, since any $\Phi \in \hat{S}(\hat{E},\hat{K})$ annihilates $\hat{E}$, we obtain $\Phi^*(\exp(s\pounds_v)E_\alpha) = 0$ and $\Phi^*(\exp(s\pounds_v)E_a) = 0$. Now, $\exp(v,s)$ has the form

$$(6.38)\qquad \begin{aligned} \overset{*}{x}{}^i &= J^i(x^j,\ q^\beta,\ p^b;s) \ , \\ \overset{*}{q}{}^\alpha &= J^\alpha(x^j,\ q^\beta,\ p^b;s) \ , \\ \overset{*}{p}{}^a &= \hat{J}^a(x^j,\ q^\beta,\ p^b;s) \ , \end{aligned}$$

because

$$(6.39)\qquad \begin{aligned} \frac{d\overset{*}{x}{}^i}{ds} &= \overset{*}{v}{}^i = f^i(\overset{*}{x}{}^j,\ \overset{*}{q}{}^\beta,\ \overset{*}{p}{}^b) \ , \\ \frac{d\overset{*}{q}{}^\alpha}{ds} &= \overset{*}{v}{}^\alpha = f^\alpha(\overset{*}{x}{}^j,\ \overset{*}{q}{}^\beta,\ \overset{*}{p}{}^b) \ , \\ \frac{d\overset{*}{p}{}^a}{ds} &= \overset{*}{\hat{v}}{}^a = \hat{f}^a(\overset{*}{x}{}^j,\ \overset{*}{q}{}^\beta,\ \overset{*}{p}{}^b) \end{aligned}$$

and $v \in TC(\hat{K})$, together with

$$(6.40)\qquad \begin{aligned} \overset{*}{y}{}^\alpha_i &= J^\alpha_i(x^j,\ q^\beta,\ p^b,\ y^\beta_j,\ h^b_j;s) \ , \\ \overset{*}{h}{}^a_i &= \hat{J}^a_i(x^j,\ q^\beta,\ p^\beta,\ y^\beta_j,\ h^b_j;s) \ , \end{aligned}$$

where

$$(6.41)\qquad \frac{d\overset{*}{y}{}^{\alpha}_{i}}{ds} = \overset{*}{v}{}^{\alpha}_{i} = \overset{*}{Z}_{i}(f^{\alpha} - \overset{*}{y}{}^{\alpha}_{j}f^{j}) , \qquad \frac{d\overset{*}{h}{}^{a}_{i}}{ds} = \overset{*}{\hat{v}}{}^{a}_{i} = \overset{*}{Z}_{i}(\hat{f}^{a} - \overset{*}{h}{}^{a}_{j}f^{j}) .$$

Theorem 6.6 tells us that if $q^{\alpha} = \phi^{\alpha}(x^{j})$, $p^{a} = \hat{\phi}^{a}(x^{j})$ is a solution of $\hat{E}$, then (6.38) gives

$$(6.42)\qquad \overset{*}{x}{}^{i} = J^{i}(x^{j}, \phi^{\beta}(x^{j}), \hat{\phi}^{b}(x^{j});s) ,$$

$$(6.43)\qquad \overset{*}{\phi}{}^{\alpha} = J^{\alpha}(x^{j}, \phi^{\beta}(x^{j}), \hat{\phi}^{b}(x^{j});s) , \qquad \overset{*}{\hat{\phi}}{}^{a} = \hat{J}^{a}(x^{j}, \phi^{\beta}(x^{j}), \hat{\phi}^{b}(x^{j});s) ,$$

and these equations define new solutions $\overset{*}{\Phi}{}^{\alpha}(x^{j};s)$, $\overset{*}{\hat{\phi}}{}^{a}(x^{j};s)$ once the variables (x^{j}) are eliminated between (6.42) and (6.43). Further, these new solutions have the property that they also satisfy (6.40), that is

$$(6.44)\qquad \frac{\partial\overset{*}{\Phi}{}^{\alpha}}{\partial\overset{*}{x}{}^{i}} = J^{\alpha}_{i}(x^{j}, \phi^{\beta}(x^{j}), \hat{\phi}^{b}(x^{j}), \frac{\partial\phi^{\beta}}{\partial x^{j}}, \frac{\partial\hat{\phi}^{b}}{\partial x^{j}};s) , \qquad \frac{\partial\overset{*}{\hat{\Phi}}{}^{a}}{\partial\overset{*}{x}{}^{i}} = \hat{J}^{a}_{i}(x^{j}, \phi^{\beta}(x^{j}), \hat{\phi}^{b}(x^{j}), \frac{\partial\phi^{\beta}}{\partial x^{j}}, \frac{\partial\hat{\phi}^{b}}{\partial x^{j}};s) .$$

When these results are viewed in the original space K of the original balance system E , we obtain

$$(6.45)\qquad \overset{*}{x}{}^{i} = J^{i}(x^{j}, \phi^{\beta}(x^{j}), \hat{\phi}^{b}(x^{j});s) , \qquad \overset{*}{\phi}{}^{\alpha} = J^{\alpha}(x^{j}, \phi^{\beta}(x^{j}), \hat{\phi}^{b}(x^{j});s) ,$$

$$(6.46)\qquad \frac{\partial\overset{*}{\Phi}{}^{\alpha}}{\partial\overset{*}{x}{}^{i}} = \frac{\partial\overset{*}{\phi}{}^{\alpha}}{\partial x^{j}}\frac{\partial x^{j}}{\partial\overset{*}{x}{}^{i}} = J^{\alpha}_{i}(x^{j}, \phi^{\beta}, \hat{\phi}^{b}, \frac{\partial\phi^{\beta}}{\partial x^{j}}, \frac{\partial\hat{\phi}^{b}}{\partial x^{j}};s)$$

in which the M functions $\{\phi^{b}(x^{j})\}$ are arbitrary to within satisfaction of $\Phi^{*}E_{a} = 0$, $\Phi^{*}C^{a} = 0$, $a = 1,\ldots,M$. These latter conditions are easily satisfied, however, for we have chosen the system (E_a, C^a) to be a linear system. Thus, if we fix $\{\phi^\beta(x^j)\}$ so that it yields a solution of (E_α, C^α), then we obtain a solution of (E_α, C^α) from (6.45) for each value of s and for each solution $\{\phi^b(x^j)\}$ of (E_a, C^a). This added generality is ephemeral, however, if J^i and J^α, and hence J^α_i happen to be independent of the variables $\phi^b(x^j)$. However, $v \rfloor C^\alpha = f^\alpha - y^\alpha_i f^i$, in which case (6.39) shows that $\hat{\partial}_a(v \rfloor C^\alpha) \not\equiv 0$ is sufficient in order that at least one of the quantities $\{d\overset{*}{x}{}^i/ds, d\overset{*}{q}{}^\alpha/ds\}$ shall depend on the variables $\{p^a, h^a_i\}$. In this event, at least one of the quantities $\{J^i, J^\alpha\}$ will depend on the $\{p^a\}$. Theorem 6.5 has shown, however, that $\hat{\partial}_a(v \rfloor C^\alpha) \not\equiv 0$ is sufficient in order that exp(v,s) restricted to K is not a contact transformation, in which case, (6.46) constitutes a system of Bäcklund transformations that generate solutions of (E_α, C^α) from any one solution of (E_α, C^α) and any solution of the linear system (E_a, C^a).

THEOREM 6.8. *If* $\hat{E}$ *admits an isovector* v *such that* $\hat{\partial}_a(v \rfloor C^\alpha) \not\equiv 0$, *then* $\Phi_v(s)$ *generates solutions of* (E_α, C^α) *from any one solution of* (E_α, C^α) *and any solution of the linear system* (E_a, C^a). *These solutions are nontrivial whenever the solution of* (E_a, C^a) *is nontrivial, even if the one solution of* (E_α, C^α) *from which the solutions arise is the trivial solution, and the process of generation of solutions arises through the Bäcklund transformations* (6.46) *whose integral manifolds are given by* (6.45).

It is clear from this result, that the method depends upon finding a $v \in T(\hat{E}, \hat{K})$ such that $\hat{\partial}_b(v \rfloor C^\alpha) \not\equiv 0$. Now, we only need one such vector field, and hence we have the possibility of obtaining one directly by making a choice of the constants $(\alpha^i_{ab}, \beta^{ij}_{ab}, \gamma_{ab}, \delta^j_{ab})$ that occur in (6.25) so that we may solve $\pounds_v E_\alpha \subset \hat{E}$, $\pounds_v E_a \subset \hat{E}$. Again, although this method will not work for every balance system (E_α, C^α) , the freedom in the choice of $(\alpha^i_{ab}, \beta^{ij}_{ab}, \gamma_{ab}, \delta^j_{ab})$ provides a sufficient number of adjustments that a large class of balance systems can be solved by this method.

CHAPTER 7

COMPUTER ASSISTED OPERATOR CALCULATIONS

The outstanding drawback with the methods presented in the previous chapters is the shear weight of the calculations that is involved. For example, if we have four independent variables, x, y, z, t and only one dependent variable $\phi(x,y,z,t)$, then kinematic space, with the coordinates x, y, z, t, q, y_1, y_2, y_3, y_4, is nine dimensional and the balance form that defines the partial differential equation that ϕ is to satisfy has degree four. A calculation of the isovectors of the balance system will thus involve $\binom{9}{4} = 126$ independent equations, each of which must be identically satisfied in the y's. This staggering amount of calculation is what has relegated isovector methods to problems with few unknowns and even fewer independent variables, the usual case being two independent variables.

There has been a recent development in computer software that offers a way around the drudgery of the extensive calculations that are demanded by isovector methods. "Algebraic programming systems" are now available that work directly with symbolic expressions rather than with values. One such system, "REDUCE 2", that was written by Dr. Anthony C. Hearn of the University of Utah, is commercially available. Since the REDUCE 2 system implements all of the standard operations with functions and operators (addition, subtraction, multiplication, division, substitution, partial differentiation, and integration), it is only necessary to add programs to the system that implement the operations peculiar to the exterior calculus. The following sections give program listings in the REDUCE 2 system that accomplish this task.

The program given in Section 7.1 implements the basis operators for exterior forms and the operations of exterior multiplication, "EXTP", and exterior differentiation, "EXTD". The implementation has been restricted to forms of degree less than six since there will generally be only the four independent variables x, y, z, t in problems so that computations of Lie derivatives of 4-forms will involve forms of degree five at most. This program is considered the basic program to which the programs given in Sections 7.2 and 7.3 may be adjoined sequentially.

Section 7.2 gives a program that implements a basis for the tangent space and the operations of Lie multiplication of two vector fields, "LIEP(,)", and Lie differentiation of a form W with respect to a vector field V , "LIED(V,W)". Section 7.3 gives the program that implements the linear homotopy operator, "HOM", and the exact part and antiexact part operators, "EP" and "AP", respectively.

It is hoped that the use of these programs with the material given in the previous chapters will provide more ready access to the intricacies of nonlinear partial differential equations and to equations of balance. The interested reader is invited to experiment with the programs and to improve on their efficiency, for much improvement is possible. To give one instance, the exterior product should have been implemented as an "nary infix" operator at the "LISP" sublanguage level and incorporated directly in the REDUCE 2 program.

7.1 THE BASIC PROGRAM FOR IMPLEMENTATION OF THE EXTERIOR CALCULUS

The following program is written in the REDUCE 2 system. All symbols are treated as operators. Exterior forms are realized as linear combinations of basis operators. For the case where the underlying space has dimension three, a 1-form is written as

F(X,Y,Z)*B1(1) + G(X,Y,Z)*B1(2) + H(X,Y,Z)*B1(3)

and a 2-form would be written as

F(X,Y,Z)*B2(1,2) + G(X,Y,Z)*B2(2,3) + H(X,Y,Z)*B2(1,3) .

Thus, $dX^1 \wedge dX^2$ becomes B2(1,2), $dX^1 \wedge dX^3$ becomes B2(1,3) and $dX^3 \wedge dX^1$ becomes -B2(1,3) . The basis operators are always sorted into canonical form in which the indices are in increasing order. Thus, if you enter B3(2,1,4) , the program will print -B3(1,2,4).

The exterior product is implemented as a binary prefix operator, EXTP(,) . If the statement EXTP(B2(1,3),B2(2,4)) is input, the program will give -B4(1,2,3,4) as the output, etc.

The exterior derivative is realized as a simple prefix operator, EXTD( ) . Thus, if $W:=X^1*X^2*B2(2,3)$, then EXTD(W) will evaluate to $X^2*B3(1,2,3)$. The only exception occurs with forms of degree zero. In this instance, the operator B0(1) must be introduced as a basis for 0-forms and the exterior derivative of a function F is then written in the form EXTD(F*B0(1)) .

```
COMMENT  THIS PROGRAM IMPLEMENTS THE EXTERIOR PRODUCT AND
     DERIVATIVE$

WRITE "B1(J), J = 1,...   IS A BASIS FOR 1-FORMS"$
WRITE "B2(J,K), J < K,   IS A BASIS FOR 2-FORMS"$
WRITE "B3(J,K,L)  IS A BASIS FOR 3-FORMS, ETC."$

WRITE "IMPLEMENTATION OF FORMS OF DEGREE > 5 IS LEFT TO THE USER"$

WRITE "THE EXTERIOR PRODUCT OF TWO FORMS M AND W IS WRITTEN"$
WRITE "          EXTP(M,W)"$
```

```
WRITE "THE EXTERIOR DERIVATIVE OF A FORM W IS WRITTEN"$
WRITE "          EXTD(W)"$

WRITE "IF  F  IS A SCALAR, THEN YOU MUST WRITE EXTD(F*BO(1))"$


OPERATOR EXTP$

FOR ALL X LET
     EXTP(X,0)=0,
     EXTP(0,X)=0$
FOR ALL X,Y LET
     EXTP(-X,Y)=-EXTP(X,Y),
     EXTP(X,-Y)=-EXTP(X,Y)$
FOR ALL X,Y,Z LET
     EXTP(X,Y+Z)=EXTP(X,Y)+EXTP(X,Z),
     EXTP(X,Y/Z)=EXTP(X,Y)/Z,
     EXTP(X/Z,Y)=EXTP(X,Y)/Z,
     EXTP(X+Y,Z)=EXTP(X,Z)+EXTP(Y,Z)$


OPERATOR BO$
FOR ALL X LET DF(BO(1),X)=0$
COMMENT  BO(1) IS THE IDENTITY$
FOR ALL X LET
     EXTP(BO(1),X)=X,
     EXTP(X,BO(1))=X$
FOR ALL X,Y LET
     EXTP(X*BO(1),Y)=X*Y,
     EXTP(X,Y*BO(1))=X*Y$

OPERATOR B1$
FOR ALL X,J LET DF(B1(J),X)=0$
FOR ALL X,Y,J LET
     EXTP(X*B1(J),Y)=X*EXTP(B1(J),Y),
     EXTP(Y,X*B1(J))=X*EXTP(Y,B1(J))$

OPERATOR B2$
FOR ALL X,J,K LET DF(B2(J,K),X)=0$
FOR ALL J,K LET
     EXTP(B1(J),B1(K))=B2(J,K)$
FOR ALL J,K SUCH THAT J=K LET
     B2(J,K)=0$
FOR ALL J,K SUCH THAT J<K LET
     B2(K,J)=-B2(J,K)$
FOR ALL X,Y,J,K LET
     EXTP(X*B2(J,K),Y)=X*EXTP(B2(J,K),Y),
     EXTP(Y,X*B2(J,K))=X*EXTP(Y,B2(J,K))$
```

```
OPERATOR B3$
FOR ALL X,J,K,L LET DF(B3(J,K,L),X)=0$
FOR ALL J,K,L LET
     EXTP(B1(J),B2(K,L))=B3(J,K,L),
     EXTP(B2(J,K),B1(L))=B3(J,K,L)$
FOR ALL J,K,L SUCH THAT J=K OR K=L LET
     B3(J,K,L)=0$
FOR ALL J,K,L SUCH THAT J<K LET
     B3(K,J,L)=-B3(J,K,L),
     B3(L,K,J)=-B3(L,J,K)$
FOR ALL X,Y,J,K,L LET
     EXTP(X*B3(J,K,L),Y)=X*EXTP(B3(J,K,L),Y),
     EXTP(Y,X*B3(J,K,L))=X*EXTP(Y,B3(J,K,L))$

OPERATOR B4$
FOR ALL X,J,K,L,M LET DF(B4(J,K,L,M),X)=0$
FOR ALL J,K,L,M LET
     EXTP(B1(J),B3(K,L,M))=B4(J,K,L,M),
     EXTP(B2(J,K),B2(L,M))=B4(J,K,L,M),
     EXTP(B3(J,K,L),B1(M))=B4(J,K,L,M)$
FOR ALL J,K,L,M SUCH THAT J=K OR K=L OR L=M LET
     B4(J,K,L,M)=0$
FOR ALL J,K,L,M SUCH THAT J<K LET
     B4(K,J,L,M)=-B4(J,K,L,M),
     B4(L,K,J,M)=-B4(L,J,K,M),
     B4(L,M,K,J)=-B4(L,M,J,K)$
FOR ALL X,Y,J,K,L,M LET
     EXTP(X*B4(J,K,L,M),Y)=X*EXTP(B4(J,K,L,M),Y),
     EXTP(Y,X*B4(J,K,L,M))=X*EXTP(Y,B4(J,K,L,M))$

OPERATOR B5$
FOR ALL X,J,K,L,M,N LET DF(B5(J,K,L,M,N),X)=0$
FOR ALL J,K,L,M,N LET
     EXTP(B1(J),B4(K,L,M,N))=B5(J,K,L,M,N),
     EXTP(B2(J,K),B3(L,M,N))=B5(J,K,L,M,N),
     EXTP(B3(J,K,L),B2(M,N))=B5(J,K,L,M,N),
     EXTP(B4(J,K,L,M),B1(N))=B5(J,K,L,M,N)$
FOR ALL J,K,L,M,N SUCH THAT J=K OR K=L OR L=M OR M=N LET
     B5(J,K,L,M,N)=0$
FOR ALL J,K,L,M,N SUCH THAT J<K LET
     B5(K,J,L,M,N)=-B5(J,K,L,M,N),
     B5(L,K,J,M,N)=-B5(L,J,K,M,N),
     B5(L,M,K,J,N)=-B5(L,M,J,K,N),
     B5(L,M,N,K,J)=-B5(L,M,N,J,K)$
FOR ALL X,Y,J.K,L,M,N LET
     EXTP(X*B5(J,K,L,M,N),Y)=X*EXTP(B5(J,K,L,M,N),Y),
     EXTP(Y,X*B5(J,K,L,M,N))=X*EXTP(Y,B5(J,K,L,M,N))$
```

```
OPERATOR XX,EXTD$
FOR ALL W LET
     EXTD(W)=
             FOR K:=1:DIM SUM EXTP(B1(K),DF(W,XX(K)))$

SYMBOLIC OPERATOR XREAD$
LET DATA=
BEGIN
        WRITE "ENTER DIMENSION AND HIT ESC:"$
        WRITE "DIM:="$
        DIM:=XREAD()$
        WRITE "ENTER VARIABLES AND HIT ESC:"$
        FOR I:=1:DIM DO
        <<WRITE "X(",I,"):="$
        XX(I):=XREAD()$
        WRITE " ">>$
        WRITE "THANK YOU"$
        WRITE " "$
        END$
WRITE "TYPE THE WORK 'DATA' AND HIT ESC:"$
END$

FACTOR B1,B2,B3,B4,B5;

END  EXTERIOR PRODUCT AND DERIVATIVE PROGRAM$
```

7.2 IMPLEMENTATION OF VECTOR FIELDS AND LIE DERIVATIVES

The following program is written in the REDUCE 2 system. It assumes that the program listed in Section 7.1 is already in memory. All symbols are treated as operators. The program prints instructions that should be adequate in its use. It is worth noting, however, that the innerproduct of a vector field V with a form W , usually written $V \rfloor W$, is realized as a binary prefix operator INN(V,W) .

```
COMMENT  THIS PROGRAM IMPLEMENTS TANGENT VECTORS, INNERPRODUCTS
     AND LIE DERIVATIVES$

OPERATOR TE$

FOR ALL X,J LET DF(TE(J),X)=0$

WRITE "TE(J), J=1,2,...,  IS A BASIS FOR THE TANGENT SPACE"$
```

```
OPERATOR EE$
FOR ALL J LET EE(J,J)=1$
FOR ALL J,K SUCH THAT J NEQ K LET EE(J,K)=0$


OPERATOR INN$

WRITE "THE INNER PRODUCT OF A VECTOR, V, AND A FORM, W,
     IS WRITTEN "$
WRITE "          INN(V,W)   ."$
WRITE " "$

FOR ALL X,Y LET
     INN(-X,Y)=-INN(X,Y),
     INN(X,-Y)=-INN(X,Y)$

FOR ALL X,Y,Z LET
     INN(X,Y+Z)=INN(X,Y)+INN(X,Z),
     INN(X,Y/Z)=INN(X,Y)/Z,
     INN(X/Z,Y)=INN(X,Y)/Z,
     INN(X+Y,Z)=INN(X,Z)+INN(Y,Z)$

FOR ALL Y LET
     INN(Y,B0(1))=0,
     INN(0,Y)=0,
     INN(Y,0)=0$

FOR ALL X,Y LET
     INN(Y,B0(1)/X)=0,
     INN(Y,X*B0(1))=0$

FOR ALL X,Y,J LET
     INN(X*TE(J),Y)=X*INN(TE(J),Y),
     INN(Y,X*B1(J))=X*INN(Y,B1(J))$

FOR ALL X,Y,J,K LET
     INN(Y,X*B2(J,K))=X*INN(Y,B2(J,K))$

FOR ALL X,Y,J,K,L LET
     INN(Y,X*B3(J,K,L))=X*INN(Y,B3(J,K,L))$

FOR ALL X,Y,J,K,L,M LET
     INN(Y,X*B4(J,K,L,M))=X*INN(Y,B4(J,K,L,M))$

FOR ALL X,Y,J,K,L,M,N LET
     INN(Y,X*B5(J,K,L,M,N))=X*INN(Y,B5(J,K,L,M,N))$

FOR ALL J,K LET
     INN(TE(J),B1(K))=EE(J,K)*B0(1)$
```

```
FOR ALL J,K,L LET
     INN(TE(J),B2(K,L))=EE(J,K)*B1(L)-EE(J,L)*B1(K)$

FOR ALL J,K,L,M LET
     INN(TE(J),B3(K,L,M))=
          EE(J,K)*B2(L,M)-EE(J,L)*B2(K,M)+EE(J,M)*B2(K,L)$

FOR ALL J,K,L,M,N LET
     INN(TE(J),B4(K,L,M,N))=
          EE(J,K)*B3(L,M,N)-EE(J,L)*B3(K,M,N)+
          EE(J,M)*B3(K,L,N)-EE(J,N)*B3(K,L,M)$

FOR ALL J,K,L,M,N,R LET
     INN(TE(J),B5(K,L,M,N,R))=
          EE(J,K)*B4(L,M,N,R)-EE(J,L)*B4(K,M,N,R)+
          EE(J,M)*B4(K,L,N,R)-EE(J,N)*B4(K,L,M,R)+
          EE(J,R)*B4(K,L,M,N)$

WRITE "THE LIE DERIVATIVE OF A FORM  W  WITH RESPECT TO A VECTOR
     IS WRITTEN "$
WRITE "           LIED(V,W) ."$

OPERATOR LIED$

FOR ALL V,W LET
     LIED(V,W)=INN(V,EXTD(W))+EXTD(INN(V,W))$


WRITE "THE LIE PRODUCT OF TWO VECTORS U AND V IS WRITTEN "$
WRITE "           LIEP(U,V)"$

OPERATOR LIEP$

FOR ALL U,V LET
     LIEP(U,V)=
     FOR J:=1:DIM SUM
     INN(U,EXTD(INN(V,B1(J))))*TE(J)/B0(1)-
     INN(V,EXTD(INN(U,B1(J))))*TE(J)/B0(1)$

END PROGRAM FOR TANGENT VECTORS AND LIE DERIVATIVES$
```

7.3 IMPLEMENTATION OF THE HOMOTOPY OPERATOR

The following program is written in the REDUCE 2 system. It assumes that the programs listed in Sections 7.1 and 7.2 are already in memory. All symbols are treated as operators and the program prints instructions that should be adequate in its use.

```
COMMENT  THIS PROGRAM IMPLEMENTS THE HOMOTOPY OPERATOR$

WRITE "THE HOMOTOPY OPERATOR IS WRITTEN"$
WRITE "           HOM(W)"$

OPERATOR XXO$

FOR K:=1:DIM DO XXO(K):=0$


WRITE "THE DEGREE OF A FORM MUST BE GIVEN BEFORE ITS HOMOTOPY
     IS EVALUATED BY WRITING"$
WRITE "           DEG:=    "$

WRITE "THE CENTER IS INITIATED TO BE THE ORIGIN."$
WRITE "IF A DIFFERENT CENTER IS DESIRED, MAKE THE CHANGES BY
     ENTERING"$
WRITE "XXO(1):=  , XXO(2):=  ,..."$

WRITE "THE INDEPENDENT VARIABLES MUST BE ENTERED AS
     Y(1),Y(2),..."$
WRITE "IN ANY FORM WHOSE HOMOTOPY IS TO BE COMPUTED."$
WRITE "VARIABLES MAY BE EXCLUDED FROM THE HOMOTOPY BY NOT
     WRITTING THEM"$
WRITE "IN TERMS OF Y(1), Y(2),..."$


OPERATOR Y,YY,RAD$

LET RAD= FOR K:=1:DIM  SUM (XX(K)-XXO(K))*TE(K)$

OPERATOR TILDA,JNT,JNTJ,JTJ,HOM$

FOR ALL K LET TILDA(K)=XXO(K)+LAM*(YY(K)-XXO(K))$

FOR ALL W LET
     JNT(W)= BEGIN
     WW:=W$
     FOR K:=1:DIM DO
          WW:=SUB(Y(K)=TILDA(K),WW)$
     FOR K:=1:DIM DO
          WW:=SUB(YY(K)=XX(K),WW)$
     RETURN WW$
     END SUBFUNCTION$

FOR ALL W LET
     JNTJ(W)=INN(RAD,JNT(W))*LAM^(DEG-1)$

FOR ALL W LET
     JTJ(W)=INT(JNTJ(W),LAM)$
```

```
FOR ALL W LET
     HOM(W)=SUB(LAM=1,JTJ(W))-SUB(LAM=0,JTJ(W))$

WRITE "THE EXACT PART OF A FORM IS WRITTEN"$
WRITE "           EP(W)"$

OPERATOR EP$

FOR ALL W LET
     EP(W)=EXTD(HOM(W))$

WRITE "THE ANTIEXACT PART OF A FORM IS WRITTEN"$
WRITE "           AP(W)"$

OPERATOR AP$

FOR ALL W LET
     AP(W)=W-EP(W)$

END OF HOMOTOPY PROGRAM$
```

7.4 EXAMPLES

The simplest way of seeing just what assistance is provided by the programs given in the previous Sections is to apply them to the problem of computing isovectors of the equations of a well established problem. For this purpose, we take the equations for shallow water waves:

(7.1) $$\partial_T U + U\,\partial_X U + V\,\partial_Y U + G\,\partial_T H = 0\ ,$$

(7.2) $$\partial_X U + \partial_Y V = 0\ ,$$

(7.3) $$\partial_T H + C(H)\,\partial_X H = 0\ ,$$

(7.4) $$\partial_Y H = 0\ .$$

Here, X, Y, T are the independent variables, U, V, H are the dependent variables and C(H) is a function of the single variable H .

The first thing that must be done is to transcribe the system

(7.1) through (7.4) into an equivalent system of balance forms on the space G_6 with coordinate functions (X,Y,T,U,V,H). It is a trivial matter to see that this transcription is given by

$$(7.5)\qquad E1 = dU\wedge dX\wedge dY + U\, dU\wedge dY\wedge dT - V\, dU\wedge dX\wedge dT + G\, dH\wedge dY\wedge dT ,$$

$$(7.6)\qquad E2 = dU\wedge dY\wedge dT - dV\wedge dX\wedge dT ,$$

$$(7.7)\qquad E3 = dH\wedge dX\wedge dY + C(H)\, dH\wedge dY\wedge dT ,$$

$$(7.8)\qquad E4 = dH\wedge dX\wedge dT .$$

Since the system (7.1)-(7.4) is of first order, there are no contact forms to be adjoined to (7.5)-(7.8). A direct calculation shows that

$$(7.9)\qquad d\begin{Bmatrix} E1 \\ E2 \\ E3 \\ E4 \end{Bmatrix} = \begin{pmatrix} 0 & -dU & 0 & 0 \\ 0 & 0 & 0 & 0 \\ 0 & 0 & 0 & 0 \\ 0 & 0 & 0 & 0 \end{pmatrix} \begin{Bmatrix} E1 \\ E2 \\ E3 \\ E4 \end{Bmatrix}$$

so that the ideal $I\{E1, E2, E3, E4\}$ is closed. In fact, it is easily seen that each of the forms $E1 + U\cdot E2$, $E2$, $E3$, $E4$ is exact so that we could write the first integrals

$$E1 + U\cdot E2 = d\alpha , \quad E2 = d\beta , \quad E3 = d\gamma , \quad E4 = d\rho .$$

Unfortunately, this does us little good in securing solutions because the only characteristic vector of the ideal generated by the balance forms is the zero vector.

The next thing to be done is to transcribe the problem so that the programs given in the previous sections may be used. To this end, set DIM:=6 and

```
X(1):=X,  X(2):=Y,  X(3):=T,  X(4):=U,  X(5):=V,  X(6):=H
```

after the DATA statement. A direct translation of the system (7.5)-(7.8) is then obtained through use of the basis operators B3(J,K,L) for $\Lambda^3(G_6)$ and the assignment statements

(7.10) E1:=B3(4,1,2) + U*B3(4,2,3) - V*B3(4,1,3) + G*B3(6,2,3) ;

(7.11) E2:=B3(4,2,3) - B3(5,1,3) ,

(7.12) E3:=B3(6,1,2) + C(H)*B3(6,2,3) ,

(7.13) E4:=B3(6,1,3) ,

where C(H) has been declared an operator.

It now remains only to implement the generic form of the isovector in order to perform the bulk of the calculations. Since TE(K) are the symbolic operator basis for the tangent space, we simply declare J to be an operator and make the following assignment statement

```
VEC:= FOR K:=1:6 SUM J(K)*TE(K); .
```

the REDUCE system will then print out

```
(7.14)      VEC:= J(1)*TE(1) + J(2)*TE(2) + J(3)*TE(3)

                 + J(4)*TE(4) + J(5)*TE(5) + J(6)*TE(6) .
```

About 80% of the work is then done automatically by the assignment statements

```
(7.15)      ISO1:= LIED(VEC,E1) - A11*E1 - A12*E2 - A13*E3 - A14*E4;

(7.16)      ISO2:= LIED(VEC,E2) - A21*E1 - A22*E2 - A23*E3 - A24*E4;
```

(7.17) ISO3:= LIED(VEC,E3) - A31*E1 - A32*E2 - A33*E3 - A34*E4;

(7.18) ISO4:= LIED(VEC,E4) - A41*E1 - A42*E2 - A43*E3 - A44*E4;

for (7.15)-(7.18) are just the requirements that VEC be an isovector of the closed ideal I{E1, E2, E3, E4} . Once the system has printed out ISO1, ISO2, ISO3, ISO4, it is a relatively simple matter to eliminate all 16 of the quantities A11, A12, ..., A43, A44 through the use of LET statements, since such statements set up global equivalences. We now resolve each of the four forms ISO1, ISO2, ISO3, ISO4 on the basis elements B3(J,K,L) , $1 \leq J < K < L \leq 6$ in order to obtain the partial differential equations that the six operators J(1),...,J(6) must satisfy in order that (7.14) define an isovector of the balance ideal. This too can be accomplished by use of LET statements and with the advantage that all resulting equations will be independent due to the fact that LET statements are global equivalences. For the problem a. hand, about 95% of the work has been done by the computer. The output of all of this is the actual system of defining partial differential equations

(7.19) $$\partial_V J(1) = \partial_U J(1) = \partial_Y J(1) = \partial_H J(1) = 0 ,$$

(7.20) $$\partial_U J(2) = \partial_V J(2) = \partial_H J(2) = 0 ,$$

(7.21) $$\partial_U J(3) = \partial_V J(3) = \partial_H J(3) = \partial_Y J(3) = 0 ,$$

(7.22) $$\partial_V J(4) = 0 ,$$

(7.23) $$\partial_V J(6) = \partial_U J(6) = \partial_Y J(6) = 0 ,$$

(7.24) $$(\partial_T + C(H)\partial_X) J(6) = 0 ,$$

(7.25) $$\frac{dC}{dH} J(6) = (\partial_T + C(H)\partial_X)(J(1) - C(H)J(3)) ,$$

$$(7.26)\qquad \partial_H J(4) = -G\partial_X J(3) ,$$

$$(7.27)\qquad \partial_X J(4) = -\partial_Y J(5) ,$$

$$(7.28)\qquad \partial_U J(4) = \partial_H J(6) + \partial_T J(3) + 2C(H)\partial_X J(3) - \partial_X J(1) ,$$

$$(7.29)\qquad J(4) = C(H)\partial_T J(3) - U\partial_T J(3) + C(H)^2\partial_X J(3) - C(H)\partial_X J(1) - U^2\partial_X J(3) + U\partial_X J(1) + \frac{dC}{dH} J(6) ,$$

$$(7.30)\qquad \partial_T J(4) = -G\partial_X J(6) - V\partial_Y J(4) + U\partial_Y J(5) ,$$

$$(7.31)\qquad \partial_V J(5) = \partial_U J(4) + \partial_Y J(2) + U\partial_X J(3) - \partial_X J(1) ,$$

$$(7.32)\qquad \partial_U J(5) = \partial_X J(2) - V\partial_X J(3) ,$$

$$(7.33)\qquad J(5) = V\partial_Y J(2) + \partial_T J(2) - V\partial_T J(3) + U\partial_X J(2) - UV\partial_X J(3) .$$

The general solution of this system is given by

$$(7.34)\qquad \begin{aligned} J(1) &= k_2 + k_3 X + k_4 + k_5 T , \\ J(2) &= \alpha(X,T) + (k_1 - k_3 + k_4 - k_7)Y , \\ J(3) &= k_6 + k_7 T , \\ J(4) &= k_5 + (k_3 - k_7)U , \\ J(5) &= (k_1 - k_3 + k_4 - 2k_7)V + \partial_T\alpha(X,T) + U\partial_X\alpha(X,T) , \\ J(6) &= -2(k_3 + k_7)H \end{aligned}$$

where $\alpha(X,T)$ is an arbitrary function of its arguments and either

(7.35) $C(H) = r + k_8 H^m$, $r = k_5/(k_7-k_3)'$, $m = \dfrac{k_7-k_3}{2(k_3+k_7)}$

or $C(H)$ an arbitrary function of H but $k_3 = k_5 = k_7 = 0$.

For the case of 1-dimensional nonlinear diffusion,

(7.36) $F(Q)\, \partial_T Q = \partial_X \partial_X Q + G(Q)(\partial_X Q)^2$,

we set DIM:=5 and X(1):=X, X(2):=T, X(3):=Q, X(4):=Y1, X(5):=Y2 after the DATA statement. The contact form and the balance form of the problem are assigned by

(7.37)
```
CON:=B1(3) - Y1*B1(1) - Y2*B1(2) ;
```

(7.38)
```
E1:= F(Q)*B2(3,1) + B2(4,2) + G(Q)*Y1*B2(3,2) ;
```

respectively, where F and G have been declared operators. With J declared an operator, the general forms of the isovectors, VEC, of the contact ideal are defined by the statements

(7.39)
```
VE:= FOR K:=1:3 SUM J(K)*TE(K);
```

(7.40)
```
SIM:= INN(VE,CON);
```

(7.41)
```
Z1:= DF(SIM,X) + Y1*DF(SIM,Q);
```

(7.42)
```
Z2:= DF(SIM,T) + Y2*DF(SIM,Q);
```

(7.43)
```
VEC:= VE + Z1*TE(4) + Z2*TE(5);
```

where DF(A,B) denotes the partial derivative of A with respect to B in the REDUCE 2 system.

Since E1 belongs to Λ^2, the equations for the isovectors of the balance ideal require the computation of the intersection of Λ^2 with the closed contact ideal. This is obtained by the assignments

(7.44) R:= FOR K:=1:5 SUM L(K)*B1(K);

(7.45) S:= L(6)*EXTD(CON) + EXTP(R,CON);

where L has been declared an operator (i.e., S=L(6)d(CON) + $R \wedge CON$ with R a 1-form in general position). Thus, S + L(7)*E1 is the intersection of Λ^2 with the balance ideal and L(1) through L(7) are undetermined coefficients that are to be algebraically eliminated from the conditions that VEC be an isovector of the balance ideal. These latter conditions are obtained from annihilating the 2-form ISO that is computed by the assignment statement

(7.46) ISO:= -S - L(7)*E1 + LIED(VEC,E1);

which will print out the required equations resolved on the 10 dimensional basis B2(1,2) through B2(4,5) .

The elimination of the seven quantities L(1) through L(7) is now a straightforward computation that can be performed directly by use of the COEFF operator in the REDUCE 2 system. Since J(1) through J(3) , that define the isovectors, are independent of the variables Y1 and Y2 , the commands ORDER Y1, Y2; FACTOR Y1, Y2; lead to listings of the results as polynomials in Y1, Y2 that we must force to vanish identically. Less than an hour's work at a terminal thus gives the following system of equations that J1:=J(1), J2:=J(2), J3:=J(3) must satisfy:

$$\frac{dG}{dQ}J3 + G\,\partial_Q J3 + \partial_Q\partial_Q J3 = 0, \tag{7.47}$$

$$2G\,\partial_X J3 + F\,\partial_T J1 - \partial_X\partial_X J1 + 2\,\partial_Q\partial_X J3 = 0, \tag{7.48}$$

$$F\,\partial_T J2 - 2\,F\,\partial_X J1 - \frac{dF}{dQ}J3 = 0, \tag{7.49}$$

$$F\,\partial_T J3 - \partial_X\partial_X J3 = 0, \tag{7.50}$$

(7.51) $\partial_Q J2 = 0, \quad \partial_X J2 = 0,$

(7.52) $\partial_Q J1 = 0.$

If we set $F = 1$, $G = 0$, this system admits the solutions for the isovectors of the linear diffusion equation given in Chapter 3. On the other hand, for dF/dQ not idencially constant and $G(Q)$ not identically zero, there are solutions of the form

(7.53) $J1 = C_2 + C_3 X + C_4 X^2 , \quad J2 = C_5 - 4C_1 T ,$

(7.54) $J3 = \exp(-\int G(Q)dQ)g(Q)(\frac{1}{2}C_3 + C_4 X + C_1)$

provided $F(Q)$ and $G(Q)$ are related by

(7.55) $F(Q) = C_6 \; g(Q)^{-4} , \quad g(Q) = \int \exp(\int G(Q)dQ)dQ \; .$

Problems with more Q's and more independent variables can be handled with straightforward modifications of the procedures given above. The user is cautioned to use unevaluated operators rather than arrays when dealing with several Q's and their associated Y's in view of the internal structure of REDUCE 2. For example, if we have three Q's, then designate them by Q(J) and their associated Y's by Y1(J), Y2(J),... or by Y(L,J) . The associated contact forms would then be designated by CON(J) .

APPENDIX

THE EXTERIOR DIFFERENTIAL CALCULUS

CHAPTER I

PRELIMINARY CONSIDERATIONS

1-1. POINTS, COORDINATES AND THE SUMMATION CONVENTION

As with any mathematical discipline, we start with the familiar and build from there. In this case, the familiar is taken to be n-dimensional number space, E_n, with the age-old Cartesian coordinates together with all of its formal and intuitively obvious properties. It is the one fixed entity with which we have to deal, for almost everything else will have an amorphous or gelatinous aspect that will become more or less concrete or real as we impose constraints upon its structure.

Let U be an open set of E_n (with respect to the Euclidean topology) and let P be a generic point in U. The underlying Cartesian coordinate cover of E_n is used to assign any point P in U its coordinates $x^1(P), x^2(P), \ldots, x^n(P)$. This is conveniently written $P:(x^1,\ldots,x^n)$, or more briefly

$P:(x^i)$ where it is understood that i runs from 1 through n. In like manner, if f denotes a function on E_n with values on the real line $\mathbb{R}$, we write

$$f(x^i) \quad \text{for} \quad f(x^1,x^2,\ldots,x^n)$$

and sometimes even $f(x)$ when it is clear that x stands for $(x^i) = (x^1(P),x^2(P),\ldots,x^n(P))$. Similarly, since $x^1 = \phi^1(t)$, $x^2 = \phi^2(t)$, ..., $x^n = \phi^n(t)$ defines a curve in E_n (a map from the real line, $\mathbb{R}$, with coordinate t into E_n), we write

$$\Phi:\mathbb{R} \longrightarrow E_n | x^i = \phi^i(t)$$

where it is understood that i runs from 1 through n.

There will be many irstances in which we will have to write quantities such as

$$\sum_{i=1}^{n} f^i(x^k)\ \omega_i(x^k) .$$

This can be significantly simplified by adopting the Einstein summation convention: *if an upstairs and a downstairs index are the same letter, then summation over the range of the index is implied.* Thus,

$$f^i(x^k)\ \omega_i(x^k) \equiv \sum_{i=1}^{n} f^i(x^k)\ \omega_i(x^k)$$

and

$$T^{ij}_{kr}\ h^k_j \equiv \sum_{j=1}^{n} \sum_{k=1}^{n} T^{ij}_{kr}\ h^k_j = S^i_r .$$

Fortunately, use of the exterior calculus will minimize the use of such outlandish looking expressions.

1-2. COMPOSITIONS OF MAPPINGS

We have already encountered two kinds of mappings: mappings from the real line into E_n (curves), and mappings from E_n into the real line. These two constructs combine to yield composition mappings, a notion that will be fundamental throughout our studies.

Let Φ be a mapping from the real line, $\mathbb{R}$, into E_n, which we write as

$$(1\text{-}2.1) \qquad \Phi:\mathbb{R} \longrightarrow E_n \mid x^i = \phi^i(t)$$

where $\mathbb{R}$ or a subset of $\mathbb{R}$ is the common domain of definition of the n functions $\{\phi^i(t)\}$ that define the mapping and E_n contains the range of the mapping. In what follows, script $\mathcal{R}$ is used to designate *range* and script $\mathcal{D}$ to designate *domain*. Thus, $\mathcal{D}(\Phi)$ denotes the subset of $\mathbb{R}$ on which each of the n functions $\{\phi^i(t)\}$ is defined and $\mathcal{R}(\Phi)$ denotes the (1-dimensional) subset of E_n whose points satisfy $x^i = \phi^i(t)$ for t in $\mathcal{D}(\Phi)$. We assume here and throughout that the n functions $\{\phi^i(t)\}$ are very smooth functions of the variable t for all t in $\mathcal{D}(\Phi)$. This will usually be taken to mean that each of the n functions $\{\phi^i(t)\}$ possesses continuous derivatives of all orders for all t in $\mathcal{D}(\Phi)$. The graphic situation is shown in Fig. 1.

Let F be a map from E_n to $\mathbb{R}$, which we write as

$$(1\text{-}2.2) \qquad F:E_n \longrightarrow \mathbb{R} \mid \tau = f(x^i)$$

where $\mathcal{D}(F)$ is the subset of E_n on which the function $f(x^i)$ is defined and $\mathbb{R}$ contains $\mathcal{R}(F)$. The function $f(x^i)$ is likewise assumed to be very smooth; that is, it possesses continuous partial derivatives of all orders for all $P:(x^i)$ in $\mathcal{D}(F)$. The graphic situation is depicted in Fig. 2, where we also show $f^{-1}(\tau_1)$ that is defined to be the subset of $\mathcal{D}(F)$ that is mapped onto the point in $\mathcal{R}(F)$ with coordinate τ_1.

Now, simply observe that the range of Φ and the domain of F are both contained in E_n. We can thus *compose* the maps Φ and F on $\mathcal{R}(\Phi) \cap \mathcal{D}(F)$ if this intersection is nonempty. When the composition is effected, we have

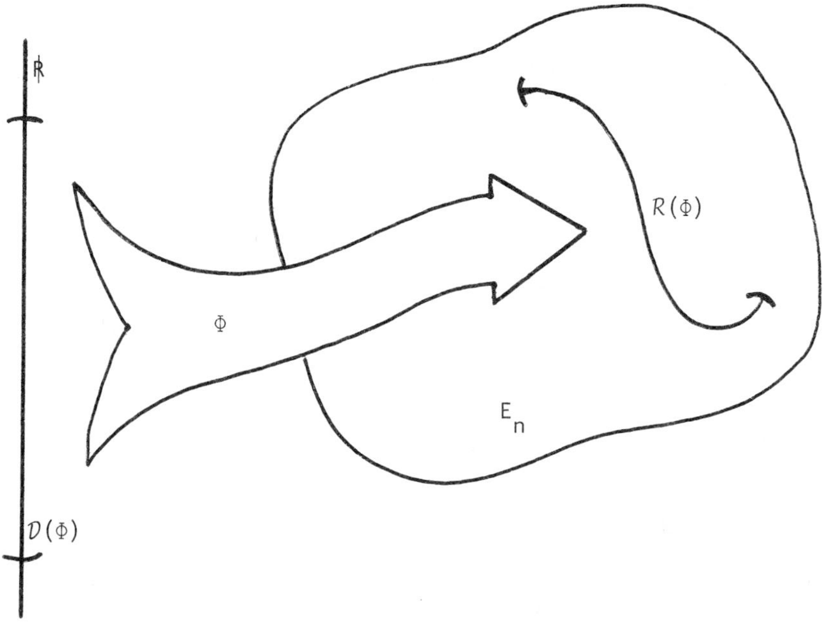

Figure 1. The Map Φ

$$\text{(1-2.3)} \qquad F \circ \Phi \;=\; f(\phi^1(t),\ldots,\phi^n(t)) \;=\; f(\phi^i(t)) \stackrel{\text{def}}{=} \tau \,,$$

namely a function with values in ℝ. The composition of the maps Φ and F thus yields the *composition map*

$$\text{(1-2.4)} \qquad T \;=\; F \circ \Phi : \mathbb{R} \longrightarrow \mathbb{R} \;|\; \tau \;=\; f(\phi^i(t)) \,.$$

Here, the domain of T consists of all those points in $\mathcal{D}(\Phi)$ whose images under the map Φ are contained in $\mathcal{R}(\Phi) \cap \mathcal{D}(F)$, and we have the situation depicted in Fig. 3. For the case where Φ and F are defined by

$$x^1 \;=\; 2t - 4 \,, \quad x^2 \;=\; t^2 - 6t \,, \quad x^3 \;=\; t\,e^{-3t} + 5 \,,$$

$$f(x^i) \;=\; (x^1)^2 - x^1 + 3(x^2)^2 - 4\,e^{-5x^3} \,,$$

the composition map $T = F \circ \Phi$ is given by

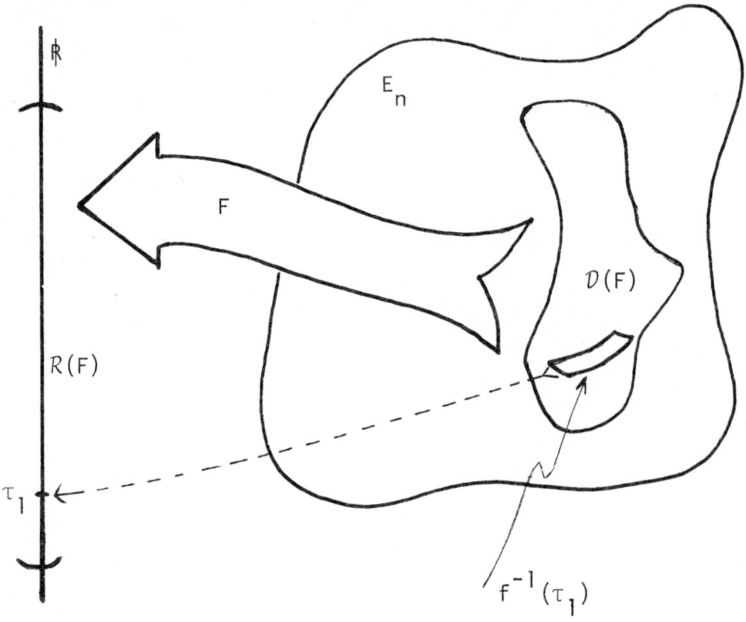

Figure 2. The Map F

$$\tau = (2t-4)^2 - 2t + 4 + 3(t^2 - 6t)^2 - 4\, e^{-5(t\, e^{-3t} + 5)} .$$

The situation just discussed is somewhat special in view of the restrictions that the domain of Φ to be contained in ℝ and that the range of F to be contained in ℝ . These restrictions are easily dispensed with, however. Let Φ be a map from a space A to a space B ,

$$(1\text{-}2.5) \qquad \Phi : A \longrightarrow B \mid y^\alpha = \phi^\alpha(x^i) ,$$

and let Ψ be a map from a space B to a space C ,

$$(1\text{-}2.6) \qquad \Psi : B \longrightarrow C \mid z^\alpha = \psi^\alpha(y^\beta) .$$

If $\mathcal{R}(\Phi) \cap \mathcal{D}(\Psi)$ is nonempty, then the composition map $T = \Psi \circ \Phi$ is a map from A to C ,

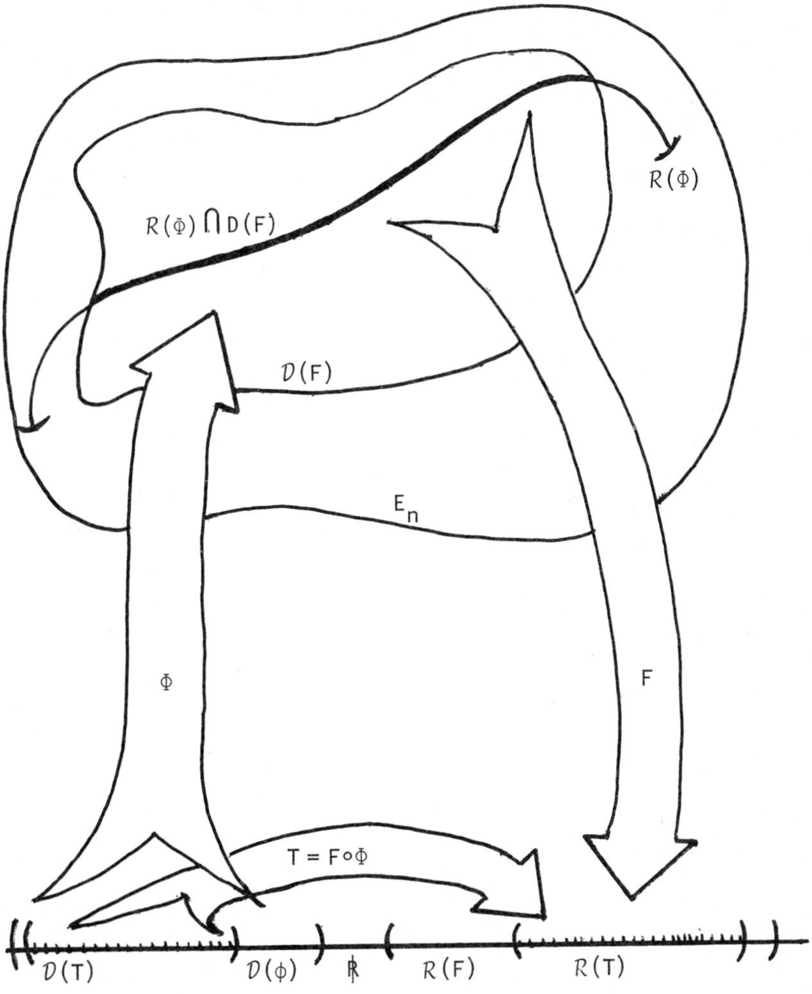

Figure 3. The Composition Map $T = F \circ \Phi$

$$T = \Psi \circ \Phi : A \longrightarrow C \mid z^{\alpha} = \psi^{\alpha}(\phi^{\beta}(x^{i})) , \tag{1-2.7}$$

with $\mathcal{D}(T) = \overset{-1}{\Phi}(\mathcal{R}(\Phi) \cap \mathcal{D}(F))$.

Here, we have used the notation $\overset{-1}{\Phi}(\rho)$ to designate the totality of points in $\mathcal{D}(\Phi)$ that is mapped into the set ρ . This notation will be used throughout.

1-3. THE IMPLICIT FUNCTION THEOREM

We now collect together several existence theorems that will be required throughout our studies. These theorems will only be stated since it is assumed that the reader is familiar with their proofs. In all instances, the proofs, in one form or another, arise directly from the renowned fixed point theorem:

Let T *be a map of a complete metric space,* S *, into itself that satisfies*

$$(1\text{-}3.1)\qquad \rho(T\phi,\ T\psi) < k\,\rho(\phi,\psi)\ ,\qquad k<1\ ,$$

where ρ *is the distance function on* S *. Then there is a unique fixed point of* T *; that is* $T\phi_o = \phi_o$ *for one and only one* $\phi_o \varepsilon S$ *. Further, this fixed point is given by*

$$(1\text{-}3.2)\qquad \phi_o = \lim_{n\to\infty} T^n\phi$$

for any $\phi \varepsilon S$ *, where* $T^n\phi \overset{def}{=} TT^{n-1}\phi$ *so that* T^n *is the* n^{th} *iteration of* T *.*

The implicit function theorem tells us when we can solve a system of k equations in $k+n$ variables for k of the variables in terms of the remaining n variables.

IMPLICIT FUNCTION THEOREM *Let* $F^1, \ldots, F^k$ *be* k *functions of the variables* $y^1, \ldots, y^k, x^1, \ldots, x^n$ *that are defined in some neighborhood,* N *, of the origin of* $E_k \times E_n$ *and are continuous together with their partial derivatives of all orders up to and including order* r *on* N *. Let*

$$(1\text{-}3.3)\qquad F^i(0,\ldots,0;0,\ldots,0) = 0\ ,\qquad i = 1,\ldots,k\ ,$$

and suppose that

$$(1\text{-}3.4)\qquad \det(\partial F^i/\partial y^j)(0,\ldots,0;0,\ldots,0) \neq 0\ .$$

Then, in a sufficiently small neighborhood of the origin in E_n *,* *there exist functions* $\phi^1(x^m), \phi^2(x^m), \ldots, \phi^k(x^m)$ *of the* n *variables* $x^1,\ldots,x^n$ *that are uniquely determined by the equations*

$$F^i(\phi^1(x^m),\phi^2(x^m),\ldots,\phi^k(x^m);\ x^1,x^2,\ldots,x^n) = 0,\quad i=1,\ldots,k. \tag{1-3.5}$$

$$\phi^i(0,\ldots,0) = 0\ ,\quad i=1,\ldots,k\ , \tag{1-3.6}$$

Further, the functions $\phi^i(x^m)$ *that serve to define the* y's *as functions of the* x's *on* N *through the relations*

$$y^i = \phi^i(x^1,\ldots,x^n)\ ,\quad i = 1,\ldots,k \tag{1-3.7}$$

are continuous together with their partial derivatives of all orders up to and including r *on* N.

This theorem is extremely powerful. All we need to know is that we have k functions F^i of $k+n$ variables such that (1) $F^i = 0$, $i = 1,\ldots,k$ are satisfied for at least one point $(y_o^i;\ x_o^m)$ of $E_k \times E_n$; (2) the functions F^i are of class C^r, and (3) that $\det(\partial F^i/\partial y^j)(y_o^\ell;\ x_o^m) \neq 0$. We then simply set $\bar{x}^m = x^m - x_o^m$, $\bar{y}^i = y^i - y_o^i$ and all of the hypotheses of the theorem are satisfied. From this we conclude that there exist k functions $\{\phi^i(\bar{x}^m)\}$ on some neighborhood of the origin in E_n such that the equations $F^i(\bar{y}^j;\ \bar{x}^m) = 0$ define the $\bar{y}$'s as functions of the $\bar{x}$'s through the relations

$$\bar{y}^i = \phi^i(\bar{x}^m)\ .$$

Particular note should be taken of the fact that the determinental condition (1-3.2) only has to be verified at the single point $(y_o^i,\ x_o^m)$ of $E_k \times E_n$ in order to apply the theorem. It then follows that

$$\det(\partial F^i/\partial y^j)(\phi^k(x^m);x^m) \neq 0$$

holds at all points of the neighborhood N over which the

functions ϕ^i are defined.

Although we now know that such functions $\{\phi^i(x)\}$ exist on N, the actual finding of these functions explicitly is another matter altogether. This state of affairs is the rule rather than the exception, for existence theorems very rarely provide a specific algorithm for the actual solution of a problem. All they tell us is that the search for a solution is not in vain.

1-4. THE INVERSE MAPPING THEOREM

A special case of the implicit function theorem is the inverse mapping theorem which will be of considerable importance in our studies

INVERSE MAPPING THEOREM. *Suppose that we are given a mapping*

$$\Phi : E_n \longrightarrow {}^{`}E_n \mid y^i = F^i(x^j) , \quad i = 1,\dots,n \tag{1-4.1}$$

where the functions $F^i(x^j)$ *are defined and of class* C^r *in some neighborhood* N *of* (x_o^i) *in* E_n *and*

$$\det(\partial F^i/\partial x^j)(x_o^k) \neq 0 . \tag{1-4.2}$$

Then there exists a neighborhood ${}^{`}N$ *of* $y_o^i = F^i(x_o^j)$ *in* ${}^{`}E_n$ *and a uniquely determined collection of* n *functions* $\{G^i(y^j)\}$ *of class* C^r *on* ${}^{`}N$ *such that* $G^i(y_o^j) = x_o^i$ *and*

$$F^i(G^j(y^k)) = y^i , \quad i = 1,\dots,n \tag{1-4.3}$$

for all (y^i) *in* ${}^{`}N$. *Thus, if we restrict* Φ *according to*

$$\Phi_N : N \subset E_n \longrightarrow {}^{`}N \subset {}^{`}E_n \mid y^i = F^i(x^j) , \quad i = 1,\dots,n \tag{1-4.4}$$

then Φ_N *has an inverse mapping* $\overset{-1}{\Phi}_N$ *that is defined by*

$$\overset{-1}{\Phi}_N : {}^{`}N \subset {}^{`}E_n \longrightarrow N \subset E_n \mid x^i = G^i(y^j) , \quad i = 1,\dots,n . \tag{1-4.5}$$

The graphic situation is shown in Fig. 4. Note that

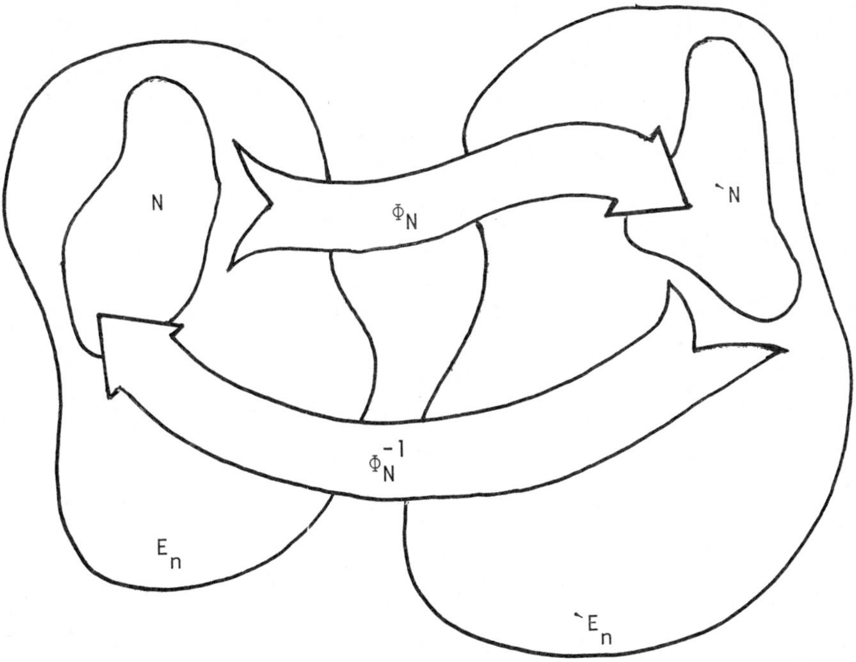

Figure 4. Inverse Mapping

$$\overset{-1}{\Phi}_N \circ \Phi_N : N \longrightarrow N$$

is the identity map on N and that

$$\Phi_N \circ \overset{-1}{\Phi}_N : {}^{\backprime}N \longrightarrow {}^{\backprime}N$$

is the identity map on ${}^{\backprime}N$. Again, we have existence of the inverse mapping, but not necessarily any explicit realization. There are, however, simple situations in which the explicit solution can be extracted. An important case of this is that of linear mappings

$$\Phi : E_n \longrightarrow {}^{\backprime}E_n \mid y^i = A^i_j\, x^j \, , \quad \det(A^i_j) \neq 0$$

in which case we have

$$\overset{-1}{\Phi} : {}^{\backprime}E_n \longrightarrow E_n \mid x^i = \overset{-1}{A}{}^i_j\, y^j$$

where $((\overset{-1}{A}{}^i_j))$ is the inverse of the nonsingular matrix $((A^i_j))$. In this important case, $N = E_n$ and $\grave{}N = \grave{}E_n$. We also note that if $x^i = G^i(y^j)$ defines the inverse of the map $y^i = F^i(x^j)$, then the matrices $\{(\partial G^i/\partial y^j)\}$ and $\{(\partial F^i/\partial x^j)\}$ are inverses of each other, either considered as functions of the x's or as functions of the y's.

1-5. SYSTEMS OF AUTONOMOUS FIRST ORDER DIFFERENTIAL EQUATIONS

An essential aspect of the discussion will center around systems of autonomous first order ordinary differential equations; namely equations of the form

$$dx^i(t)/dt = f^i(x^1(t),\ldots,x^n(t)) , \quad i = 1,\ldots,n .$$

The name autonomous derives from the fact that the right-hand sides of these equations do not contain the independent variable t explicitly; that is, the values of the derivatives are uniquely determined by the values of the dependent variables irrespective of the value of the independent variable t.

EXISTENCE THEOREM FOR AUTONOMOUS ORDINARY DIFFERENTIAL EQUATIONS. *Let* $f^1(x^1,\ldots,x^n),\ldots,f^n(x^1,\ldots,x^n)$ *be* C^r *functions in some neighborhood,* N, *of the origin of* E_n *with* $r > 1$. *Then a neighborhood* U *of the origin of* E_n *and a neighborhood* I *of* 0 *in* $\mathbb{R}$ *can be found such that there exist unique functions* $\phi^1(t;x^i_o),\ldots,\phi^n(t;x^i_o)$, *for any* (x^i_o) *in* U *and for all* t *in* I, *such that*

$$(1\text{-}5.1) \qquad d\phi^i/dt = f^i(\phi^1,\ldots,\phi^n)$$

and

$$(1\text{-}5.2) \qquad \phi^i(0;x^j_o) = x^i_o = x^i(0) .$$

Further, the functions $\{\phi^i\}$ *are of class* C^{r+1} *for all* t *in*

I *and of class* C^r *in the initial data* $\{x_o^i\}$ *for all* $\{x_o^i\}$ *in* U .

In most applications, the given functions $\{f^i\}$ will be of class C^∞ , in which case the functions ϕ^i will be of class C^∞ in t and of class C^∞ in the initial data. In this case there is a representation of the functions ϕ^i that proves to be of significant use. Let us first observe that evaluation of (1-5.1) at $t = 0$ together with (1-5.2) yields

$$\left.\frac{d\phi^i}{dt}\right|_{t=0} = f^i(x_o^j) .$$

If we differentiate (1-5.1) with respect to t and then use (1-5.1) again, we obtain

$$\frac{d^2\phi^i}{dt^2} = \frac{\partial f^i}{\partial \phi^j}\frac{d\phi^j}{dt} = \frac{\partial f^i}{\partial \phi^j} f^j ,$$

and similarly

$$\frac{d^3\phi^i}{dt^3} = \frac{\partial^2 f^i}{\partial \phi^j \partial \phi^k} f^j f^k + \frac{\partial f^i}{\partial \phi^j}\frac{\partial f^j}{\partial \phi^k} f^k .$$

Thus, an evaluation at $t = 0$ and use of (1-5.2) shows that

$$\left.\frac{d^2\phi^i}{dt^2}\right|_{t=0} = \frac{\partial f^i(x_o^m)}{\partial x_o^j} f^j(x_o^m) ,$$

$$\left.\frac{d^3\phi^i}{dt^3}\right|_{t=0} = \frac{\partial^2 f^i(x_o^m)}{\partial x_o^j \partial x_o^k} f^j(x_o^m) f^k(x_o^m)$$

$$+ \frac{\partial f^i(x_o^m)}{\partial x_o^j}\frac{\partial f^j(x_o^m)}{\partial x_o^k} f^k(x_o^m) .$$

Now, Taylor's theorem with remainder tells us that in a sufficiently small neighborhood of $t = 0$, we may write

$$\phi^i(t;x_o^m) = x_o^i + t\left(\frac{d\phi^i}{dt}\right)\Big|_{t=0} + \frac{t^2}{2!}\left(\frac{d^2\phi^i}{dt^2}\right)\Big|_{t=0} + \dots .$$

When the above expressions are used to evaluate the various derivatives at $t = 0$, the results can be written as

$$\phi^i(t;x_o^m) = \left[\exp\left(t\, f^j(z^m)\frac{\partial}{\partial z^j}\right) z^i\right]_{z^m = x_o^m} ,$$

where $\exp(\xi)$ stands for the corresponding power series $\sum_{k=0}^{\infty} \frac{1}{k!}\xi^k$. We may thus represent the solution of the system (1-5.1) in the useful form

(1-5.3) $$\phi^i(t;x_o^m) = \exp\left(t\, f^j(x_o^m)\frac{\partial}{\partial x_o^j}\right) x_o^i .$$

for t in some sufficiently small neighborhood of the origin. Written in slightly different notation, we thus have the following result.

If the n *functions* $\{f^i(x^m)\}$ *are of class* C^∞ *in some neighborhood of the origin of* E_n *, then, for* t *in a sufficiently small neighborhood of* 0 *, the system of autonomous ordinary differential equations*

(1-5.4) $$\frac{d\bar{x}^i(t)}{dt} = f^i(\bar{x}^m(t)) , \quad i = 1,\dots,n$$

subject to the initial data

(1-5.5) $$\bar{x}^i(0) = x^i$$

has unique solutions that may be represented by

(1-5.6) $$\bar{x}^i(t) = \exp\left(t\, f^j(x^m)\frac{\partial}{\partial x^j}\right) x^i .$$

This form of writing the result provides us with a significant amount of new information when proper care is used with the known formula $\exp(a+b) = \exp(a)\exp(b)$.

The essential new aspect of these considerations is that the argument of the exponential is not a number or even a function; rather, it is an operator

$$t\,f \stackrel{\text{def}}{=} t\,f^j(x^m)\,\frac{\partial}{\partial x^j}\,.$$

Thus, $\exp(t\,f)$ is not defined if it does not operate on a function of the x's. It is for this reason that we must write $\exp(t\,f)x^i$ in (1-5.6), for otherwise there would be just nonsense.

1-6. FINITE DIMENSION VECTOR SPACES, DUAL SPACES AND ALGEBRAS

The reader is assumed to be familiar with vector spaces of finite dimension and with the concepts of linear algebra. However, just to be certain that all mean the same thing by the well-worn words, and as a gentle reminder, the definitions will be given.

Let V be a collection of elements $u, v, w, \ldots$ $(u, v, w, \ldots \in V)$ and let $a, b, c, \ldots$ be elements of $\mathbb{R}$. Suppose that we are given an operation $+$,

$$+: \quad (V,V) \longrightarrow V \mid w = u + v\,,$$

of $V \times V$ to V, and an operation $\cdot$,

$$\cdot: \quad (\mathbb{R},V) \longrightarrow V \mid w = a \cdot u\,,$$

of $\mathbb{R} \times V$ to V. Then V is a *vector space* over the real number field, $\mathbb{R}$, if and only if V and the operations $+$ and $\cdot$ satisfy the following conditions:

V1. $u + (v + w) = (u + v) + w$ for all $u,v,w \in V$,

V2. $u + v = v + u$ for all $u,v \in V$,

V3. There exists an element 0 of V such that $0 + v = v$ for all $v \in V$,

V4. For each $u \in V$ there exists one and only one $v \in V$ such that $u + v = 0$,

V5. $(ab) \cdot u = a \cdot (b \cdot u)$ for all $a, b \in \mathbb{R}$ and all $u \in V$,

V6. $(a + b) \cdot u = a \cdot u + b \cdot u$ for all $a, b \in \mathbb{R}$ and all $u \in V$,

V7. $a \cdot (u + v) = a \cdot u + a \cdot v$ for all $a \in \mathbb{R}$ and all $u, v \in V$,

V8. $1 \cdot u = u$ for all $u \in V$.

A set of elements $u_1, u_2, \ldots, u_k \in V$ is said to be *linearly independent* if and only if the only values of the constants $c_1, c_2, \ldots, c_k \in \mathbb{R}$ that satisfy

$$c_1 \cdot u_1 + c_2 \cdot u_2 + \ldots + c_k \cdot u_k = 0$$

are $c_1 = c_2 = \ldots = c_k = 0$.

A vector space V is said to be n-dimensional if and only if the maximal number of linearly independent elements of V is n . In this event, we write V_n . If V is n-dimensional, then any system of n linearly independent elements $v_1, \ldots, v_n$ is a *basis* for V_n , and any $u \in V_n$ can be expressed as $u = c_1 \cdot v_1 + c_2 \cdot v_2 + \ldots + c_n \cdot v_n$, where the constants $c_1, \ldots, c_n \in \mathbb{R}$ are uniquely determined by u and $v_1, \ldots, v_n$.

Notice that there is nothing in the definition of a vector space that requires it to be of finite dimension. A particularly important example of a nonfinite dimensional vector space is the collection $\Lambda^0(E_n)$ of *all* C^∞ mappings of E_n into $\mathbb{R}$, where $+$ is defined by

$$(f + g)(x^i) = f(x^i) + g(x^i) \tag{1-6.1}$$

and $\cdot$ is defined by

$$(a \cdot f)(x^i) = a\, f(x^i) \ . \tag{1-6.2}$$

Clearly, the sum of two C^∞ functions is a C^∞ function and a numerical multiple of a C^∞ function is a C^∞ function, so that $\Lambda^0(E_n)$ is closed under the operations $+$ and $\cdot$. It is then a trivial matter to verify that axioms V1 through V8 are indeed satisfied.

We now come to two concepts that may not be quite so familiar. Let u^* be a mapping from a vector space V to $\mathbb{R}$,

$$u^*: V \longrightarrow \mathbb{R} \mid r = \langle u^*, v \rangle , \quad v \in V . \tag{1-6.3}$$

Such mappings are called *functionals* in order to distinguish them from ordinary functions. A functional is said to be *linear* if and only if

$$\langle u^*, a \cdot v + b \cdot w \rangle = a \langle u^*, v \rangle + b \langle u^*, w \rangle . \tag{1-6.4}$$

The collection of all linear functionals on a vector space V is referred to as the *dual space* of V and is usually designated by V^* . Since the range of any functional is the real line, it is immediate that the definitions

$$\langle u^* + v^*, w \rangle = \langle u^*, w \rangle + \langle v^*, w \rangle , \tag{1-6.5}$$

$$\langle b \cdot u^*, w \rangle = b \langle u^*, w \rangle \tag{1-6.6}$$

makes V^* into a vector space. Thus, *the dual of any vector space is a vector space*. Further, if V is n-dimensional then V^* is also n-dimensional. The reason that dual spaces are important, beside their obvious computational efficiency, is that they almost invariably have nicer properties than the vector space with which one starts. This is particularly true whenever the original vector space is not finite dimensional.

The second concept is that of an *algebra* over the real number field. This is simply a vector space V over the real number field together with a binary operation $(\,|\,)$ on $V \times V$ to V such that

$$(a \cdot u + b \cdot v | w) = a \cdot (u|w) + b \cdot (v|w) , \tag{1-6.7}$$

$$(1\text{-}6.8) \qquad (u|a\cdot v + b\cdot w) = a\cdot(u|v) + b\cdot(u|w) \; ;$$

that is, it is linear in both the first and the second arguments. For example, the vector space $\Lambda^o(E_n)$ of all C^∞ functions on E_n becomes an algebra under the definition

$$(1\text{-}6.9) \qquad (f|g)(x^i) = f(x^i)\, g(x^i) \; .$$

Since the product of two C^∞ functions is a C^∞ function, (1-6.9) is on $\Lambda^o(E_n)\times\Lambda^o(E_n)$ to $\Lambda^o(E_n)$. It is then a simple matter to verify that (1-6.9) satisfies (1-6.7) and (1-6.8). An algebra is said to be *commutative* if and only if

$$(1\text{-}6.10) \qquad (u|v) = (v|u)$$

for all $u, v \in V$. If (1-6.10) does not hold for all $u, v \in V$ then the algebra is said to be *noncommutative*. An algebra is said to be *associative* if and only if

$$(1\text{-}6.11) \qquad (u|(v|w)) = ((u|v)|w)$$

for all $u,v,w \in V$. The situation that is most familiar is that of an algebra that is both associative and commutative, such as $\Lambda^o(E_n)$ with the definition (1-6.9). However, noncommutative, nonassociative algebras will be of central importance and will prove to be instrumental in the solutions of a number of real and important problems.

CHAPTER II

VECTOR FIELDS

2-1. THE TANGENT TO A CURVE AT A POINT

There is a convention that will save us many repetitions; namely, any function that appears is assumed to be as many times differentiable as required. In fact, we will actually go one step further and assume that every function is of class C^∞ unless noted to the contrary.

Let J be an open interval of the real line, $\mathbb{R}$, that contains the point $t = 0$. The map

$$(2\text{-}1.1) \qquad \Gamma : J \longrightarrow E_n \mid \bar{x}^i = \gamma^i(t)$$

of J into E_n defines a smooth (C^∞) curve in E_n if all of the n functions $\{\gamma^i(t)\}$ are smooth (C^∞) functions on the common domain J. Such a curve Γ provides the information that is required in order to compute the "tangent vector" to the curve: simply apply the operation d/dt to each of the n functions $\{\gamma^i(t)\}$ so as to obtain the n quantities $\{d\gamma^i(t)/dt\}$. Although this is indeed a simple operation, it can lead to a great deal of confusion, for $\{d\gamma^i(t)/dt\}$, in its most elementary interpretation, defines the directed line segment that starts at the point in E_n with coordinates $\bar{x}^i = \gamma^i(t)$. Thus, if $\gamma^i(t_1) \neq \gamma^i(t_2)$ for any one value of the index i, then the quantities $\{d\gamma^i(t_1)/dt\}$ and $\{d\gamma^i(t_2)/dt\}$ have their points of origination at $P_1 : (\gamma^i(t_1))$ and $P_2 : (\gamma^i(t_2))$, respectively. The geometric situation is depicted in Fig. 5. There is thus no way that $\{d\gamma^i(t)/dt\}$ can be combined with the operations of a vector space for vectors can only be added if they have a common origin. It should thus be clear that we have to start over from scratch if we are to have any success in associating a vector space with the intuitive notion of tangents to curves.

Let P denote a generic point of E_n and let the coordinates of P be (x_o^i) with respect to the established Cartesian coordinate cover of E_n, $P : (x_o^i)$. Again, let J be an open interval of the real line that contains the point $t = 0$, but this time, let the map Γ be defined by

$$(2\text{-}1.2) \qquad \Gamma : J \longrightarrow E_n \mid \bar{x}^i = x_o^i + \gamma^i(t) , \quad \gamma^i(0) = 0 .$$

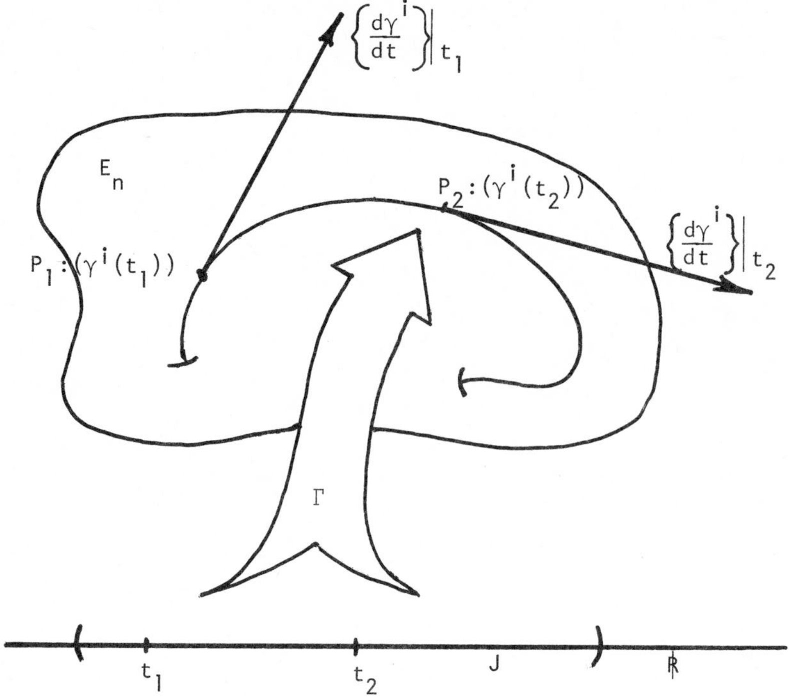

Figure 5. The Curve Γ and Its Tangents

We then have, on applying d/dt to the functions $\bar{x}^i = x_o^i + \gamma^i(t)$,

$$d\bar{x}^i/dt = d\gamma^i(t)/dt ,$$

and hence

$$\left.\frac{d\bar{x}^i}{dt}\right|_{t=0} = v^i , \quad i = 1,\ldots,n , \tag{2-1.3}$$

where the n numbers $\{v^i\}$ are given by

$$v^i = \left.\frac{d\gamma^i(t)}{dt}\right|_{t=0} . \tag{2-1.4}$$

Thus, for example, if $\bar{x}^i = x_o^i + a^i t + o(t)$, then the numbers v^i are given by $v^i = a^i$.

The important thing to note here is that the point P in E_n is considered fixed. With this in mind, we consider the totality of all curves Γ that pass through the point P; that is, we allow the functions $\{\gamma^i(t)\}$ in (2-1.2) to range through the collection of all n-tuples of functions with domain J such that $\gamma^i(0) = 0$. Now, all such functions look like $\gamma^i(t) = v^i t + o(t)$, where the n-tuples of numbers $\{v^i\}$ range through the collection of all n-tuples of real numbers. Applying d/dt to the functions $x_o^i + \gamma^i(t) = x_o^i + v^i t + o(t)$, we obtain the n-tuples

$$v^i = \left.\frac{d\bar{x}^i(t)}{dt}\right|_{t=0} . \tag{2-1.5}$$

As the n-tuple of real numbers $\{v^i\}$ ranges through the collection of all n-tuples of real numbers, we obtain the collection of all tangents at the point P to all curves through the point P. Thus, if we identify the n-tuple $\{v^i\}$ with the components of a vector in an n-dimensional vector space V, we may consider this vector space as attached to the space E_n at the point P. Under this interpretation, we write $T_P(E_n)$ for the n-dimensional vector space V that is attached to E_n at the point P and refer to $T_P(E_n)$ as the *tangent space to* E_n *at* P. Stated another way, $d/dt|_{t=0}$ establishes a 1-to-1 correspondence (via. (2-1.5)) between the totality of all curves through the point P and the elements of the vector space $T_P(E_n)$. The geometric notions are shown in Fig. 6 where, for simplicity, we have considered only E_2 realized as a surface in 3-dimensional Euclidean space.

2-2. THE TANGENT SPACE $T(E_n)$ AND VECTOR FIELDS ON E_n

We have just seen that the vector space $T_P(E_n)$ is attached to E_n at the generic point P of E_n in a natural way by allowing $d/dt|_{t=0}$ to act on all curves through the point $P:(x_o^i)$ such that $\bar{x}^i = \gamma^i(t)$ with $\gamma^i(0) = x_o^i$. On the other hand, the point P of E_n was generic in these considerations.

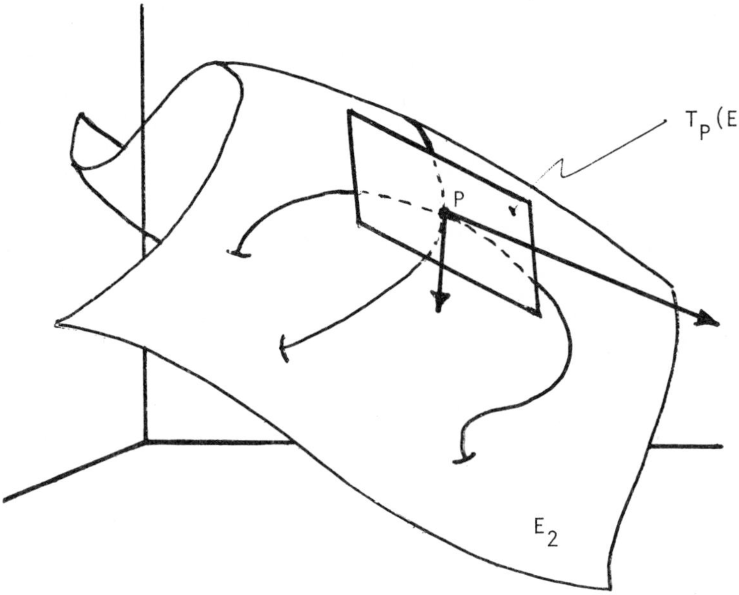

Figure 6. The Space $T_P(E_2)$

Thus, if we allow P to range throughout E_n or a smooth subset S of E_n, we obtain a vector space attached to each point of E_n or to each point of S. The union of these n-dimensional vector spaces, as P ranges over E_n or a subset S of E_n, is referred to as the *tangent space* of E_n or of S. The notation $T(E_n)$ or $T(S)$ is now standard for tangent spaces, so we write

$$T(E_n) = \bigcup_{P \varepsilon E_n} T_P(E_n) \tag{2-2.1}$$

and

$$T(S) = \bigcup_{P \varepsilon S} T_P(E_n) \quad \text{for} \quad S \subset E_n . \tag{2-2.2}$$

The reason why we have to consider $T(E_n)$ as the ensemble of n-dimensional vector spaces attached to each of the points of E_n is so that we can make sense out of the notion of the tangents to a curve in E_n and of a vector field in E_n. Let us first return to the curve Γ in E_n that is specified by (2-1.1):

$$\Gamma : J \longrightarrow E_n \mid \bar{x}^i = \gamma^i(t) .$$

If we apply the operator d/dt to the functions $\{\gamma^i(t)\}$, we obtain the n-tuples of functions $\{d\gamma^i(t)/dt\}$. For each value of t in the interval J, say t_1, the n-tuple $d\gamma^i(t_1)/dt = v^i$ specifies the selection of an element from the vector space $T_P(E_n)$ with $P:(\gamma^i(t_1))$. Thus, as t ranges through J, $d\gamma^i(t)/dt = v^i(t)$ specifies a selection of an elements from $T(\Gamma)$, and of course, this collection of vectors is only defined on the curve Γ itself. The *tangent field to a curve* Γ *is thus a map*

$$(2\text{-}2.3) \qquad T\Gamma : J \longrightarrow T(\Gamma)$$

that selects the element $v^i(t) = d\gamma^i(t)/dt$ *for each* $t \in J$ *from the vector space* $T_{P:(\gamma^i(t))}(E_n)$ *that belongs to*

$$T(\Gamma) = \bigcup_{t \in J} T_{P:(\gamma^i(t))}(E_n) .$$

A straightforward generalization of the notion of the tangent field to a curve yields the following definition of a general vector field.

<u>Definition.</u> A *vector field* v on a region S of E_n is a smooth (C^∞) map

$$(2\text{-}2.4) \qquad v : S \subset E_n \longrightarrow T(S) \mid v^i = v^i(x^j) , \quad i = 1,\dots,n$$

such that if $P:(x^j) \in S$ then $\{v^i(x^j)\} \in T_P(E_n)$.

A vector field thus has nothing whatsoever to do with just one vector space. Rather, it is an assignment of an element from each of a large collection of vector spaces; indeed, there are as many vector spaces involved as there are points in the domain S

of the vector field.

The prejudices that naturally grow up with our first exposure to Euclidean space with its Cartesian coordinate cover tend to be a stumbling block to a full understanding of the notion of a vector field. We always have in the back of our minds the certainty of the ability to measure distances and angles in Euclidean space and the almost self evident notion of the rigidity and permanence of these distance and angle measurements; that is, Euclidean space is truly cast in concrete (with the implied pun). On the other hand, for most of our studies, and for a large number of important problems, we have just an n-dimensional number space E_n that is not equipped in any specified manner with an ability to measure distances and angles. Stated another way, E_n without further restrictions is to be taken as an amorphous or gelatinous arena in which we are only able to label its points by coordinate labels, but not assign these coordinate labels their concrete meanings that would be assigned if there were Cartesian coordinates of an n-dimensional Euclidean space. Granted, this is somewhat unsettling, but we do indeed have something of permanence to deal with, namely the vector spaces $T_P(E_n)$ and the tangent space $T(E_n)$ itself. Indeed, without the ability to attach the n-dimensional vector space structure $T_P(E_n)$ to each point of E_n, which is what $T(E_n)$ actually is, there would be little hope of any sort of quantification in such gelatinous circumstances. Stated another way, we can speak of vector fields and all of their attendant consequences and computational efficacies without having to make specific assumptions about the geometry of the domain space E_n, and this is often a distinct advantage in specific problems.

2-3. OPERATOR REPRESENTATIONS OF VECTOR FIELDS

The formation of $T_P(E_n)$ and of $T(E_n)$, although conceptually clear, give vector fields as n-tuples of functions

$\{v^i(x^j)\}$ that specify the selections of the appropriate elements from $T(E_n)$ as $P:(x^i)$ ranges over E_n. This is cumbersome to work with. It is therefore preferable to proceed along different, but equivalent lines so that one works with scalar quantities (i.e., quantities that have a 1-dimensional range).

Let us return to the construct of the tangent to a curve through a point in E_n. Let

$$(2\text{-}3.1) \qquad F:E_n \longrightarrow R \mid r = f(x^1,\ldots,x^n) = f(x^i) ,$$

where f is a smooth function whose domain contains the range of the map

$$(2\text{-}3.2) \qquad \Gamma:J \longrightarrow E_n \mid \bar{x}^i = \gamma^i(t) , \quad \gamma^i(0) = x_o^i$$

that defines the curve of interest through the point $P:(x_o^i)$. Under these conditions, we may compose the maps F and Γ to obtain the map

$$(2\text{-}3.3) \qquad T = F\circ\Gamma : \mathbb{R} \longrightarrow \mathbb{R} \mid \tau = F(t) = f(\gamma^i(t)) .$$

Now, the chain rule of the calculus shows that

$$(2\text{-}3.4) \qquad dF(t)/dt = \frac{d\gamma^i(t)}{dt} \left.\frac{\partial f(x^j)}{\partial x^i}\right|_{x^i=\gamma^i(t)}$$

The important thing to be observed here is the occurrence of the coefficients $\{d\gamma^i(t)/dt\}$ on the right-hand side of (2-3.4). This suggests that vectors and vector fields can be defined in a manner similar to (2-3.4) provided we take advantage of the freedom in the choice of the function $f(x^j)$ that defines the map F. For example, if we make the choice $f(x^j) = x^k$, which is in fact a C^∞ function on E_n, then (2-3.4) yields

$$dF(t)/dt = \frac{d\gamma^i(t)}{dt} \left.\frac{\partial x^k}{\partial x^i}\right|_{x=\gamma(t)} = \frac{d\gamma^k(t)}{dt} .$$

We can thus recover the n-tuple $\{d\gamma^i(t)/dt\}$ provided we take a sufficiently large collection of functions for the f's and evaluate the corresponding $dF(t)/dt$'s .

We now have to put these ideas on a firm foundation. The essential ingredient in the above considerations is that of the functions f that generate the maps from E_n to $\mathbb{R}$. Clearly, these should be as smooth as possible for otherwise spurious complications may be introduced by discontinuities in the functions or some of their derivatives. The ideal candidate for the choice of the collection from which the f's are picked has already been encountered; namely the algebra $\Lambda^0(E_n)$ of C^∞ functions on E_n . This algebra has the vector space operations

$$(f+g)(x^j) = f(x^j) + g(x^j) , \quad (b \cdot f)(x^j) = b\, f(x^j)$$

and the multiplication law

$$(f|g)(x^k) = f(x^k)\, g(x^k) .$$

Since each element of $\Lambda^0(E_n)$ is a C^∞ function, the partial derivatives of any element of $\Lambda^0(E_n)$ is also a C^∞ function. (It is primarily for this reason that we take $\Lambda^0(E_n)$, for a collection such as all functions of class C^{12} would be mapped into the collection of all functions of class C^{11} by partial differentiation.)

If we start with any element $f(x^i)$ of $\Lambda^0(E_n)$, the action of an operator of the form

$$H = h^i(x^j) \frac{\partial}{\partial x^i}$$

is well defined when it acts on $f(x^j)$;

$$H(f) = h^i(x^j) \frac{\partial}{\partial x^i} f(x^j) = g(x^j) .$$

If each of the n functions $h^i(x^j)$ belongs to $\Lambda^0(E_n)$, then $H(f) = g$ gives back an element of $\Lambda^0(E_n)$. Further, it is clear from the calculus that

$$H(af + bg) = aH(f) + bH(g) , \quad H(fg) = fH(g) + gH(f) .$$

The first of these says that H is a linear operator on $\Lambda^o(E_n)$ and the second that H satisfies the rule of Leibniz, and they completely delineate how the operator H combines with the vector space and multiplication operations of the algebra $\Lambda^o(E_n)$.

Definition. An operation H on an algebra A is a *derivation* if and only if H maps A to A and satisfies the conditions

$$H(a \cdot f + b \cdot g) = a \cdot H(f) + b \cdot H(g) \tag{2-3.5}$$

$$H((f|g)) = (H(f)|g) + (f|H(g)) \tag{2-3.6}$$

for all $f,g \in A$ and all $a,b \in \mathbb{R}$.

With these facts in hand, we can now give a new definition of a vector field on E_n.

Definition. A (C^∞) *vector field* on E_n is a derivation

$$v = v^i(x^j) \frac{\partial}{\partial x^i} \tag{2-3.7}$$

on the algebra $\Lambda^o(E_n)$ of C^∞ functions on E_n.

This definition turns out to be both practical and powerful. The first thing to be noted is what happens when a vector field acts on the simplest possible functions other than constants; namely on the functions $x^1, x^2, \ldots, x^n$. It follows directly from (2-3.7) that

$$v(x^k) = v^i(x^j) \frac{\partial x^k}{\partial x^i} = v^k(x^j) , \tag{2-3.8}$$

and hence any derivation on $\Lambda^o(E_n)$ reproduces its coefficients. This can be stated in the following equivalent fashion.

LEMMA 2-3.1. *If* v *is a vector field on* E_n, *then* v *is uniquely determined by its action on the* n *functions* $x^1, x^2, \ldots, x^n$;

$$(2\text{-}3.9) \qquad v = v(x^i)\frac{\partial}{\partial x^i}\,.$$

Lemma 2-3.1 implies that *every derivation on the algebra* $\Lambda^o(E_n)$ *is a vector field.* Now simply observe that the collection of all vector fields on E_n is just the tangent space $T(E_n)$, so that we obtain the following equivalent definition.

Definition. The *tangent space*, $T(E_n)$, is the collection of all derivations on the algebra $\Lambda^o(E_n)$.

Let us go back and recall that $T(E_n)$ is also the union over $P \in E_n$ of $T_P(E_n)$. Now, each of the vector spaces $T_P(E_n)$ being n-dimensional says that each $T_P(E_n)$ has various collections of systems of n vectors that constitute a basis for $T_P(E_n)$. The obvious question is whether we can choose a basis in each $T_P(E_n)$ so that they vary smoothly as P sweeps out E_n; that is, can we choose a basis for all of $T(E_n)$. It turns out that we can, as the next Lemma shows.

LEMMA 2-3.2. *The* n *vector fields* $\partial/\partial x^1, \partial/\partial x^2, \ldots, \partial/\partial x^n$ *constitute a basis for* $T(E_n)$.

Proof. Each of the quantities $\partial/\partial x^i$, $i = 1,\ldots,n$, is a derivation on $\Lambda^o(E_n)$, so each is a vector field on E_n. Lemma 2-3.1 then shows that any element of $T(P)$ is a linear combination of these n vector fields: $v = v(x^i)\partial/\partial x^i = v^i(x^j)\partial/\partial x^i$. It thus remains only to show that these n vectors are linearly independent. However,

$$\frac{\partial}{\partial x^i}x^j = \delta^j_i = \begin{cases} 1 & \text{if } i = j \\ 0 & \text{if } i \neq j \end{cases},$$

and the n-by-n matrix with entries δ^i_j is defined at every point of E_n, constant on E_n, and nonsingular throughout E_n. Thus, since the n functions $x^1, \ldots, x^n$ belong to $\Lambda^o(E_n)$, the n vector fields $\{\partial/\partial x^i\}$ are linearly independent throughout E_n;

i.e., $0 = c^i \frac{\partial}{\partial x^i} x^j = c^j$ for $j = 1,\ldots,n$ implies $c^i(\partial/\partial x^i)f = 0$ for all $f \in \Lambda^0(E_n)$ if and only if $c^i = 0$, $i=1,\ldots,n$.

Definition. The basis $\{\partial/\partial x^i\}$ of $T(E_n)$ is the *natural basis* of $T(E_n)$ with respect to the (x^i) coordinate cover of E_n .

We shall see later that the natural basis will change when we change coordinate covers of E_n , so the qualification is essential. For the time being, however, the coordinate cover of E_n is fixed.

The existence of a global basis for $T(E_n)$ makes everything much simpler. Although we may still not add vectors in $T_{P_1}(E_n)$ to $T_{P_2}(E_n)$ for P_1 different from P_2 , verfication of the vector space properties of $T(E_n)$ is now significantly simplified. In fact we do it once and for all for each P in E_n simultaneously. Simply observe if u and v belong to $T(E_n)$,

$$u = u^i(x^j)\partial/\partial x^i , \quad v = v^i(x^j)\partial/\partial x^i ,$$

then we define the vector space operations + and • as would be done with respect to any basis of a finite dimensional vector space

$$\text{(2-3.10)} \quad u + v = (u^i(x^j) + v^i(x^j))\partial/\partial x^i ,$$

$$\text{(2-3.11)} \quad b \cdot u = b\, u^i(x^j)\partial/\partial x^i .$$

It is then a trivial matter to verify that postulates V1 through V8 of a vector space are satisfied; indeed, the calculations are identical to those for an ordinary n-dimensional vector space of ordered n-tuples of real numbers. We note in passing that $u^i(x^j) + v^i(x^j)$ and $b\, u^i(x^j)$ on the right-hand sides of (2-3.10) and (2-3.11) are well defined, throughout E_n , from the vector space properties of $\Lambda^0(E_n)$, so that $T(E_n)$ is closed under the vector space operations, as indeed it must. On the other hand, a $v \in T(E_n)$ only has a value at a point in E_n when it acts on a specific element of $\Lambda^0(E_n)$, in which case it has a single, unique value; *not* the n values of its coefficients.

2-4. ORBITS

We started our considerations with the notion of the tangents to a curve, and have wound up with a rather abstract notion of a vector field as a derivation on an algebra. The full circle can be closed if we now show that a vector field leads in a natural manner to a unique system of curves. Not only is this useful from the conceptual point of view, it also provides the basis for solving a number of important problems, among which is that of solving linear first order partial differential equations with variable coefficients.

The association in question comes about through the following definition.

Definition. The n functions $\{\bar{x}^i(t)\}$ define a curve in E_n, by the map

$$(2\text{-}4.1) \qquad \Gamma : J \longrightarrow E_n \mid x^i = \bar{x}^i(t) ,$$

that is an *orbit* of the vector field

$$(2\text{-}4.2) \qquad v = v(x^i)\partial/\partial x^i = v^i(x^j)\partial/\partial x^i$$

if and only if

$$(2\text{-}4.3) \qquad d\bar{x}^i(t)/dt = v(\bar{x}^i(t)) = v^i(\bar{x}^j(t))$$

holds for all t in the open interval J of $\mathbb{R}$.

The only thing that might give some hesitation about this definition would be whether there are any functions that satisfy the system of nonlinear first order differential equations

(2-4.3). This is easily dispensed with, for the system (2-4.3) is an autonomous system and the functions $v^i(\bar{x}^j)$ are of class C^∞ because $v \in T(E_n)$. The basic existence theorem for such systems that was given in the previous Chapter thus guarantees the existence of a solution of (2-4.3) for some interval J such that $\{\bar{x}^i(t)\}$ passes through any point P in E_n at $t = 0$.

These remarks bring up a very important point, namely that we are interested in not just one orbit of a given $v \in T(E_n)$; rather, we would like to know all orbits of v so that we have a unique curve through each point P of E_n that is associated with v. This goal is easily achieved, for the basic existence theorem tells us that the solutions of (2-4.3) are C^∞ functions of the initial data because the functions $v^i(\bar{x}^j(t))$ are C^∞ functions of the $\bar{x}$'s. We therefore give the more useful definition.

Definition. The *orbits* of a vector field

$$(2\text{-}4.4)\qquad v = v(x^i)\partial/\partial x^i = v^i(x^j)\partial/\partial x^i$$

are all solutions of the autonomous system of differential equations

$$(2\text{-}4.5)\qquad d\bar{x}^i(t)/dt = v(\bar{x}^i(t)) = v^i(\bar{x}^j(t))$$

subject to the initial data

$$(2\text{-}4.6)\qquad \bar{x}^i(0) = x^i .$$

THEOREM 2-4.1. *The orbits of a vector field*

$$(2\text{-}4.7)\qquad v = v(x^i)\partial/\partial x^i = v^i(x^j)\partial/\partial x^i$$

have the form

$$(2\text{-}4.8)\qquad \bar{x}^i(t) = \exp(tv)x^i = \exp(t\, v^j(x^k)\partial/\partial x^j)x^i$$

for all t *in some open interval* J *that contains* $t = 0$ *. Thus, a vector field determines and is determined by its orbits.*

Proof. The representation (2-4.8) follows directly from application of the result (1-5.6) to the autonomous system (2-4.5) subject to the initial data (2-4.6) for all t in some open interval J containing $t = 0$. Thus, a vector field determines its orbits. Conversely, if we write (2-4.8) in the form

$$\bar{x}^i(t;x^j) = \exp(tv)x^i ,$$

then

$$d\bar{x}^i/dt = v(\exp(tv)x^i) .$$

An evaluation for $t = 0$ then gives $\bar{x}^i(0) = x^i$ and

$$d\bar{x}^i/dt\Big|_{t=0} = v(x^i) = v^i(x^j) ,$$

and hence the orbits of v determine v .

The proof of the last theorem has given us some important additional information, for it tells us how to compute the vector field from its orbits. In order to do this, we simply differentiate with respect to t and then evaluate at the origin $t = 0$. When the answer is written in terms of the initial data, we have

$$v^i(x_o^j) = \frac{d}{dt}\bar{x}^i(t)\Big|_{t=0} .$$

There is a very simple way of visualizing the orbits of a vector field, and, in fact, the vector field itself. Suppose that we have a fluid that fills up E_n and at time $t = 0$ we paint the molecules of this fluid red. If the fluid undergoes a steady flow with velocity vector field v , then $\bar{x}^i(t) = \exp(tv)x^i$ defines the coordinates of the fluid molecule at time t that was at the point with coordinates $\{x^i\}$ at time $t = 0$. The orbits of the velocity vector field thus become the trajectories of the fluid molecules that undergo the steady flow with the given velocity vector field v .

A few examples may help to clarify what is going on. If we start with the vector field

$$v = -2y\,\partial/\partial x + x\,z\,\partial/\partial z$$

in E_3 with coordinates (x,y,z) , we have to solve the system of differential equations

$$d\bar{x}/dt = -2\bar{y}\ , \quad d\bar{y}/dt = 0\ , \quad d\bar{z}/dt = \bar{x}\bar{z}$$

subject to the initial data

$$\bar{x}(0) = x\ , \quad \bar{y}(0) = y\ , \quad \bar{z}(0) = z\ .$$

This is quite an easy problem, so we simply write down the solution:

$$\bar{x}(t) = x - 2yt\ , \quad \bar{y}(t) = y\ , \quad \bar{z}(t) = z\,\exp(xt - yt^2)\ .$$

A differentiation with respect to t followed by an evaluation at $t = 0$ then gives

$$d\bar{x}(0)/dt = -2y\ , \quad d\bar{y}(0)/dt = 0\ , \quad d\bar{z}(0)/dt = xz\ ,$$

from which we recover the components $v^i(x^j)$ of our vector field; that is, $v^1(x^j) = -2y$, $v^2(x^j) = 0$, $v^3(x^j) = xz$ with $x^1 = x$, $x^2 = y$, $x^3 = z$.

In the second example, we suppose that we are given the vector field

$$v = y(1-2x)\partial/\partial x + (3z^2-5)\partial/\partial y$$

in E_4 with coordinates (x,y,z,r) . The orbits are determined by solving

$$\frac{d\bar{x}}{dt} = \bar{y}(1-2\bar{x})\ , \quad \frac{d\bar{y}}{dt} = 3\bar{z}^2 - 5\ , \quad \frac{d\bar{z}}{dt} = 0\ , \quad \frac{d\bar{r}}{dt} = 0$$

subject to the initial data

$$\bar{x}(0) = x, \quad \bar{y}(0) = y, \quad \bar{z}(0) = z, \quad \bar{r}(0) = r.$$

The solution of this problem is given by

$$\bar{x}(t) = (x - \tfrac{1}{2})\exp(-2yt - (3z^2-5)t^2) + \tfrac{1}{2},$$

$$\bar{y}(t) = (3z^2-5)t + y, \quad \bar{z}(t) = z, \quad \bar{r}(t) = r.$$

Again, differentiation with respect to t and evaluation at $t=0$ reproduces the given vector field v. It should also be noted that this solution can be obtained by summing the operator series representations that are given by

$$\bar{x}(t) = \exp(t\, y(1-2x)\partial/\partial x + t(3z^2-5)\partial/\partial y)x,$$

$$\bar{y}(t) = \exp(t\, y(1-2x)\partial/\partial x + t(3z^2-5)\partial/\partial y)y,$$

$$\bar{z}(t) = \exp(t\, y(1-2x)\partial/\partial x + t(3z^2-5)\partial/\partial y)z,$$

$$\bar{r}(t) = \exp(t\, y(1-2x)\partial/\partial x + t(3z^2-5)\partial/\partial y)r.$$

Explicit solutions rather than the operator series representations are preferable whenever the explicit solutions can be obtained. Obtaining explicit solutions is not always an easy matter, however, and so it is often the case that the operator series representations are all that can be given.

2-5. THE METHOD OF CHARACTERISTICS FOR LINEAR AND QUASILINEAR FIRST ORDER PARTIAL DIFFERENTIAL EQUATIONS

A linear (homogeneous) first order partial differential equation for the determination of a function $f(x^j)$ on E_n is a partial differential equation of the form

$$\text{(2-5.1)} \qquad v^i(x^j)\frac{\partial f}{\partial x^i} = 0,$$

where the n coefficient functions $\{v^i(x^j)\}$ are a given system of n functions on E_n. If we use the given functions $\{v^i(x^j)\}$ to construct the vector field

$$(2\text{-}5.2) \qquad v = v^i(x^j)\partial/\partial x^i ,$$

then we can rewrite (2-5.1) in the equivalent form

$$(2\text{-}5.3) \qquad v(f) = 0 .$$

Thus, the problem of finding all solutions to the partial differential equation (2-5.1) is equivalent to the problem of finding all elements of $\Lambda^o(E_n)$ that are mapped into 0 by the vector field v. This does not solve the problem, it only gives a restatement of it. However, this restatement in terms of a vector field makes available to us all of the information we already know concerning vector fields.

Of particular importance is the system of orbital equations associated with the vector field v that is defined by the partial differential equation (2-5.1):

$$(2\text{-}5.4) \qquad d\bar{x}^i(t)/dt = v(\bar{x}^i) = v^i(\bar{x}^j(t)) ,$$

$$(2\text{-}5.5) \qquad \bar{x}^i(0) = x^i .$$

Suppose that $g(x^j) = c$ is a given surface in E_n. This surface is taken into itself by the orbits of the vector fields if and only if $g(\bar{x}^i(t)) = c$ holds for all t in some open interval J of $t = 0$. When this expression is differentiated with respect to t and we make use of (2-5.4), we have

$$0 = dc/dt = dg(\bar{x}^j(t))/dt = \frac{\partial g}{\partial \bar{x}^i}\frac{d\bar{x}^i(t)}{dt} = v^i(\bar{x}^j)\frac{\partial g}{\partial \bar{x}^i} ;$$

that is $v(g) = 0$. Thus, we come full circle to our original partial differential equation. This intimate relation between the partial differential equation (2-5.1) and the vector field

v is codified in the following way.

Definition. The vector field

$$(2\text{-}5.6)\qquad v = v^i(x^j)\partial/\partial x^i$$

is said to be a *characteristic* vector field of the linear partial differential equation

$$(2\text{-}5.7)\qquad 0 = v(f) = v^i(x^j)\frac{\partial f}{\partial x^i}$$

and the orbits of the vector field v are the *characteristic curves* (characteristics) of the partial differential equation $v(f) = 0$.

We have seen that the characteristics (orbits) of a vector field can be written in the form

$$(2\text{-}5.8)\qquad \bar{x}^i(t) = \exp(tv)x^i .$$

Suppose that we can solve one of these equations for t as a function of the x's and the $\bar{x}$ that occurs in that equation. This can be used to eliminate t from the remaining $n-1$ $\bar{x}$'s so that we wind up with $n-1$ functions of the x's and the $\bar{x}$'s that do not involve the independent variable t explicitly. This in turn shows that we can find $n-1$ functions $g_1,\ldots,g_{n-1}$ so that the $n-1$ relations that obtain amongst the x's and the $\bar{x}$'s can be written as

$$(2\text{-}5.9)\qquad g_a(x^j) = g_a(\bar{x}^j) , \quad a = 1,\ldots,n ,$$

for otherwise putting $\bar{x}^i = \bar{x}^i(t)$ would give an explicit dependence on the independent variable t. However, we also have

$$0 = dg_a(x^j)/dt$$

because the x's are initial data and hence independent of t, and hence (2-5.9) yields

$$(2\text{-}5.10)\quad 0 = dg_a(x^j)/dt = \frac{\partial g_a(\bar{x}^j(t))}{\partial \bar{x}^i(t)} \frac{d\bar{x}^i(t)}{dt} = v(g_a(\bar{x}^j(t))) .$$

Thus, each of the $n-1$ functions $g_a(\bar{x}^j)$ that is defined in this way satisfies $v(g_a(\bar{x}^j)) = 0$. Finally, if we set

$$(2\text{-}5.11)\quad f = \Psi(g_1(x^j), \ldots, g_{n-1}(x^j)) ,$$

then f satisfies $v(f) = 0$ provided the function Ψ is of class C^1 in its $n-1$ arguments; simply observe that

$$v(f) = v(\Psi) = \sum_{a=1}^{n-1} \frac{\partial \psi}{\partial g_a} v(g_a) = 0$$

because $v(g_a) = 0$ for $a = 1,\ldots,n-1$. This method of solving $v(f) = 0$ is known as the *method of characteristics* and the solutions obtained from (2-5.11) as Ψ ranges over all possible C^1 functions of its arguments generates all C^1 solutions of the given partial differential equation.

For example, suppose that we are given the partial differential equation

$$x\, z \frac{\partial f}{\partial z} - 2y \frac{\partial f}{\partial x} = 0 .$$

The characteristic vector field of this equation is obviously

$$v = -2y\, \partial/\partial x + x\, z\, \partial/\partial z$$

and the equations for the characteristics are

$$d\bar{x}(t)/dt = -2\bar{y}(t) , \quad d\bar{y}(t)/dt = 0 , \quad d\bar{z}(t)/dt = \bar{x}(t)\, \bar{z}(t) ,$$

subject to the initial data $\bar{x}(0) = x$, $\bar{y}(0) = y$, $\bar{z}(0) = z$. We have already solved this problem in a previous example:

$$\bar{x}(t) = x - 2yt , \quad \bar{y}(t) = y , \quad \bar{z}(t) = z \exp(xt - yt^2)$$

since the first of these relations yields

$$t = \frac{x-\bar{x}}{2y} .$$

We obtain

$$\bar{z} = z \exp\left(\frac{x-\bar{x}}{2y} x - y\left(\frac{x-\bar{x}}{2y}\right)^2\right)$$

$$= z \exp\left(\frac{x-\bar{x}}{2y} \frac{x+\bar{x}}{2}\right) = z \exp\left(\frac{x^2-\bar{x}^2}{4y}\right)$$

$$= z \exp\left(\frac{x^2}{4y} - \frac{\bar{x}^2}{4\bar{y}}\right) = z \exp\left(\frac{x^2}{4y}\right)\exp\left(-\frac{\bar{x}^2}{4\bar{y}}\right)$$

since $y = \bar{y}$. Thus

$$y = \bar{y} , \quad \bar{z} \exp\left(\frac{\bar{x}^2}{4\bar{y}}\right) = z \exp\left(\frac{x^2}{4y}\right)$$

and the two functions g_1 and g_2 are

$$g_1 = y , \quad g_2 = z \exp\left(\frac{x^2}{4y}\right) .$$

The general solution of $v(f) = 0$ is accordingly given by

$$f = \psi\left(y, z \exp\left(\frac{x^2}{4y}\right)\right) .$$

We now turn to somewhat more complicated first order partial differential equations; namely, those referred to as quasilinear.

Definition. A first order partial differential equation of the form

(2-5.12) $$v(f) = g(x^j;f) , \quad v = v^i(x^j)\partial_i$$

is referred to as *quasilinear*.

Let us look for a solution of (2-5.12) that is defined implicitly by means of the equation

(2-5.13) $$F(x^j;f(x^j)) = C = \text{constant}.$$

If we take the total partial derivative of (2-5.13), we obtain

$$\frac{\partial F(x^j;f(x^j))}{\partial x^i} = \partial_i F(x^j;f(x^j)) + \frac{\partial F}{\partial f}\,\partial_i f = 0 .$$

When the ith of these equations is multiplied by $v^i(x^j)$ and i is summed from 1 through n, the result is

$$(2\text{-}5.14) \qquad v^i\,\partial_i F + \frac{\partial F}{\partial f}\,v^i\,\partial_i f = 0 .$$

Thus, if $f(x^j)$ satisfies (2-5.12) then the function $F(x^j,f)$ satisfies

$$v^i\,\partial_i F + g(x^j,f)\,\frac{\partial F}{\partial f} = 0 ;$$

that is

$$(2\text{-}5.15) \qquad (v + g(x^j,f)\partial_f)(F) = 0 ,$$

where ∂_f stands for $\partial/\partial f$ where the $n+1$ quantities $(x^1, x^2, \ldots, x^n, f)$ are considered as the independent coordinates of a point in E_{n+1}. Conversely, if $F(x^j,f)$, considered as an element of $\Lambda^0(E_{n+1})$, satisfies (2-5.15), then $F(x^j,f) = C$ defines f implicitly as a function of (x^j) provided $\partial F/\partial f = 0$. However, (2-5.14) shows that $F(x^j,f(x^j)) = C$ satisfies the identity

$$(2\text{-}5.16) \qquad (v + v^i\,\partial_i f\,\partial_f)(F) = 0 .$$

An elimination of the common term $v(F)$ between (2-5.15) and (2-5.16) thus gives

$$(2\text{-}5.17) \qquad (v^i\,\partial_i f - g(x^j,f))\partial_f F = 0 ,$$

from which we conclude that $f(x^j)$ defined by $F(x^j,f(x^j)) = C$ satisfies the original equation (2-5.12) whenever $\frac{\partial F}{\partial f} = 0$. This result can be summarized in the following convenient fashion:

The general solution of the quasilinear partial differential equation

(2-5.18) $v(f) = g(x^j, f(x^j))$

is obtained in the implicit form

(2-5.19) $F(x^j; f) = C$

by finding the general solution of the linear partial differential equation

(2-5.20) $(v + g(x^j, f)\partial_f)(F) = 0$.

where $v + g(x^j, f)\partial_f$ *is considered as an element of* $T(E_{n+1})$ *and* E_{n+1} *has the coordinate cover* $(x^1, x^2, \ldots, x^n, f)$.

This new formulation has replaced E_n by E_{n+1} with coordinate cover $(x^1, x^2, \ldots, x^n, f)$ and v by $v + g(x^j, f)\partial/\partial f$ which is an element of $T(E_{n+1})$. The restriction that $v \in T(E_n)$ is thus not essential. If we consider a general $\hat{u} \in T(E_{n+1})$,

$$\hat{u} = u^i(x^j, f)\partial/\partial x^i + g(x^j, f)\partial/\partial f ,$$

then any solution of $\hat{u}(F) = 0$ defines a solution of

$$u^i(x^j, f)\partial_i f = g(x^j, f)$$

through the relation $F(x^j, f) = C =$ constant. The method of characteristics is thus applicable to general quasilinear first order partial differential equations of the form

$$u^i(x^j, f)\partial f/\partial x^i = g(x^j, f) ,$$

where both the function g and the n functions $\{u^i\}$ are allowed to depend on the unknown function f as well as the n independent variables $\{x^i\}$.

A simple but illustrative example is provided by

(2-5.21) $(2x\ \partial/\partial x - 3x\ \partial/\partial y)(f) = (f)^2$.

In this case, we have to solve the linear equation

(2-5.22) $\quad (2x\, \partial/\partial x - 3y\, \partial/\partial y + f^2\, \partial/\partial f)(F) = 0$

whose characteristic vector field is

(2-5.23) $\quad \hat{v} = 2x\, \partial/\partial x - 3y\, \partial/\partial y + f^2\, \partial/\partial f\,.$

The orbital equations of this vector field are

(2-5.24) $\quad d\bar{x}/dt = 2\bar{x}\,, \quad d\bar{y}/dt = -3\bar{y}\,, \quad d\bar{f}/dt = (\bar{f})^2$

subject to the initial data

(2-5.25) $\quad \bar{x}(0) = x\,, \quad \bar{y}(0) = y\,, \quad \bar{f}(0) = f\,.$

Since the general solutions of these equations are

$$\bar{x}(t) = x\, e^{2t}\,, \quad \bar{y}(t) = y\, e^{-3t}\,, \quad -(\bar{f}(t))^{-1} = t - (f)^{-1}\,,$$

using the third to eliminate t from the first two yields

$$\bar{x}/x = e^{-2/\bar{f}}\, e^{2/f}\,, \quad \bar{y}/y = e^{3/\bar{f}}\, e^{-3/f}\,.$$

The invariant functions are thus given by

$$g_1 = \bar{x}\, e^{2/\bar{f}} = x\, e^{2/f}\,, \quad g_2 = \bar{y}\, e^{-3/\bar{f}} = y\, e^{-3/f}\,,$$

and hence the general solution of (2-5.22) is

(2-5.26) $\quad F(x,y,f) = \Psi(x\, e^{2/f}, y\, e^{-3/f})\,.$

The general solution of the original equation (2-5.21) is thus given in implicit form by

(2-5.27) $\quad \Psi(x\, e^{2/f}, y\, e^{-3/f}) = C$

when $\Psi(p,q)$ ranges over all C^1 functions of its arguments. The solution may also be written as

$$Q(y^2x^3, \frac{1}{f} - \frac{1}{2} \ln|x|) = \ {}^{`}C$$

since $y^2x^3 = \bar{y}^2\bar{x}^3$ and $\frac{1}{f} - \frac{1}{2} \ln|x| = \frac{1}{\bar{f}} - \frac{1}{2} \ln|\bar{x}|$ are also invariant functions of the orbits of (2-5.24).

2-6. THE NATURAL LIE ALGEBRA OF $T(E_n)$

We have already seen that $T(E_n)$ has a natural vector space structure that is inherited through its construction as the union over $P \in E_n$ of the vector spaces $T_P(E_n)$. The question thus arises as to whether a binary operation $(\ |\ )$ can be defined over $T(E_n) \times T(E_n)$ in such a way that $T(E_n)$ becomes an algebra. The easiest way of answering this question is to use the realization of $T(E_n)$ as the collection of all derivations

(2-6.1) $$v = v^i(x^j)\, \partial/\partial x^i$$

on the algebra $\Lambda^0(E_n)$. Before doing this, let us introduce a notational convenience that will simplify the writing of many of the required expressions. Specifically, let us set

(2-6.2) $$\partial_i \equiv \frac{\partial}{\partial x^i} .$$

The natural basis for $T(E_n)$ is then given by the n vector fields $\{\partial_i,\ i = 1,\ldots,n\}$ and (2-6.1) becomes

(2-6.3) $$v = v^i(x^j)\partial_i .$$

We begin by noting that $v \in T(E_n)$ being a derivation on $\Lambda^0(E_n)$ implies that

(2-6.4) $$v(f)(x^j) = g(x^j)$$

with $g(x^j) \in \Lambda^0(E_n)$. Thus, if

(2-6.5) $$u = u^i(x^j)\partial_i$$

is any other element of $T(E_n)$, we may apply u to $g(x^j)$. When (2-6.3) through (2-6.5) are used, we have

$$(2\text{-}6.6)\qquad u(v(f)) = u^i\partial_i(v^j\partial_j f) = u^i(\partial_i v^j\ \partial_j f + v^j\ \partial_i\partial_j f)$$
$$= u^i\ (\partial_i v^j)\ \partial_j f + u^i\ v^j\ \partial_i\partial_j f$$

where $\partial_i\partial_j f = \partial^2 f/\partial x^i \partial x^j$ so that

$$(2\text{-}6.7)\qquad \partial_i\partial_j f = \partial_j\partial_i f$$

for all $f \varepsilon \Lambda^o(E_n)$. An inspection of the right-hand side of (2-6.6) shows that the first term, $u^i\ (\partial_i v^j)\ \partial_j f$ is again a derivation on $\Lambda^o(E_n)$; simply put $w^j(x^k) = u^i(x^k)\partial_i v^j(x^k)$ to obtain $u^i\ (\partial_i v^j)\ \partial_j f = w^j\ \partial_j f$. The second term on the right-hand side of (2-6.7) is not a derivation due to the occurrence of the second derivatives $\partial_i\partial_j f$. It thus follows that $u(v(f))$ does not define a binary operation on $T(E_n) \times T(E_n)$ *to* $T(E_n)$. The work thus far is not in vain, however, if we can eliminate the disruptive set of terms $u^i\ v^j\ \partial_i\partial_j f$.

Let us apply u and v in the opposite order. This gives, in direct analogy with (2-6.6),

$$(2\text{-}6.8)\qquad v(u(f)) = v^i\ (\partial_i u^j)\ \partial_j f + v^i\ u^j\ \partial_i\partial_j f .$$

If we now subtract (2-6.8) from (2-6.6), the result is

$$(2\text{-}6.9)\qquad u(v(f)) - v(u(f)) = (u^i\ \partial_i v^j - v^i\ \partial_i u^j)\ \partial_j f$$
$$+ (u^i\ v^j - u^j\ v^i)\ \partial_i\partial_j f .$$

If we set

$$(2\text{-}6.10)\qquad S^{ij} = u^i\ v^j - u^j\ v^i ,\qquad H_{ij} = \partial_i\partial_j f$$

then

$$(2\text{-}6.11)\qquad S^{ij} = -\ S^{ji} ,\qquad H_{ij} = H_{ji} ,$$

and

$$(2\text{-}6.12)\qquad (u^i v^j - u^j v^i)\,\partial_i\partial_j f = S^{ij} H_{ij}\ .$$

The following Lemma shows that this sum vanishes identically.

LEMMA 2-6.1. *If* $S^{ij} = -S^{ji}$ *and* $H_{ij} = H_{ji}$ *then the sum* $W = S^{ij} H_{ij}$ *vanishes identically.*

Proof. We simply note that the symmetry of H_{ij} and the anti-symmetry of S^{ij} yields

$$W = S^{ij} H_{ij} = -S^{ji} H_{ij} = -S^{ji} H_{ji} = -W$$

so that W vanishes identically.

The desired result is now at hand, for we have shown that

$$(2\text{-}6.13)\qquad u(v(f)) - v(u(f)) = (u^i\,\partial_i v^j - v^i\,\partial_i u^j)\,\partial_j f$$

defines a binary operation on $T(E_n)\times T(E_n)$ to $T(E_n)$. The first step in defining a multiplication on $T(E_n)$ with the required properties is thus accomplished.

LEMMA 2-6.2. *The operation* $[u,v]$ *defined by*

$$(2\text{-}6.14)\qquad [u,v](f) = u(v(f)) - v(u(f))$$

for all $f \in \Lambda^0(E_n)$ *has the representation*

$$(2\text{-}6.15)\qquad [u,v] = (u^j\,\partial_j v^i - v^j\,\partial_j u^i)\partial_i$$

for every u, v *in* $T(E_n)$. *Thus* $[u,v]$ *defines a map of* $T(E_n)\times T(E_n)$ *to* $T(E_n)$.

All that now remains is to check that the binary operation $[\ ,\ ]$ satisfies the conditions (1-6.7) and (1-6.8) in order to conclude that $T(P)$ equipped with the multiplication $[\ ,\ ]$ forms an algebra over the real number field. This task is simplified by the noting that the following result is a direct consequence of (2-6.14).

LEMMA 2-6.3. *The binary operation* [,] *satisfies the identity*

$$(2\text{-}6.16) \qquad [u,v] = -[v,u]$$

for all $u,v \in T(E_n)$.

The final argument goes as follows:

LEMMA 2-6.4. *The binary operation* [,] *satisfies the identities*

$$(2\text{-}6.17) \qquad [u,av+bw] = a[u,v] + b[u,w]$$

$$(2\text{-}6.18) \qquad [au+bv,w] = a[u,w] + b[v,w]$$

for all $u,v,w \in T(E_n)$.

Proof. Use of Lemma 2-6.3 shows that (2-6.18) is a consequence of (2-6.17). Thus, we need only establish (2-6.17). This, however, is straightforward, for the definition (2-6.14) gives

$$\begin{aligned}
[u,av+bw](f) &= u(\{av+bw\}(f)) - \{av+bw\}(u(f)) \\
&= u(av(f) + bw(f)) - av(u(f)) - bw(u(f)) \\
&= au(v(f)) + bu(w(f)) - av(u(f)) - bw(u(f)) \\
&= a\{u(v(f)) - v(u(f))\} + b\{u(w(f)) - w(u(f))\} \\
&= a[u,v](f) + b[u,w](f)
\end{aligned}$$

for all $f \in \Lambda^0(E_n)$.

The multiplication that is defined on $T(E_n)$ by (2-6.14) occurs so often that it has been given a special name.

Definition. The multiplication, [,] , that is defined on

$T(E_n)$ by

(2-6.19) $$[u,v](f) = u(v(f)) - v(u(f))$$

for all $f \varepsilon \Lambda^o(E_n)$ is called the *Lie product.*

We can thus state our principal result in the following form.

THEOREM 2-6.1. *The tangent space,* $T(E_n)$, *equipped with the Lie product,* [,] , *is an algebra over the real number field.*

The algebra induced on $T(E_n)$ by the Lie product possesses certain properties that makes it quite different from the algebras that you may have already encountered. We already know that $T(E_n)$ is an *anticommutative* algebra, for Lemma 2-6.3 has shown that $[u,v] = -[v,u]$. This is not too unusual, for the classic vector product of two vectors in 3-dimensional Euclidean space exhibits this property: $\vec{a} \times \vec{b} = -\vec{b} \times \vec{a}$. The telling result is that $T(E_n)$ is a nonassociative algebra, as the following Lemma demonstrates.

LEMMA 2-6.5. *The Lie product satisfies the Jacobi identity*

(2-6.20) $$[u,[v,w]] + [v,[w,u]] + [w,[u,v]] = 0 .$$

The proof of this result is a straightforward, but lengthy calculation based on the definition (2-6.15). After each of the three terms is evaluated by this formula, it is found that the sum given by (2-6.20) leads to a collection of terms all of which cancel in pairs. The details of this calculation are left to the reader as an exercise.

The lack of associativity is now easily established. If we use the anticommutativity of the Lie product in the thrid term in (2-6.20), we have

$$[u,[v,w]] + [v,[w,u]] - [[u,v],w] = 0 ,$$

and hence

$$(2\text{-}6.21)\qquad [u,[v,w]] - [[u,v],w] = -[v,[w,u]] .$$

On the other hand, the Lie product is associative only if

$$[u,[v,w]] - [[u,v],w]$$

vanishes, which is definitely not the case here. Particular care must thus be exercised in the use of the algebra $T(E_n)$ for it is something of a revelation just how many familiar results cease to hold as soon as the multiplication becomes nonassociative.

Anticommutative algebras with the Jacobi identity occur so frequently that they have been given a special name that derives from their historical origins in the work of S. Lie in continuous groups.

Definition. An algebra whose binary operation $(\mid)$ satisfies

$$(2\text{-}6.22)\qquad (u|v) = -(v|u) ,$$

$$(2\text{-}6.23)\qquad (u|(v|w)) + (v|(w|u)) + (w|(u|v)) = 0$$

is a *Lie algebra.*

The basic Theorem 2-6.1 can thus be restated in the following equivalent form.

THEOREM 2-6.2. *The tangent space,* $T(E_n)$, *equipped with the Lie product is a Lie algebra.*

The following result is recorded for future reference.

LEMMA 2-6.6. *If* $p(x^j)$ *and* $q(x^j)$ *belong to* $\Lambda^0(E_n)$, *then*

$$(2\text{-}6.24)\qquad [pu,qv] = pq[u,v] + pu(q)v - qv(p)u .$$

Proof. The result follows directly from (2-6.14):

$$[pu,qv](f) = pu(qv(f)) - qv(pu(f))$$

$$= pu(q)v(f) + pqu(v(f)) - qv(p)u(f) - qpv(u(f))$$

$$= pq\{u(v(f)) - v(u(f))\} + pu(q)v(f) - qv(p)u(f)$$

$$= pq[u,v](f) + \{pu(q)v - qv(p)u\}(f) .$$

2-7. LIE SUBALGEBRA OF $T(E_n)$ AND SYSTEMS OF LINEAR FIRST ORDER PARTIAL DIFFERENTIAL EQUATIONS

The concept of a Lie subalgebra of a Lie algebra is possibly even more useful than the corresponding concept of a subspace of a vector space.

Definition. A *Lie subalgebra* of a Lie algebra A is vector subspace of A, considered as a vector space over the real number field, that is closed under the multiplication operation of the algebra.

Suppose that $u_1,\ldots,u_r$ belong to $T(E_n)$ and let S denote the vector subspace of $T(E_n)$ that is spanned by the elements $u_1,\ldots,u_r$. For simplicity, let us write

$$\{u_1, \ldots, u_r\} = \{u_a\}$$

where it is understood that the index a runs from 1 through r. The definition of a Lie subalgebra shows that the subspace S forms a Lie subalgebra only if S is closed under $[\ ,\ ]$; that is, the Lie product of every two elements of S belongs to S. Since S is spanned by $\{u_a\}$, this can be the case if and only if $[u_a, u_b]$ can be expressed as a linear combination of the u_a's.

LEMMA 2-7.1. *If S is a subspace of $T(E_n)$ that is spanned by the r elements $\{u_1,\ldots,u_r\} = \{u_a, a=1,\ldots,r\}$, then S is a Lie subalgebra of $T(E_n)$ if and only if there exist* r^3

elements $\{c^e_{ab}(x^j)\}$ *of* $\Lambda^o(E_n)$ *such that*

$$[u_a, u_b] = c^e_{ab}\, u_e , \tag{2-7.1}$$

$$c^e_{ab} = - c^e_{ab} , \tag{2-7.2}$$

and

$$0 = \{u_a(c^h_{bc}) + u_b(c^h_{ca}) + u_c(c^h_{ab}) + c^e_{bc}c^h_{ae} + c^e_{ca}c^h_{be} + c^e_{ab}c^h_{ce}\}u_h(f) \tag{2-7.3}$$

for all $f \in \Lambda^o(E_n)$.

Proof. We have already seen that S can be a subalgebra of $T(E_n)$ only if $[u_a, u_b]$ can be expressed as a linear combination of the u_a's, and this is exactly what (2-7.1) says. Since $[u_a, u_b] = - [u_b, u_a]$, we must also have (2-7.2). Similarly, the Jacobi identity, (2-6.20) gives the conditions

$$\begin{aligned} 0 &= [u_a,[u_b, u_c]] + [u_b,[u_c, u_a]] + [u_c,[u_a, u_b]] \\ &= [u_a, c^e_{bc}\, u_e] + [u_b, c^e_{ca}\, u_e] + [u_c, c^e_{ab}\, u_e] \\ &= u_a(c^e_{bc})u_e + u_b(c^e_{ca})u_e + u_c(c^e_{ab})u_e \\ &\quad + c^e_{bc}[u_a, u_e] + c^e_{ca}[u_b, u_e] + c^e_{ab}[u_c, u_e] ; \end{aligned}$$

that is, with e set equal to h in the first three terms,

$$0 = \{u_a(c^h_{bc}) + u_b(c^h_{ca}) + u_c(c^h_{ab}) + c^e_{bc}\, c^h_{ae} + c^e_{ca}\, c^h_{be} + c^e_{ab}\, c^h_{ce}\}u_h(f)$$

must hold for all $f \in \Lambda^o(E_n)$.

This Lemma shows that a Lie subalgebra is a much more constrained structure than a subspace; simply look at the additional requirements (2-7.1) through (2-7.3). For example, the subspace of $T(E_3)$ that is spanned by the elements

$$u_1 = x\,\partial_y - y\,\partial_x\,, \quad u_2 = y\,\partial_z - z\,\partial_y$$

gives

$$[u_1, u_2] = x\,\partial_z - z\,\partial_x \overset{\text{def}}{=} u_3\,.$$

However, $x\,u_2 - y\,u_3 + z\,u_1 \equiv 0$ and the coefficient functions in this expression all vanish only when all of the u's vanish, namely at $x = y = z = 0$. Hence the subspace spanned by u_1 and u_2 forms a Lie subalgebra of $T(E_3)$. On the other hand, for the subspace spanned by

$$v_1 = x\,\partial_y - y\,\partial_x + \partial_z\,, \quad v_2 = \partial_y$$

we have

$$[v_1, v_2] = \partial_x \overset{\text{def}}{=} v_3\,.$$

Since $\alpha\,v_1 + \beta\,v_2 + \gamma\,v_3 = 0$ only when $\alpha = \beta = \gamma = 0$, the subspace spanned by v_1 and v_2 does not form a Lie subalgebra of $T(E_3)$. We must thus consider the enlarged set v_1, v_2, v_3. However, $\partial_z = v_1 + y\,v_3 - x\,v_2$ and hence the subspace of $T(E_3)$ that is spanned by v_1, v_2, v_3 is also spanned by $\partial_x, \partial_y, \partial_z$; the completion of the subspace spanned by v_1, v_2 to a Lie subalgebra gives the original Lie algebra $T(E_3)$.

It should be noted in passing that, *so long as the* $\{C^e_{ab}\}$ *in* (2-7.1) *are not restricted to be constants, we can form linear combinations of the elements of* $T(E_n)$ *whose coefficients are themselves elements of* $\Lambda^0(E_n)$; that is, the vector subspaces may be formed by terms of the form

$$\alpha_1(x^j)u_1 + \alpha_2(x^j)u_2 + \dots\,.$$

The formation of a Lie subalgebra is then invariant for such replacements of constants by functions in forming linear combinations. This follows from (2-6.24),

$$[\alpha u_a, \beta u_b] = \alpha\beta[u_a, u_b] + \alpha u_a(\beta)u_b - \beta u_b(\alpha)u_a$$

which shows that if the subspace spanned by $u_1,\ldots,u_r$ forms a Lie subalgebra, then the space spanned by $\alpha_1^a(x^j)u_a,\ldots,\alpha_r^a(x^j)u_a$ also forms a Lie subalgebra.

Suppose that we start with a subspace S_1 of $T(E_n)$ that is spanned by the elements $u_1,\ldots,u_r$ that does not form a Lie subalgebra. In this event, there will be certain Lie products of the elements $u_1,\ldots,u_r$ that can not be expressed as linear combinations of the set $u_1,\ldots,u_r$. If there are s such Lie products, each of them is a new element of $T(E_n)$. Designating these new elements by $u_{r+1},\ldots,u_{r+s}$, we form the new set of elements $u_1,\ldots,u_r$, $u_{r+1},\ldots,u_{r+s}$. Designate the subspace of $T(E_n)$ that is spanned by these $r+s$ elements as S_3. Now either S_2 forms a Lie subalgebra or it does not. If it does, then we are through. If it does not, we adjoin all new elements of $T(E_n)$ that are formed by taking Lie products to obtain a new set of spanning elements for the new subspace S_3. Clearly, this process can be continued indefinitely. When it is performed, we say that the original subspace S_1 has been *completed* to a Lie subalgebra S.

The notion of the completion of a subspace of $T(E_n)$ to a Lie subalgebra is useful in many contexts. One of the simplest is that associated with the solvability of a system of first order linear partial differential equations. Suppose that we are given r elements $\{u_a\}$ of $T(E_n)$, and we need to find all functions $f(x^j)$ on E_n that satisfy the system of r first order linear partial differential equations

$$(2\text{-}7.4) \qquad u_a(f) = 0 , \quad a = 1,\ldots,r .$$

Each of these first order equations can be solved, at least in principle, by the method of characteristics that was given in

Section 2.5. What we now have to do is to determine whether there exist solutions of any one of them that are also solutions of the remaining $r-1$ equations of the system (2-7.4); that is, whether the general solution

$$f = \Psi(g_{1,a}(x^j), \ldots, g_{n-1,a}(x^j))$$

of the ath equation can be forced to satisfy the remaining $r-1$ equations by an appropriate choice of the function Ψ.

Suppose that f is a solution of $u_a(f) = 0$ and also a solution of $u_b(f) = 0$. It then follows that f will satisfy $u_a(u_b(f)) = 0$ and $u_b(u_a(f)) = 0$. These conditions contain second derivatives of the function f. The dependence on second derivatives is easily eliminated, however, for we need only form

$$(2\text{-}7.5) \qquad [u_a, u_b](f) = u_a(u_b(f)) - u_b(u_a(f)) = 0 .$$

Thus, *if* f *is to satisfy the system* $u_a(f) = 0$, $a = 1,\ldots,r$, *then it must also satisfy the linear first order partial differential equations*

$$(2\text{-}7.6) \qquad [u_a, u_b](f) = 0 .$$

If the subspace S_1 of $T(E_n)$ that is spanned by $\{u_a\}$ is a Lie subalgebra of $T(E_n)$, then

$$(2\text{-}7.7) \qquad [u_a, u_b](f) = c^e_{ab}\, u_e(f) ,$$

in which case satisfaction of the conditions (2-7.6) is implied by satisfaction of the original equations $u_a(f) = 0$, $a = 1,\ldots,r$. On the other hand, if the subspace S_1 does not form a Lie subalgebra of $T(E_n)$ then we have to adjoin to our original system of new equations that come from the conditions (2-7.6). This yields a subspace S_2 of $T(E_n)$ that is spanned by $u_1,\ldots,u_r$, $u_{r+1},\ldots,u_{r+s}$ if there are s new equations. The process just described now has to be repeated starting with the subspace S_2;

that is, we have to complete the subspace S_1 to a Lie subalgebra of $T(E_n)$.

<u>THEOREM 2-7.1</u>. *Let* S_1 *be a subspace of* $T(E_n)$ *that is spanned by a given system of* r *elements* $u_1,\ldots,u_r$ *and let* S *be the completion of* S_1 *to a Lie subalgebra of* $T(E_n)$. *Necessary conditions for the existence of a solution to the system of first order linear partial differential equations*

$$(2\text{-}7.8) \qquad u_a(f) = 0 , \quad a = 1,\ldots,r$$

is that f *satisfy*

$$(2\text{-}7.9) \qquad u(f) = 0$$

for every u *in the subspace* S . *Thus, a solution of* (2-7.8) *exists only if there is at least one function on* E_n *that is annihilated by the Lie subalgebra that is the completion of the subspace* S_1 .

The above theorem shows that there is quite a difference between a single first order partial differential equation and a system of two or more such equations. It is also clear that everything depends on the structure of the Lie subalgebra S of $T(E_n)$ that is obtained by completion of the subspace S_1 . Suppose that S has a basis that is comprised of k elements

$$v_1, v_2, \ldots, v_k$$

with $k \geq r$ necessarily. We then have the problem of solving the system of k equations $v_b(f) = 0$, $b = 1,\ldots,k$. On the other hand, for any P in E_n , $\dim(T_P(E_n)) = n$, so that $k \leq n$. We thus have two cases, $k < n$ and $k = n$. For the case $k = n$, we have n linearly independent elements $\{v_1,\ldots,v_n\} = \{v_i\}$ with

$$v_i = v_i^j(x^k)\,\partial_j , \quad \det(v_i^j(x^k)) \neq 0$$

on some open set N in E_n. Thus, for all P in N, we can solve

$$v_i(f) = v_i^j \partial_j f = 0$$

for $\partial_j f$ and obtain $\partial_j f = 0$, $j = 1,\dots,n$. In this event, the only solution is the trivial solution f = constant.

COROLLARY 2-7.1. *Let S_1 be a subspace of $T(E_n)$ that is associated with a given system of first order linear partial differential equations, as in Theorem 2-7.1, and let S be the completion of S_1 to a Lie subalgebra. If S has an n-dimensional basis on some neighborhood N of E_n then the only solution of the given system of partial differential equations of N is the trivial solution $f(x^j)$ = constant.*

The following result will simply be stated, in the interests of completeness since the proof is fairly technical and significantly more complicated than the corresponding results that be obtained after we have mastered the exterior calculus.

COROLLARY 2-7.2. *Let S_1 and S be as in Corollary 2-7.1 and let S have a k-dimensional basis on a neighborhood N of E_n with $k < n$. Then the given system of partial differential equations has exactly $n-k$ independent integrals $h_1(x^j), \dots, h_{n-k}(x^j)$ on N and the general solution is given by $\Psi(h_1,\dots,h_{n-k})$.*

In fact, the substance of the proof is to show that the integration of any one of the k equations (by the method of characteristics) will give $f = \eta(g_1,\dots,g_{n-1})$, in which case the unknown function η will satisfy a complete system of equations in the variables $g_1,\dots,g_{n-1}$ whose dimension is exactly $k-1$. A continuation of this process will ultimately yield $f = \Psi(h_1(x^j), \dots, h_{n-k}(x^j))$, where the $n-k$ functions $\{h_m(x^j)\}$ are simultaneous integrals of the completed system.

We have seen that the subspace of $T(E_3)$ spanned by

$$u_1 = x\,\partial_y - y\,\partial_x , \qquad u_2 = y\,\partial_z - z\,\partial_y$$

forms a Lie subalgebra of $T(E_3)$. The method of characteristics applied to $u_1(f) = 0$ yields $f = \Psi(x^2 + y^2, z)$, while $u_2(f) = 0$ yields $f = \phi(x, y^2 + z^2)$. Thus, the common integrals of $u_1(f) = u_2(f) = 0$ are given by $f = \rho(x^2 + y^2 + z^2)$, namely an arbitrary C^1 function of the $3-2 = 1$ argument $x^2 + y^2 + z^2$. On the other hand, the completion of the subspace spanned by

$$v_1 = x\,\partial_y - y\,\partial_x + \partial_z , \qquad v_2 = \partial_y$$

to a Lie subalgebra gives all of $T(E_3)$. Thus, f satisfies $v_1(f) = v_2(f) = 0$ only if $\partial_1 f = \partial_2 f = \partial_3 f = 0$, in which case the only solutions are the trivial ones f = constant.

Another occurrence of a Lie subalgebra, and one which arises in almost any physical discipline is in the theory of Lie groups. In fact, the whole theory actually started there. *A Lie subalgebra of* $T(E_n)$, *that is spanned by a finite number of elements* $v_1, \ldots, v_r$ *such that*

$$[v_a, v_b] = C^e_{ab}\, v_e , \qquad \partial_i\, C^e_{ab} = 0$$

generates a Lie group that can be realized as a group of point transformations acting on E_n (a group of transformations that moves the points of E_n along the orbits of all linear combinations of the vector fileds $v_1, \ldots, v_r$ with constant coefficients). In this event, the C's are referred to as the *structure constants* of the group.

The careful reader will have perceived that we started with vector spaces over the real number field, $\mathbb{R}$, and have progressed to vector spaces over the associative algebra $\Lambda^0(E_n)$ of C^∞ functions on E_n. A case in point is $T(E_n)$, for it is closed under multiplication of its elements by elements of $\Lambda^0(E_n)$ rather than just by constants (elements of $\mathbb{R}$); if $v \in T(E_n)$ then $f(x^j)v \in T(E_n)$ for any $f(x^j) \in \Lambda^0(E_n)$. The meaning of linear dependence in a vector space over $\mathbb{R}$ and in a vector space over $\Lambda^0(E_n)$ are quite different, however.

Definition. A collection $\{v_a, a = 1,\ldots,r\}$ of elements of $T(E_n)$ is said to be *linearly dependent over* $\mathbb{R}$ (linearly dependent with constant coefficients) if and only if

$$c_1 v_1 + c_2 v_2 + \ldots + c_r v_r = 0$$

and

$$\partial_i c_a = 0 , \quad a = 1,\ldots,r$$

can be satisfied simultaneously at every point of E_n without all of the c's being zero. A collection of elements of $T(E_n)$ is said to be *linearly independent over* $\mathbb{R}$ if it is not linearly dependent over $\mathbb{R}$.

Definition. A collection $v_1, \ldots, v_r$ of elements of $T(E_n)$ is said to be *linearly dependent* over $\Lambda^0(E_n)$ if and only if

$$f_1(x^j)v_1 + f_2(x^j)v_2 + \ldots + f_r(x^j)v_r = 0 ,$$

can be satisfied simultaneously at every point of E_n by a collection of r elements $\{f_a(x^j)\}$ of $\Lambda^0(E_n)$ not all of which vanish identically over E_n. A collection of elements of $T(E_n)$ is said to be *linearly independent over* $\Lambda^0(E_n)$ if and only if it is not linearly dependent over $\Lambda^0(E_n)$.

We first note that there can be at most n elements in a linearly independent system over $\Lambda^0(E_n)$, for each $T_P(E_n)$ for each $P \in E_n$ is n-dimensional. This is also seen from the fact that any element of $T(E_n)$ is a linear combination over $\Lambda^0(E_n)$ of the natural basis $\{\partial_i\}$. On the other hand, there can be significantly more than n elements of $T(E_n)$ that are linearly independent over $\mathbb{R}$. For example,

$$\partial_x,\ \partial_y,\ \partial_z,\ x\,\partial_y - y\,\partial_x\ ,\ y\,\partial_z - z\,\partial_y\ ,\ z\,\partial_x - x\,\partial_z$$

is a system of six elements of $T(E_3)$ that is linearly independent over $\mathbb{R}$. Simply note that $x\,\partial_y - y\,\partial_x$ can not be written as a linear combination of $\partial_x, \partial_y, \partial_z$ with constant coefficients for all $(x,y,z) \in E_3$.

2-8. BEHAVIOR UNDER MAPPINGS

Our considerations have been confined thus far to a domain space E_n with a fixed (Cartesian) coordinate cover. The question thus naturally arises as to what happens when we change the coordinate cover. As with most things, the obvious question is not necessarily the easiest to answer, so we pose a less obvious but easier question to answer first: what happens when we perform mappings from E_n to E_m.

As before, we assume that E_n is covered by the (x^i) coordinate system. Let E_m be covered by the (y^A) coordinate system with capital Latin indices ranging over the integers from 1 to m. If Φ is a smooth map from E_n to E_m that is represented by the m functions $\{\phi^A(x^j)\}$, then we write

$$\Phi : E_n \longrightarrow E_m \mid y^A = \phi^A(x^j)\ . \tag{2-8.1}$$

When we recall that $T(E_n)$ is the collection of all derivations on the algebra $\Lambda^0(E_n)$, it is clear that the behavior of elements of $T(E_n)$ is to a great extent controlled by the relationship between the algebras $\Lambda^0(E_n)$ and $\Lambda^0(E_m)$ that is induced by the mapping Φ. We therefore take up this matter first.

We know that any element $f(x^j)$ of $\Lambda^0(E_n)$ can be viewed as generating a mapping F from E_n to $\mathbb{R}$,

$$F : E_n \longrightarrow \mathbb{R} \mid \tau = f(x^j)\ , \tag{2-8.2}$$

and likewise, any element $g(y^A)$ of $\Lambda^0(E_m)$ can be viewed as

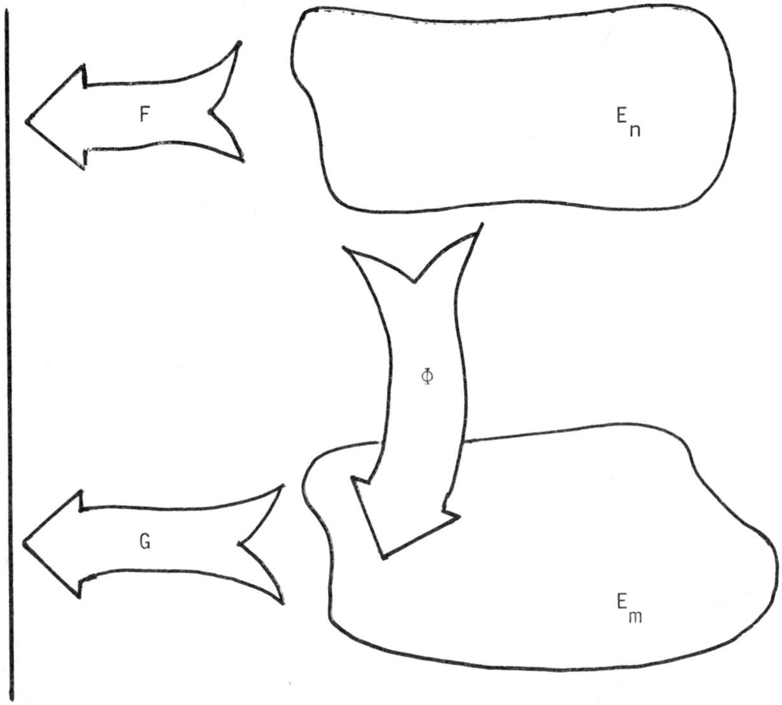

Figure 7. The Maps ϕ, F and G

generating a mapping G from E_m to $\mathbb{R}$,

$$(2\text{-}8.3) \qquad G:E_m \longrightarrow \mathbb{R} \mid \tau = g(y^A) .$$

Combining these three mappings generates the graphic situation shown in Fig. 7. It is clear from this figure, or from the fact that $\mathcal{R}(\Phi) = \mathcal{D}(G) = E_m$, that we may compose the mappings G and Φ . This, however, yields the map

$$(2\text{-}8.4) \qquad H = G\circ\Phi:E_n \longrightarrow \mathbb{R} \mid \tau = g(\phi^A(x^j)) .$$

When this is done for every g in $\Lambda^o(E_m)$, a map from $\Lambda^o(E_m)$ to $\Lambda^o(E_m)$ is obtained, that is commonly denoted by Φ^* ;

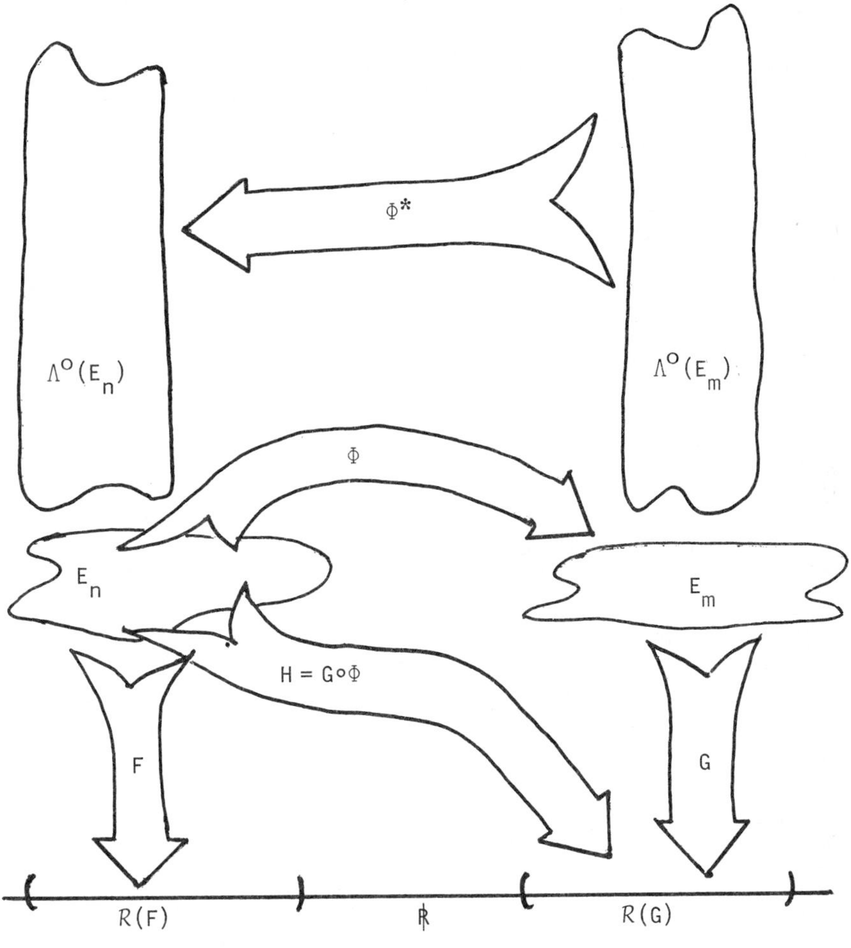

Figure 8. The Induced Map Φ^*

$$\text{(2-8.5)} \quad \Phi^*:\Lambda^o(E_m) \longrightarrow \Lambda^o(E_n) \mid H = G\circ\Phi \; ; \quad h(x^j) = g(\phi^A(x^j)) \; .$$

Thus, any map Φ from E_n to E_m induces a map Φ^* of any element of $\Lambda^o(E_m)$ onto an element of $\Lambda^o(E_n)$; that is, Φ induces the map Φ^* that is in the *opposite* direction to that of Φ . This gives rise to the graphic situation that is shown in Fig. 8. The important thing to be noted here is that there is a unique element $h(x^j) = g(\phi^A(x^j))$ of $\Lambda^o(E_n)$, for any given element $g(y^A)$ of $\Lambda^o(E_m)$, that is induced by the given map Φ.

We are now in a position to determine how a vector field behaves under the map Φ of E_n to E_m that is given by (2-8.1). Let v be an element of $T(E_n)$, so that we have

$$(2\text{-}8.6) \qquad v = v^i(x^j)\,\partial_i$$

for some given n-tuple of functions $\{v^i(x^j)\}$. Since v acts on all elements of $\Lambda^o(E_n)$, it certainly acts on those elements $h(x^j) = g(\phi^A(x^j))$ that are induced by Φ^*, and we have

$$(2\text{-}8.7) \quad v(h) = v^i(x^j)\,\partial_i g(\phi^A(x^j)) = v^i(x^j)\,\partial_i\phi^A\,\partial_A g\Big|_{y=\phi(x)} .$$

Now, set

$$(2\text{-}8.8) \qquad v^A(x^j) = v^i(x^j)\,\partial\phi^A(x^j)/\partial x^i ,$$

in which case (2-8.7) becomes

$$(2\text{-}8.9) \qquad v(h) = v^A(x^j)\,\partial_A g(y^B)\Big|_{y=\phi(x)} .$$

On the other hand, a generic element of $T(E_m)$ looks like

$$(2\text{-}8.10) \qquad \grave{v}(g) = \grave{v}^A(y^B)\,\partial_A g(y^B)$$

for any $g \in \Lambda^o(E_m)$. A comparison of (2-8.9) and (2-8.10) shows that we will have

$$(2\text{-}8.11) \qquad v(h) = \grave{v}(g) , \quad h = g\circ\Phi$$

if and only if we may achieve satisfaction of the relations

$$(2\text{-}8.12) \qquad \grave{v}^A(y^B) = v^A(x^j) = v^i(x^j)\,\partial\phi^A(x^j)/\partial x^i$$

as a consequence of the relations $y^A = \phi^A(x^j)$ that define the map Φ. In other words we have to be able to solve the equations

$$(2\text{-}8.13)\qquad y^A - \phi^A(x^j) = 0$$

for the x's as functions of the y's. The implicit function theorem (Section 1.5) tells us that this is only possible if $m = n$, in which case $E_m = {}^\backprime E_n$ with ${}^\backprime E_n$ denoting a replica of E_n. Further, the inverse mapping theorem (Section 1.6) also requires satisfaction of the condition $\det(\partial_i \phi^A(x^j)) \neq 0$ in order to be able to determine the inverse mapping

$$(2\text{-}8.14)\qquad \overset{-1}{\Phi}: {}^\backprime E_n = E_m \longrightarrow E_n \mid x^i = \overset{-1}{\phi}{}^i(y^B) .$$

THEOREM 2-8.1. *If* Φ *is an invertible mapping of* E_n *to* ${}^\backprime E_n$, *then* Φ *induces the map*

$$(2\text{-}8.15)\qquad \Phi_*: T(E_n) \longrightarrow T({}^\backprime E_n)$$

such that

$$(2\text{-}8.16)\qquad v(h) = {}^\backprime v(g) , \qquad h = g \circ \Phi$$

for all $g \in \Lambda^0({}^\backprime E_n)$, *in which case*

$$(2\text{-}8.17)\qquad v = v^i(x^j)\, \partial_i , \qquad {}^\backprime v = {}^\backprime v^A(y^B)\, \partial_A$$

are related by

$$(2\text{-}8.18)\qquad {}^\backprime v^A(y^B) = \{v^i\, \partial_i \phi^A\} \circ \overset{-1}{\Phi} .$$

The geometric situation is shown in Fig. 9.

The "fly in the ointment" in these considerations is obviously the need to express the coefficient functions $v^A(x^j)$ *as functions of the* y's so that $v^A(x^j)\,\partial_A$ becomes a derivation ${}^\backprime v^A(y^B)\,\partial_A$ on $\Lambda^0(E_m)$. In other words, implementation of the requirement that we map whole vector fields to vector fields. If we back off of this requirement, then everything becomes much simpler. A case in the extreme is the tangent field to a specific

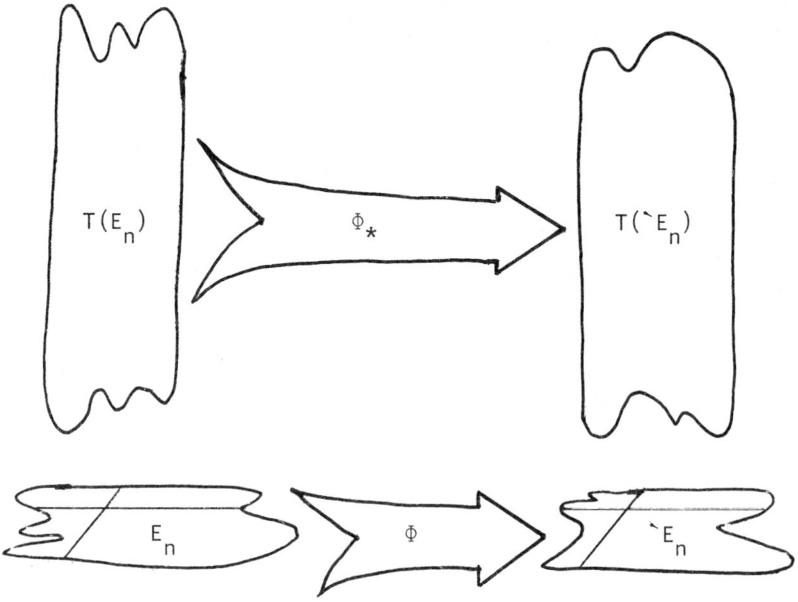

Figure 9. The Map Φ_*

curve in E_n . In this instance, we have the mappings

(2-8.19) $\Gamma : J \longrightarrow E_n \mid x^i = \gamma^i(t)$,

(2-8.20) $\Phi : E_n \longrightarrow E_m \mid y^A = \phi^A(x^i)$,

and the induced composite map (see Fig. 10)

(2-8.21) $`\Gamma = \Phi \circ \Gamma : J \longrightarrow E_m \mid y^A = \phi^A(\gamma^i(t))$.

It is then a simple matter to see that

(2-8.22) $T\Gamma : J \longrightarrow T_{\Gamma(J)}(E_n) \mid v^i(t) = d\gamma^i(t)/dt$,

(2-8.23) $`(T\Gamma) : J \longrightarrow T_{\Gamma(J)}(E_m) \mid `v^A(t) = \frac{d}{dt}\, \phi^A(\gamma^i(t))$,

and hence

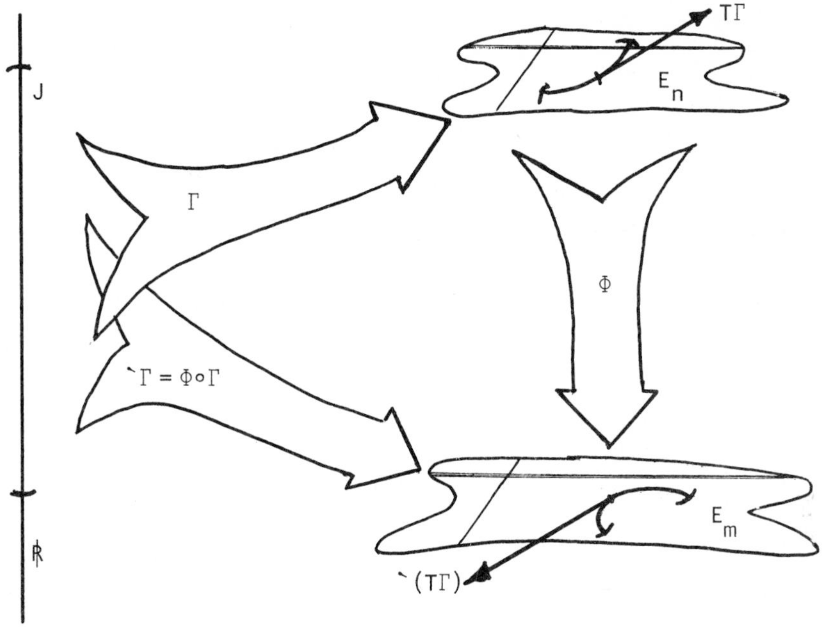

Figure 10. The Map TΦ of the Tangent Field to a Curve

$$(2\text{-}8.24)\quad {}^{\backprime}v^A(t) = \left.\frac{\partial\phi^A(x^j)}{\partial x^k}\right|_{x=\gamma} d\gamma^k(t)/dt = \left.\frac{\partial\phi^A(x^k)}{\partial x^k}\right|_{x=\gamma(t)} v^k(t)$$

Thus, the map Φ induces a map of $T\Gamma$ onto $(T\Gamma)$ that is given by (2-8.24). It thus follows that a map Φ from any E_n to any E_m induces a map of any one orbit of a given vector field on E_n onto the tangent field to the image in E_m of the given orbit. What we can not do, for $n \neq m$ is to map all orbits of a vector field on E_n onto curves in E_m in such a fashion that a vector field is induced on E_m. From the practical point of view, what happens is that the images in E_m of all orbits of a vector field in E_n result in a system of curves that have a large number of intersections or do not fill E_m. If there is even one intersection of the images of two different orbits, say at $\bar{y}^A_o$ in E_m, then there are two different tangents to two different curves through $\bar{y}^A_o$ and it becomes impossible to describe such curves by an autonomous system of equations $d\bar{y}^A(t) = v^A(\bar{y}^B(t))$, $\bar{y}^A(0) = \bar{y}^A_o$. Thus, our results are consistent.

Let us now restrict attention to the case in which Φ is invertible. We then have the maps

$$(2\text{-}8.25) \quad \Phi : E_n \longrightarrow {}^{\backprime}E_n \mid y^A = \phi^A(x^j), \quad A = 1, \ldots, n, \quad \det(\partial_i \phi^A) \neq 0,$$

$$(2\text{-}8.26) \quad \Phi_* : T(E_n) \longrightarrow T({}^{\backprime}E_n) \mid {}^{\backprime}v^A(y^B) = \{v^i \, \partial_i \phi^A\} \circ \overset{-1}{\Phi} ,$$

where ${}^{\backprime}E_n$ is a replica of E_n . If we were not told beforehand to the contrary, we could equally well read the equations $y^A = \phi^A(x^j)$, $A = 1, \ldots, n$ as a system of equations that assigns new coordinates (y^A) to the point P of E_n that had coordinates (x^i) in the given coordinate cover of E_n . That is, the equations (2-8.25) may be viewed as a change of the coordinate cover of E_n , in which case (2-8.27) gives the law of transformation for the components of vector fields with respect to the bases $\{\partial_i\}$ and $\{\partial_A\}$, respectively. Perhaps, it becomes a little easier to follow if we set ${}^{\backprime}x^A = y^A$ and then replace the upper case indices by lower case indices:

THEOREM 2-8.2. *The components of a vector field undergo the induced transformation*

$$(2\text{-}8.27) \quad (\Phi C)_* : T(E_n) \longrightarrow T(E_n) \mid {}^{\backprime}v^i({}^{\backprime}x^k) = \{v^j \, \partial_j \phi^i\} \circ (\overset{-1}{\Phi} C)$$

for a change of coordinate cover

$$(2\text{-}8.28) \quad \Phi C : {}^{\backprime}x^k = \phi^k(x^j) ,$$

where

$$v = v^i \partial_i , \quad {}^{\backprime}v = {}^{\backprime}v^i \, {}^{\backprime}\partial_i , \quad {}^{\backprime}\partial_i \equiv \frac{\partial}{\partial {}^{\backprime}x^i} ,$$

and we have

(2-8.29) $v(h) = \grave{}v(\grave{}h)\ ,\quad h = \grave{}h\circ\Phi = (\Phi C)^{*}\grave{}h\ .$

A restriction of enormous severity would be imposed if we were only able to work with problems involving invertible mappings of E_n to $\grave{}E_n$. It is for this reason that every effort is made to shift as many considerations as possible to the *dual space* of $T(E_n)$, for this space possesses very nice properties under mappings from E_n to E_m with $n \neq m$.

There is another situation in which a map $\Phi : E_n \longrightarrow E_m$ has a well defined induced action on $T(E_n)$, and this comes about through the notion of a coordinate transformation. Let

(2-8.30) $\Phi : E_n \longrightarrow E_m \mid y^A = \phi^A(x^j)$

be a given map with $m > n$. In this case, the range of Φ in E_m can not sweep out all of E_m. What it does sweep out is an n-dimensional surface in E_m if

(2-8.31) $\text{rank}(\partial_i \phi^A(x^j)) = n$

for all points $P:(x^j)$ in some neighborhood (possibly all) of E_n. Simply observe that $y^A = \phi^A(x^j)$, for $x^1 = c^1$, $x^2 = c^2$, ..., $x^{k-1} = c^{k-1}$, $x^{k+1} = c^{k+1}$, ..., $x^n = c^n$ defines a curve in E_m whose tangent vector field has components $\{\partial_k \phi^A\}$. Thus, each of the n vectors $\{\partial_1 \phi^A \partial_A\}, \ldots, \{\partial_n \phi^A \partial_A\}$ is tangent to the range of Φ at the image of $P:(x^j)$ in E_m. The condition (2-8.31) then tells us that these n vectors are linearly independent and hence $\mathcal{R}(\Phi)$ is an n-dimensional surface in E_m. The important thing to realize here is that the map Φ may be considered as inducing $(x^1, \ldots, x^n)$ as new coordinates on $\mathcal{R}(\Phi)$. If $v = v^i(x^j)\partial_i \in T(E_n)$ then $v(h) = v^i(x^j)\ \partial_i \phi^A(x^j)\ \partial_A g$ for $h = g\circ\Phi = \Phi^* g$. However, $\{\partial_j \phi^A \partial_A,\ j = 1, \ldots, n\}$ are n linearly independent vector fields tangent to $\mathcal{R}(\Phi)$ and hence constitute a basis for $T(\mathcal{R}(\Phi))$. We may thus define the action of Φ_* as this induced map from $T(E_n)$ to $T(\mathcal{R}(\Phi))$.

<u>THEOREM 2-8.3</u>. *If* $\Phi : E_n \longrightarrow E_m \mid y^A = \phi^A(x^j)$ *is a given map such that*

(2-8.32) $\quad m > n \; ,$

(2-8.33) $\quad \operatorname{rank}(\partial_i \phi^A(x^j)) \;=\; n \; ,$

then the induced map

(2-8.34) $\quad \Phi_* : T(E_n) \longrightarrow T(R(\Phi)) \mid v(h) \;=\; {}^{\backprime}v(g) \; , \quad h \;=\; \Phi^* g$

is well defined with respect to the induced basis

(2-8.35) $\quad {}^{\backprime}\partial_i \overset{\text{def}}{=} \partial_i \phi^A \, \partial_A$

of $T(R(\Phi))$. *In this case we have*

(2-8.36) $\quad {}^{\backprime}\partial_i \;=\; \partial_i \phi^A \, \partial_A$

considered as operators on $\Lambda^0(R(\Phi))$ *whose elements consist of the composition of all elements of* $\Lambda^0(E_m)$ *with the map* Φ .

In all instances where Φ_* is well defined, it induces a vector field on the range of Φ ; that is, it pulls a vector field from the domain of Φ up to the range of Φ . This direct graphic property is reinforced with the following definition.

<u>Definition</u>. If Φ is a map for which Φ_* is well defined, then $\Phi_* v$ is referred to as the *pull forward by* Φ and Φ_* is referred to as the *pull forward* map.

All that now remains is to codify how Φ_* combines with the Lie algebra operations of $T(E_n)$.

<u>THEOREM 2-8.4</u>. *The pull forward map has the following properties whenever the maps* Φ *and* Ψ *are such that* Φ_*, Ψ_* *and* $(\Psi \circ \Phi)_*$ *are well defined:*

PF1. $\quad \Phi_*(fu + gv) \;=\; \overset{-1}{\Phi}{}^{*} f \; \Phi_* u + \overset{-1}{\Phi}{}^{*} g \; \Phi_* v \; ;$

PF2. $\Phi_*[u, v] = [\Phi_* u, \Phi_* v]$;

PF3. $(\Psi \circ \Phi)_* = \Psi_* \Phi_*$.

Proof. We have already seen that Φ induces the map Φ^* of $\Lambda^0(\mathcal{R}(\Phi))$ to $\Lambda^0(\mathcal{D}(\Phi))$. Hence $\overset{-1}{\Phi}{}^*$ maps $\Lambda^0(\mathcal{D}(\Phi))$ into $\Lambda^0(\mathcal{R}(\Phi))$. Property PF1 is thus a direct consequence that Φ_* maps $T(\mathcal{D}(\Phi))$ into $T(\mathcal{R}(\Phi))$ and that $T(\mathcal{R}(\Phi))$ is a vector space over $\Lambda^0(\mathcal{R}(\Phi))$. By definition, $[u, v](h) = u(v(h)) - v(u(h))$. When this is combined with $u(h) = \grave{u}(g)$, $v(h) = \grave{v}(g)$, $h = \Phi^* g$, we obtain

$$[\grave{u}, \grave{v}](g) = \grave{u}(\grave{v}(g)) - \grave{v}(\grave{u}(g))$$

$$= \grave{u}(v(h)) - \grave{v}(u(h))$$

$$= u(v(h)) - v(u(h)) = [u, v](h) .$$

This establishes property PF2 . Property PF3 is a direct consequence of Fig. 11.

The following direct proof of PF2 may prove informative. We have

$$\grave{}[u, v] = (u^k \partial_k v^j - v^k \partial_k u^j) \partial_j \phi^B \, \partial_B ,$$

and (2-6.15) and (2-8.18). However

$$u^k (\partial_k v^j) \partial_j \phi^B = u^k \partial_k (v^j \partial_j \phi^B) - u^k v^j \partial_k \partial_j \phi^B$$

$$= u^k \partial_k \phi^A \partial_A (v^j \partial_j \phi^B) - u^k v^j \partial_k \partial_j \phi^B$$

and hence

$$\grave{}[u, v] = \{u^k \partial_k \phi^A \partial_A (v^j \partial_j \phi^B) - v^k \partial_k \phi^A \partial_A (u^j \partial_j \phi^B)\} \partial_B$$

$$- \{u^k v^j \partial_k \partial_j \phi^B - v^k u^j \partial_j \partial_k \phi^B\} \partial_B =$$

$$= \{u^k \partial_k \phi^A \partial_A (v^j \partial_j \phi^B) - v^k \partial_k \phi^A \partial_A (u^j \partial_j \phi^B)\} \partial_B$$

$$= \{\grave{}u^A \partial_A \grave{}v^B - \grave{}v^A \partial_A \grave{}u^B\} \partial_B$$

$$= [\grave{}u, \grave{}v] .$$

CHAPTER III

EXTERIOR FORMS

3-1. THE DUAL SPACE $T^*(E_n)$

Let E_n be our now familiar n-dimensional number space with the given Cartesian coordinate cover (x^i) and let $T(E_n)$ be the tangent space of E_n that was studied in the last Chapter. Thus, $T(E_n)$ is a Lie algebra over $\Lambda^o(E_n)$ with the globally defined basis $\{\partial_i, i = 1,\ldots,n\}$. Since $T(E_n)$ is the union over $P \varepsilon E_n$ of the vector spaces $T_P(E_n)$, the concept of a dual space that was introduced in Section 1-6 can be extended to that of the dual of $T(E_n)$ as the union over $P \varepsilon E_n$ of the dual vector spaces $T^*_P(E_n)$.

Definition. The *dual space*, $T^*(E_n)$, is the union over $P \varepsilon E_n$ of the dual vector spaces $T^*_P(E_n)$:

$$(3\text{-}1.1) \qquad T^*(E_n) = \bigcup_{P \varepsilon E_n} T^*_P(E_n) .$$

This definition shows that $T^*(E_n)$ inherits a vector space structure from the vector space structure of each of its constituent $T^*_P(E_n)$'s. Thus, if $\omega \varepsilon T^*(E_n)$ and $v \varepsilon T(E_n)$, the notation introduced in Section 1-6 yields

$$(3\text{-}1.2) \qquad \omega: T(E_n) \longrightarrow \mathbb{R} \mid r = \langle \omega, v \rangle , \quad \forall\, v \varepsilon T(E_n)$$

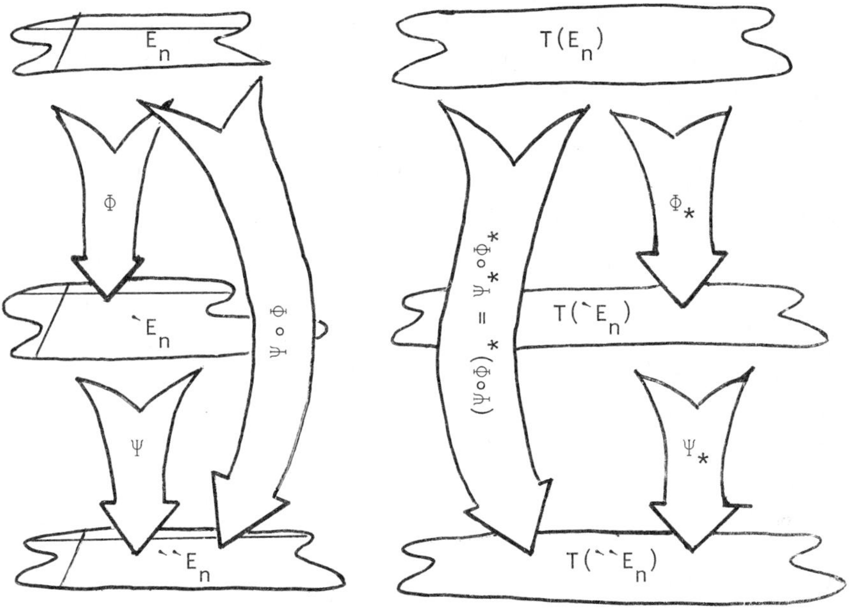

Figure 11. Pull Forward by Composition Maps

$$(3\text{-}1.3) \qquad \langle \omega, fu + gv \rangle = f\langle \omega, u \rangle + g\langle \omega, v \rangle ,$$

$$(3\text{-}1.4) \qquad \langle f\omega + g\rho, v \rangle = f\langle \omega, v \rangle + g\langle \rho, v \rangle .$$

LEMMA 3-1.1. *The dual space* $T^*(E_n)$ *is a vector space over* $\Lambda^0(E_n)$.

We now need to clothe this abstract result in a realizable and computationally useful form. This is most easily accomplished by the introduction of a natural global basis.

Definition. A collection of n elements $\{\theta^i\}$ of $T^*(E_n)$ is a *natural global basis* for $T^*(E_n)$ (with respect to the (x^i) coordinate cover) if and only if

$$(3\text{-}1.5) \qquad \langle \theta^i, \partial_j \rangle = \delta^i_j .$$

The fact that we can define something does not necessarily

mean that we can use it. In the present case, the definition of a natural global basis gives us the equations (3-1.5) to work with, but we have no guarantee of either the existence or the uniqueness of such a collection of elements of $T^*(E_n)$.

LEMMA 3-1.2. *If a solution of the system* (3-1.5) *exists, then it is unique.*

Proof. Suppose that there were two natural bases, $\{\theta^i\}$ and $\{\rho^i\}$; $\langle\theta^i, \partial_j\rangle = \delta^i_j$, $\langle\rho^i, \partial_j\rangle = \delta^i_j$. Subtracting the corresponding equations of these systems gives us

$$\langle\theta^i, \partial_j\rangle - \langle\rho^i, \partial_j\rangle = \langle\theta^i - \rho^i, \partial_j\rangle = 0 .$$

Thus, for every n-tuple of C^∞ functions $\{v^i(x^k)\}$, we have

$$v^j\langle\theta^i - \rho^i, \partial_j\rangle = \langle\theta^i - \rho^i, v\rangle = 0$$

for all $v = v^j(x^k)\partial_j \in T(E_n)$. It then follows from the uniqueness of the zero element o^* of $T^*(E_n)$ that $\theta^i - \rho^i = o^*$, and uniqueness is established.

We now turn to the question of existence. The simplest way to establish existence is to exhibit a solution explicitly. For any $f \in \Lambda^0(E_n)$ and any $v = v^i\partial_i \in T(E_n)$ we have $v(f) = v^i \frac{\partial f}{\partial x^i}$ takes its values in $\mathbb{R}$. Thus, $v(f)$ may be viewed as a linear functional on $T(E_n)$ since it has the required linearity properties: $v(f+g) = v(f) + v(g)$, $(u+v)(f) = u(f) + v(f)$. *The symbol* df *, that is defined by*

$$(3\text{-}1.6) \qquad \langle df, v\rangle = v(f) , \quad \forall\, v \in T(E_n)$$

is an element of $T^*(E_n)$ *for each* $f \in \Lambda^0(E_n)$.

The result we have just obtained sometimes gives a good bit of trouble the first time around, so the following comments may prove useful. The key to the situation is that $v = v^i(x^j)\partial_i$ realizes any element v of $T(E_n)$ as an *operator*, and thus does not possess values, *per se*. On the other hand, when v acts on an element $f(x^j)$ of $\Lambda^0(E_n)$, we obtain the new element $v(f) = v^i(x^j)\partial_i f(x^j) = g(x^j)$ of $\Lambda^0(E_n)$, that does have a value for any $P:(x^j)$ in E_n. Finally observe that the values of $g(x^j)$ belong to $\mathbb{R}$, and hence it is consistent to view the full quantity $v(f)$ as a mapping from $T(E_n)$ to $\mathbb{R}$; namely, a functional on $T(E_n)$. This, however, is exactly what (3-1.6) says if we call the element of $T^*(E_n)$ that is thus defined the symbol df.

LEMMA 3-1.3. *The unique natural dual basis of* $T^*(E_n)$ *is given by the* n *elements* $\{dx^i, i=1,\ldots,n\}$. *Thus, any element* ω *of* $T^*(E_n)$ *can be written uniquely as*

$$\omega = \omega_i(x^j)\, dx^i \tag{3-1.7}$$

where the coefficients $\omega_i(x^j)$ *are determined by*

$$\omega_i(x^j) = \langle \omega, \partial_i \rangle \tag{3-1.8}$$

and hence belong to $\Lambda^0(E_n)$.

Proof. The quantities $\{dx^i\}$ are defined by

$$\langle dx^i, v\rangle = v(x^i)$$

and hence $\langle dx^i, \partial_j\rangle = \partial_j x^i = \delta^i_j$. This establishes existence of the n quantities $\{dx^i\}$ that satisfy (3-1.5). Lemma 3-1.2 then establishes their uniqueness. Incidentally, $\langle dx^i, \partial_j\rangle = \delta^i_j$ also shows that $\{dx^i\}$ is a linearly independent system. Since $T^*(E_n)$ is a vector space over $\Lambda^0(E_n)$, any $\omega \in T^*(E_n)$ can be written in the form (3-1.7). Hence, $\langle \omega, \partial_j\rangle = \langle \omega_i dx^i, \partial_j\rangle = \omega_i(x^k)\langle dx^i, \partial_j\rangle = \omega_i(x^k)\delta^i_j = \omega_j(x^k)$ and (3-1.8) is established.

All that is now needed in order to make these ideas complete is a concrete realization of the abstract symbol df. If we apply df to the basis $\{\partial_i\}$, (3-1.6) gives

$$\langle df, \partial_i\rangle = \partial_i f. \tag{3-1.9}$$

On the other hand, $df \in T^*(E_n)$ and Lemma 3-1.3 shows that

$$df = \gamma_i(x^k)\, dx^i \tag{3-1.10}$$

with

$$\gamma_i(x^k) = \langle df, \partial_i \rangle = \partial_i f(x^k) \tag{3-1.11}$$

by (3-1.8) and (3-1.9). Thus, (3-1.10) yields

$$df = \partial_i f(x^k)\, dx^i . \tag{3-1.12}$$

This, however, is the familiar total differential of the function f of the n variables $x^1, x^2, \ldots, x^n$.

LEMMA 3-1.4. *The abstract element* df *of* $T^*(E_n)$ *that is defined by* (3-1.6) *for any* $f \in \Lambda^0(E_n)$ *has the realization*

$$df = \partial_i f(x^k)\, dx^i \tag{3-1.13}$$

as the total differential of the function $f(x^k)$.

There is still some degree of ambiguity, for (3-1.13) defines df in terms of the "differentials" $dx^1, dx^2, \ldots, dx^n$. Thus, either we interpret the quantities $\{dx^i\}$ as abstract basis elements - the notion of the differential of a coordinate function is a primitive in the logical sense - or else dx^i is interpreted as the differential of the coordiante function x^i and (3-1.13) evaluates something in terms of itself. The reader should keep this carefully in mind, for it will not be resolved until later in the next chapter when we introduce the operation of exterior differentiation.

There is another interpretation that can be attached to (3-1.13) that is important if for no other reason than to show consistency with what has gone before. Suppose we divide both sides of (3-1.13) by dt so as to obtain

$$df(x^k)/dt = \partial_i f(x^k)\, dx^i/dt .$$

Read in this way, (3-1.13) may be viewed as providing an approximation to the change in the value of the function $f(x^k)$ that obtains from evaluation at $P:(x^j)$ and at $P:(x^j + v^j t + o(t))$ with $v^j = dx^j/dt|_{t=0}$. We are thus back to the familiar ground of the differentiation of the composition of a map from $\mathbb{R}$ to E_n with a map from E_n to $\mathbb{R}$. There is thus no possibility of inconsistency, as should be evident from the various possible interpretations of the symbols on the two sides of (3-1.6).

This same view, when applied to quite an arbitrary element

$$(3\text{-}1.14) \qquad \omega = \omega_i(x^j)\, dx^i$$

of $T^*(E_n)$ shows something quite different. If we again divide both sides by dt, the result is

$$(3\text{-}1.15) \qquad \omega/dt = \omega_i(x^j)\, dx^i/dt\ .$$

The right-hand side of this expression is well defined whenever the x's are functions of the variable t, but the left-hand side, ω/dt, makes no sense (is not d/dt of any single function) unless we are sufficiently charitable to interpret ω as an infinitesimal of the same order as dt whenever the x's are functions of the variable t. This should not be too surprising, for it is hoped that everyone has heard of an inexact differential. It is simply an expression of the form

$$(3\text{-}1.16) \qquad \omega = \omega_i(x^j)\, dx^i$$

that is not the derivative with respect to t of any function $f(x^j)$ when the x's become functions of the variable t. Quantities such as (3-1.16) are referred to as *differential forms* or Pfaffian forms. In fact, it is this terminology, namely differential forms, that is the most widely used today. Accordingly, the sumbol $\Lambda^1(E_n)$ is commonly used instead of $T^*(E_n)$. The following definition is used in the approach given here in order to codify this situation.

Definition. $\Lambda^1(E_n)$, the space of *differential forms* of degree one on E_n , coincides with the dual space $T^*(E_n)$. The elements of $\Lambda^1(E_n)$ have the general form

$$\omega = \omega_i(x^j)\, dx^i \tag{3-1.17}$$

in terms of the natural basis $\{dx^i\}$, and are referred to as *differential forms of degree one*. $\Lambda^0(E_n)$ is referred to as the *space of forms of degree zero*.

There is one additional and very useful result that obtains as a direct consequence of Lemmas 3-1.1 through 3-1.4.

LEMMA 3-1.5. *Let* $v = v^i(x^k)\partial_i$ *be a given element of* $T(E_n)$ *and let* $\omega = \omega_i(x^k)\, dx^i$ *be a given element of* $\Lambda^1(E_n)$. *Then* ω *applied to* v *has the evaluation*

$$\langle \omega, v \rangle = \omega_i(x^k)\, v^i(x^k) \ . \tag{3-1.18}$$

Proof. $\langle \omega, v \rangle = \langle \omega_i dx^i, v^j \partial_j \rangle = \omega_i v^j \langle dx^i, \partial_j \rangle$

$$= \omega_i v^j \delta^i_j = \omega_i v^i \ .$$

The simplest physical example of an exterior form arises in mechanics. Let E_3 designate 3-dimensional Euclidean space with the Cartesian coordinate cover (x^1, x^2, x^3) . By a field of force in E_3 we mean a collection of three functions $F_1(x^i)$, $F_2(x^i)$, $F_3(x^i)$ such that a particle at $P:(x^i)$ will experience the force that is specified by evaluating the three functions $\{F_i(x^j)\}$. With this information, let us define the differential form

$$W = F_i(x^j)\, dx^i \ . \tag{3-1.19}$$

If we view $\{dx^i\}$ as the components of an infinitesimal displacement of the particle along its trajectory in E_3 through the point $P:(x^j)$, then the differential form W that is defined by

(3-1.19) is the work done in this infinitesimal displacement. Further, if $v = v^i(t)\partial_i$ is the velocity vector field of the trajectory $x^i = \phi^i(t)$ of the particle, then $dx^i = v^i dt$ at the point $P:(\phi^i(t))$ and W becomes $W = F_i(\phi^j(t))\, v^i\, dt$. This same result can also be obtained by allowing W to act on v

$$\langle W, v\rangle = F_i(\phi^j(t))\, v^i$$

and this expression is simply the instantaneous rate at which work is done on the particle by the force field. In particular, we have

$$W = F_i(\phi^j(t))\, v^i\, dt = \langle W, v\rangle dt$$

in this particular case. A little reflection on this example will quickly convince you that each of the various interpretations of elements $\Lambda^1(E_n)$ that we have obtained above is of use, depending on the nature of the problem and the kind of questions being asked.

3-2. THE EXTERIOR OR "VECK" PRODUCT

We have already seen that the vector space $T(E_n)$ equipped with the Lie product becomes an algebra. It is thus natural to ask whether $\Lambda^1(E_n)$ can be made into an algebra. The answer is unequivocally no, so we must proceed along other lines.

The vector space structure of $\Lambda^1(E_n)$ is completely delineated by the fact that $\Lambda^1(E_n)$ has the natural basis $\{dx^i\}$ as a vector space over $\Lambda^0(E_n)$; every element of $\Lambda^1(E_n)$ has the form $\omega_i(x^j)\, dx^i$ with $\omega_i(x^j) \in \Lambda^0(E_n)$. A rule of combination can thus be defined by defining it for the elements of the basis $\{dx^i\}$ and requiring that it behave properly with respect to the vector space operations.

<u>Definition</u>. The *exterior* or *veck* product $\wedge$ is defined on $\Lambda^1(E_n) \times \Lambda^1(E_n)$ by the requirements

(3-2.1) $$dx^i \wedge dx^j = -dx^j \wedge dx^i \;;$$

(3-2.2) $$dx^i \wedge f(x^k)dx^j = f(x^k)\, dx^i \wedge dx^j$$

for all $f(x^k) \in \Lambda^0(E_n)$;

(3-2.3) $$\alpha \wedge (\beta + \gamma) = \alpha \wedge \beta + \alpha \wedge \gamma$$

for all $\alpha, \beta, \gamma \in \Lambda^1(E_n)$.

Let us first show that this law of combination is actually defined for any pair of elements $\alpha = \alpha_i\, dx^i$ and $\beta = \beta_i\, dx^i$ of $\Lambda^1(E_n)$. Now, $\alpha \wedge \beta$ is defined by the relation

$$\alpha \wedge \beta = (\alpha_i\, dx^i) \wedge (\beta_j\, dx^j)$$

$$= (\alpha_1\, dx^1 + \ldots + \alpha_n\, dx^n) \wedge (\beta_1\, dx^1 + \ldots + \beta_n\, dx^n) \;.$$

Thus, use of (3-2.3) followed by (3-2.2) yields

$$\alpha \wedge \beta = (\alpha_1\, dx^1 + \ldots + \alpha_n\, dx^n) \wedge \beta_1\, dx^1 + \ldots$$

$$+ (\alpha_1\, dx^1 + \ldots + \alpha_n\, dx^n) \wedge \beta_n\, dx^n$$

$$= \beta_1(\alpha_1\, dx^1 + \ldots + \alpha_n\, dx^n) \wedge dx^1 + \ldots$$

$$+ \beta_n(\alpha_1\, dx^1 + \ldots + \alpha_n\, dx^n) \wedge dx^n \;.$$

However, (3-2.1) and (3-2.3) imply that

$$(\alpha + \beta) \wedge \gamma = -\gamma \wedge (\alpha + \beta) = -\gamma \wedge \alpha - \gamma \wedge \beta = \alpha \wedge \gamma + \beta \wedge \gamma$$

and hence

(3-2.4) $$\alpha \wedge \beta = \beta_1\alpha_1\, dx^1 \wedge dx^1 + \beta_1\alpha_2\, dx^2 \wedge dx^1 + \ldots + \beta_n\alpha_1\, dx^1 \wedge dx^n + \beta_n\alpha_2\, dx^2 \wedge dx^n + \ldots + \beta_n\alpha_n\, dx^n \wedge dx^n$$

$$= \alpha_i\beta_j\, dx^i \wedge dx^j \;.$$

However, $dx^i \wedge dx^j = - dx^j \wedge dx^i$, by (3-2.1) and hence (3-2.4) also gives

(3-2.5) $$\alpha \wedge \beta = - \alpha_i \beta_j \, dx^j \wedge dx^i = - \alpha_j \beta_i \, dx^i \wedge dx^j .$$

The following result is then obtained by adding (3-2.4) and (3-2.5).

LEMMA 3-2.1. *The exterior product of* $\alpha = \alpha_i \, dx^i$ *and* $\beta = \beta_j \, dx^j$ *has the evaluation*

(3-2.6) $$2 \alpha \wedge \beta = (\alpha_i \beta_j - \alpha_j \beta_i) \, dx^i \wedge dx^j .$$

We certainly can, at least in principle, form the veck product of all pairs of elements of $\Lambda^1(E_n)$. If we form all linear combinations of the quantities thus obtained with coefficients from $\Lambda^0(E_n)$, we obtain a vector space over $\Lambda^0(E_n)$ that is usually denoted by $\Lambda^2(E_n)$. Since $dx^i \wedge dx^j = - dx^j \wedge dx^i$ it follows that this vector space has the basis $\{dx^i \wedge dx^j, \; i < j\}$ so it is a vector space of dimension $n(n-1)/2$.

Definition. The $n(n-1)/2$ dimensional vector space $\Lambda^2(E_n)$ over $\Lambda^0(E_n)$ with the basis

(3-2.7) $$\{dx^i \wedge dx^j , \quad i < j\}$$

is the vector space of exterior 2-forms over E_n . The elements of $\Lambda^2(E_n)$ are referred to as 2-*forms*, or *exterior forms of degree* 2.

Definition. The elements of $\Lambda^0(E_n)$ are referred to as *exterior forms of degree* 0. The elements of $\Lambda^1(E_n)$ are referred to as *exterior forms of degree* 1.

With these definitions at hand, it is permissible to say that the veck product, $\wedge$, defines a map of $\Lambda^1(E_n) \times \Lambda^1(E_n)$ to $\Lambda^2(E_n)$.

What we have done once, we can certainly do again. The

vector space $\Lambda^3(E_n)$ of *forms of degree* 3 is thus generated by forming all linear combinations of all veck products of the basis elements dx^i taken three at a time. Since there are now three products involved, it is clearly necessary to say something about how such multiple multiplications associate. The simplest thing is to require that the multiplication be associative

$$(3\text{-}2.8) \qquad \alpha \wedge (\beta \wedge \gamma) \quad = \quad (\alpha \wedge \beta) \wedge \gamma \ .$$

From now on, we simply continue this process and obtain the sequence of vector spaces $\Lambda^4(E_n)$, $\Lambda^5(E_n)$, ... We note, however, that (3-2.1) implies $dx^i \wedge dx^i = 0$. Thus, when we come to $\Lambda^n(E_n)$, we can only form the single independent element

$$dx^1 \wedge dx^2 \wedge dx^3 \wedge \ldots \wedge dx^{n-1} \wedge dx^n \ .$$

Further, if we veck this with any dx^i , the result is

$$dx^i \wedge dx^1 \wedge dx^2 \wedge \ldots \wedge dx^n \quad = \quad 0$$

since dx^i occurs twice in any such product. The process thus terminates with $\Lambda^n(E_n)$.

Definition. The space $\Lambda^k(E_n)$, $0 < k \leq n$, of *exterior forms of degree* k is the vector space of dimension $\binom{n}{k} = \frac{n!}{k!(n-k)!}$ over $\Lambda^0(E_n)$ with the natural basis

$$(3\text{-}2.9) \qquad \{dx^{i_1} \wedge dx^{i_2} \wedge \ldots \wedge dx^{i_k} \ , \quad i_1 < i_2 < \ldots < i_k\} \ .$$

If $\alpha \in \Lambda^k(E_n)$ then we write

$$(3\text{-}2.10) \qquad \deg(\alpha) \quad = \quad k \ .$$

The value of $\binom{n}{k}$ for the dimension of Λ^k comes directly from (3-2.9), namely n distinct things taken k at a time. The following table summarizes these results.

Space	Basis	Dimension
$\Lambda^0(E_n)$	1	$1 = \binom{n}{0}$
$\Lambda^1(E_n)$	$\{dx^k\}$	$n = \binom{n}{1}$
$\Lambda^2(E_n)$	$\{dx^k \wedge dx^m,\ k<m\}$	$\frac{n(n-1)}{2} = \binom{n}{2}$
$\Lambda^3(E_n)$	$\{dx^k \wedge dx^m \wedge dx^r,\ k<m<r\}$	$\frac{n(n-1)(n-2)}{3!} = \binom{n}{3}$
$\vdots$	$\vdots$	$\vdots$
$\Lambda^n(E_n)$	$dx^1 \wedge dx^2 \wedge \ldots \wedge dx^n$	$1 = \binom{n}{n}$

The sequence of vector spaces $\Lambda^0(E_n),\ldots,\Lambda^n(E_n)$ allows us to collect together the properties of the veck product in a more elegant formulation.

LEMMA 3-2.2. *The operation* $\wedge$ *of exterior multiplication generates a map*

$$(3\text{-}2.11) \qquad \wedge : \Lambda^r(x^j) \times \Lambda^s(x^j) \longrightarrow \Lambda^{r+s}(x^j)$$

that exhibits the following properties:

P1. $\alpha \wedge (\beta + \gamma) = \alpha \wedge \beta + \alpha \wedge \gamma$;

P2. $\alpha \wedge \beta = (-1)^{\deg(\alpha)\deg(\beta)}\ \beta \wedge \alpha$;

P3. $\alpha \wedge (\beta \wedge \gamma) = (\alpha \wedge \beta) \wedge \gamma$.

Proof. The only thing that hasn't already been verified is that P2 implies (3-2.2). Since $f(x^k)$ belongs to $\Lambda^0(E_n)$, $\deg(f) = 0$, and P2 implies $dx^i \wedge f(x^k) = (-1)^{\deg(f)\deg(dx^i)}$ $f(x^k) \wedge dx^i = f(x^k)dx^i$. Thus, *in exterior multiplications by elements of* $\Lambda^0(E_n)$ *we may omit the symbol* $\wedge$. The associative property P3 then gives $dx^i \wedge f \wedge dx^j = f \wedge dx^i \wedge dx^j = f\, dx^i \wedge dx^j$.

The sequence of vector spaces $\Lambda^0(E_n)$, $\Lambda^1(E_n)$, ..., $\Lambda^n(E_n)$ thus possess the following properties:

(i) each $\Lambda^k(E_n)$ is a vector space over $\Lambda^0(E_n)$ with the operations of addition and multiplication by elements of $\Lambda^0(E_n)$, but elements from $\Lambda^k(E_n)$ and $\Lambda^m(E_n)$ with $k \neq m$ can not be added;

(ii) any element of $\Lambda^k(E_n)$ can be vecked with any element from $\Lambda^m(E_n)$, but the answer belongs to $\Lambda^{k+m}(E_n)$.

These results are similar in structure to what obtains with the algebra of polynomials. There we can add two polynomials of the same degree together to obtain a polynomial of the same degree, and we can also multiply polynomials of degrees k and m together to obtain a polynomial of degree $k+m$. Now the theory of polynomials is significantly simplified by introducing an all-encompassing space P of all polynomials that is *graded* by the degrees of the polynomials that constitute its subspaces. We do the same thing here.

Definition. The *graded exterior algebra of differential forms*, is the direct sum

$$\Lambda(E_n) = \Lambda^0(E_n) \oplus \Lambda^1(E_n) \oplus \Lambda^2(E_n) \oplus \ldots \oplus \Lambda^n(E_n) \tag{3-2.12}$$

with the vector space operations of each $\Lambda^k(E_n)$ together with the exterior product $\wedge$ as a map from $\Lambda(E_n) \times \Lambda(E_n)$ to $\Lambda(E_n)$.

The definition of each $\Lambda^k(E_n)$ shows that the natural basis for $\Lambda(E_n)$ consists of the direct sum of the natural bases for each of its constituent graded subspaces. The basis is thus given by

$$1 \oplus \{dx^i\} \oplus \{dx^i \wedge dx^j,\ i<j\} \oplus \ldots \oplus dx^1 \wedge dx^2 \wedge \ldots \wedge dx^n ,$$

and thus $\dim(\Lambda) = \sum_{k=0}^{n} \binom{n}{k}$. However $(1+t)^n = \sum_{k=0}^{n} \binom{n}{k} t^k$, so that

$$2^n = \sum_{k=0}^{n} \binom{n}{k} \text{ and hence}$$

(3-2.13) $$\dim(\Lambda(E_n)) = 2^n .$$

There is one word of caution that must be sounded when it comes to writing an element of $\Lambda^k(E_n)$, for there are actually two conventions in current use. We know that $\Lambda^k(E_n)$ has the basis

$$\{dx^{i_1} \wedge dx^{i_2} \wedge \ldots \wedge dx^{i_k} , \quad i_1 < i_2 < \ldots < i_k\}$$

and hence any $\omega \in \Lambda^k(E_n)$ is uniquely written as

(3-2.14) $$\omega = \sum_{i_1<i_2<\ldots<i_k} \omega_{i_1 i_2 \ldots i_k}(x^m) dx^{i_1} \wedge dx^{i_2} \wedge \ldots \wedge dx^{i_k}$$

in terms of the $\binom{n}{k}$ functions

(3-2.15) $$\{\omega_{i_1 i_2 \ldots i_k}(x^m) , \quad i_1 < i_2 < \ldots < i_k\} .$$

Now, we can extend the $\binom{n}{k}$ functions (3-2.15) to n^k functions $\{\omega_{i_1 i_2 \ldots i_n}(x^m)\}$ for all values of the indices $i_1, \ldots, i_k$ by requiring the new functions to be antisymmetric in every pair of indices. If we then sum over all k indices from 1 through n we will repeat each term of (3-2.15) $k!$ times. Thus, we must write

(3-2.16) $$\omega = \frac{1}{k!} \omega_{i_1 i_2 \ldots i_k}(x^m) \, dx^{i_1} \wedge dx^{i_2} \wedge \ldots \wedge dx^{i_k}$$

with the operatable summation convention in order that the result agree with (3-2.14). Please note the $1/k!$ that arises in this alternative form of writing the answer, and the fact that *the new* n^k *coefficient function* $\{\omega_{i_1 i_2 \ldots i_k}\}$ *are now antisymmetric in every pair of indices.* This situation has already been in evidence in the process of exterior multiplication, as the factor of two on the left-hand side of (3-2.6) clearly shows. The first time through, the situation appears unduly complex, for there are actually two different representations involved; representations in terms of the natural basis, and representations that use the convenience of the summation convention. A little practice with the manipulations, however, easily sets the mind at rest.

3-3. ALGEBRAIC RESULTS

One of the uses of forms of degree higher than one is the characterization of subspaces of $\Lambda^1(E_n)$. This is accomplished by use of the notion of simple elements of $\Lambda^k(E_n)$.

<u>Definition.</u> An element ω of $\Lambda^k(E_n)$ is said to be *simple* if and only if there exist k elements $\{\eta^1, \eta^2, \ldots, \eta^k\} = \{\eta^a\}$ of $\Lambda^1(E_n)$ such that

$$\omega = \eta^1 \wedge \eta^2 \wedge \ldots \wedge \eta^k . \tag{3-3.1}$$

We have already encountered simple elements of $\Lambda^k(E_n)$; namely, the basis elements of $\Lambda^k(E_n)$. In fact, the representation

$$\omega = \sum_{i_1<i_2<\ldots<i_k} \omega_{i_1 i_2 \ldots i_k}(x^m)\, dx^{i_1} \wedge dx^{i_2} \wedge \ldots \wedge dx^{i_k} \tag{3-3.2}$$

shows that *every element of* $\Lambda^k(E_n)$ *consists of linear combination* of simple elements of $\Lambda^k(E_n)$ with coefficients from $\Lambda^0(E_n)$.

<u>LEMMA 3-3.1</u>. *The collection of all simple elements of* $\Lambda^k(E_n)$ *is closed under multiplication by elements of* $\Lambda^0(E_n)$. *The collection of all simple elements of* $\Lambda(E_n)$ *is closed under exterior multiplication. Every* $\omega \in \Lambda^n(E_n)$ *is simple.*

<u>Proof.</u> Any simple element of $\Lambda^k(E_n)$ has the form (3-3.1), and hence

$$f\,\omega = f\,\eta^1 \wedge \eta^2 \wedge \ldots \wedge \eta^k .$$

However, $f\,\eta^1 = \gamma$ belongs to $\Lambda^1(E_n)$ so that

$$f\,\omega = \gamma \wedge \eta^2 \wedge \ldots \wedge \eta^k$$

and hence $f\,\omega$ is also simple. If $\omega \in \Lambda^k(E_n)$ and $\beta \in \Lambda^m(E_n)$ are simple, then

$$\omega \wedge \beta = \eta^1 \wedge \ldots \wedge \eta^k \wedge \rho^1 \wedge \ldots \wedge \rho^m$$

with $\beta = \rho^1 \wedge \ldots \wedge \rho^m$. This shows that the exterior product of any two simple elements of $\Lambda(E_n)$ is also a simple element of $\Lambda(E_n)$. Finally, if $\omega \in \Lambda^n(E_n)$, then

$$\omega = f(x^m)\,dx^1 \wedge \ldots \wedge dx^n$$

since $\Lambda^n(E_n)$ is 1-dimensional, and hence every element of $\Lambda^n(E_n)$ is simple.

LEMMA 3-3.2. *If* $\omega \in \Lambda^k(E_n)$ *with* k *an odd integer, then*

(3-3.3) $$\omega \wedge \omega = 0\,.$$

Proof. If k is an odd integer, then $\deg(\omega) = 2r+1$, and property P2 gives

$$\omega \wedge \omega = (-1)^{(2r+1)^2}\,\omega \wedge \omega = -\,\omega \wedge \omega$$

and the result is established.

Note. If $\omega \in \Lambda^k(E_n)$ with k an even integer, then $\omega \wedge \omega$ need not vanish. Simply observe that

$$(dx \wedge dy + dz \wedge dt) \wedge (dx \wedge dy + dz \wedge dt) = 2\,dx \wedge dy \wedge dz \wedge dt \neq 0\,.$$

LEMMA 3-3.3. *Let* $\omega^1, \ldots, \omega^k$ *be* k *given elements of* $\Lambda^1(E_n)$ *and construct the simple* k-*form*

(3-3.4) $$\Omega = \omega^1 \wedge \omega^2 \wedge \ldots \wedge \omega^k\,.$$

A necessary and sufficient condition that the k *given* 1-*forms be linearly dependent is*

(3-3.4) $\quad \Omega = 0 \, .$

A necessary and sufficient condition that the k *given* 1-*forms be linearly independent is*

(3-3.5) $\quad \Omega \neq 0 \, .$

Proof. If $\omega^1, \ldots, \omega^k$ are linearly dependent, then one of them, say the first, can be expressed as a linear combination of the remaining ones:

(3-3.6) $$\omega^1 = f_2 \, \omega^2 + \ldots + f_k \, \omega^k \, .$$

Now simply veck (3-3.6) with $\omega^2 \wedge \omega^3 \wedge \ldots \wedge \omega^k$ to obtain

$$\Omega = f_2 \, \omega^2 \wedge \omega^2 \wedge \omega^3 \wedge \ldots \wedge \omega^k + \ldots + f_k \, \omega^k \wedge \omega^2 \wedge \omega^3 \wedge \ldots \wedge \omega^k \, .$$

Each of the k-1 terms on the right-hand side contains two identical 1-form factors, so that each term vanishes by Lemma 3-3.1. Conversely, consider the equation

(3-3.7) $$g_1 \, \omega^1 + g_2 \, \omega^2 + \ldots + g_k \, \omega^k = 0 \, .$$

If we veck this equation with the simple element $\omega^2 \wedge \omega^3 \wedge \ldots \wedge \omega^k$ the only term on the left-hand side that does not contain two identical 1-form factors is $g_1 \, \omega^1 \wedge \omega^2 \wedge \ldots \wedge \omega^k$, and hence

$$g_1 \Omega = 0$$

if (3-3.7) is to be satisfied. In like manner, vecking (3-3.7) with each of the remaining k-1 possible simple elements of $\Lambda^{k-1}(E_n)$ that can be formed from $\omega^1, \ldots, \omega^k$ gives the requirements

(3-3.8) $$g_m \Omega = 0 \, , \quad m = 1, \ldots, k$$

in order that (3-3.7) be satisfied. Thus, if $\Omega \neq 0$, then

(3-3.7) can be satisfied only by $g_1 = g_2 = \dots = 0$ and the result is established.

LEMMA 3-3.7. *If* $\eta^1, \dots, \eta^k$ *are* k *elements of* $\Lambda^1(E_n)$ *and another collection of* k *elements of* $\Lambda^1(E_n)$ *is defined by*

$$(3\text{-}3.9) \qquad \omega^a = K^a_b \, \eta^b ,$$

then

$$(3\text{-}3.10) \qquad \omega^1 \wedge \dots \wedge \omega^k = \det(K^a_b)\, \eta^1 \wedge \dots \wedge \eta^k .$$

Proof. The simplest way of establishing this result is by induction on the number of 1-forms. For one 1-form, we obviously have $\omega^1 = K^1_1 \eta^1 = \det(K^1_1)\, \eta^1$, and for two,

$$\omega^1 \wedge \omega^2 = K^1_a K^2_b \, \eta^a \wedge \eta^b = (K^1_1 K^2_2 - K^1_2 K^2_1)\, \eta^1 \wedge \eta^2$$

$$= \det(K^a_b)\, \eta^1 \wedge \eta^2 .$$

Let us assume that the result is true for $r-1$ 1-forms:

$$(3\text{-}3.11) \qquad \omega^1 \wedge \dots \wedge \omega^r = \det(K^p_q)\, \eta^1 \wedge \dots \wedge \eta^r ,$$

where (K^p_q) is an $(r-1)$-by-$(r-1)$ matrix. In this event, we have

(3-3.12)

$$\omega^1 \wedge \dots \wedge \omega^{r-1} \wedge \omega^r = K^1_{a_1} K^2_{a_2} \dots K^{r-1}_{a_{r-1}} K^r_{a_r} \, \eta^{a_1} \wedge \dots \wedge \eta^{a_{r-1}} \wedge \eta^{a_r}$$

$$= K^1_{a_1} K^2_{a_2} \dots K^{r-1}_{a_{r-1}} K^r_1 \, \eta^{a_1} \wedge \dots \wedge \eta^{a_{r-1}} \wedge \eta^1$$

$$+ K^1_{a_1} K^2_{a_2} \dots K^{r-1}_{a_{r-1}} K^r_2 \, \eta^{a_1} \wedge \dots \wedge \eta^{a_{r-1}} \wedge \eta^2 + \dots$$

$$+ K^1_{a_1} K^2_{a_2} \dots K^{r-1}_{a_{r-1}} K^r_r \, \eta^{a_1} \wedge \dots \wedge \eta^{a_{r-1}} \wedge \eta^r .$$

Now, observe that none of the indices $a_1, \dots, a_{r-1}$ in the first term can be equal to 1 for otherwise η^1 would appear twice as a factor and the answer would be zero. Likewise, none of $a_1, \ldots, a_{r-1}$ in the second term can be 2, ..., and none of $a_1, \ldots, a_{r-1}$ in the last term can be equal to r. Thus, if we write (3-3.12) in the equivalent form

$$\omega^1 \wedge \ldots \wedge \omega^r = K^r_1 (K^1_{a_1} K^2_{a_2} \ldots K^{r-1}_{a_{r-1}} \eta^{a_1} \wedge \ldots \wedge \eta^{a_{r-1}}) \wedge \eta^1 + \ldots$$

$$+ K^r_r (K^1_{a_1} K^2_{a_2} \ldots K^{r-1}_{a_{r-1}} \eta^{a_1} \wedge \ldots \wedge \eta^{a_{r-1}}) \wedge \eta^r$$

then reordering the factors η^p and using the induction hypothesis gives

$$= K^r_1 C^r_1 \eta^1 \wedge \eta^2 \wedge \ldots \wedge \eta^r + \ldots$$

$$K^r_r C^r_r \eta^1 \wedge \eta^2 \wedge \ldots \wedge \eta^r = \det(K^a_b)\, \eta^1 \wedge \ldots \wedge \eta^r$$

where C^r_b is the cofactor of K^r_b.

This last result is quite useful in the study of determinants. For example, suppose that $\eta^b = L^b_c \rho^c$ for another collection $\rho^1, \ldots, \rho^k$ of 1-forms. We then have

$$\omega^a = K^a_b L^b_c \rho^c$$

from (3-3.9). Using Lemma 3-3.4 then gives

$$\omega^1 \wedge \ldots \wedge \omega^r = \det(K^a_b)\, \eta^1 \wedge \ldots \wedge \eta^r ,$$

$$\eta^1 \wedge \ldots \wedge \eta^r = \det(L^a_b)\, \rho^1 \wedge \ldots \wedge \rho^r ,$$

$$\omega^1 \wedge \ldots \wedge \omega^r = \det(K^a_b L^b_c)\, \rho^1 \wedge \ldots \wedge \rho^r ,$$

from which we may conclude that

$$\det(K^a_b L^b_c) = \det(K^a_b) \det(L^c_e) .$$

All of the necessary results are now at hand in order to

obtain the basic characterization of subspaces of $\Lambda^1(E_n)$.

<u>THEOREM 3-3.1</u>. *There exists a nonzero simple r-form Ω for each r-dimensional subspace U_r of $\Lambda^1(E_n)$, that is determined by U_r up to a nonzero scalar factor, such that α belongs to U_r if and only if*

$$(3\text{-}3.13) \qquad \alpha \wedge \Omega = 0 .$$

Let U_{r_1} and U_{r_2} be two subspaces of $\Lambda^1(E_n)$ with characteristic r_1- and r_2-forms Ω_1 and Ω_2, respectively. A necessary and sufficient condition that U_{r_1} be contained in U_{r_2} is that there exists an (r_2-r_1)-form γ such that $\Omega_2 = \Omega_1 \wedge \gamma$. A necessary and sufficient condition that U_{r_1} and U_{r_2} have only the zero 1-form in common is $\Omega_1 \wedge \Omega_2 \neq 0$. If this condition is satisfied, then $\Omega_1 \wedge \Omega_2$ is a characteristic simple (r_1+r_2)-form of the subspace $U_{r_1} \bigcup U_{r_2}$.

<u>Proof</u>. If $\omega^1,\ldots,\omega^{r_1}$ is a basis for U_{r_1}, then set $\Omega_1 = \omega^1 \wedge \ldots \wedge \omega^{r_1}$. If $\eta^1,\ldots,\eta^{r_1}$ is any other basis for U_{r_1}, then $\eta^a = K^a_b\,\omega^b$ with $\det(K^a_b) \neq 0$. Lemma 3-3.4 shows that $\eta^1 \wedge \ldots \wedge \eta^{r_1} = \det(K^a_b)\Omega_1$, and hence Ω_1 is determined by U_{r_1} up to a scalar factor. An $\alpha \in \Lambda^1(E_n)$ belongs to U_{r_1} if and only if α is a linear combination of the basis elements of U_{r_1}, in which case α, $\omega^1,\ldots,\omega^{r_1}$ are linearly dependent. Lemma 3-3.3 shows that a necessary and sufficient condition for this to hold is that $0 = \alpha \wedge \omega^1 \wedge \ldots \wedge \omega^{r_1} = \alpha \wedge \Omega_1$. A necessary and sufficient condition that U_{r_1} be contained in U_{r_2} is that a basis $\omega^1,\ldots,\omega^{r_1}$ of U_{r_1} can be extended to a basis $\omega^1,\ldots,\omega^{r_1}$, $\omega^{r_1+1},\ldots,\omega^{r_2}$ of U_{r_2}. In this event, $\Omega_1 = \omega^1 \wedge \ldots \wedge \omega^{r_1}$,

$$\Omega_2 = \omega^1 \wedge \ldots \wedge \omega^{r_1} \wedge \omega^{r_1+1} \wedge \ldots \wedge \omega^{r_2} = \Omega_1 \wedge \omega^{r_1+1} \wedge \ldots \wedge \omega^{r_2} = \Omega_1 \wedge \gamma$$

with $\gamma = \omega^{r_1+1} \wedge \ldots \wedge \omega^{r_2} \in \Lambda^{r_2-r_1}(E_n)$. A necessary and sufficient condition that $U_{r_1} \bigcup U_{r_2}$ be the zero element of $\Lambda^1(E_n)$ is that $\omega^1,\ldots,\omega^{r_1}$, $\beta^1,\ldots,\beta^{r_2}$ be independent, where

$\omega^1,\ldots,\omega^{r_1}$ is a basis for U_{r_1} and $\beta^1,\ldots,\beta^{r_2}$ is a basis for U_{r_2}. This is the case if and only if $\Omega_1 \wedge \Omega_2 = \omega^1 \wedge \ldots \wedge \omega^{r_1} \wedge \beta^1 \wedge \ldots \wedge \beta^{r_2} \neq 0$, by Lemma 3-3.3. If this condition is satisfied, then $\Omega_1 \wedge \Omega_2$ is a simple characteristic (r_1+r_2)-form of $U_{r_1} \cup U_{r_2}$.

The formation of forms of degree higher than one is indeed a useful construct. This is particularly true when we realize that most algebraic problems revolve around characterizations of subspaces of solutions, etc. In this regard, the following Lemma of E. Cartan proves to be quite useful.

LEMMA 3-3.5. *Let* $\omega^1,\ldots,\omega^r$ *be* r *linearly independent 1-forms and suppose that the* r *1-forms* $\gamma^1,\ldots,\gamma^r$ *satisfy*

$$\omega^1 \wedge \gamma^1 + \omega^2 \wedge \gamma^2 + \ldots + \omega^r \wedge \gamma^r = 0 . \tag{3-3.14}$$

Then each of the 1-forms $\gamma^1,\ldots,\gamma^r$ *belongs to the subspace spanned by* $\omega^1,\ldots,\omega^r$. *Thus, there exists a matrix* $((A^a_b))$ *such that*

$$\gamma^a = A^a_b\, \omega^b , \tag{3-3.15}$$

and, in addition, we have

$$A^a_b = A^b_a , \quad a,b = 1,\ldots,r . \tag{3-3.16}$$

Proof. Since $\omega^1,\ldots,\omega^r$ are linearly independent,

$$\Omega = \omega^1 \wedge \omega^2 \wedge \ldots \wedge \omega^r \neq 0 . \tag{3-3.17}$$

Thus, if we veck (3-3.14) with $\omega^2 \wedge \omega^3 \wedge \ldots \wedge \omega^r$, we obtain

$$\gamma^1 \wedge \Omega = 0 . \tag{3-3.18}$$

Indeed, proceeding in like fashion, we see that

$$\gamma^a \wedge \Omega = 0 , \quad a = 1,\ldots,r \tag{3-3.19}$$

and hence each of the γ's is linearly dependent on the ω's. This establishes (3-3.15). A substitution of (3-3.15) back into (3-3.14) yields the additional condition (3-3.16).

3-4. INNER MULTIPLICATION

The graded exterior algebra $\Lambda(E_n)$ can be considered as generated from $\Lambda^o(E_n)$ and $\Lambda^1(E_n)$ by the exterior product, $\wedge$, together with the vector space properties that are thereby implied. On the other hand, the exterior product is only an ascending operator,

$$\wedge : \Lambda^k(E_n) \times \Lambda^m(E_n) \longrightarrow \Lambda^{k+m}(E_n) ,$$

and hence we can not come back down the ladder once we have reached the top-most collection $\Lambda^n(E_n)$. Serious difficulties would thus arise if we did not define an operation that lowers the degrees of forms.

Definition. *Inner multiplication*, or the *pull down* is a map

$$(3\text{-}4.1) \qquad \rfloor : T(E_n) \times \Lambda^k(E_n) \longrightarrow \Lambda^{k-1}(E_n)$$

with the following properties:

I1. $\quad v \rfloor f = 0$

for all $v \in T(E_n)$ and all $f \in \Lambda^o(E_n)$;

I2. $\quad v \rfloor \omega = \langle \omega, v \rangle$

for all $v \in T(E_n)$ and all $\omega \in \Lambda^1(E_n)$;

I3. $\quad v \rfloor (\alpha + \beta) = v \rfloor \alpha + v \rfloor \beta$

for all $\alpha, \beta \in \Lambda^k(E_n)$, $k = 1, \ldots, n$ and all $v \in T(E_n)$;

I4. $\quad v \rfloor (\alpha \wedge \beta) = (v \rfloor \alpha) \wedge \beta + (-1)^{\deg(\alpha)} \alpha \wedge (v \rfloor \beta)$

for all $\alpha, \beta \in \Lambda(E_n)$ and all $v \in T(E_n)$.

We have taken a somewhat different tack here in that $\rfloor$ is defined in terms of its axioms I1 through I4, rather than define the operation directly and then establish its properties. It is thus incumbent upon us to show that I1 through I4 result in $\rfloor$ being well defined.

<u>LEMMA 3-4.1</u>. *The operation* $\rfloor$ *that satisfies axioms* I1 *through* I4 *is well defined on* $T(E_n) \times \Lambda(E_n)$ *and yields a map of* $T(E_n) \times \Lambda^k(E_n)$ *to* $\Lambda^{k-1}(E_n)$. *It also satisfies*

I5. $$(fu + gv)\rfloor\omega = fu\rfloor\omega + gv\rfloor\omega .$$

<u>Proof</u>. Axiom I1 defines $\rfloor$ for all forms of degree zero and axiom I2 defines $\rfloor$ for all forms of degree 1. In particular,

$$(3\text{-}4.2) \qquad v = v^i \partial_i , \qquad \omega = w_i\, dx^i ,$$

then

$$(3\text{-}4.3) \qquad v\rfloor\omega = \langle\omega, v\rangle = w_i\, v^i \in \Lambda^0(E_n) .$$

Since $\langle\omega, v\rangle$ is a linear functional, I5 is clearly satisfied for forms of degree zero and one. On the other hand, any form of degree k is the sum of simple forms of degree k , and hence it is sufficient to show that $\rfloor$ is well defined for a simple k-form by axiom I3. However, axioms I2 and I4 yield

$$(3\text{-}4.4) \quad v\rfloor(\omega^1\wedge\omega^2\wedge\ldots\wedge\omega^k) = (v\rfloor\omega^1)\wedge\omega^2\wedge\ldots\wedge\omega^k$$

$$- \omega^1\wedge(v\rfloor(\omega^2\wedge\omega^3\wedge\ldots\wedge\omega^k))$$

$$= \langle\omega^1, v\rangle\omega^2\wedge\ldots\wedge\omega^k$$

$$- \omega^1\wedge\{(v\rfloor\omega^2)\wedge\omega^3\wedge\ldots\wedge\omega^k - \omega^2\wedge(v\rfloor(\omega^3\wedge\ldots\wedge\omega^k))\}$$

$$= \langle \omega^1, v\rangle \omega^2 \wedge \ldots \wedge \omega^k - \langle \omega^2, v\rangle \omega^1 \wedge \omega^3 \wedge \ldots \wedge \omega^k$$

$$+ \langle \omega^3, v\rangle \omega^1 \wedge \omega^2 \wedge \omega^4 \wedge \ldots \wedge \omega^k$$

$$+ \ldots + (-1)^{k-1} \langle \omega^k, v\rangle \omega^1 \wedge \ldots \wedge \omega^{k-1}$$

and this belongs to $\Lambda^{k-1}(E_n)$ since each term belongs to $\Lambda^{k-1}(E_n)$. It is also clear from (3-4.4) that property I5 is likewise satisfied since $\langle \omega^k, v\rangle$ is a linear functional for each value of k.

LEMMA 3-4.2. *We have*

I6. $$u \rfloor (v \rfloor \omega) = - v \rfloor (u \rfloor \omega)$$

for all $u, v \in T(E_n)$ *and all* $\omega \in \Lambda(E_n)$, *and hence*

(3-4.5) $$v \rfloor (v \rfloor \omega) = 0 .$$

Proof. Since any element of $\Lambda^k(E_n)$ is the sum of simple elements, it is sufficient to establish property I6 for simple elements of $\Lambda^k(E_n)$. In this event, simply allow $u \rfloor$ to act on both sides of (3-4.4), reverse the roles of u and v and then add the results. The answer is zero which is the same thing as I6, and this implies (3-4.5).

It does prove useful to go through a direct proof of Lemma 3-4.2 for basis elements of $\Lambda^2(E_n)$ in order to improve our ability to work with the pull down operation:

$$v \rfloor (dx^i \wedge dx^j) = (v \rfloor dx^i) \wedge dx^j - dx^i \wedge (v \rfloor dx^j)$$

$$= \langle dx^i, v\rangle dx^j - \langle dx^j, v\rangle dx^i = v^i dx^j - v^j dx^i ;$$

$$u \rfloor (v \rfloor (dx^i \wedge dx^j)) = v^i u \rfloor dx^j - v^j u \rfloor dx^i = v^i u^j - v^j u^i ;$$

$$v \rfloor (u \rfloor (dx^i \wedge dx^j)) = u^i v^j - u^j v^i = - u \rfloor (v \rfloor (dx^i \wedge dx^j)) .$$

We have already seen that a linear first order partial differential equation can be written in the form

$$(3\text{-}4.6) \qquad v(f) = 0$$

for a given $v \in T(E_n)$. If we further recall that df is an element of $\Lambda^1(E_n)$ that is defined by

$$(3\text{-}4.7) \qquad \langle df, v \rangle = v(f) ,$$

then the definition of $\rfloor$ shows that

$$(3\text{-}4.8) \qquad v \rfloor df = \langle df, v \rangle = v(f) .$$

The partial differential equation (3-4.6) can thus be written in the equivalent form

$$(3\text{-}4.9) \qquad v \rfloor df = 0 ,$$

and the vector field v is the characteristic vector field of this partial differential equation. The equivalence of the formulations (3-4.6) and (3-4.9) clearly provide a means of viewing partial differential equations in terms of exterior forms. In this vein, the following definition proves to be useful.

Definition. A vector field $v \in T(E_n)$ is said to be a *characteristic* vector field of the given exterior form ω if and only if

$$(3\text{-}4.10) \qquad v \rfloor \omega = 0 .$$

LEMMA 3-4.3. *If a given exterior form* ω *has one nonzero characteristic vector field, then the collection of all characteristic vector fields of* ω *forms a subspace,* $\omega T(E_n)$, *of* $T(E_n)$.

Proof. If $v \neq 0$ is a characteristic vector field of ω then $f(v \rfloor \omega) = (fv) \rfloor \omega = 0$ and hence fv is a characteristic vector field of ω . The exterior form ω thus has at least a 1-dimensional subspace of characteristic vector fields. If u and v are characteristic vector fields, then

$$(fu+gv)\rfloor\omega = fu\rfloor\omega + gv\rfloor\omega = 0$$

and hence the collection of all characteristic vector fields of ω is closed under the vector space operations of $T(E_n)$; that is $\omega T(E_n)$ is a subspace of $T(E_n)$. This lemma will prove to be instrumental when we come to the problem of solving exterior equations. The following result is useful in this regard.

LEMMA 3-4.4. *If*

$$\omega = w_i(x^j)\, dx^i \tag{3-4.11}$$

is a 1-*form, then each of the* $n-1$ *vector fields*

$$v_k = w_1(x^j)\partial_k - w_k(x^j)\partial_1 , \quad k = 2,\dots,n \tag{3-4.12}$$

is a characteristic vector field of ω . *Thus, if* $w_1(x^j) \neq 0$ *on some neighborhood* N *in* E_n , *then* ω *has an* $(n-1)$-*dimensional characteristic subspace on* N .

Proof. If we use (3-4.12) to form $v_k\rfloor\omega$, we obtain

$$v_k\rfloor\omega = w_1\partial_k\rfloor\omega - w_k\partial_1\rfloor\omega = w_1w_k - w_kw_1 = 0 .$$

Since it is clear by inspection of (3-4.12) that these $n-1$ vectors are linearly independent whenever $w_1(x^j) \neq 0$, the result follows.

The important thing to be realized here, is that the $n-1$ characteristic vector fields need not mesh on the neighborhood N so that they are tangent to a family of surfaces of dimension $n-1$. We come back to this question later when we have more definite information concerning the possible forms that a 1-form may take.

Some example may prove helpful at this point, If

$$\omega = \frac{1}{2} a_{ij} \, dx^i \wedge dx^j , \quad a_{ij} = - a_{ji}$$

then

$$v \rfloor \omega = \frac{1}{2} a_{ij} \; v \rfloor (dx^i \wedge dx^j) = \frac{1}{2} a_{ij} \{ (v \rfloor dx^i) \wedge dx^j - dx^i \wedge (v \rfloor dx^j) \}$$

$$= \frac{1}{2} a_{ij} \{ v^i dx^j - v^j dx^i \} = \frac{1}{2} a_{ij} v^i dx^j - \frac{1}{2} a_{ij} v^j dx^i$$

$$= \frac{1}{2} a_{ji} v^j dx^i - \frac{1}{2} a_{ij} v^j dx^i$$

$$= \frac{1}{2} (a_{ji} - a_{ij}) v^j dx^i = v^j a_{ji} dx^i \;;$$

and if

$$\omega = \frac{1}{3!} a_{ijk} \, dx^i \wedge dx^j \wedge dx^k ,$$

$$a_{ijk} = - a_{jik} , \quad a_{ijk} = - a_{ikj} , \quad a_{ijk} = - a_{kji}$$

then

$$v \rfloor \omega = \frac{1}{3!} a_{ijk} (v^i dx^j \wedge dx^k - v^j dx^i \wedge dx^k + v^k dx^i \wedge dx^j)$$

$$= \frac{1}{3!} (a_{ijk} v^i dx^j \wedge dx^k - a_{jik} v^i dx^j \wedge dx^k + a_{jki} v^i dx^j \wedge dx^k)$$

$$= \frac{1}{3!} v^i (a_{ijk} - a_{jik} + a_{jki}) dx^j \wedge dx^k$$

$$= \frac{1}{3!} v^i (a_{ijk} + a_{ijk} + a_{ijk}) dx^j \wedge dx^k$$

$$= \frac{1}{2} v^i \; a_{ijk} \, dx^j \wedge dx^k .$$

Thus, in general, if

$$\omega = \frac{1}{k!} a_{i_1 i_2 \ldots i_k} dx^{i_1} \wedge dx^{i_2} \wedge \ldots \wedge dx^{i_k} \tag{3-4.13}$$

with $\{a_{i_1 i_2 \dots i_k}\}$ antisymmetric in all pairs of indices, then

$$(3\text{-}4.14) \qquad v \rfloor \omega = \frac{1}{(k-1)!} v^i a_{i i_2 \dots i_k} dx^{i_2} \wedge dx^{i_3} \wedge \dots \wedge dx^{i_k} .$$

An immediate use of the pull down operator $\rfloor$ is that it provides a means of testing whether a given exterior form is simple or not.

LEMMA 3-4.5. *Let* ω *be a given nonzero element of* $\Lambda^k(E_n)$ *and let* $(\omega|v_1,v_2,\dots,v_{k-1})$ *be the* 1-*form that is defined by*

$$(3\text{-}4.15) \quad (\omega|v_1,v_2,\dots,v_{k-1}) = v_{k-1}\rfloor v_{k-2}\rfloor \dots \rfloor v_2 \rfloor v_1 \rfloor \omega .$$

The k-*form* ω *is simple if and only if*

$$(3\text{-}4.16) \qquad \omega \wedge (\omega|v_1, v_2, \dots, v_{k-1}) = 0$$

for all (k-1)-*tuples of elements* $v_1,\dots,v_{k-1}$ *of* $T(E_n)$.

Proof. The proof will be given for the case $k=2$ since it is representative of the general case. If $\omega \varepsilon \Lambda^2(E_n)$ is simple, then

$$(3\text{-}4.17) \qquad \omega = \omega_1 \wedge \omega_2$$

with ω_1 and ω_2 linearly independent 1-forms. Pick a basis $\{u^1,u^2,\dots u^n\}$ of $T(E_n)$ such that

$$(3\text{-}4.18) \; u^1\rfloor\omega_1 = 1 , \quad u^1\rfloor\omega_2 = 0 , \quad u^2\rfloor\omega_1 = 0 , \quad u^2\rfloor\omega_2 = 1 ;$$

$$u^b\rfloor\omega_1 = 0 , \quad u^b\rfloor\omega_2 = 0 , \quad b = 3,\dots,n ,$$

then any $v \varepsilon T(E_n)$ may be expressed as

$$v = p_1 u^1 + p_2 u^2 + p_3 u^3 + \dots + p_n u^n .$$

Now, any v belonging to the subspace of $T(E_n)$ that is spanned by $\{u_3,u_4,\dots u_n\}$ has the property that

$$v \rfloor \omega = v \rfloor \omega_1 \wedge \omega_2 = (v \rfloor \omega_1)\omega_2 - (v \rfloor \omega_2)\omega_1 = 0 .$$

Hence, we need consider only those v's that belong to the sub-space S_2 that is spanned by $\{u_1, u_2\}$:

(3-4.19) $$v = r_1 u^1 + r_2 u^2 .$$

But,

$$(\omega | u^1) = u^1 \rfloor \omega = u^1 \rfloor (\omega_1 \wedge \omega_2) = \omega_2$$

$$(\omega | u^2) = u^2 \rfloor \omega = u^2 \rfloor (\omega_1 \wedge \omega_2) = -\omega_1$$

by (3-4.18), and hence

(3-4.20) $$(\omega | v) = r_1 \omega_2 - r_2 \omega_1$$

for v in S_2. A combination of (3-4.17) with (3-4.20) now gives

$$\omega \wedge (\omega | v) = (\omega_1 \wedge \omega_2) \wedge (r_1 \omega_1 - r_2 \omega_2) = 0$$

for all r_1, r_2 and hence for all v in S_2. Thus, if ω is simple then (3-4.16) is satisfied. On the other hand, a general $\omega \in \Lambda^2(E_n)$ is given by

(3-4.21) $$\omega = \frac{1}{2} w_{ij} \, dx^i \wedge dx^j , \quad w_{ij} = -w_{ji} ,$$

so that any $v = v^m \partial_m$ yields

(3-4.22) $$(\omega | v) = v^m w_{mk} \, dx^k$$

by (3-4.14). Thus,

(3-4.23) $$\omega \wedge (\omega | v) = \frac{1}{2} w_{ij} \, dx^i \wedge dx^j \wedge (v^m w_{mk} \, dx^k)$$

has to vanish for all possible values of the n quantities $\{v^m\}$ if (3-4.16) is to be satisfied for all $v \in T(E_n)$:

$$(3\text{-}4.24) \qquad w_{mk}\, w_{ij}\, dx^k \wedge dx^i \wedge dx^j = 0\ , \qquad m = 1,\ldots,n\ .$$

This is a complicated system of $n\binom{n}{3}$ *quadratic equations* for the determination of the $\{w_{ij}\}$ due to the antisymmetry of $dx^k \wedge dx^i \wedge dx^j$ in each pair of indices. We will not actually obtain the solution to this system. Suffice it to say that the general solution is given by

$$(3\text{-}4.25) \qquad w_{ij} = \frac{1}{2}(\omega_{1i}\, \omega_{2j} - \omega_{1j}\, \omega_{2i})\ ,$$

in which case $\omega = \omega_1 \wedge \omega_2$ with $\omega_1 = \omega_{1i}\, dx^i$, $\omega_2 = \omega_{2i}\, dx^i$.

Note. For a form $\omega = \frac{1}{r!} w_{i_1 \ldots i_r}\, dx^{i_1} \wedge \ldots \wedge dx^{i_r}$ of degree r , (3-4.16) yields the *quadratic* conditions

$$(3\text{-}4.26) \qquad w_{i_1 \ldots i_{r-1} j_1}\, w_{j_2 \ldots j_{r+1}}\, dx^{j_1} \wedge \ldots \wedge dx^{j_{r+1}} = 0\ .$$

The condition that an r-form be simple is always a nonlinear requirement on its coefficients. This nonlinearity is what makes a number of problems very difficult.

3-5. TOP DOWN GENERATION OF BASES

The whole graded exterior algebra $\Lambda(E_n)$ has been generated from $\Lambda^0(E_n)$ and $\Lambda^1(E_n)$ by constructing sums of exterior products. In fact, the basis for $\Lambda^k(E_n)$ is constructed by forming the maximal set of linearly independent k-fold exterior products of the basis elements $\{dx^i\}$ of $\Lambda^1(E_n)$. Now that we have the pull down operator $\lrcorner$ that lowers the degree of an exterior form by one, it might be expected that we could generate bases by starting at the top with $\Lambda^n(E_n)$. This is indeed the case as we now show.

Definition. The natural basis of $\Lambda^n(E_n)$ is referred to as the *volume element* of E_n and is denoted by μ :

$$(3\text{-}5.1) \qquad \mu = dx^1 \wedge dx^2 \wedge \ldots \wedge dx^n\ .$$

LEMMA 3-5.1. *The collection of* (n-1)-*forms*

(3-5.2) $$\mu_i = \partial_i \rfloor \mu , \quad i = 1,\ldots,n$$

has the following properties: (1) *the set* $\{\mu_i , \ i=1,\ldots,n\}$ *forms a basis for* $\Lambda^{n-1}(E_n)$ *so that any* $\alpha \in \Lambda^{n-1}(E_n)$ *can be uniquely expressed in the form*

(3-5.3) $$\alpha = \alpha^i \mu_i ;$$

(2) *we have*

(3-5.4) $$dx^i \wedge \mu_j = \delta^i_j \mu ;$$

(3) *if* $v = v^i \partial_i \in T(E_n)$, *then*

(3-5.5) $$v \rfloor \mu = v^i \mu_i ;$$

(4) *if we write* μ *in the equivalent form*

(3-5.6) $$\mu = \frac{1}{n!} e_{i_1 i_2 \ldots i_n} dx^{i_1} \wedge dx^{i_2} \wedge \ldots \wedge dx^{i_n} ,$$

then

(3-5.7) $$\mu_j = \frac{1}{(n-1)!} e_{j i_2 \ldots i_n} dx^{i_2} \wedge dx^{i_3} \wedge \ldots \wedge dx^{i_n} ;$$

(5) *if we write* $\alpha \in \Lambda^{n-1}(E_n)$ *in the equivalent form*

(3-5.8) $$\alpha = \frac{1}{(n-1)!} \alpha_{i_2 i_3 \ldots i_n} dx^{i_2} \wedge dx^{i_3} \wedge \ldots \wedge dx^{i_n} ,$$

then equality of (3-5.3) *and* (3-5.7) *imply*

(3-5.9) $$\alpha_{i_2 i_3 \ldots i_n} = \alpha^j e_{j i_2 i_3 \ldots i_n} .$$

Proof. We first establish (3-5.4). Since μ is a basis for $\Lambda^n(E_n)$, $dx^i \wedge \mu \in \Lambda^{n+1}(E_n)$ and hence $dx^i \wedge \mu = 0$. It thus follows from property I4 that

$$0 = \partial_j \rfloor (dx^i \wedge \mu) = (\partial_j \rfloor dx^i)\mu - dx^i \wedge (\partial_j \rfloor \mu)$$

$$= \delta^i_j \mu - dx^i \wedge \mu_j .$$

If the equation $0 = c^j \mu_j$ is to be satisfied, we must also satisfy $0 = dx^i \wedge (c^j \mu_j) = c^j dx^i \wedge \mu_j = c^j \delta^i_j \mu = c^j \mu$ for all $j = 1,\ldots,n$. However, $\mu \neq 0$ and hence $0 = c^j \mu_j$ is satisfied if and only if $c^1 = c^2 = \ldots = c^n = 0$; that is, $\{\mu_i, i=1,\ldots,n\}$ are n linearly independent elements of $\Lambda^{n-1}(E_n)$. Thus, since $\Lambda^{n-1}(E_n)$ is n-dimensional, $\{\mu_i, i=1,\ldots,n\}$ is a basis for $\Lambda^{n-1}(E_n)$ and (3-5.3) holds. If we veck both sides of (3-5.3) with dx^j , the result is $dx^j \wedge \alpha = \alpha^i dx^j \wedge \mu_i = \alpha^i \delta^j_i \mu = \alpha^j \mu$ by (3-5.4) and hence

$$\text{(3-5.10)} \qquad \alpha^j \mu = dx^j \wedge \alpha .$$

Thus, since $dx^j \wedge \alpha \in \Lambda^n(E_n)$ and $\Lambda^n(E_n)$ is 1-dimensional with basis μ , (3-5.10) serves to uniquely determine $\{\alpha^i\}$. The result (3-5.5) follows directly from the definition of μ_i , for $v \rfloor \mu = v^i \partial_i \rfloor \mu = v^i \mu_i$. The remaining results are straightforward calculations that are left to the reader.

The almost trivial result given by (3-5.5) is the basis for the solution of a number of important problems when (3-5.5) is interpreted properly.

LEMMA 3-5.2. *Resolution of* $\Lambda^{n-1}(E_n)$ *on the basis* $\{\mu_i\}$ *induces a 1-to-1 correspondence between* $T(E_n)$ *and* $\Lambda^{n-1}(E_n)$. *In particular, any* $v = v^i \partial_i \in T(E_n)$ *induces the* (n-1)-*form* $\alpha = v^i \mu_i = v \rfloor \mu$ *and any* $\alpha = \alpha^i \mu_i \in \Lambda^{n-1}(E_n)$ *induces the tangent vector field* $v = \alpha^i \partial_i$.

This whole process can now be repeated starting with the (n-1)-forms $\{\mu_i\}$. The proof of the following results follow exactly the lines given for the proof of Lemma 3-5.1.

LEMMA 3-5.3. *The collection of* (n-2)-*forms*

(3-5.11) $\mu_{ji} = \partial_j \rfloor \mu_i$

has the following properties: (1)

(3-5.12) $\mu_{ji} = - \mu_{ij}$;

(2) *the set* $\{\mu_{ij} \mid i < j\}$ *forms a basis for* $\Lambda^{n-2}(E_n)$ *and hence any* $\alpha \in \Lambda^{n-2}(E_n)$ *can be uniquely expressed in the form*

(3-5.13) $\alpha = \frac{1}{2} \alpha^{ij} \mu_{ij} \quad \alpha_{ij} = - \alpha_{ji}$;

(3) *we have*

(3-5.14) $dx^i \wedge \mu_{jk} = \delta^i_j \mu_k - \delta^i_k \mu_j$;

(4) *if* $v = v^i \partial_i$ *is any element of* $T(E_n)$ *and* $\beta = \beta^i \mu_i$ *is any element of* $\Lambda^{n-1}(E_n)$, *then*

(3-5.15) $v \rfloor \beta = v^i \beta^j \mu_{ij}$.

3-6. IDEALS OF $\Lambda(E_n)$

There is a construct that uses the full scope of $\Lambda(E_n)$ as a graded algebra and which will prove to be of considerable importance in our later studies; namely, an ideal. Most people, on seeing the definition of an ideal for the first time are somewhat perplexed. This may well be the case right now, but bear with the discussion for it will prove to provide almost instantaneous answers to certain questions.

Definition. An *ideal*, I , of an algebra A over $\Lambda^0(E_n)$ is a subset of elements of A such that (1) if $\alpha \in I$ and $\beta \in I$, then $\alpha + \beta \in I$ if the sum is defined; (2) if $\alpha \in I$ then $f\alpha \in I$ for all $f \in \Lambda^0(E_n)$; (3) if $\alpha \in I$, then $(\alpha|\gamma) \in I$ for every $\gamma \in A$.

This definition, simply put, says that an ideal of an algebra is a vector subspace of the algebra such that the multiplication of any element of the algebra by an element of the ideal gives back an element of the ideal.

Definition. An ideal I is daid to be *generated* by the elements $\alpha_1, \alpha_2, \ldots, \alpha_k$ if every element of I is sum of terms, each of which contains at least one of the elements of the set $\alpha_1, \ldots, \alpha_k$ as a factor. If I is generated by $\alpha_1, \ldots, \alpha_k$, then we write $I\{\alpha_1, \ldots, \alpha_k\}$.

These ideas can be applied directly to the graded algebra $\Lambda(E_n)$. Thus, if I is an ideal of $\Lambda(E_n)$ that is generated by the exterior forms $\alpha_1, \ldots, \alpha_k$, then $\beta \in I\{\alpha_1, \ldots, \alpha_k\}$ if and only if

$$(3\text{-}6.1) \qquad \beta = \gamma_1 \wedge \alpha_1 + \gamma_2 \wedge \alpha_2 + \ldots + \gamma_k \wedge \alpha_k$$

for some $\gamma_1, \gamma_2, \ldots, \gamma_k \in \Lambda(E_n)$, such that for each $\gamma_r \neq 0$ $\deg(\alpha_r) + \deg(\gamma_r) = s$, $r = 1, \ldots, k$, with $1 \leq s \leq n$. The restriction that the sums of the degrees be a constant is clearly necessary in order that the additions be defined, in which case β belongs to $\Lambda^s(E_n)$. For example, let E_{2n+1} have coordinates $(q^1, \ldots, q^n, y^1, \ldots, y^n, t)$ and define the collection $\{C^i\}$ of n 1-forms by

$$(3\text{-}6.2) \qquad C^i = dq^i - y^i \, dt \ , \quad i = 1, \ldots, n \ .$$

The form $\beta \in \Lambda^3(E_{2n+1})$ belongs to the ideal $I\{C^1, \ldots, C^n\}$ if and only if there exist 2-forms $\gamma^1, \ldots, \gamma^n$ such that

$$(3\text{-}6.3) \qquad \beta = \gamma^1 \wedge C^1 + \gamma^2 \wedge C^2 + \ldots + \gamma^n \wedge C^n \ .$$

The 1-forms (3-6.2) are an essential ingredient in the study of particle mechanics, where they are called contact forms and the ideal $I\{C^1, \ldots, C^n\}$ is the contact ideal.

One of the primary uses of ideals is that it allows us to talk about forms that are determined up to sums of products of given forms in a meaningful manner. Suppose that α is determined by the relation

$$(3\text{-}6.4) \qquad \alpha = \rho + \gamma^1 \wedge C^1 + \gamma^2 \wedge C^2 + \ldots + \gamma^n \wedge C^n$$

where ρ does not belong to $I\{C^1,\ldots,C^n\}$, and the forms $\gamma^1, \ldots, \gamma^n$ are not known. It is still meaningful to say that α is determined by ρ to within an element of $I\{C^1,\ldots,C^n\}$. This statement can be made even more precise if we write (3-6.4) in the equivalent form

$$(3\text{-}6.5) \qquad \alpha - \rho \in I\{C^1,\ldots,C^n\} .$$

In order to recover an actual equation, it is usual to write

$$(3\text{-}6.6) \qquad \alpha - \rho = 0 \bmod I\{C^1,\ldots,C^n\} .$$

The nice thing about this notation is that any such equation remains true under exterior multiplications just as zero multiplied by anything is still zero. Simply observe that $\gamma \wedge (\alpha-\rho) \in I\{C^1,\ldots,C^n\}$ if $(\alpha-\rho) \in I\{C^1,\ldots,C^n\}$ since I is an ideal. Thus,

$$\alpha - \rho = 0 \bmod I \longrightarrow \gamma \wedge (\alpha-\rho) = 0 \bmod I \quad \forall \gamma \in \Lambda(E_n) ,$$

and significant computational simplifications result.

Although the concept of an ideal is an algebraic one, the principle use of ideals of $\Lambda(E_n)$ arises in conjunction with the calculus of exterior forms. A full realization of the utility of ideals can only be shown after we have mastered the material of the next chapter. Suffice it to say that we will use the notion of an ideal over and over again in the applications.

There is an example that can be given at this point that is in some respects typical of the use of ideals. Suppose that $\omega^1, \ldots, \omega^k$ are k given elements of $\Lambda^1(E_n)$ and we wish to determine all vector fields $v \in T(E_n)$ that are simultaneous characteristic vector fields of all of the given 1-forms;

$$(3\text{-}6.7)\qquad v \rfloor \omega^a = 0 , \quad a = 1,\dots,k .$$

In this case, we say that v is a *characteristic vector field of the exterior system* $\{\omega^a, a=1,\dots,k\}$.

LEMMA 3-6.1. *A vector field* v *is a characteristic vector field of an exterior system* $\{\omega^a, a=1,\dots,k\}$ *if and only if* $v \rfloor \rho$ *belongs to the ideal* $I\{\omega^1,\dots,\omega^k\}$ *for every* $\rho \in I\{\omega^1,\dots,\omega^k\}$.

Proof. If $\rho \in I\{\omega^1,\dots,\omega^k\}$, then

$$(3\text{-}6.8)\qquad \rho = \beta_a \wedge \omega^a ,$$

and hence

$$(3\text{-}6.9)\qquad v \rfloor \rho = (v \rfloor \beta_a) \wedge \omega^a + (-1)^{\deg(\beta_a)} \beta_a \wedge (v \rfloor \omega^a)$$

$$= (-1)^{\deg(\beta_a)} (v \rfloor \omega^a) \beta_a \mod I .$$

Thus, if v is a characteristic vector field of the given exterior system, then $v \rfloor \rho$ belongs to $I\{\omega^1,\dots,\omega^k\}$ for every $\rho \in I\{\omega^1,\dots,\omega^k\}$. Conversely, if $v \rfloor \rho = 0 \mod I\{\omega^1,\dots,\omega^k\}$ for all $\rho \in I\{\omega^1,\dots,\omega^k\}$ then it must be true for all $\rho = K_a \omega^a$; that is

$$(3\text{-}6.10)\qquad K_a v \rfloor \omega^a = 0 \mod I\{\omega^1,\dots,\omega^k\} .$$

However, I is generated by $\omega^1, \dots, \omega^k$ so that the only 0-form in the ideal is the zero function:

$$(3\text{-}6.11)\qquad K_a v \rfloor \omega^a = 0 .$$

Since this must hold for all possible choices of K_a , we obtain

$$(3\text{-}6.12)\qquad v \rfloor \omega^a = 0 \quad a = 1,\dots,k .$$

LEMMA 3-6.2. *The set of all characteristic vector fields of an exterior system* $\{\omega^a, a=1,\dots,k\}$ *forms a subspace* S *of* $T(E_n)$

over $\Lambda^{o}(E_n)$. *If*

$$(3\text{-}6.13) \qquad \Omega = \omega^1 \wedge \ldots \wedge \omega^k \neq 0 ,$$

so that the k *given* 1-*forms are linearly independent, then* S *has dimension* $n-k$.

Proof. If u and v are characteristic vector fields of $\{\omega^a, a=1,\ldots,k\}$, then $(fu+gv) \rfloor \rho = fu \rfloor \rho + gv \rfloor \rho = 0 + 0 \bmod I\{\omega^1,\ldots,\omega^k\}$ for all $\rho \in I\{\omega^a\}$ and hence characteristic vector fields form a subspace of $T(E_n)$ over $\Lambda^{o}(E_n)$. Under satisfaction of (3-6.13), the k given ω's are linearly independent. Complete them to a basis $\omega^1, \ldots, \omega^k, \omega^{k+1}, \ldots, \omega^n$ for $\Lambda^1(E_n)$. The basis for $T(E_n)$ that is dual to this one is given by the system of n vector fields $\{V_i\}$ such that

$$(3\text{-}6.14) \qquad V_i \rfloor \omega^j = \langle \omega^j, V_i \rangle = \delta^j_i .$$

Thus, for $i = k+1,\ldots,n$ and $j = 1,\ldots,k$ we have

$$(3\text{-}6.15) \qquad V_i \rfloor \omega^j = 0 .$$

The $n-k$ vector fields $V_{k+1}, \ldots, V_n$ are thus linearly independent characteristic vector fields of the given exterior system. On the other hand, $V_1 \rfloor \omega^1 = V_2 \rfloor \omega^2 = \ldots V_k \rfloor \omega^k = 1$, and hence $V_{k+1}, \ldots, V_n$ span the subspace S .

3-7. BEHAVIOR UNDER MAPPINGS

Up to this point, our discussions of exterior forms have taken place in E_n with the fixed coordinate cover, namely the (x^i) coordinate cover. In point of fact, this coordinate cover has been all pervasive, for the basis elements of $\Lambda^k(E_n)$ and for all of $\Lambda(E_n)$ have been written in terms of exterior products of the differentials of these coordinate functions, $\{dx^i\}$. It is thus essential that we study how elements of the graded exterior algebra behave under changes of the coordinate cover and under mappings in general, for otherwise our results would be restricted to that one paltry case of E_n with only the (x^i) coordinate system. We shall follow the same procedure as given in Section 2-8 and for the same reasons; that is, we first study the general mapping properties of exterior forms.

Let E_m be an m-dimensional number space with coordinate functions (y^A) and let Φ be a given map from E_n to E_m,

$$\text{(3-7.1)} \qquad \Phi : E_n \longrightarrow E_m \mid y^A = \phi^A(x^i) .$$

It was shown in Section 2-8 that Φ induces a map Φ^* of $\Lambda^0(E_m)$ to $\Lambda^0(E_n)$ by composition,

$$\text{(3-7.2)} \quad \Phi^* : \Lambda^0(E_m) \longrightarrow \Lambda^0(E_n) \mid h(x^i) = g \circ \Phi = g(\phi^A(x^i)) ;$$

that is, Φ^* maps in the direction opposite to that of Φ. Now, $\Lambda^0(E_n)$ is the first entry of the sequence $\Lambda^0(E_n)$, $\Lambda^1(E_n)$, ..., $\Lambda^n(E_n)$ that makes up $\Lambda(E_n)$, and any form of degree k is the sum of terms, each of which is an element of $\Lambda^0(E_n)$ multiplied by a k-fold exterior product of the elements dx^1, ..., dx^n. It thus follows that the behavior of $\Lambda(E_n)$ under the map Φ will be determined once we obtain the mapping properties of the elements dx^1, dx^2, ..., dx^n.

The starting point for our considerations is obviously E_m since the domain of Φ^* is $\Lambda^0(E_m)$. Since each of the coordinate functions $\{y^A\}$ is also an element of $\Lambda^0(E_m)$, we can actually combine (3-7.1) and (3-7.2) by writing

$$\text{(3-7.3)} \qquad \Phi^* y^A = \phi^A(x^j) , \quad A = 1,\ldots,m$$

since $\phi^A(x^j)$ is an element of $\Lambda^0(E_n)$. On the other hand, the base elements $\{dy^A\}$ of $\Lambda^1(E_m)$ are simply the differentials of the functions $\{y^A\}$, while the differentials of $\Phi^* y^A$ are

$$(3\text{-}7.4)\qquad d\Phi^* y^A = d\phi^A(x^j) = \frac{\partial\phi^A(x^j)}{\partial x^i}\, dx^i$$

because $\Phi^* y^A \in \Lambda^o(E_n)$. However, the right-hand side of (3-7.4) is a linear combination of the base elements $\{dx^i\}$ of $\Lambda^1(E_n)$, and hence $d\Phi^* y^A$ belongs to $\Lambda^1(E_n)$. In other words, we just apply the chain rule of the calculus. These considerations do, nevertheless, establish the consistency of the following definition of the action of Φ^* on the basis $\{dy^A\}$:

$$(3\text{-}7.5)\qquad \Phi^* dy^A \overset{\text{def}}{=} d\Phi^* y^A = (\partial_i \phi^A(x^j))\, dx^i .$$

An arbitrary 1-form in E_m has the structure

$$(3\text{-}7.6)\qquad {}^{\backprime}\beta = b_A(y^B)\, dy^A$$

where $b_A(y^B)$ are an m-tuple of elements of $\Lambda^o(E_m)$. Since we know how such functions behave under Φ^* and how the dy^A behave under Φ^*, we would know how ${}^{\backprime}\beta$ behaves under Φ^* if we require Φ^* of products to be equal to the products of Φ^* acting on each factor separately:

$$(3\text{-}7.7)\qquad \Phi^* {}^{\backprime}\beta = \Phi^* b_A\ \Phi^* dy^A .$$

A combination of (3-7.7) with (3-7.4) now gives

$$(3\text{-}7.8)\qquad \Phi^* {}^{\backprime}\beta = b_A(\phi^B(x^j))\, \frac{\partial\phi^A(x^j)}{\partial x^i}\, dx^i$$

so that $\Phi^* {}^{\backprime}\beta$ certainly belongs to $\Lambda^1(E_n)$. Thus, if we denote the image of ${}^{\backprime}\beta$ in $\Lambda^1(E_n)$ by $\beta = b_i(x^j)dx^i$, then (3-7.8) shows that

$$(3\text{-}7.9)\qquad {}^{\backprime}\beta = b_A(y^B)dy^A ,\quad \beta = b_i(x^j)dx^i ,\quad \Phi^* {}^{\backprime}\beta = \beta ,$$

$$(3\text{-}7.10)\qquad b_i(x^j) = b_A(\phi^B(x^j))\, \frac{\partial\phi^A(x^j)}{\partial x^i} .$$

It now remains only to check that these definitions are

consistent with the fact that elements of $\Lambda^1(E_m)$ are linear functionals on $T(E_m)$ and that elements of $\Lambda^1(E_n)$ are linear functionals on $T(E_n)$. Let

$$(3\text{-}7.11) \qquad v = v^i \partial_i$$

be an arbitrary element of $T(E_n)$ and suppose that Φ is such that Φ_* is well defined on $T(E_n)$ to $T(E_m)$. In this case we have

$$(3\text{-}7.12) \qquad \Phi_* v = v^i (\partial\phi^A/\partial x^i)\, \partial_A = {}^\backprime v^A\, \partial_A .$$

If ${}^\backprime\beta$ and $\Phi_*{}^\backprime\beta$ are as given in (3-7.9), then

$$\langle {}^\backprime\beta, \Phi_* v\rangle = b_A\, {}^\backprime v^A = b_A \frac{\partial\phi^A}{\partial x^i} v^i ,$$

$$\langle \Phi^{*}{}^\backprime\beta, v\rangle = b_i v^i = b_A \frac{\partial\phi^A}{\partial x^i} v^i ,$$

and hence

$$(3\text{-}7.13) \qquad \langle {}^\backprime\beta, \Phi_* v\rangle = \langle \Phi^{*}{}^\backprime\beta, v\rangle ;$$

that is, *the value of the linear functional* $\langle \beta, v\rangle$ *is invariant under the mappings induced by* Φ.

The result also goes the other way. Suppose that (3-7.13) holds for all $v \in T(E_n)$ and that the action of Φ_* on $T(E_n)$ is well defined. If we then write

$$(3\text{-}7.14) \qquad {}^\backprime\beta = b_A(y^B)\, dx^A , \qquad \Phi^{*}{}^\backprime\beta = b_i(x^j)\, dx^i ,$$

then (3-7.13) gives

$$(3\text{-}7.15) \qquad 0 = \langle {}^\backprime\beta, \Phi_* v\rangle - \langle \Phi^{*}{}^\backprime\beta, v\rangle = b_A(y^B)\, {}^\backprime v^A - b_i(x^j) v^i$$

$$= b_A(\phi^B(x^j)) \frac{\partial\phi^A(x^j)}{\partial x^i} v^i - b_i(x^j) v^i$$

for all n-tuples $v^1, \ldots, v^n$. This is the case, however, if and only if $b_i = b_A \, \partial\phi^A/\partial x^i$, namely (3-7.10) holds. It should now be clear that the action of Φ^* can be extended to all of $\Lambda(E_m)$ by the requirement $\Phi^*(\alpha \wedge \beta) = \Phi^*\alpha \wedge \Phi^*\beta$.

THEOREM 3-7.1. *A smooth map*

$$\Phi : E_n \longrightarrow E_m \mid y^A = \phi^A(x^j) \tag{3-7.16}$$

induces the smooth map

$$\Phi^* : \Lambda(E_m) \longrightarrow \Lambda(E_n) \tag{3-7.17}$$

by the requirements

$$\Phi^* dy^A = d\Phi^* y^A , \quad A = 1,\ldots,m , \tag{3-7.18}$$

$$\Phi^*(\alpha \wedge \beta) = \Phi^*\alpha \wedge \Phi^*\beta , \quad \Phi^*(\alpha + \beta) = \Phi^*\alpha + \Phi^*\beta . \tag{3-7.19}$$

Proof. All that remains to be shown is that (3-7.18) and (3-7.19) define Φ^* on all of $\Lambda(E_n)$. Now, any element of Λ is the sum of simple elements and hence $\Phi^*(\alpha + \beta) = \Phi^*\alpha + \Phi^*\beta$ shows that we need only consider simple elements. On the other hand, simple elements are exterior products of 1-forms and hence $\Phi^*(\alpha \wedge \beta) = \Phi^*\alpha \wedge \Phi^*\beta$ shows that we need only establish the result for 1-forms. This has already been done for arbitrary 1-forms $\beta = b_A(y^B)dy^A$, and so the result is established.

We make particular note of the fact that Φ^* is defined for any smooth map Φ . This is in strong contrast to the situation with vector fields, for there Φ_* was only well defined for certain kinds of maps. There is no restriction that $m = n$ or that Φ be invertible for the construction of the map Φ^* . Exterior forms thus have very nice behavior for all smooth mappings of E_n to E_m no matter what the values of n and m are. This pleasant circumstance is one of the reasons why we will use exterior forms rather than vector fields whenever possible.

A detailed examination of the mapping properties of each of

the vector spaces $\Lambda^1(E_n)$, $\Lambda^2(E_n)$, ..., $\Lambda^n(E_n)$ is most helpful in understanding the full import of the map Φ^*. For 1-forms, let us write $\grave{\alpha} = \grave{\alpha}_A(y^B)dy^A$ and $\alpha = \Phi^*\grave{\alpha} = \alpha_i(x^j)dx^i$, then the previous results give

$$\text{(3-7.20)} \qquad \alpha_i(x^j) = \frac{\partial\phi^A(x^j)}{\partial x^i}\,\grave{\alpha}_A(\phi^B(x^j)) .$$

For 2-forms, we write

$$\text{(3-7.21)} \quad \grave{\beta} = \frac{1}{2}\grave{\beta}_{AB}(y^C)dy^A \wedge dy^B, \quad \beta = \Phi^*\grave{\beta} = \frac{1}{2}\beta_{ij}(x^k)dx^i \wedge dx^j .$$

The relations between the coefficients are obtained directly from how elements of $\Lambda^0(E_m)$ and the dy^A's behave under Φ^*:

$$\Phi^*\grave{\beta} = \frac{1}{2}\Phi^*(\grave{\beta}_{AB}dy^A \wedge dy^B) = \frac{1}{2}\Phi^*\grave{\beta}_{AB}\;\Phi^*dy^A \wedge \Phi^*dy^B$$

$$= \frac{1}{2}\grave{\beta}_{AB}(\phi^C(x^j))\,\frac{\partial\phi^A}{\partial x^i}dx^i \wedge \frac{\partial\phi^B}{\partial x^j}dx^j$$

$$= \frac{1}{2}\grave{\beta}_{AB}(\phi)\,\frac{\partial\phi^A}{\partial x^i}\frac{\partial\phi^B}{\partial x^j}dx^i \wedge dx^j = \frac{1}{2}\beta_{ij}(x^k)dx^i \wedge dx^j ,$$

so that

$$\text{(3-7.22)} \qquad \beta_{ij}(x^k) = \frac{\partial\phi^A(x^k)}{\partial x^i}\frac{\partial\phi^B(x^k)}{\partial x^j}\,\grave{\beta}_{AB}(\phi^C(x^k)) .$$

In the case of 3-forms, we have

$$\text{(3-7.23)} \qquad \beta_{ijk} = \frac{\partial\phi^A}{\partial x^i}\frac{\partial\phi^B}{\partial x^j}\frac{\partial\phi^C}{\partial x^k}\,\grave{\beta}_{ABC}(\phi(x)) ,$$

and so forth. The graphic situation is shown in Fig. 12. An inspection of this figure establishes the following result.

THEOREM 3-7.2. *If* $\Phi: E_n \longrightarrow E_m$ *and* $\Psi: E_m \longrightarrow E_r$ *are smooth and composable, then their composition induces the map* $(\Psi\circ\Phi)^*$ *with*

$$\text{(3-7.24)} \qquad (\Psi\circ\Phi)^* = \Phi^*\Psi^* .$$

Let us look in more detail at two specific cases. In the

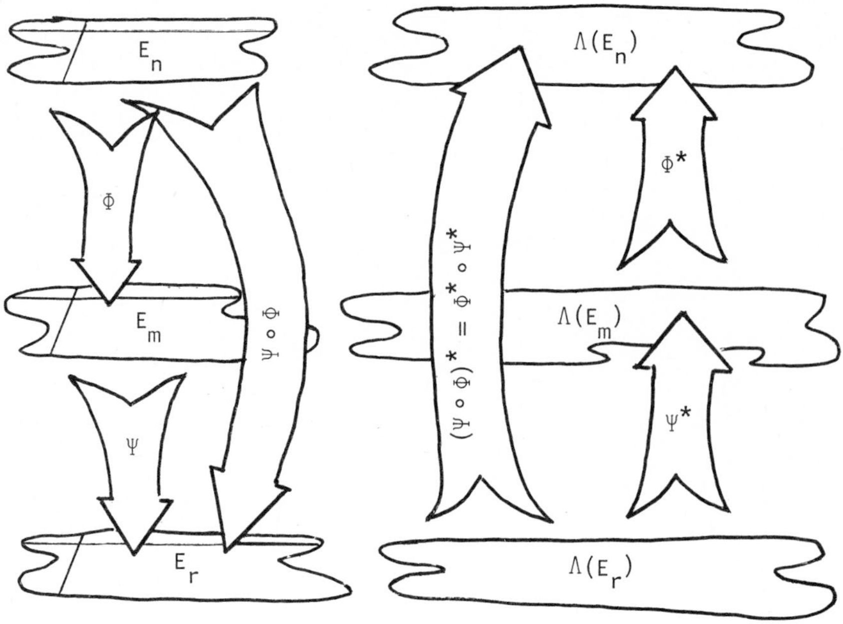

Figure 12. The induced maps of Λ

first, $\Phi : \mathbb{R} \longrightarrow E_3 \mid y^A = \phi^A(t)$ so that Φ^* is a map from $\Lambda(E_3)$ to $\Lambda(\mathbb{R}) = \Lambda^0(\mathbb{R}) \oplus \Lambda^1(\mathbb{R})$. Suppose that ${}^\backprime\beta = f(x,y,z)dx + g(x,y,z)dy + h(x,y,z)dz$, $\in \Lambda^1(E_3)$ then $\Phi^{*\backprime}\beta = \{f(\phi^1,\phi^2,\phi^3) d\phi^1/dt + g(\phi)d\phi^2/dt + h(\phi)d\phi^3/dt\}dt$. Further, if ${}^\backprime\gamma \in \Lambda^2(E_3)$, then $\Phi^*\gamma = 0$ because every 2-form on the 1-dimensional space $\mathbb{R}$ vanishes identically. In general *if* $\Phi : E_n \longrightarrow E_m$, $n < m$ *then* $\Phi^*\omega = 0$ *for every* $\omega \in \Lambda^{n+r}(E_m)$ *with* $r > 0$. However $\Phi : E_n \longrightarrow E_m \mid y^A = \phi^A(x^1,\ldots,x^n)$, $A = 1,\ldots,m$ with $n < m$ defines an n-dimensional surface in E_m, and hence $\Phi^*\alpha = 0$ if $\deg(\alpha) \leq n$ may be viewed as what it means to "solve" an exterior equation $\alpha = 0$ when α does not vanish identically on E_n.

Definition. Let $\alpha \in \Lambda^r(E_m)$ be a given r-form on E_m that does not vanish identically on E_m. An exterior equation

(3-7.25) $\alpha = 0$

is said to be *solved* in parametric form by any map

(3-7.26) $\Phi : E_k \longrightarrow E_m \mid y^A = \phi^A(x^j)$

with $k \leq r$ if any only if

(3-7.27) $\Phi^* \alpha = 0$

is satisfied identically on the domain of Φ.

Thus, a solution of $\alpha = 0$, $\alpha \in \Lambda^r(E_m)$ can be a curve in E_m, a two-dimensional surface in $E_m, \ldots,$ or a surface of dimension r in E_m. This does not mean that there can not be solutions of $\alpha = 0$, $\alpha \in \Lambda^r(E_m)$ that constitute surfaces of dimension greater than r; rather, such solutions are not parametric. Indeed, there can be solutions of $\alpha = 0$ that are surfaces whose dimension is as large as $m-1$, it is just that such surfaces are not conveniently written in parametric form. We note that $\Phi^* \alpha \in \Lambda^r(E_k)$ and hence $\Phi^* \alpha$ vanishes identically if $k < r$; in which case we have a *trivial* parametric solution: *nontrivial parametric solutions are possible only if* $k \geq r$. We will have much more to say about this important topic later.

An example will probably make the situation as clear as it is going to be at this point. Let α be given by

$$\alpha = x \, dy - 3 y \, dx$$

and let $\Phi : R \longrightarrow E_2 \mid x = \phi(t)$, $y = \psi(t)$, where $\phi(t)$ and $\psi(t)$ are as yet unknown. We then have

$$\Phi^* \alpha = \left\{ \phi(t) \frac{d\psi}{dt} - 3 \psi(t) \frac{d\phi}{dt} \right\} dt ,$$

and hence $\Phi^* \alpha = 0$ throughout $\mathbb{R}$ if and only if

$$\phi \frac{d\psi}{dt} = 3 \psi(t) \frac{d\phi}{dt} .$$

Thus, $\alpha = 0$ is solved by every map $\Phi : R \longrightarrow E_2 \mid x = \phi(t)$,

$y = \psi(t)$ (i.e., every curve) for which the functions $\phi(t)$ and $\psi(t)$ stand in the relation $\phi\, d\psi/dt = 3\,\psi\, d\phi/dt$. There are obviously uncountably many such curves, for simply assign the function $\phi(t)$ and then use the differential equation to compute $\psi(t)$. The exterior equation $x\,dy - 3\,y\,dx = 0$ thus has uncountably many parametric solutions of dimension one.

Invertible maps

$$(3\text{-}7.28) \qquad \Phi : E_n \longrightarrow {}^{\backprime}E_n \mid y^A = \phi^A(x^j), \quad \det(\partial\phi^A/\partial x^i) \neq 0$$

are of particular importance, for (y^A) may be interpreted as new coordinates of a point that had coordinates (x^i). We have already seen that

$$(3\text{-}7.29) \qquad \Phi^* dy^A = \frac{\partial\phi^A(x^j)}{\partial x^i}\, dx^i ,$$

and hence

$$(3\text{-}7.30) \qquad \Phi^* {}^{\backprime}\mu = \Phi^*(dy^1 \wedge \ldots \wedge dy^n) = \Phi^* dy^1 \wedge \ldots \wedge \Phi^* dy^n .$$

The actual calculation is significantly simplified by use of Lemma 3-3.4 since (3-7.29) gives a direct relation between the 1-forms $d\Phi^* y^A$ and the 1-forms dx^i. When (3-3.10) is used, the right-hand side of (3-7.30) becomes $\det(\partial\phi^A/\partial x^i)\mu$, which can also be written as $\det(\partial y^A/\partial x^i)\mu$:

$$(3\text{-}7.31) \qquad \Phi^* {}^{\backprime}\mu = \det(\partial y^A/\partial x^i)\mu .$$

A mnemonic convenience obtains if we write

$$\mu(y) = dy^1 \wedge \ldots \wedge dy^n , \qquad \mu(x) = dx^1 \wedge \ldots \wedge dx^n ,$$

for (3-7.31) then becomes

$$(3\text{-}7.32) \qquad \Phi^* \mu(y) = \det(\partial y^A/\partial x^i)\mu(x) .$$

If $\alpha(y)$ is an n-form on ${}^{\backprime}E_n$, then

$$(3\text{-}7.33) \qquad \alpha(y) = f(y^A)\mu(y) .$$

Thus, if we write

$$(3\text{-}7.34) \qquad \alpha(x) = g(x^j)\mu(x) = \Phi^*\alpha(y) ,$$

then (3-7.32) shows that the coefficient functions f and g stand in the relation

$$(3\text{-}7.35) \qquad g(x^j) = \det(\partial y^B/\partial x^k) f(y^A(x^j)) .$$

The case of invertible maps allows us to obtain the mapping properties of the "top down" basis $\{\mu_i\}$ of Λ^{n-1}. In this regard, it is useful to introduce the notation

$$(3\text{-}7.36) \qquad \mu_i(x) = \partial_i \rfloor \mu(x) , \qquad \mu_A(y) = \partial_A \rfloor \mu(y) .$$

For $v \in T(E_n)$, Φ induces the map Φ_* so that $\Phi_* v \in T({}^\backprime E_n)$. This shows that the action of Φ^* on the pull down of a form α by v is given by

$$(3\text{-}7.37) \qquad \Phi^*(\Phi_* v \rfloor {}^\backprime\alpha) = v \rfloor \Phi^*\alpha .$$

Now,

$$(3\text{-}7.38) \qquad \Phi_* v = v^i \frac{\partial \phi^A}{\partial x^i} \partial_A ,$$

so that

$$(3\text{-}7.39) \qquad \Phi_* v \rfloor \mu(y) = v^i \frac{\partial \phi^A}{\partial x^i} \partial_A \rfloor \mu(y) = v^i \frac{\partial \phi^A}{\partial x^i} \mu_A(y) ,$$

$$(3\text{-}7.40) \qquad v \rfloor \Phi^*\mu(y) = v^i \partial_i \rfloor \det\left(\frac{\partial \phi^A}{\partial x^j}\right)\mu = \det\left(\frac{\partial \phi^A}{\partial x^j}\right) v^i \mu_i(x) ,$$

with

$$(3\text{-}7.41) \qquad \Phi^*(\Phi_* v \rfloor \mu(y)) = v \rfloor \Phi^*\mu(y)$$

for all $v \in T(E_n)$ implies

$$(3\text{-}7.42)\qquad \frac{\partial\phi^A}{\partial x^i}\,\Phi^*\mu_A(y) = \det\left(\frac{\partial\phi^B}{\partial x^k}\right)\mu_i(x)\ .$$

These results, in turn, allow us to obtain the relations between elements of $\Lambda^{n-1}(E_n)$ and $\Lambda^{n-1}({}^\backprime E_n)$ when resolved on the bases $\mu_i(x)$ and $\mu_A(y)$, respectively. If ${}^\backprime\alpha(y)$ is an element of $\Lambda^{n-1}({}^\backprime E_n)$, then

$$(3\text{-}7.43)\qquad {}^\backprime\alpha(y) = F^A(y^B)\,\mu_A(y)\ .$$

Thus, if we set

$$(3\text{-}7.44)\qquad \alpha(x) = F^i(x^j)\,\mu_i(x) = \Phi^{*\,\backprime}\alpha(y)\ ,$$

then

$$F^i(x)\,\mu_i(x) = F^A(\phi^B(x))\,\Phi^*\mu_A(y)$$

while (3-7.42) shows that

$$F^i(x)\,\mu_i(x) = F^i(x)\,\frac{\partial\phi^A}{\partial x^i}\,\frac{1}{\det\left(\frac{\partial\phi^B}{\partial x^k}\right)}\,\Phi^*\mu_A(y)\ .$$

We thus obtain the relations

$$(3\text{-}7.45)\qquad F^A(\phi^B(x)) = \frac{1}{\det\left(\frac{\partial\phi^B}{\partial x^k}\right)}\,\frac{\partial\phi^A}{\partial x^i}\,F^i(x)\ .$$

An additional insight into the intrinsic structure of the basis $\{\mu_i(x)\}$ of $\Lambda^{n-1}(E_n)$ can now be obtained. A map

$$(3\text{-}7.46)\qquad \Phi : E_{n-1} \longrightarrow E_n \mid x^i = \phi^i(u^\alpha)$$

defines an (n-1)-dimensional surface S in E_n if we require that rank $(\partial\phi^i/\partial u^\alpha) = n-1$ throughout the domain of Φ. In these circumstances, the coordinates $(u^\alpha,\ \alpha=1,\dots,n-1)$ of E_{n-1} are referred to as the surface coordinates of S in E_n. If we allow Φ_* to act on the element ∂_α, we obtain

$$\text{(3-7.47)} \qquad \Phi_* \partial_\alpha = \frac{\partial \phi^i}{\partial u^\alpha} \partial_i$$

on S , which is a vector field on S that is tangent to S in E_n for each value of $\alpha = 1,\ldots,n-1$. However (3-7.37) shows that

$$\text{(3-7.48)} \qquad \Phi^*(\Phi_* \partial_\alpha \rfloor \mu(x)) = \partial_\alpha \rfloor \Phi^* \mu(x) = 0 ,$$

since $\Phi^* \mu(x) \in \Lambda^n(E_{n-1})$ vanishes identically. A combination of (3-7.47) and (3-7.48) thus gives

$$\text{(3-7.49)} \qquad 0 = \Phi^*(\Phi_* \partial_\alpha \rfloor \mu(x)) = \Phi^*\left(\frac{\partial \phi^i}{\partial u^\alpha} \partial_i \rfloor \mu(x)\right)$$

$$= \Phi^*\left(\frac{\partial \phi^i}{\partial u^\alpha} \mu_i\right) .$$

On the other hand, $\Phi^*(\mu_i \; \partial\phi^i/\partial u^\alpha)$ can be interpreted as having values on the surface S when S is given the surface coordinates (u^α) . If we denote this restriction to S by $\mu_i|_S$, then (3-7.49) gives

$$\text{(3-7.50)} \qquad \frac{\partial \phi^i(u^\beta)}{\partial u^\alpha} \mu_i|_S = 0 , \qquad \alpha = 1,\ldots,n-1 .$$

In view of the fact that $(\partial\phi^i/\partial u^\alpha)\partial_i$ is tangent to S for $\alpha = 1,\ldots,n-1$, (3-7.50) simply states that $\{\mu_i|_S\}$ may be viewed as the *normal* element of the surface S .

A terminology has grown up in the current literature that should be noted here.

Definition. A map

$$\text{(3-7.51)} \qquad \Phi : E_m \longrightarrow E_n \mid x^i = \phi^i(u^1,\ldots,u^m)$$

with $m < n$ is called a *section* of E_n . If

$$\text{(3-7.52)} \qquad \operatorname{rank}(\partial\phi^i/\partial u^\alpha) = m ,$$

on $\mathcal{D}(\Phi)$ then the section Φ of E_n is said to be *regular*.

The definition of a regular section $\Phi: E_m \longrightarrow E_n$ shows that it defines a regular m-dimensional surface in E_n, in which case the coordinates (u^α) of E_m may be viewed as surface coordinates of the regular section. If a regular section $\Phi: E_m \longrightarrow E_n$ is such that $\Phi^*\omega = 0$ then the section is said to *solve* the exterior equation $\omega = 0$. Thus, a regular section of E_n that solves $\omega = 0$ gives a solution of $\omega = 0$ in parametric form. The important thing to note here is that the solution of $\omega = 0$ is now interpreted as a regular m-dimensional surface in E_n with the property that $\Phi^*\omega = 0$ is now realized as the evaluation of ω on the section Φ of E_n. The ability to realize solutions of exterior equations as regular sections of the original space (i.e., as surfaces in the original space) provides significant insights into problems, for it allows full use of one's geometric intuition.

CHAPTER IV

EXTERIOR DERIVATIVES, LIE DERIVATIVES AND INTEGRATION

4-1. THE EXTERIOR DERIVATIVE

The natural basis $\{dx^i\}$ of $\Lambda^1(E_n)$ with respect to the coordinate cover (x^i) was obtained in Section 3-1 by introducing the notion of the differential of a function as a primitive concept: that is, dx^i can be considered either as a basis form or as the differential of the element x^i of $\Lambda^0(E_n)$. The question thus arises as to whether an operation "d" can be introduced into the exterior algebra so that the notion of a differential is not restricted solely to forms of degree zero.

We begin by noting that any element of $\Lambda^k(E_n)$ is the sum of terms of the form $f(x^j)\, dx^{i_1} \wedge dx^{i_2} \wedge \ldots \wedge dx^{i_k}$. The operation d can thus be introduced by specifying how it combines with the operations $+$ and $\wedge$. We therefore require

$$(4\text{-}1.1) \qquad d(\alpha+\beta) = d\alpha + d\beta ,$$

$$(4\text{-}1.2) \qquad d(\alpha\wedge\beta) = (d\alpha)\wedge\beta + (-1)^{\deg(\alpha)}\,\alpha\wedge(d\beta) .$$

Consistency requires us to demand that

$$(4\text{-}1.3) \qquad df = (\partial f/\partial x^j)dx^j \qquad \forall\, f \in \Lambda^0(E_n) .$$

If d is allowed to act on both sides of (4-1.3) and (4-1.2) is used, we obtain

$$(4\text{-}1.4) \qquad ddf = d\left(\frac{\partial f}{\partial x^j}\, dx^j\right) = d\left(\frac{\partial f}{\partial x^j}\right)\wedge dx^j + \frac{\partial f}{\partial x^j}\, ddx^j$$

$$= \frac{\partial^2 f}{\partial x^k \partial x^j}\, dx^k \wedge dx^j + \frac{\partial f}{\partial x^j}\, ddx^j$$

$$= \frac{1}{2}\left(\frac{\partial^2 f}{\partial x^k \partial x^j} - \frac{\partial^2 f}{\partial x^j \partial x^k}\right) dx^k \wedge dx^j + \frac{\partial f}{\partial x^j}\, ddx^j$$

$$= \frac{\partial f}{\partial x^j}\, ddx^j$$

because $(\partial f/\partial x^j) \in \Lambda^0(E_n)$. This result shows that we may require

$$(4\text{-}1.5) \qquad ddx^i = 0 , \quad i = 1,\ldots,n$$

without contradiction. Such a requirement certainly achieves maximal simplicity, for there would only be derivatives of first order that could ever occur.

It now remains for us to show that an operator with these properties is defined on all of $\Lambda(E_n)$ and that it is unique. If $\beta \in \Lambda^k(E_n)$, then β is a sum of terms of the form $f(x^m)dx^{i_1}\wedge dx^{i_2}\wedge\ldots\wedge dx^{i_k}$. Thus, if d satisfies (4-1.1) then it is sufficient to check that d is well defined for monomials of the form

$$(4\text{-}1.6) \qquad \gamma = f(x^m)dx^{i_1}\wedge dx^{i_2}\wedge\ldots\wedge dx^{i_k} .$$

Thus, applying d to both sides of (4-1.6) and using property (4-1.2), we have

$$(4\text{-}1.7)\qquad d\gamma = df \wedge dx^{i_1} \wedge dx^{i_2} \wedge \ldots \wedge dx^{i_k}$$

$$+ f\, d\{dx^{i_1} \wedge dx^{i_2} \wedge \ldots \wedge dx^{i_k}\} .$$

Thus, (4-1.2) and (4-1.5) imply

$$(4\text{-}1.8)\qquad d\gamma = \frac{\partial f}{\partial x^j}\, dx^j \wedge dx^{i_1} \wedge \ldots \wedge dx^{i_k} ,$$

from which we conclude that d is defined for every element of $\Lambda(E_n)$ and is unique. In particular $d:\Lambda^k(E_n) \longrightarrow \Lambda^{k+1}(E_n)$. If we apply d to both sides of (4-1.8) and make use of (4-1.5), it is obvious that

$$(4\text{-}1.9)\qquad dd\gamma = 0$$

for all $\gamma \varepsilon \Lambda(E_n)$, which implies (4-1.5).

These results are conveniently summarized as follows.

THEOREM 4-1.1. *There is one and only one operator* d *on* $\Lambda(E_n)$ *with the following properties*

D1. $\qquad d(\alpha + \beta) = d\alpha + d\beta$;

D2. $\qquad d(\alpha \wedge \beta) = (d\alpha) \wedge \beta + (-1)^{\deg(\alpha)}\, \alpha \wedge (d\beta)$;

D3. $\qquad df = (\partial f/\partial x^j)dx^j$

for all $f \varepsilon \Lambda^0(E_n)$;

D4. $\qquad dd\alpha = 0$

for all $\alpha \varepsilon \Lambda(E_n)$. *If* $\alpha \varepsilon \Lambda^k(E_n)$, *then* $d\alpha \varepsilon \Lambda^{k+1}(E_n)$ *and hence* d *may be viewed as the map*

$$(4\text{-}1.10)\qquad d:\Lambda^k(E_m) \longrightarrow \Lambda^{k+1}(E_m)$$

that satisfies properties D1 *through* D4 .

Definition. The operator d , given in Theorem 4-1.1 is called *exterior differentiation.*

The actual calculation of the exterior derivative of any given exterior form follows directly from properties D1 through D4 . For example, for a 1-form

$$\alpha = \alpha_i(x^k)dx^i , \tag{4-1.11}$$

we have

$$d\alpha = d\alpha_i \wedge dx^i = \frac{\partial\alpha_i}{\partial x^j} dx^j \wedge dx^i \tag{4-1.12}$$

$$= \frac{1}{2}(\partial_j\alpha_i - \partial_i\alpha_j)dx^j \wedge dx^i .$$

Similarly, if

$$\alpha = \frac{1}{2}\alpha_{ij} \quad dx^i \wedge dx^j , \quad \alpha_{ij} = -\alpha_{ji} ,$$

then

$$d\alpha = \frac{1}{2} d\alpha_{ij} \wedge dx^i \wedge dx^j = \frac{1}{2}(\partial_k\alpha_{ij})dx^k \wedge dx^i \wedge dx^j \tag{4-1.13}$$

$$= \frac{1}{3!}(\partial_k\alpha_{ij} + \partial_i\alpha_{jk} + \partial_j\alpha_{ki})dx^k \wedge dx^i \wedge dx^j .$$

What about elements from the top of $\Lambda(E_n)$?

THEOREM 4-1.2. *If* $\omega \in \Lambda^n(E_n)$, *then* $d\omega = 0$.

Proof. Since d maps $\Lambda^k(E_n)$ into $\Lambda^{k+1}(E_n)$, d maps $\Lambda^n(E_n)$ into $\Lambda^{n+1}(E_n)$ and $\Lambda^{n+1}(E_n)$ contains the single element 0 .

Elements of $\Lambda^{n-1}(E_n)$ are most easily handled by first establishing the following result.

LEMMA 4-1.1. *The base elements* μ_i *and* μ_{ij} *satisfy*

$$d\mu_i = 0 , \quad d\mu_{ij} = 0 . \tag{4-1.14}$$

Proof. By definition (see Section 3-5)

$$\mu_i = \partial_i \lrcorner \mu = \partial_i \lrcorner (dx^1 \wedge dx^2 \wedge \dots \wedge dx^n) \tag{4-1.15}$$

so that each μ_i is the product of $n-1$ simple factors of the form dx^{i_j}. Since $ddx^{i_j} = 0$, it follows that $d\mu_i = 0$. The same argument applies to the (n-2)-forms μ_{ij}.

THEOREM 4-1.3. *If* $\alpha \in \Lambda^{n-1}(E_n)$, *then resolution of* α *on the basis* $\{\mu_i\}$ *by*

$$\alpha = \alpha^i \mu_i \tag{4-1.16}$$

gives

$$d\alpha = (\partial_i \alpha^i)\mu . \tag{4-1.17}$$

If $\alpha \in \Lambda^{n-2}$, *then resolution of* α *on the basis* $\{\mu_{ij}, i<j\}$ *by*

$$\alpha = \frac{1}{2}\alpha^{ij}\mu_{ij} , \quad \alpha^{ij} = -\alpha^{ji} \tag{4-1.18}$$

gives

$$d\alpha = (\partial_i \alpha^{ij})\mu_j . \tag{4-1.19}$$

Proof. Applying d to both sides of (4-1.16) and using properties D1 through D4 gives $d\alpha = d\alpha^i \wedge \mu_i + \alpha^i d\mu_i$. However $d\mu_i = 0$ by Lemma 4-1.1 and hence $d\alpha = (\partial_j \alpha^i) dx^j \wedge \mu_i = (\partial_j \alpha^i)\delta_i^j \mu = (\partial_i \alpha^i)\mu$ by (3-5.4). Similarly, (4-1.18) gives $d\alpha = \frac{1}{2} d\alpha^{ij} \wedge \mu_{ij} = \frac{1}{2}(\partial_k \alpha^{ij}) dx^k \wedge \mu_{ij} = \frac{1}{2}(\partial_k \alpha^{ij})(\delta_i^k \mu_j - \delta_j^k \mu_i) = \frac{1}{2}\partial_i \alpha^{ij}\mu_j - \frac{1}{2}\partial_j \alpha^{ij}\mu_i$ by (3-5.13). Thus, $\alpha^{ij} = -\alpha^{ji}$ yields $d\alpha = \frac{1}{2}\partial_i \alpha^{ij}\mu_j - \frac{1}{2}\partial_i \alpha^{ji}\mu_j = \frac{1}{2}\partial_i(\alpha^{ij} - \alpha^{ji})\mu_j = \partial_i \alpha^{ij}\mu_j$ and the result is established.

The important thing to note in the context of Theorem 4-1.3 is the emergence of the "divergence" type quantities $\partial_i \alpha^i$ and $\partial_i \alpha^{ij}$ as the coefficients of μ and μ_j, respectively. There is thus no difficulty in obtaining "divergence" type quantities in the exterior calculus provided the right choice of basis is made; namely, $\{\mu_i\}$ for $\Lambda^{n-1}(E_n)$ and $\{\mu_{ij}, i<j\}$ for $\Lambda^{n-2}(E_n)$.

Let E_3 with coordinates $(X^A) = (X^1, X^2, X^3)$ be the 3-dimensional Euclidean space of reference configurations of a deformable body and let $\grave{E}_3$ with coordinates $(x^i) = (x^1, x^2, x^3)$ be the 3-dimensional Euclidean space of current configurations of the body. If we denote the components of the Piola-Kirchhoff stress tensor by $((\sigma^{Ai}))$, we can form the collection of three 2-forms $\{\sigma^i\} \in \Lambda^2(E_3)$ by

$$\sigma^i = \sigma^{Ai} \mu_A ,$$

where $\mu_A = \partial_A \lrcorner \mu = \partial_A \lrcorner (dX^1 \wedge dX^2 \wedge dX^3)$ are the elements of oriented surface of E_3 and a basis for $\Lambda^2(E_3)$. The standard interpretation of the Piola-Kirchhoff stress tensor then shows that the traction vector T (element of $T(\grave{E}_3)$) for a unit element of bounding surface of the deformable body is given by

$$T = \sigma^i \partial_i = \sigma^{Ai} \mu_A \partial_i .$$

It is now an elementary calculation to see that

$$d\sigma^i = d(\sigma^{Ai} \mu_A) = (\partial_A \sigma^{Ai})\mu$$

is the resulting force vector that acts on an element of unit volume in the reference configuration as a consequence of the application of tractions to the boundary of the body. Since $dd\omega \equiv 0$, it follows that any $\{\sigma^i\}$ given by

$$\sigma^i = d\beta^i , \quad \beta^i = \beta^i_A \, dX^A \in \Lambda^1(E_3)$$

yields an identically zero resulting force vector per unit volume in the reference configuration; i.e., a null stress distribution. Thus,

$$\sigma^{Ai}\,\mu_A = d\beta^i = \partial_B\beta^i_A\, dX^B \wedge dX^A = \frac{1}{2}\,(\partial_B\beta^i_A - \partial_A\beta^i_B)\, dX^B \wedge dX^A$$

defines the collection of all null Piola-Kirchhoff stress distributions.

We have seen that any smooth map $\Phi : E_n \longrightarrow E_m$ induces a smooth map $\Phi^* : \Lambda(E_m) \longrightarrow \Lambda(E_n)$ simply by composition and application of the chain rule, and this came about by setting $d\Phi^* y^A = \Phi^* dy^A$. It would thus appear that d and Φ^* should commute, as indeed they do.

THEOREM 4-1.4. *If* $\Phi : E_n \longrightarrow E_m$ *is a smooth map, then*

$$(4\text{-}1.20) \qquad d(\Phi^*\omega) = \Phi^*(d\omega)$$

for all $\omega \in \Lambda(E_m)$.

Proof. Since every element of $\Lambda(E_m)$ is the sum of simple elements of $\Lambda(E_m)$, it suffices to establish (4-1.20) for a generic 1-form

$$\alpha = \alpha_A(y^B)\, dy^A \,, \qquad d\alpha = \partial_E\alpha_A(y^B)\, dy^E \wedge dy^A \,.$$

Thus, if $y^A = \phi^A(x^j)$ defines the map Φ , we have

$$\Phi^*\alpha = \alpha_A(\phi^B)\, \Phi^* dy^A \,,$$

$$\Phi^*(d\alpha) = \partial_E\alpha_A(\phi^B)\, \Phi^* dy^E \wedge \Phi^* dy^A \,,$$

$$d(\Phi^*\alpha) = \partial_i(\alpha_A(\phi^B))\, dx^i \wedge \Phi^* dy^A$$

$$= (\partial_E\alpha_A(\phi^B))\, \frac{\partial\phi^E}{\partial x^i}\, dx^i \wedge \Phi^* dy^A \,.$$

However, (3-7.4) shows that $\frac{\partial\phi^E}{\partial x^i}\, dx^i = d\Phi^* y^E = \Phi^* dy^E$, and hence

$$d(\Phi^*\alpha) = \partial_E\alpha_A(\phi^B)\Phi^* dy^E \wedge \Phi^* dy^A = \Phi^*(d\alpha) \,.$$

The importance of this last result is that the order of mappings and exterior differentiation is immaterial. Thus, in particular, the action of Φ^* is well defined for any smooth map Φ whether or not exterior differentiation is involved. For example, if $\alpha \varepsilon \Lambda^k(E_m)$ and $\beta \varepsilon \Lambda^{k+1}(E_m)$ are given, then $\gamma = d\alpha + \beta$ belongs to $\Lambda^{k+1}(E_m)$ and $\Phi^*\gamma = d\Phi^*\alpha + \Phi^*\beta \varepsilon \Lambda^{k+1}(E_n)$ is well defined.

The operation d of exterior differentiation leads to the definition of two subspaces of $\Lambda(E_n)$ in a natural way.

<u>Definition.</u> An element α of $\Lambda(E_n)$ is said to be *closed* if and only if

$$(4\text{-}1.21) \qquad d\alpha = 0 .$$

<u>Definition.</u> An element α of $\Lambda(E_n)$ is said to be *exact* if and only if there exists a $\beta \varepsilon \Lambda(E_n)$ such that

$$(4\text{-}1.22) \qquad \alpha = d\beta .$$

<u>THEOREM 4-1.5.</u> *The collection of all closed elements of* $\Lambda(E_n)$ *forms a subsapce* $C(E_n)$ *of* $\Lambda(E_n)$ *over* $\mathbb{R}$ *but not over* $\Lambda^0(E_n)$. *The collection of all exact elements of* $\Lambda(E_n)$ *forms a subspace* $E(E_n)$ *of* $\Lambda(E_n)$ *over* $\mathbb{R}$ *but not over* $\Lambda^0(E_n)$, *and*

$$(4\text{-}1.23) \qquad E(E_n) \subset C(E_n) .$$

<u>Proof.</u> If α and β are closed elements of $\Lambda^k(E_n)$, then $d\alpha = 0$, $d\beta = 0$ and

$$d(f(x)\alpha + g(x)\beta) = df \wedge \alpha + dg \wedge \beta .$$

Thus, if $df = 0$, $dg = 0$, that is $f = \text{constant}$, $g = \text{constant}$, then $d(f\alpha + g\beta) = 0$ and $C(E_n)$ is a subspace over $\mathbb{R}$. On the other hand, it is easily seen that for $df \neq 0$, $dg \neq 0$, then $d\alpha = 0$, $d\beta = 0$ does not imply $d(f\alpha + g\beta) = 0$ for all f , $g \varepsilon \Lambda^0(E_n)$. Thus, $C(E_n)$ is not a vector subspace over $\Lambda^0(E_n)$. Similarly, if α and β belong to $E(E_n)$, there exist elements

ρ and η such that $\alpha = d\rho$, $\beta = d\eta$ and $f\alpha + g\beta = f\, d\rho + g\, d\eta = d(f\rho) + d(g\eta) - df \wedge \rho - dg \wedge \eta$. The same argument then shows that $E(E_n)$ is a vector subspace over $\mathbb{R}$ but not over $\Lambda^0(E_n)$. Finally, if $\beta \varepsilon E(E_n)$, then $\beta = d\eta$ for some η and hence $d\beta = dd\eta = 0$; that is β is closed.

The inclusion of $E(E_n)$ in $C(E_n)$ is the basis by which a large number of significant problems are solved. A typical such problem is the following. Suppose that we are given a 1-form

(4-1.24) $$F = F_i(x^j)\, dx^i$$

and we would like to find an element η of $\Lambda^0(E_n)$ such that

(4-1.25) $$F = d\eta = \partial_i \eta\, dx^i .$$

A comparison of (4-1.24) and (4-1.25) shows that this is the case if and only if the function η satisfies the simultaneous system of partial differential equations

(4-1.26) $$F_i(x^j) = \partial_i \eta(x^j) ;$$

that is the force with components F_i admits a potential function η. From our new point of view, (4-1.25) says that such a function η can exist if and only if F is exact, in which case F must be closed; that is

(4-1.27) $$0 = dF = \frac{1}{2}(\partial_i F_j - \partial_j F_i)\, dx^i \wedge dx^j$$

is a system of necessary conditions for the existence of the function η; that is

(4-1.28) $$\partial_i F_j = \partial_j F_i .$$

This generalized $\vec{F} = \vec{\nabla}\eta$ only if $\vec{\nabla}\times\vec{F} = \vec{0}$ in 3 dimensions to any number of dimensions. There is quite a bit more here, however. Suppose that α is a given 2-form and we ask when can we find a 1-form β such that

$$(4\text{-}1.29) \qquad \alpha = d\beta .$$

Since (4-1.29) says that α is exact if there exists such a β, then α must likewise be closed in order that there exist such a β:

$$(4\text{-}1.30) \qquad d\alpha = 0 .$$

Thus, if $\alpha = \frac{1}{2}\alpha_{ij}\, dx^i \wedge dx^j$, then the coefficient functions α_{ij}, with $\alpha_{ij} = -\alpha_{ji}$ must satisfy

$$(4\text{-}1.31) \qquad \partial_i\alpha_{jk} + \partial_j\alpha_{ki} + \partial_k\alpha_{ij} = 0$$

(see (4-1.13)). Similarly, if $\alpha = \alpha^i\mu_i \in \Lambda^{n-1}(E_n)$ then $\alpha = d\beta$ only if $0 = d\alpha = (\partial_i\alpha^i)\mu$; that is $\partial_i\alpha^i = 0$.

Suppose that we satisfy these necessary conditions for the existence of solutions; $\alpha = d\beta$ only if $d\alpha = 0$. The question thus arises as to whether we can actually find such a β as to make the equation $\alpha = d\beta$ true for given α . This will indeed be the case if we could show that every closed form is an exact form; $d\alpha = 0$ implies a β such that $\alpha = d\beta$. A little reflection will easily show that such wishful thinking is just that, for such a result can not be true in general. For instance, suppose that we are in 3-dimensions and F satisfies $\vec{\nabla}\times\vec{F} = \vec{0}$ on a region with a hole in it. We then know that $\vec{F} = \vec{\nabla}\eta$ can not necessarily be satisfied by a single-valued function η because of the hole: the line integral of $\vec{F}\cdot d\vec{r}$ around the hole need not be zero! On the other hand, there is a partial converse of the inclusion $\mathcal{E}(E_n) \subset \mathcal{C}(E_n)$ that comes about if the region of definition is sufficiently nice. This is known as the Poincaré Lemma, which we prove in the next chapter.

LEMMA 1-4.2. (Poincaré Lemma) *If S is a region of E_n that can be shrunk to a point in a smooth way (S is starshaped with respect to one of its points), then*

(4-1.32) $\quad C(S) \subset E(S)$;

that is, if $d\alpha = 0$ *on* S *then there exists a* β *on* S *such that* $\alpha = d\beta$.

Significantly more information is actually available. In fact, the form β can be computed directly from the given form α by application of the homotopy operator H that is the subject of the next chapter.

We noted in Section 3-1 that dx^i could be given two different interpretations. The first was that of an element of the natural basis for $\Lambda^1(E_n)$. For this interpretation, let us write $(dx)^i$ since the definition of the dual basis gives $\langle (dx)^i, \partial_j \rangle = \delta^i_j$ with $\langle (dx)^i, \partial_j \rangle = \partial_j(x^i)$. The second interpretation was that of the differential of the coordinate function x^i . In order to make a clear distinction, let us write $d(x^i)$ for this differential. The considerations up to this point have been based upon the identification $(dx)^i = d(x^i)$ without a proof of consistency, as the careful reader has already realized! This situation is easily remedied, however, once the axioms D1 through D4 of the exterior derivative have been given. In order to see this, let us return to the original definition of the base elements $(dx)^i$ by means of $\langle (dx)^i, \partial_j \rangle = \delta^i_j$. The only axiom that is affected is D3 , which must now be given in the following form

D3. $\quad d(f) = \partial f/\partial x^j \, (dx)^j$.

Written in this way, D3 defines the action of the operator d on elements of $\Lambda^0(E_n)$. Now, simply observe that x^i , for given i , is an element of $\Lambda^0(E_n)$. We may thus apply D3 to obtain the desired result

$$d(x^i) = \partial x^i/\partial x^j \, (dx)^j = (dx)^i .$$

The two interpretations of dx^i are thus seen to be consistent.

Their simultaneous employment in the previous Chapter thus leads to no internal inconsistency, rather only an economy.

4-2. CLOSED IDEALS AND A CONFLUENCE OF IDEAS

The most useful ideals of the exterior algebra of forms are those that remain unchanged under exterior differentiation.

Definition. An ideal I of $\Lambda(E_n)$ is said to be *closed* if and only if $d\rho \in I$ for every $\rho \in I$, in which case we write

$$dI \subset I . \tag{4-2.1}$$

Any finitely generated ideal $I\{\omega^1,\ldots,\omega^k\}$ can be closed by formation of a new ideal $\bar{I}\{\omega^1,\ldots,\omega^k;d\omega^1,\ldots,d\omega^k\}$ called the *closure* of $I\{\omega^1,\ldots,\omega^k\}$. If $\rho \in \bar{I}$, then

$$\rho = \gamma_a \wedge \omega^a + \Gamma_a \wedge d\omega^a$$

in which case

$$d\rho = d\gamma_a \wedge \omega^a + \{(-1)^{\deg(\gamma_a)} \gamma_a + d\Gamma_a\} \wedge d\omega^a$$

and hence $d\rho \in \bar{I}$.

The really important question is when is a given ideal $I\{\omega^1,\ldots,\omega^k\}$ a closed ideal as it stands. This question is answered by the following Theorem for the case in which all of the generators of the ideal are of the same degree.

THEOREM 4-2.1. *Let* $I\{\omega^1,\ldots,\omega^k\} \overset{\text{def}}{=} I\{\omega^a\}$ *be an ideal of* $\Lambda(E_n)$ *such that each of its generators is of the same degree. Then* $I\{\omega^a\}$ *is a closed ideal if and only if there exist* k^2 *1-forms* $\{\Gamma^a_b\}$ *such that the generators satisfy*

$$d\omega^a = \Gamma^a_b \wedge \omega^b . \tag{4-2.2}$$

Proof. If $\rho \in I\{\omega^a\}$, then $\rho = \gamma_a \wedge \omega^a$ for some k-tuple of forms $\{\gamma_a\}$ with $\deg(\gamma_a) = b$ and hence

$$d\rho = d\gamma_a \wedge \omega^a + (-1)^b \gamma_a \wedge d\omega^a = (-1)^b \gamma_a \wedge d\omega^a \mod I\{\omega^k\} .$$

Thus, $d\rho \in I\{\omega^a\}$ for every $\rho \in I\{\omega^a\}$ if and only if each of the forms $d\omega^a$ belongs to $I\{\omega^a\}$. Thus, since $d\omega^a$ has degree one greater than ω^a, $d\omega^a \in I\{\omega^a\}$ for $a = 1,\ldots,n$ if and only if there exist k^2 1-forms Γ^a_b such that the generators satisfy (4-2.2).

We note in passing that the Γ^a_b in (4-2.2) can not be entirely arbitrary, for $dd\omega^a$ must vanish:

$$0 = dd\omega^a = d\Gamma^a_b \wedge \omega^b - \Gamma^a_b \wedge d\omega^b = d\Gamma^a_b \wedge \omega^b - \Gamma^a_b \wedge \Gamma^b_c \wedge \omega^c ,$$

that is,

$$(4\text{-}2.3) \qquad (d\Gamma^a_b - \Gamma^a_c \wedge \Gamma^c_b) \wedge \omega^b = 0 .$$

The question of existence and uniqueness of solutions of systems of exterior differential equations of the form (4-2.2) subject to the integrability conditions (4-2.3) will be studied in the next Chapter. It suffices at this point to simply note that there need not exist a system of k^2 1-forms Γ^a_b for which (4-2.2) can be satisfied for given $\{\omega^a\}$; *not every ideal generated by forms of the same degree is closed.*

An alternative and sometimes more useful criteria for the closure of an ideal is available when the ideal is generated by forms of degree 1. Suppose that I is generated by 1-forms $\omega^1,\ldots,\omega^r$. If these r 1-forms are not linearly independent, $\omega^1 \wedge \ldots \wedge \omega^r = 0$, we may replace the generators $\omega^1,\ldots,\omega^r$ by a basis $\gamma^1,\ldots,\gamma^k$ for the subspace of $\Lambda^1(E_n)$ that is spanned by $\omega^1,\ldots,\omega^r$ and the ideal I will be unchanged. We may therefore assume that the generators of I are linearly independent over $\Lambda^0(E_n)$ without loss of generality.

<u>THEOREM 2-4.4.</u> *Let* $I\{\omega^a\}$ *be an ideal of* $\Lambda(E_n)$ *whose generators* $\omega^1,\ldots,\omega^k$ *are 1-forms such that*

(4-2.4) $\omega^1 \wedge \omega^2 \wedge \ldots \wedge \omega^k = \Omega(I) \neq 0$,

and $k<n-1$ *then* $dI\{\omega^a\} \subset I\{\omega^a\}$ *if and only if*

(4-2.5) $\Omega(I) \wedge d\omega^a = 0$, $a = 1,\ldots,k$.

Proof. If the generators satisfy (4-2.2), then they also satisfy (4-2.5). In order to establish the converse, complete $\omega^1,\ldots,\omega^k$ to a basis $\omega^1,\ldots,\omega^k,\omega^{k+1},\ldots,\omega^n$ of $\Lambda^1(E_n)$ and label the additional 1-forms with the index α, β, ... Since $d\omega^a \in \Lambda^2(E_n)$, we may always write

$$d\omega^a = \xi^a_{ij}\,\omega^i \wedge \omega^j = \xi^a_{bc}\,\omega^b \wedge \omega^c + \xi^a_{\alpha c}\,\omega^\alpha \wedge \omega^c + \xi^a_{\alpha\beta}\,\omega^\alpha \wedge \omega^\beta ,$$

and hence

$$\Omega(I) \wedge d\omega^a = \xi^a_{\alpha\beta}\Omega(I) \wedge \omega^\alpha \wedge \omega^\beta .$$

Since $k<n-1$, $\deg(\Omega(I) \wedge \omega^\alpha \wedge \omega^\beta) \leq n$ and hence $\Omega(I) \wedge \omega^\alpha \wedge \omega^\beta$ is a simple nonzero (k+2)-form because ω^α and ω^β are independent and not members of the subspace spanned by $\omega^1,\ldots,\omega^k$. Thus, (4-2.5) is satisfied only when $\xi^a_{\alpha\beta} = 0$, $\alpha,\beta = k+1,\ldots,n$, $a = 1,\ldots,k$; that is

$$d\omega^a = (\xi^a_{bc}\omega^b + \xi^a_{\alpha c}\omega^\alpha) \wedge \omega^c = \Gamma^a_c \wedge \omega^c$$

with $\Gamma^a_c = \xi^a_{bc}\omega^b + \xi^a_{\alpha c}\omega^\alpha$.

The case $k \geq n-1$ was excluded because $\Omega(I) \wedge d\omega^a$ would have degree greater than n and hence vanish identically. This case is taken care of by the following.

THEOREM 4-2.3. *If* I *is generated by either* $n-1$ *or* n *linearly independent* 1-*forms then* $dI \subset I$.

Proof. For $k=n$, $\{\omega^a\}$ constitute a basis for $\Lambda^1(E_n)$ and hence $d\omega^a = \xi^a_{bc}\ \omega^b \wedge \omega^c$. The result follows by Theorem 4-2.1. For the case $k = n-1$, let γ be an additional 1-form such that $\omega^1, \ldots, \omega^{n-1}, \gamma$ is a basis for $\Lambda^1(E_n)$. We then have

$$d\omega^a = \xi^a_{bc}\ \omega^b \wedge \omega^c + \xi^a_c\ \gamma \wedge \omega^c$$

and the result again follows by Theorem 4-2.1.

A simple example at this point will allow us to pull together the various ideas that have been introduced and will provide the springboard for the considerations of the next section. Consider the 1-form ω on E_n that is defined by

$$(4\text{-}2.6) \qquad \omega = f(x^j)\ dg(x^j) = f\ \partial_i g\ dx^i , \quad f \neq 0 .$$

Exterior differentiation yields

$$(4\text{-}2.7) \qquad d\omega = df \wedge dg$$

so that

$$(4\text{-}2.8) \qquad \omega \wedge d\omega = 0 .$$

Theorem 4-2.2 thus shows that the ideal, $I\{\omega\}$ that is generated by ω is closed: $dI \subset I$. Indeed, since $f \neq 0$, we can rewrite (4-2.7) in the equivalent form

$$(4\text{-}2.9) \qquad d\omega = \frac{1}{f}\ df \wedge f\ dg = \frac{1}{f}\ df \wedge \omega ,$$

which is just Theorem 4-2.1.

Suppose that we need to solve the exterior equation

$$(4\text{-}2.10) \qquad \omega = 0 .$$

It follows directly from (4-2.6) that $\omega = 0$ holds on every (n-1)-dimensional surface

$$(4\text{-}2.11) \qquad g(x^j) = \text{constant.}$$

Accordingly, if the map

(4-2.12) $\Phi: E_r \longrightarrow E_n \mid x^i = \phi^i(u^1,\ldots,u^r)$

is such that

(4-2.13) $g(\phi^j(u^a)) = \text{constant},$

then

(4-2.14) $\Phi^*\omega = f(\phi^j)\, dg(\phi^j) = f(\phi^j) \frac{\partial g(\phi^j)}{\partial \phi^i} \frac{\partial \phi^i}{\partial u^a} du^a = 0$

and hence Φ gives the solutions of $\omega = 0$ in parametric form.

Starting with $\omega = \omega^1$, we can complete to a basis $\omega^1, \omega^2, \ldots, \omega^n$ of $\Lambda^1(E_n)$. The dual basis is given by

(4-2.15) $V_i \rfloor \omega^j = \delta^j_i ,$

and hence we obtain

(4-2.16) $V_\alpha \rfloor \omega = 0 , \quad \alpha = 2,\ldots,n ;$

that is, each of the $n-1$ vector fields V_α, $\alpha = 2,\ldots,n$ is a characteristic vector of ω. The characteristic subspace $\mathcal{S}(I\{\omega\})$ of the ideal $I\{\omega\}$ is thus of dimension $n-1$. When (4-2.6) is substituted into (4-2.16) and we note that $f \neq 0$, we obtain

(4-2.17) $V_\alpha \rfloor dg = V_\alpha(g) = 0 , \quad \alpha = 1,\ldots,n .$

The function g thus satisfies the system (4-2.17) of $n-1$ simultaneous first order partial differential equations. Since $g(x^j)$ is not identically constant (i.e., ω is not the zero element of $\Lambda^1(E_n)$) the system (4-2.17) is complete. This implies that

(4-2.18) $[V_\alpha, V_\beta] = C^\gamma_{\alpha\beta} V_\gamma ,$

for if one of these Lie products were not expressible in terms of the original set of $n-1$ linearly independent vector fields $\{V_\alpha\}$, we would obtain n linearly independent vector fields that would annihilate g and g would have to be identically constant. The characteristic subspace $S(I\{\omega\})$ is thus a Lie subalgebra of $T(E_n)$. Conversely, if we are given a Lie subalgebra of $T(E_n)$ that is spanned by $n-1$ linearly independent vector fields $\{V_\alpha, \alpha = 2,\ldots,n\}$, we can reverse the process by (4-2.15) and obtain the 1-form $\omega^1 = \omega$ for which the subalgebra is the subalgebra of the characteristic subspace of $S(I\{\omega\})$. This form will then satisfy $\omega \wedge d\omega = 0$ and hence there exists a 1-form Γ such that

$$(4\text{-}2.19) \qquad d\omega = \Gamma \wedge \omega .$$

Exterior differentiation then yields

$$(4\text{-}2.20) \qquad 0 = dd\omega = d\Gamma \wedge \omega - \Gamma \wedge d\omega = d\Gamma \wedge \omega - \Gamma \wedge \Gamma \wedge \omega$$

$$= d\Gamma \wedge \omega .$$

Since (4-2.19) also implies $d\omega = (\Gamma + \lambda\omega) \wedge \omega$, (4-2.20) is also equivalent to $0 = d(\Gamma + \lambda\omega) \wedge \omega$ and we can always choose λ so that $\Gamma + \lambda\omega$ is exact,

$$(4\text{-}2.21) \qquad \Gamma = d\eta - \lambda\omega .$$

In this event, (4-2.19) becomes

$$(4\text{-}2.22) \qquad d\omega = d\eta \wedge \omega$$

which is satisfied by $\omega = e^\eta dg$ for some g. The function g can then be obtained from the given ω by $dg = e^{-\eta}\omega$; that is $e^\eta = f$. The system of simultaneous complete first order differential equations $V_\alpha(g) = 0$ is thus solved by solving the equivalent problem $\omega = 0$.

The duality between $\omega = 0$, $dI\{\omega\} \subset I\{\omega\}$ and the complete system $V_\alpha(t) = 0$, $\alpha = 2,\ldots,n$ is what allows us to actually solve the complete system $V_\alpha(t) = 0$, $\alpha = 2,\ldots,n$ explicitly. The ability to obtain explicit solutions is a very valuable thing, and it is thus natural to ask whether there are similar dualities when the given data is more complicated; that is, when there are more than 1 form of degree 1 that generate the ideal I, or when we have a complete system of partial differential equations whose vector fields span a subspace of dimension less than $n-1$. This question is answered in the affirmative by the celebrated Frobenius theorem which we study in the next section.

4-3. THE FROBENIUS THEOREM

The reason why everything worked in the example given at the end of the last section was the shape of the 1-form $\omega = f(x^j)\, dg(x^j)$; namely, ω was proportional to the exact 1-form $dg(x^j)$. This idea clearly generalized to the case of more than a single 1-form

We have already seen that any given system of 1-forms can be replaced by an equivalent system of linearly independent 1-forms that spans the same subspace of $\Lambda^1(E_n)$ as was spanned by the given system. Linear independence of any given system of 1-forms may thus be assumed without loss of generality.

Definition. A collection of r linearly independent 1-forms, ω^a, $a = 1,\ldots,r$ is an *exterior system of dimension* r. The symbol D_r will be used to designate such a system and the ideal generated by $\{\omega^a\}$ will be denoted by $I\{D_r\}$.

Definition. An exterior system D_r with 1-forms $\{\omega^a\}$ is said to be *completely integrable* if and only if there exist r independent functions $\{g^a(x^j)\}$ such that each of the r 1-forms $\{\omega^a\}$ vanishes on the $(n-r)$-dimensional surface

$$(4\text{-}3.1) \qquad \{g^a(x^j) = \text{constant}, \quad a = 1,\ldots,r\}$$

for each choice of the r constants.

THEOREM 4-3.1. *An exterior system* D_r *with* 1-*forms* $\{\omega^a\}$ *is completely integrable if and only if there exists a nonsingular* r-*by*-r *matrix of functions* $((A^a_b(x^j)))$ *and* r *independent functions* $g(x^j)$ *such that*

$$\omega^a = A^a_b\, dg^b . \tag{4-3.2}$$

Proof. If $\{\omega^a\}$ are given by (4-3.2), then they are completely integrable. Suppose, therefore that $\{\omega^a\}$ is completely integrable so that r linearly independent functions $g^a(x^j)$ exist. We can then construct the ideal $I\{dg^a\}$ and this ideal is the largest closed ideal such that every one of its elements vanishes on the surface $\{g^a(x^j) = \text{constant}, a = 1,\dots,r\}$. However, complete integrability of $\{\omega^a\}$ says that each ω^a vanishes on the surface $g^a(x^j) = \text{constant}$, $a = 1,\dots,r$, and hence $I\{D_r\} \subset I\{dg^a\}$. Thus, since both the ω's and the dg's are 1-forms, the ideal inclusion is satisfied only if there exists a matrix $((A^a_b))$ such that $\omega^a = A^a_b\, dg^b$. Since the ω's are linearly independent and the g's are independent, $\omega^1 \wedge \dots \wedge \omega^r \neq 0$, $dg^1 \wedge \dots \wedge dg^r \neq 0$ and hence $\omega^1 \wedge \dots \wedge \omega^r = \det(A^a_b)\, dg^1 \wedge \dots \wedge dg^r$ implies $\det(A^a_b) \neq 0$ so that the result is established.

Let us start with (4-3.2). Exterior differentiation and $\det(A^a_b) \neq 0$ then give

$$d\omega^a = dA^a_b \wedge dg^b = dA^a_e \overset{-1}{A}{}^e_c \wedge A^c_b\, dg^b = dA^a_e \overset{-1}{A}{}^e_c \wedge \omega^c . \tag{4-3.3}$$

Theorem 4-2.1 thus shows that *the ideal* $I\{D_r\}$ *is closed if* D_r *is completely integrable.* The converse of this statement, D_r *is completely integrable if the ideal* $I\{D_r\}$ *is closed,* is the famous Frobenius theorem. We will not actually prove this theorem now since to do so would involve a number of arguments whose groundwork has not yet been prepared.

THEOREM 4-3.2. (Frobenius) *An exterior system* D_r *that is defined by* r *1-forms* $\{\omega^a\}$ *is completely integrable if and only if the ideal* $I\{D_r\}$ *is closed; that is* $dI\{D_r\} \subset I\{D_r\}$ *which is equivalent to*

$$(4\text{-}3.4) \qquad d\omega^a = \Gamma^a_b \wedge \omega^b$$

or

$$(4\text{-}3.5) \qquad \omega^1 \wedge \omega^2 \wedge \ldots \wedge \omega^r \wedge d\omega^a = 0 \,, \quad a = 1,\ldots,r \,.$$

We now turn to the question of equivalent systems of first order partial differential equations. If $\{\omega^a\}$ define a differential system D_r, then the linear independence of the ω^a, $a = 1,\ldots,r$ allow us to complete this system to a basis $\omega^1,\ldots,\omega^r,\omega^{r+1},\ldots,\omega^n$ of $\Lambda^1(E_n)$. The dual basis of $T(E_n)$ is then defined by

$$(4\text{-}3.6) \qquad V_i \lrcorner\, \omega^j = \delta^j_i \,.$$

Thus, $\{V_\alpha, \alpha = r+1,\ldots,n\}$ satisfies

$$(4\text{-}3.7) \qquad V_\alpha \lrcorner\, \omega^a = 0 \,, \quad a = 1,\ldots,r \,, \quad \alpha = r+1,\ldots,n \,,$$

so that $\{V_\alpha\}$ span the subspace of characteristic vector fields of D_r.

<u>LEMMA 4-3.1</u>. *The characteristic subspace,* $S(D_r)$, *of an exterior system* D_r *is a subspace of* $T(E_n)$ *of dimension* $n-r$ *and a basis for* $S(D_r)$ *is given by* $\{V_\alpha, \alpha = r+1,\ldots,n\}$ *where* $\{V_i, i = 1,\ldots,n\}$ *is the dual basis that is defined by* (4-3.6).

<u>Proof</u>. By definition $S(D_r)$ consists of all elements of $T(E_n)$ such that $V \lrcorner\, \rho \in I\{D_r\}$ for all $\rho \in I\{D_r\}$. The construct immediately above shows that $\{V_\alpha\}$ is a basis for $S(D_r)$ and hence $S(D_r)$ is of dimension $n-r$.

We now need a technical result in order to connect complete integrability of D_r with the corresponding notion of complete integrability of the system $V_\alpha(f) = 0$, $\alpha = 1,\ldots,r$.

LEMMA 4-3.2. *If* η *is any element of* $\Lambda^1(E_n)$ *and* u,v *are any elements of* $T(E_n)$, *then*

$$u(v\rfloor\eta) - v(u\rfloor\eta) = [u,v]\rfloor\eta + v\rfloor u\rfloor d\eta \, . \tag{4-3.8}$$

Proof. Set $\eta = \eta_i dx^i$, $u = u^i\partial_i$, $v = v^i\partial_i$, then $u\rfloor\eta = u^i\eta_i$, $v\rfloor\eta = v^i\eta_i$ and hence

$$\begin{aligned} u(v\rfloor\eta) - v(u\rfloor\eta) &= \{u(v^i) - v(u^i)\}\eta_i \\ &\quad + v^i u(\eta_i) - u^i v(\eta_i) \\ &= [u,v]\rfloor\eta^i + v^i u(\eta_i) - u^i v(\eta_i) \, . \end{aligned}$$

On the other hand,

$$\begin{aligned} u\rfloor d\eta &= u\rfloor(d\eta_i \wedge dx^i) = (u\rfloor d\eta_i)dx^i - (u\rfloor dx^i)d\eta_i \\ &= u(\eta_i)dx^i - u^i d\eta_i \end{aligned}$$

so that

$$v\rfloor u\rfloor d\eta = u(\eta_i)v^i - u^i v(\eta_i) \, ;$$

$$u(v\rfloor\eta) - v(u\rfloor\eta) = [u,v]\rfloor\eta + v\rfloor u\rfloor d\eta \, .$$

THEOREM 4-3.3. *An exterior system* D_r *that is defined by the* r *1-forms* $\{\omega^a\}$ *is completely integrable if and only if its characteristic subspace* $S(D_r)$ *is a Lie subalgebra of* $T(E_n)$

Proof. Let $\{V_\alpha, \alpha = r+1,\ldots,n\}$ be a basis for $S(D_r)$, so that

$$V_\alpha\rfloor\omega^a = 0 \, , \quad a = 1,\ldots,r \, , \quad \alpha = r+1,\ldots,n \tag{4-3.9}$$

and let Greek indices have the range $r+1,\ldots,n$. A direct application of (4-3.8) yields

(4-3.10) $$V_\alpha(V_\beta \rfloor \omega^a) - V_\beta(V_\alpha \rfloor \omega^a) = [V_\alpha, V_\beta] \rfloor \omega^a + V_\beta \rfloor V_\alpha \rfloor d\omega^a$$

and hence

(4-3.11) $$[V_\alpha, V_\beta] \rfloor \omega^a = V_\alpha \rfloor V_\beta \rfloor d\omega^a .$$

If D_r is completely integrable, then $d\omega^a = \Gamma^a_b \wedge \omega^b$, $V_\beta \rfloor d\omega^a = (V_\beta \rfloor \Gamma^a_b)\omega^b - \Gamma^a_b \, V_\beta \rfloor \omega^b = (V_\beta \rfloor \Gamma^a_b)\omega^b$, and $V_\alpha \rfloor V_\beta \rfloor d\omega^a = (V_\beta \rfloor \Gamma^a_b) \, V_\alpha \rfloor \omega^b = 0$. The right-hand side of (4-3.11) thus vanishes and hence $[V_\alpha, V_\beta] \rfloor \omega^a = 0$ shows that $[V_\alpha, V_\beta] \in S(D_r)$. Since $\{V_\alpha, \alpha = r+1, \ldots, n\}$ is a basis for $S(D_r)$, we have

(4-3.12) $$[V_\alpha, V_\beta] = C^\gamma_{\alpha\beta} \, V_\gamma$$

and hence $S(D_r)$ is a Lie subalgebra of $T(E_n)$. Conversely, if $S(D_r)$ is a Lie subalgebra of $T(E_n)$, (4-3.12) holds and hence $[V_\alpha, V_\beta] \rfloor \omega^a = C^\gamma_{\alpha\beta} \, V_\gamma \rfloor \omega^a = 0$. In this event, (4-3.11) implies

(4-3.13) $$V_\alpha \rfloor V_\beta \rfloor d\omega^a = 0 .$$

Thus, if we write $(V_\alpha \rfloor \omega^\beta = \delta^\beta_\alpha, \; V_\alpha \rfloor \omega^a = 0)$

$$d\omega^a = \xi^a_{bc} \, \omega^b \wedge \omega^c + \xi^a_{\alpha c} \, \omega^\alpha \wedge \omega^c + \xi^a_{\alpha\beta} \, \omega^\alpha \wedge \omega^\beta$$

then (4-3.13) can be satisfied only if $\xi^a_{\alpha\beta} = 0$; that is

(4-3.14) $$d\omega^a = (\xi^a_{bc} \, \omega^b + \xi^a_{\alpha c} \, \omega^\alpha) \wedge \omega^c = \Gamma^a_c \wedge \omega^c .$$

The Frobenius theorem then shows that D_r is completely integrable.

COROLLARY 4-3.1. *Any complete system*

(4-3.15) $$V_\alpha(f) = 0 , \quad \alpha = r+1, \ldots, n$$

of linear partial differential equations in E_n *defines and is defined by a completely integrable exterior system* D_r *whose*

1-forms $\{\omega^a, a = 1,\ldots,r\}$ *satisfy the requirements*

$$(4\text{-}3.16) \qquad V_i \rfloor \omega^j = \delta^j_i ,$$

where $\{V_i\}$ *is a completion of* $\{V_\alpha\}$ *to a basis for* $T(E_n)$. *Thus, the functions* $\{g^a(x^j)\}$ *that occur in*

$$(4\text{-}3.17) \qquad \omega^a = A^a_b \, dg^b$$

are r *independent solutions of* (4-3.15) *and the general solution of* (4-3.15) *is given by*

$$(4\text{-}3.18) \qquad f = \Psi(g^1(x^j),\ldots,g^r(x^j)) .$$

Proof. The only thing that has not yet been proven is $V_\alpha(g^a) = 0$. However, $V_\alpha \rfloor \omega^a = A^a_b \, V_\alpha \rfloor dg^b = A^a_b \, V_\alpha(g^b) = 0$ because each V_α is a characteristic vector of D_r . Thus, since $\det(A^a_b) \neq 0$ we indeed have $V_\alpha(g^a) = 0$. It is then a trivial matter to see that (4-3.18) constitutes the general solution of (4-3.15); $V_\alpha(f) = V_\alpha(\Psi) = \frac{\partial\Psi}{\partial g^a} V_\alpha(g^a) = 0$, $\alpha = r+1,\ldots,n$.

From the practical point of view, Corollary 4-3.1 is the important result, for it tells us how to find the independent integrals of a complete system of linear partial differential equations. The construction of the corresponding exterior system is also fundamental whenever we have to deal with mappings. This should be obvious since exterior forms behave very nicely under mappings while vector fields do not unless the mapping is invertible.

It is clear that not every exterior system D_r is completely integrable, for $dI\{D_r\}$ need not be contained in $I\{D_r\}$. The question thus naturally arises as to what can be said for not completely integrable exterior systems.

Definition. Let D_r be an exterior system with linearly independent 1-forms $\{\omega^a, a = 1,\ldots,r\}$ and let $\{V_\alpha, \alpha = r+1,\ldots,n\}$ be a basis for the characteristic subspace, $S(D_r)$, of the exterior system. If $\bar{S}(D_r)$ is the completion of $S(D_r)$ to a Lie subalgebra of $T(E_n)$, then D_r is said to be *subintegrable* if and only if the dimension of $\bar{S}(D_r)$ is less than n. If $\{\bar{V}_\alpha, \alpha = r-s, \ldots, n\}$ is a basis for $\bar{S}(D_r)$, and $\{\gamma^1, \ldots, \gamma^{r-s}\}$ is a basis for the subspace of $\Lambda^1(E_n)$ whose characteristic subspace is $\bar{S}(D_r)$, then $\{\gamma^1, \ldots, \gamma^{r-s}\}$ is the *maximal subexterior* system of the exterior system D_r that is completely integrable.

The concept of the maximal subexterior system of a subintegrable exterior system provides certain information that is quite often useful in the solution of explicit problems. Needless to say, subintegrable exterior systems are much harder to deal with than completely integrable exterior systems, while exterior systems that are not subintegrable are well-nigh impossible.

The closure of an ideal of $\Lambda(E_n)$ and the complete integrability of its system of characteristic vector fields is not restricted to the case where the ideal is generated by 1-forms, as the Cartan theorem shows.

THEOREM 4-3.4 (Cartan). *Let I be an ideal of $\Lambda(E_n)$ whose characteristic system forms a subspace S of $T(E_n)$ with constant dimension. If I is closed then its characteristic system is completely integrable.*

Proof. By definition,

$$S = \{v \in T(E_n) \mid v \rfloor I \subset I\} .$$

Since S is of fixed dimension, say r, there exist r linearly independent elements v_α, $\alpha = 1, \ldots, r$ of $T(E_n)$ that span S. The definition of the Lie derivative given in Section 4-5 yields

$$v \rfloor \pounds_u \omega = v \rfloor u \rfloor d\omega + v \rfloor d(u \rfloor \omega)$$

and
$$\pounds_u (v \rfloor \omega) = [u,v] \rfloor \omega + v \rfloor \pounds_u \omega$$
$$= u \rfloor d(v \rfloor \omega) + d(u \rfloor v \rfloor \omega) .$$

When these relations are combined, we obtain the basic relation

$$(4\text{-}3.19)\qquad [u,v]\rfloor\omega = u\rfloor d(v\rfloor\omega) - v\rfloor d(u\rfloor\omega) - v\rfloor u\rfloor d\omega + d(u\rfloor v\rfloor\omega)$$

that is the generalization of (4-3.8). Thus, if $\omega\in I$, we have

$$(4\text{-}3.20)\qquad [v_\alpha,v_\beta]\rfloor\omega = v_\alpha\rfloor d(v_\beta\rfloor\omega) - v_\beta\rfloor d(v_\alpha\rfloor\omega) - v_\beta\rfloor v_\alpha\rfloor d\omega + d(v_\alpha\rfloor v_\beta\rfloor\omega) .$$

However, $v_\alpha\rfloor I\subset I$ and $dI\subset I$ show that each of the terms on the right-hand side of (4-3.20) is contained in I. Thus, $[v_\alpha,v_\beta]\rfloor\omega$ is contained in I and hence $[v_\alpha,v_\beta]$ belongs to S. Accordingly, S forms a Lie subalgebra of $T(E_n)$ and S is completely integrable.

The Cartan theorem provides the basis for a simple and direct proof of the Frobenius theorem.

Proof of the Frobenius theorem. Let D_r denote the exterior system that is defined by r 1-forms $\{\omega^a\}$ and let $I\{D_r\}$ denote the ideal of $\Lambda(E_n)$ that is generated by D_r. We need to show that $dI\{D_r\}\subset I\{D_r\}$ implies that there exist a nonsingular r-by-r matrix $((A^a_b))$ and r independent elements $\{g^a(x^j)\}$ of $\Lambda^0(E_n)$ such that $\omega^a = A^a_b\, dg^b$. Let $\{v_\alpha,\ \alpha=r+1,\dots,n\}$ be a basis for the characteristic subspace $S(D_r)$ of D_r. The Cartan theorem shows that $S(D_r)$ forms a Lie subalgebra of $T(E_n)$ and hence there exist r independent elements $\{g^a(x^j)\}$ of $\Lambda^0(E_n)$ such that

$$(4\text{-}3.21)\qquad v_\alpha(g^a) = v_\alpha\rfloor dg^a = 0,\quad a=1,\dots,r,\quad \alpha=r+1,\dots,n ,$$

$$(4\text{-}3.22)\qquad \Omega = dg^1\wedge dg^2\wedge\dots\wedge dg^r \neq 0 .$$

These equations show that $S(D_r)$ is a characteristic system of the system of r 1-forms $\{dg^a\}$ that spans an r-dimensional subspace S^* of $\Lambda^1(E_n)$ characterized by the simple r-form Ω. However, $S(D_r)$ is also the characteristic subspace of D_r, so that $v_\alpha \rfloor \omega^a = 0$, and any $v \in T(E_n)$ that does not belong to $S(D_r)$ gives $v \rfloor \omega^a \neq 0$. Thus, each of the ω's belongs to S^*, that is, $\omega^a \wedge \Omega = 0$, and hence $\omega^a = A^a_b\, dg^b$ for some r-by-r matrix $((A^a_b))$. It thus follows that

$$\omega^1 \wedge \omega^2 \wedge \ldots \wedge \omega^r = \det(A^a_b)\Omega$$

and hence the independence of the ω's and (4-3.22) shows that $\det(A^a_b) \neq 0$.

Particular note should be taken of the fact that the A's and the g's are not unique. Simply observe that if $\{F^a(g^1,\ldots,g^r)\}$ is any system of functions such that $\det(\partial F^a/\partial g^b) \neq 0$, then dF^a is another system of r independent elements of $\Lambda^0(E_n)$ such that $v_\alpha \rfloor dF^a = 0$. We could then write $\omega^a = B^a_b\, dF^b$, in which case $\omega^a = A^a_b\, dg^b$ yields the relations $A^a_b = B^a_c\, \partial F^c/\partial g^b$.

The Cartan theorem also leads to the following result that is often useful in specific problems.

<u>THEOREM 4-3.5</u>. *If* I *is a closed ideal of* $\Lambda(E_n)$ *whose characteristic subspace is of dimension* $n-r$, *then there exist* r *independent elements* $\{g^a(x^i),\ a=1,\ldots,r\}$ *of* $\Lambda^0(E_n)$ *such that* I *is contained in the closed ideal that is generated by the* r *elements* $\{dg^1,\ldots,dg^r\}$.

<u>Proof</u>. Since I is closed, its characteristic subspace $S(I)$ forms a Lie subalgebra of $T(E_n)$ and this subalgebra is of dimension $n-r$ by hypothesis. There thus exist $n-r$ elements $\{v^\alpha\}$ of $T(E_n)$ that form a basis for $S(I)$ and satisfy

$$[v^\alpha, v^\beta] = C^{\alpha\beta}_\gamma\, v^\gamma .$$

There are thus r independent first integrals $\{g^a,\ a=1,\ldots,r\}$ of the system $v^\alpha(g) = 0$, so that $v^\alpha \rfloor dg^a = 0$. The ideal of

$\Lambda(E_n)$ that is generated by $\{dg^1,\ldots,dg^r\}$ is a closed ideal that admits $S(I)$ as its characteristic subspace. Thus, since each of the generators of $\{dg^1,\ldots,dg^r\}$ is of lowest possible degree, namely one, while some of the generators of I may be of degree greater than one, I is contained in the closed ideal generated by $\{dg^1,\ldots,dg^r\}$.

A simple example is useful here. Consider the ideal I of $\Lambda(E_4)$ that is generated by

$$\omega^1 = dx^1 + x^2\, dx^3 + dx^4 \;, \quad \omega^2 = x^2\, dx^2 \wedge dx^3 \;.$$

Since $\omega^2 = x^2\, d\omega^1$ and $d\omega^2 = 0$, I is a closed ideal of $\Lambda(E_4)$. The characteristic subspace of I is 1-dimensional and is spanned by $v^1 = \partial_1 - \partial_4$, and $v^1(g) = 0$ admits the three independent first integrals

$$g^1 = x^1 + x^4 \;, \quad g^2 = x^2 \;, \quad g^3 = x^3 \;.$$

We thus consider the ideal J generated by the 1-forms $dg^1 = dx^1 + dx^4$, $dg^2 = dx^2$, $dg^3 = dx^3$. Since $\omega^1 = dg^1 + g^2\, dg^2$, $\omega^2 = g^2\, dg^2 \wedge dg^3$, I is contained in J. Further, I is a proper subideal of J for J contains terms of the form $f(x^i)dx^2 = f(x^i)dg^2$ while I does not.

Theorem 4-3.5 provides the means whereby subsets of E_n can be found on which a given ideal vanishes.

<u>THEOREM 4-3.6</u>. *Let* I *be a given ideal of* $\Lambda(E_n)$ *and construct its closure* $\bar{I} = I \cup dI$. *If the dimension of the characteristic subspace,* $S(\bar{I})$, *of* $\bar{I}$ *is* $n-r \neq 0$, *then there exist* $(n-r)$*-dimensional subsets*

$$(4\text{-}3.23) \qquad g^a(x^i) = k^a \;, \quad a = 1,\ldots,r$$

of E_n *on which every element of* I *and of* $\bar{I}$ *vanish. The* r *functions* $\{g^a(x^j)\}$ *are the primitive integrals of the characteristic system* $v^\alpha(g) = 0$, $a = r+1,\ldots,n$, *where* $\{v^\alpha,\ \alpha = r+1,\ldots,n\}$

is a basis for the Lie subalgebra $S(\bar{I})$ *of* $T(E_n)$.

Proof. Since $\bar{I}$ is a closed ideal of $\Lambda(E_n)$, Theorem 4-3.5 shows that $\bar{I}$ is contained in the ideal J that is generated by $\{dg^1,\ldots,dg^r\}$. Noting that every element of J vanishes on the (n-r)-dimensional subset (4-3.23) for any values of the constants k^a, the result is established.

We observe that I is a proper subideal of $\bar{I}$, unless I is itself a closed ideal, and $\bar{I}$ is a proper subideal of J. It is thus clear that there will in general be subsets of E_n of dimension greater than $n-r$ on which each element of I will vanish. The problem of finding all subsets of E_n on which each element of I vanishes for dI not contained in I is a very difficult one and is very sensitive to the structure of the elements of $\Lambda(E_n)$ that generate I. In most instances, Theorem 4-3.6 is the best that can be obtained without entering into the specifics of the generators of I.

4-4. THE DARBOUX CLASS OF A 1-FORM AND THE DARBOUX THEOREM

We now turn to the question of how to obtain an intrinsic characterization and a general representation of an arbitrary 1-form. As with most such problems, the first task is to obtain the characterization.

Let ω be a given 1-form. Starting with ω we construct the following sequence of exterior forms of increasing degree:

$$I_1 = \omega, \quad I_2 = d\omega. \tag{4-4.1}$$

$$I_{2k+1} = \omega \wedge I_{2k} = I_1 \wedge I_{2k}, \tag{4-4.2}$$

$$I_{2k+2} = dI_{2k+1} = I_2 \wedge I_{2k}. \tag{4-4.3}$$

This sequence is finite for all forms of degree higher than n vanish identically so that $I_{n+1} \equiv 0$. For each point $P:(x^i)$

in an n-dimensional subset S of E_n we can find an integer $K(x^i)$ such that

$$I_{K(x^i)} \neq 0 \ , \quad I_{K(x^i)+m} = 0 \tag{4-4.4}$$

for all $m > 0$.

Definition. The *Darboux class*, $K(\omega,S)$, of the 1-form ω relative to the set S is the integer

$$K(\omega,S) = \max_{(x^j)\in S} K(x^j) \ . \tag{4-4.5}$$

Definition. If $K(x^j) = K(\omega,S)$ then $P:(x^j)$ is said to be a *regular point* of ω relative to S . If $K(x^j) < K(\omega,S)$ then $P:(x^j)$ is said to be a *critical point* of ω relative to S .

The important thing to note here is that the class of a 1-form, its regular points and its critical points can be determined in a finite number of steps simply by constructing the sequence given by (4-4.1) through (4-4.3).

Definition. The *rank* of ω relative to S is the even integer that is the greatest even integer less than or equal to $K(\omega,S)$, and is written as $2p(\omega,S)$. The *index* of ω relative to S is given by

$$\varepsilon(\omega,S) = K(\omega,S) - 2p(\omega,S) \ . \tag{4-4.6}$$

The Darboux theorem gives the following representation.

THEOREM 4-4.1. (Darboux) *Let* ω *be a 1-form with Darboux class* $K(\omega,S)$ *on an n-dimensional subset* S *of* E_n . *There exist* $p(\omega,S)$ *positive, scalar-valued functions* $\{u_a(x^j)\}$ *and* $p(\omega,S) + \varepsilon(\omega,S)$ *scalar-valued functions* $\{v_b(x^j)\}$ *that constitute a system of* $K(\omega,S)$ *functionally independent functions at every regular point of* S *such that*

$$(4\text{-}4.7)\qquad \omega = \sum_{a=1}^{p(\omega,S)} u_a\, dv_a + \varepsilon(\omega,S)\, dv_{p(\omega,S)+1}\ .$$

A detailed proof of this theorem will not be given. There are, however, certain aspects of the proof that are worth mentioning since they give a direct method for obtaining the functions u_i and v_i . One first uses the Frobenius theorem to show that there exists a scalar-valued function η such that

$$(4\text{-}4.8)\qquad I_{K(\omega)} = d\eta \wedge I_{K(\omega)-1}$$

at all regular points of $\omega(x^\beta)$. Two classes of transformations are then constructed. The first consists of similarity transformations. A 1-form ω is said to undergo a *similarity transformation* to an image form $`\omega$ if

$$(4\text{-}4.9)\qquad `\omega = \sigma\omega\ ,\quad \sigma > 0\ .$$

If we construct the sequence $`I_k$ from the image form $`\omega$, it follows readily that

$$(4\text{-}4.10)\qquad `I_{2k} = \sigma^k I_{2k} + k\,\sigma^{k-1} d\sigma \wedge I_{2k-1},\quad `I_{2k+1} = \sigma^{k+1} I_{2k+1}\ .$$

Accordingly, if $K(\omega) = 2p(\omega)$, (4-4.8) can be used to obtain

$$`I_{2p(`\omega)} = \sigma^k\, d\eta \wedge I_{2k-1} + k\,\sigma^{k-1}\, d\sigma \wedge I_{2k-1},\quad k = p(\omega)\ .$$

The choice $\sigma = e^{-\eta/k} = e^{-\eta/p(\omega)}$ then reduces the class of the image form $`\omega$ to $2p(\omega)-1$ . A 1-form $\omega$ is said to undergo a *gradient transformation* to an image form $`\omega$ if

$$(4\text{-}4.11)\qquad `\omega = \omega + d\lambda\ .$$

This gives

$$(4\text{-}4.12)\qquad `I_{2k} = I_{2k}\ ,\quad `I_{2k+1} = I_{2k+1} + d\lambda \wedge I_{2k}\ .$$

Thus if $K(\omega)$ is an odd integer $=2p(\omega)+1$, then use of (4-4.8) gives

$$\tilde{I}_{K(\tilde{\omega})} = I_{K(\omega)} + d\lambda \wedge I_{2p(\omega)} = d\eta \wedge I_{2p(\omega)} + d\lambda \wedge I_{2p(\omega)} .$$

The choice $\lambda = -\eta$ thus reduces the class of ω to $2p(\omega)$. A succession of gradient transformations and similarity transformations will thus reduce the class of any form ω to one. The functions u_i and v_i are then constructed from the σ's and the λ's by inverting the sequence of similarity and gradient transformations, on noting that a 1-form of class 1 is an exact 1-form.

We note the lack of uniqueness in the representation given by the Darboux theorem. In particular, since $u_i dv_i = d(u_i v_i) - v_i du_i$, (4-4.7) can also be written as

$$(4\text{-}4.13) \qquad \omega = -\sum_{i=1}^{p(\omega)} v_i du_i + d\{\varepsilon(\omega) v_{2p(\omega)+1} + \sum_{i=1}^{p(\omega)} u_i v_i\} .$$

There are thus many different representations of $\omega(x)$ that involve the same total number of functions, and for each, the function whose exterior derivative occurs with unity coefficient will be different.

THEOREM 4-4.2. *If* η *is a closed 2-form on an n-dimensional region* S *of* E_n *that is starshaped, then* $\eta = d\omega$ *and* η *has the canonical form*

$$(4\text{-}4.14) \qquad \eta = \sum_{a=1}^{p(\omega,S)} du_a \wedge dv_a$$

at all regular points of ω *in* S.

Proof. Since η is closed and S is starshaped, the Poincaré Lemma implies the existence of an $\omega \in \Lambda^1(S)$ such that $\eta = d\omega$. The Darboux theorem applied to ω then gives (4-4.7) and hence an exterior differentiation yields (4-4.14).

The Darboux theorem provides explicit results in connection with the problem of solving $\omega = 0$. We simply substitute (4-4.7) for ω and equate the result to zero. Now observe that for

$\varepsilon(\omega,S) = 0$ we may always divide by u_1 since $u_1 \neq 0$ throughout S. The problem of solving $\omega = 0$ can thus always be reduced to solving

$$(4\text{-}4.15) \qquad \bar{\omega} = dv_1 + u_2\, dv_2 + \dots + u_r\, dv_r = 0$$

with all of the u's positive. Solutions of (4-4.15) are obviously given by the intersection of the surfaces in S that are defined by

$$(4\text{-}4.16) \qquad v_1 = c_1 , \quad v_2 = c_2 , \quad \dots, \quad v_r = c_r .$$

There are other solutions, but these obtain by relations between the u's and the v's. For example, suppose that we set

$$(4\text{-}4.17) \qquad v_1 = \phi(v_2,\dots,v_r) .$$

We than have

$$\bar{\omega} = (u_2 + \partial\phi/\partial v_2)dv_2 + \dots + (u_r + \partial\phi/\partial v_r)dv_r$$

and hence $\bar{\omega} = 0$ can be satisfied by (4-4.17) and

$$(4\text{-}4.18) \qquad u_2 = -\,\partial\phi/\partial v_2, \quad \dots, \quad u_r = -\,\partial\phi/\partial v_r .$$

Definition. A 1-form ω is said to have the *inaccessibility* property if and only if a sufficiently small neighborhood of any point P contains a point Q that can not be reached by a path from P that satisfies $\omega = 0$. If ω does not have the inaccessibility property then it has the *accessibility* property.

THEOREM 4-4.3. *A 1-form ω has the inaccessibility property on an n-dimensional set S of E_n if and only if the Darboux class of ω relative to S is less than three. If $K(\omega,S) \geq 3$ then ω has the accessibility property on S.*

Proof. If $K(\omega,S) = 1$, then $\omega = dv_1$ and hence $\omega = 0$ holds only on (n-1)-dimensional surfaces $v_1 = c_1$. If P belongs to

$v_1 = c_1$, then simply take Q to belong to $v_1 = c_1 + \delta$ for sufficiently small δ in order to obtain a point in the neighborhood of P that can not be reached by any path in S for which $\omega = 0$. If $K(\omega,S) = 2$, then $\omega = u_1 \, dv_1$ and $\omega = 0$ is equivalent to $dv_1 = 0$ which is the previous case. For $K(\omega,S) = 3$, $\omega = 0$ becomes

$$(4\text{-}4.19) \qquad 0 = dv_1 + u_2 \, dv_2 .$$

The cases with $K(\omega,S) > 3$ proceed along the same lines as that for $K(\omega,S) = 3$, so we will establish accessibility for (4-4.19) only. It is clear that v_1 = constant, v_2 = constant define surfaces of dimension $n - 2$ of mutually accessible points. Thus, suppose that $P:(x_o^i)$ and $Q:(x_1^i)$ and set

$$v_1(x_o^i) = v_1^o , \quad v_2(x_o^i) = v_2^o , \quad v_1(x_1^i) = v_1^1 , \quad v_2(x_1^i) = v_2^1 .$$

Accessibility will be established if we can find a path in E_n that connects the values (v_1^o, v_2^o) with (v_1^1, v_2^1) . To this end, consider the path for which the induced functions $v_a(x^i(t)) = \bar{v}_a(t)$ are such that

$$(4\text{-}4.20) \qquad \bar{v}_2(t) = v_2^o + (v_2^1 - v_2^o)t , \quad \bar{u}_2(t) = u_2^o + h\,t .$$

This leads to the requirement

$$(4\text{-}4.21) \qquad d\bar{v}_1/dt = (v_2^o - v_2^1)(u_2^o + h\,t)$$

and hence

$$(4\text{-}4.22) \qquad \bar{v}_1(t) = v_1^o + (v_2^o - v_2^1)(u_2^o + h\,t/2)t .$$

Thus, if we set $t = 1$, then $\bar{v}_2(1) = v_2^1$ and h can be chosen so as to secure

$$\bar{v}_1(1) = v_1^1 = v_1^o + (v_2^o - v_2^1)(u_2^o + h/2) .$$

This trivial calculation establishes accessibility for $K(\omega,S) = 3$ since v_1, v_2 and u_2 are functionally independent functions.

The whole discipline of thermodynamics can be based upon the concept of inaccessibility, as shows in the pioneering work of Carathéodory. There, physics dictates that the 1-form Q of heat addition must be such that there are thermodynamic states that are inaccessible from any given state by paths for which $Q = 0$ (adiabatic processes). This implies that $K(Q,S) = 2$; that is $Q = u_1\, dv_1$ from which the entropy and thermodynamic temperature of the system may be inferred.

4-5. FINITE DEFORMATIONS AND LIE DERIVATIVES

If $v = v^i(x^j)\partial_i$ is a given element of $T(E_n)$, then the orbits of v are given by the solutions of

$$\frac{d\bar{x}^i(t)}{dt} = v^i(\bar{x}^j(t)) \tag{4-5.1}$$

subject to the initial data

$$\bar{x}^i(t) = x^i . \tag{4-5.2}$$

We have also seen that v may be considered as a derivation on $\Lambda^o(E_n)$. Thus, since $\Lambda^o(E_n)$ is the first element in the sequence $\Lambda^o(E_n)$, $\Lambda^1(E_n)$, $\Lambda^2(E_n)$, ... , $\Lambda^n(E_n)$, it is natural to ask whether v can be extended to a derivation on all of $\Lambda(E_n)$ in such a way that it agrees with v when applied to elements of Λ^o . This would indeed be a useful construct if one could also extend in such a fashion that $[u,v]f = u(v(f)) - v(u(f))$ is also extended to apply to all of $\Lambda(E_n)$.

The underlying precept from which we can construct such an extension is the operator series representation of the solutions of (4-5.1), (4-5.2):

$$\bar{x}^i(t) = \exp(tv)x^i . \tag{4-5.3}$$

These equations can be thought of as defining a 1-parameter family of maps

$$\Psi_t : E_n \longrightarrow E_n \mid \bar{x}^i(t) = \exp(tv)x^i \tag{4-5.4}$$

that moves any point $P:(x^i)$ at $t = 0$ into the point $P_t:(\exp(tv)x^i)$ at time t. It is, however, conceptually simpler to consider Ψ_t as a map from the space E_n into a replica $E_{n,t}$ of E_n where the replicas are parametrized by t ; that is, we shift consideration to $\mathbb{R}\times E_n$. This alternative view gives the graphic situation shown in Fig. 13.

If we label points in $\mathbb{R}\times E_n$ by $(t;x^1,\ldots,x^n)$ there is the obvious projection map

$$\Pi : \mathbb{R}\times E_n \longrightarrow E_n \mid (t;x^1,\ldots,x^n) \longrightarrow (x^1,\ldots,x^n) . \tag{4-5.5}$$

Let ω be a k-form on E_n. Since the map Π induces the map Π^* on k-forms that maps in the opposite direction to Π, the k-form $\Pi^*\omega$ is induced on every replica of E_n in $\mathbb{R}\times E_n$ and this collection of k-forms is independent of the value of the parameter t since Π projects all values of t onto $t = 0$. From now on, we simply assume that ω is defined on every replica $E_{n,t}$ of E_n in $\mathbb{R}\times E_n$ and drop the Π^* map.

Let $Q:(\exp(\tau v)x^i)$ be the image in $E_{n,\tau}$ of the point $P:(x^i)$ in $E_{n,0}$ under the map Ψ_τ. We can then evaluate ω at Q in $E_{n,\tau}$ and at P in $E_{n,0}$, but we can not compare them since forms at different points or forms at different points in different spaces can not be added (we can not add an element of $T_P^*(E_n)$ to an element of $T_Q^*(E_n)$ for P different from Q). The form ω at Q in $E_{n,\tau}$ can however be mapped into a form in $E_{n,0}$ at P be the action of Ψ_τ^*. This will result in a new k-form

$$\overset{m}{\omega}(\tau) = \Psi_\tau^*\omega = \exp(\tau v)^*\omega \tag{4-5.6}$$

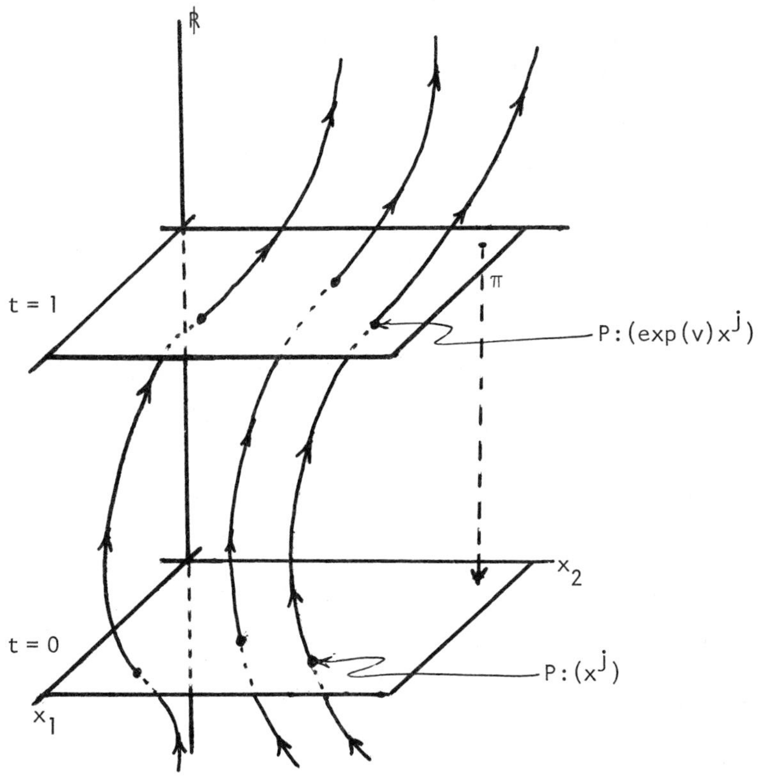

Figure 13. The Family of Maps Ψ_t

on $E_{n,0}$ that can be evaluated (at P) and compared with ω at P. The definition (4-5.6) is expressed in the following graphic way: *the new field* $\overset{m}{\omega}(\tau)$ *at the old point* P *is equal to the old field at the new point* Q *pulled back to* P *by the map* $\exp(\tau v)^*$.

Definition. The *finite deformation*, $\Delta_v(\tau)\omega$ of a given $\omega \varepsilon \Lambda^k(E_n)$, that is induced by the vector field v, is given by

(4-5.7) $$\Delta_v(\tau)\omega = \exp(\tau v)^*\omega - \omega .$$

Definition. The *Lie derivative*, $\pounds_v \omega$, of a given $\omega \varepsilon \Lambda^k(E_n)$ with respect to the vector field v is given by

$$(4\text{-}5.8) \qquad \pounds_v \omega = \lim_{\tau \to 0} \left(\frac{\exp(\tau v)^* \omega - \omega}{\tau} \right) .$$

Clearly $\pounds_v \omega \varepsilon \Lambda^k(E_n)$ for $\omega \varepsilon \Lambda^k(E_n)$. The following theorem provides a direct parallel between $\bar{x}^i(t)$ and $\overset{m}{\omega}(t)$.

THEOREM 4-5.1. *If* $\omega \varepsilon \Lambda^k(E_n)$ *and* $v \varepsilon T(E_n)$, *then*

$$(4\text{-}5.9) \qquad \overset{m}{\omega}(t) = \exp(t\pounds_v)\omega .$$

Proof. The definition of the derivative and (4-5.6) yield

$$\frac{d\overset{m}{\omega}(t)}{dt} = \lim_{\tau \to 0} \frac{\exp((t+\tau)v)^* \omega - \exp(tv)^* \omega}{\tau} .$$

But $\exp((t+\tau)v) = \exp(tv) \cdot \exp(\tau v)$ because tv and τv commute, and hence

$$\exp((t+\tau)v)^* = \{\exp(tv) \cdot \exp(\tau v)\}^*$$

$$= \exp(\tau v)^* \exp(tv)^* ;$$

that is

$$\frac{d\overset{m}{\omega}(t)}{dt} = \lim_{\tau \to 0} \left(\frac{\exp(\tau v)^* - 1}{\tau} \right) \exp(tv)^* \omega .$$

However, $\exp(tv)^* \omega = \overset{m}{\omega}(t)$ belongs to $\Lambda^k(E_n)$ and hence (4-5.8) gives

$$(4\text{-}5.10) \qquad \frac{d\overset{m}{\omega}(t)}{dt} = \pounds_v \overset{m}{\omega}(t) .$$

Thus, since (4-5.6) also implies

$$\overset{m}{\omega}(0) = \omega ,$$

the series solution of (4-5.10) subject to this initial data gives (4-5.9).

The Lie derivative would seem to be a candidate for the

extension of v to an operator on forms of arbitrary degree. The following lemma provides the explicit evaluation whereby the requisite properties are easily established.

LEMMA 4-5.1. *The Lie derivative of a k-form* ω *with respect to a* $v \in T(E_n)$ *has the explicit evaluation*

$$\pounds_v \omega = v \rfloor d\omega + d(v \rfloor \omega) . \tag{4-5.11}$$

Proof. Since the proof is the same for forms of any degree, we confine attention to forms of degree one in the interests of computational simplicity. If $\omega = \omega_i dx^i$, $v = v^i \partial_i$, then

$$\begin{aligned} \exp(\tau v)^* \omega &= \omega_i(\exp(\tau v)x^j) d\{\exp(\tau v)x^i\} \\ &= \omega + \tau\left(\frac{\partial \omega_i}{\partial x^j} v^j + \omega_j \frac{\partial v^j}{\partial x^i}\right) dx^i + o(\tau) . \end{aligned}$$

The definition (4-5.8) thus gives

$$\pounds_v \omega = \left(\frac{\partial \omega_i}{\partial x^j} v^j + \omega_j \frac{\partial v^j}{\partial x^i}\right) dx^i . \tag{4-5.12}$$

If we add and subtract $\frac{\partial \omega_j}{\partial x^i} v^j dx^i$, we obtain

$$\pounds_v \omega = \left\{\left(\frac{\partial \omega_i}{\partial x^j} - \frac{\partial \omega_j}{\partial x^i}\right)v^j + \frac{\partial}{\partial x^i}(\omega_j v^j)\right\} dx^i$$

from which (4-5.11) follows directly.

THEOREM 4-5.2. *The Lie derivative is a map from* $\Lambda^k(E_n)$ *to* $\Lambda^k(E_n)$, $k = 0,1,\ldots,n$, *that exhibits the properties*

L1: $\pounds_v f = v(f) \quad \forall f \in \Lambda^0(E_n)$,

L2: $\pounds_v(\alpha + \beta) = \pounds_v \alpha + \pounds_v \beta$,

L3: $\pounds_v(\alpha \wedge \beta) = (\pounds_v \alpha) \wedge \beta + \alpha \wedge \pounds_v \beta$,

L4: $\mathcal{L}_v d\alpha = d\,\mathcal{L}_v \alpha$,

L5: $\mathcal{L}_{fv}\alpha = f\,\mathcal{L}_v\alpha + df \wedge (v \rfloor \alpha)$,

L6: $\mathcal{L}_{u+v}\alpha = \mathcal{L}_u\alpha + \mathcal{L}_v\alpha$,

and hence $\mathcal{L}_v$ *is a derivation on the graded algebra* $\Lambda(E_n)$ *that agrees with the derivation* v *on* $\Lambda^0(E_n)$.

<u>Proof.</u> That $\mathcal{L}_v$ is on Λ^k to Λ^k follows from

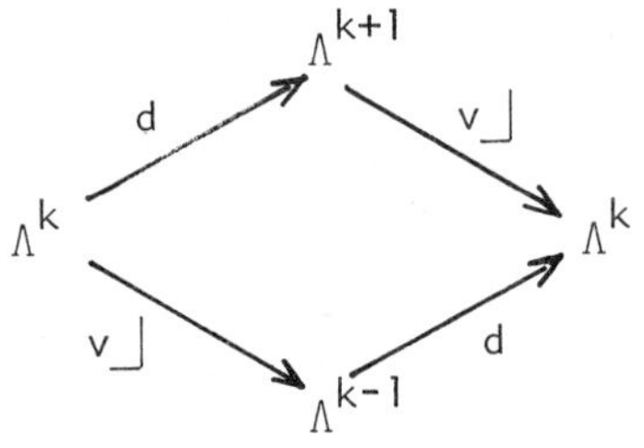

and (4-5.13). Properties L2 and L6 are immediate consequences of the linearity of $\mathcal{L}_v\alpha$ in v for fixed α and in α for fixed v , as shown by the definition (4-5.13). For L4 , we note that

$$\mathcal{L}_v d\alpha = v \rfloor dd\alpha + d(v \rfloor d\alpha) = d(v \rfloor d\alpha)$$

$$d\mathcal{L}_v\alpha = d\{v \rfloor d\alpha + d(v \rfloor \alpha)\} = d(v \rfloor d\alpha) ,$$

while L1 follows from

$$\mathcal{L}_v f = v \rfloor df + d(v \rfloor f) = v \rfloor df = v(f) .$$

By definition

$$\mathcal{L}_{fv}\alpha = fv \rfloor d\alpha + d(fv \rfloor \alpha) = fv \rfloor d\alpha + fd(v \rfloor \alpha) + df \wedge (v \rfloor \alpha)$$

$$= f\mathcal{L}_v\alpha + df \wedge (v \rfloor \alpha) ,$$

which is just property L5 . Again, be definition, with $a = \deg(\alpha)$

$$\begin{aligned}
\pounds_v(\alpha \wedge \beta) &= v \rfloor d(\alpha \wedge \beta) + d(v \rfloor (\alpha \wedge \beta)) \\
&= v \rfloor (d\alpha \wedge \beta + (-1)^a \alpha \wedge d\beta) + d((v \rfloor \alpha) \wedge \beta + (-1)^a \alpha \wedge (v \rfloor \beta)) \\
&= (v \rfloor d\alpha) \wedge \beta + (-1)^{a+1} d\alpha \wedge (v \rfloor \beta) + (-1)^a (v \rfloor \alpha) \wedge d\beta \\
&\quad + (-1)^{2a} \alpha \wedge (v \rfloor d\beta) + d(v \rfloor \alpha) \wedge \beta + (-1)^{a-1} (v \rfloor \alpha) \wedge d\beta \\
&\quad + (-1)^a d\alpha \wedge (v \rfloor \beta) + (-1)^{2a} \alpha \wedge d(v \rfloor \beta) \\
&= \{(v \rfloor d\alpha) + d(v \rfloor \alpha)\} \wedge \beta + \alpha \wedge \{v \rfloor d\beta + d(v \rfloor \beta)\} .
\end{aligned}$$

Theorem 4-5.2 provides a full description of the action of $\pounds_v$ on exterior forms. The question now arises as to what happens when $\pounds_v$ acts on $T(E_n)$. We first note that $v \rfloor \omega$ is a well defined exterior form for any given exterior form ω , and hence $\pounds_u(v \rfloor \omega)$ is well defined. Now, property L3 shows that $\pounds_u$ is a derivation on $\Lambda(E_n)$, so it is consistent to define the action of $\pounds_u$ on $v \in T(E_n)$ by requiring $\pounds_u$ to act as a derivation on $v \rfloor \omega$ for all $\omega \in \Lambda(E_n)$, that is

$$\pounds_u(v \rfloor \omega) = (\pounds_u v) \rfloor \omega + v \rfloor \pounds_u \omega .$$

Definition. The action of $\pounds$ on $T(E_n)$ is defined by the requirement that

$$(\pounds_u v) \rfloor \omega = \pounds_u(v \rfloor \omega) - v \rfloor \pounds_u \omega \tag{4-5.12}$$

hold for all $\omega \in \Lambda(E_n)$.

LEMMA 4-5.2. *If* $u, v \in T(E_n)$, *then the action of* $\pounds_u$ *on* v *has the explicit evaluation*

$$\pounds_u v = [u,v] . \tag{4-5.13}$$

Thus, $\pounds_u$ *is a map from* $T(E_n)$ *to* $T(E_n)$ *and*

(4-5.14) $$\pounds_u v = -\pounds_v u$$

Proof. Since (4-5.12) must hold for all ω, putting $\omega = df \in \Lambda^1$ and $v \rfloor df = v(f)$ give

$$\begin{aligned}(\pounds_u v) \rfloor df &= (\pounds_u v)(f) \\ &= \pounds_u(v(f)) - v \rfloor \pounds_u df = \pounds_u(v(f)) - v \rfloor d(u(f)) \\ &= u \rfloor d(v(f)) - v \rfloor d(u(f)) = u(v(f)) - v(u(f)) .\end{aligned}$$

The definition of the Lie product thus gives

$$(\pounds_u v)(f) = [u,v](f)$$

for all $f \in \Lambda^0(E_n)$ and hence (4-5.13) is established. The relation (4-5.14) is an immediate consequence of the properties of the Lie product.

It is instructive to see that (4-5.13) agrees with a direct definition of the Lie derivative similar to (4-5.8). Let u and v be given elements of $T(E_n)$ and define $\bar{x}^i(s)$ by $\bar{x}^i(s) = \exp(su)x^i$ so that $\bar{P}:(\bar{x}^i(s))$ is the image of $P:(x^i)$ under transport along the orbits of the vector field u. Since $\exp(su)_*$ maps in the same direction as $\exp(su)$, the appropriate definition of the Lie derivative is

(4-5.15) $$\pounds_u v = \lim_{s \to 0} \left\{ \frac{\bar{v} - (\exp(su))_* v}{s} \right\},$$

where $\bar{v}$ denotes the evaluation of the vector field v at $\bar{P}:(\bar{x}^i(s))$; that is, $\bar{v} = v^i(\bar{x}^j(s))\bar{\partial}_i$ with $\bar{\partial}_i = \partial/\partial\bar{x}^i$. It is now an easy matter to see that

$$\begin{aligned}\bar{v} - (\exp(su))_* v &= v^i(\exp(su)x^j)\bar{\partial}_i - (\exp(su))_*(v^k(x^j)\partial_k) \\ &= s\{u^j\, \partial_j v^i - v^j\, \partial_j u^i\}\bar{\partial}_i + o(s) ,\end{aligned}$$

and hence $\pounds_u v = [u,v]$. The respective definitions of the Lie derivative show that $\exp(s\pounds_u)\omega$ is similar to a description in terms of Eulerian variables while $\exp(s\pounds_u)v$ is similar to a description in terms of Lagrangian variables.

Now that we know that $\pounds_u v = [u,v]$, we turn to the computation of $\pounds_{[u,v]}$. By definition,

$$\pounds_{[u,v]}\Omega = \pounds_{\pounds_u v}\Omega = (\pounds_u v)\rfloor d\Omega + d((\pounds_u v)\rfloor\Omega) . \tag{4-5.16}$$

Use of (4-5.12), first with $\omega = d\Omega$ and then with $\omega = \Omega$, yield

$$(\pounds_u v)\rfloor d\Omega = \pounds_u(v\rfloor d\Omega) - v\rfloor(\pounds_u d\Omega) ,$$

$$(\pounds_u v)\rfloor \Omega = \pounds_u(v\rfloor \Omega) - v\rfloor(\pounds_u \Omega) .$$

When these results are put back into (4-5.16), we have

$$\begin{aligned}\pounds_{[u,v]}\Omega &= \pounds_u(v\rfloor d\Omega) - v\rfloor(\pounds_u d\Omega) + d[\pounds_u(v\rfloor\Omega) - v\rfloor\pounds_u\Omega]\\ &= \pounds_u(v\rfloor d\Omega) - v\rfloor d\pounds_u\Omega + \pounds_u d(v\rfloor\Omega) - d\{v\rfloor\pounds_u\Omega\}\\ &= \pounds_u\{v\rfloor d\Omega + d(v\rfloor\Omega)\} - v\rfloor d\pounds_u\Omega - d(v\rfloor\pounds_u\Omega)\\ &= \pounds_u(\pounds_v\Omega) - \pounds_v(\pounds_u\Omega) = (\pounds_u\pounds_v - \pounds_v\pounds_u)\Omega .\end{aligned}$$

THEOREM 4-5.3. *The Lie derivative satisfies*

$$\pounds_{[u,v]}\omega = (\pounds_u\pounds_v - \pounds_v\pounds_u)\omega \tag{4-5.17}$$

and hence the Lie algebra of $T(E_n)$ *as operators on* $\Lambda^0(E_n)$ *extends to a Lie algebra of operators on* $\Lambda(E_n)$ *under the correspondence*

$$[u,v] \longrightarrow \pounds_u\pounds_v - \pounds_v\pounds_u . \tag{4-5.18}$$

If $\{u_a, a=1,\dots,r\}$ *form a Lie subalgebra of* $T(E_n)$ *with*

$$[u_a,u_b] = C^e_{ab}\, u_e , \tag{4-5.19}$$

then

(4-5.20) $$(\pounds_{u_a}\pounds_{u_b} - \pounds_{u_b}\pounds_{u_a})\omega = C^e_{ab}\,\pounds_{u_e}\omega + dC^e_{ab}\wedge(u_e\rfloor\omega)\ .$$

Thus a Lie subalgebra of $T(E_n)$ *extends to a Lie subalgebra of operators on* Λ *only if*

(4-5.21) $$dC^e_{ab} = 0\ ;$$

that is, the Lie subalgebra is the Lie subalgebra of a Lie group.

Proof. We have already established (4-5.17). If the collection $\{u_a\}$ satisfy (4-5.19), then (4-5.17) gives

$$\pounds_{[u_a,u_b]}\omega = (\pounds_{u_a}\pounds_{u_b} - \pounds_{u_b}\pounds_{u_a})\omega$$

$$= \pounds_{C^2_{ab}u_e}\omega = C^e_{ab}\,\pounds_{u_e}\omega + dC^2_{ab}\wedge(u_2\rfloor\omega)$$

which is just (4-5.20). The remaining conclusions are obvious.

Lie derivatives are very useful in many contexts. One of particular utility arises in connection with ideals of $\Lambda(E_n)$.

Definition. A vector field $v\in T(E_n)$ is an *isovector* of an ideal I of $\Lambda(E_n)$ if and only if

(4-5.22) $$\Delta_v(t)\rho\subset I$$

for all $\rho\in I$ and all t in some neighborhood of $t=0$.

LEMMA 4-5.3. A $v\in T(E_n)$ *is an isovector of* I *if and only if*

(4-5.23) $$\pounds_v\rho\subset I$$

for all $\rho\in I$.

Proof. Be definition, (4-5.7), $\Delta_v(t)\rho = \exp(tv)^*\rho - \rho = \overset{m}{\rho}(t) - \rho$, and hence (4-5.9) gives

$$(4\text{-}5.24) \qquad \Delta_v(t)\rho = \{\exp(t£_v) - 1\}\rho .$$

Thus $\Delta_v(t)\rho \subset I$ for all $\rho \subset I$ if and only if $£_v\rho \subset I$.

THEOREM 4-5.4. *The isovectors of an ideal* I *of* $\Lambda(E_n)$ *form a Lie subalgebra of* $T(E_n)$ *that is the Lie algebra of a Lie group.*

Proof. Lemma 4-5.3 shows that v is an isovector of I if and only if

$$(4\text{-}5.25) \qquad £_v\rho = 0 \bmod I$$

for all $\rho \varepsilon I$. Suppose u and v are isovectors of I. We then have

$$(£_u£_v - £_v£_u)\rho = £_u(£_v\rho) - £_v(£_u\rho) = 0 \bmod I$$

and hence

$$£_{[u,v]}\rho = (£_u£_v - £_v£_u)\rho = 0 \bmod I .$$

Thus $[u,v]$ is an isovector whenever u and v are isovectors and the set of all isovectors is closed under the Lie product. However

$$£_{u+v}\rho = £_u\rho + £_v\rho = 0 \bmod I$$

but

$$£_{fu}\rho = f£_u\rho + df \wedge (u \rfloor \rho) = df \wedge (u \rfloor \rho) \bmod I ,$$

so that all isovectors of I form a subspace of $T(E_n)$ over $\mathbb{R}$ but not over $\Lambda^0(E_n)$; i.e. $£_{fu}\rho = 0 \bmod I$ only if $df = 0$ since isovectors are not necessarily characteristic vectors of I. The vector subspace property over $\mathbb{R}$, but not over $\Lambda^0(E_n)$ then implies

$$[u_a, u_b] = C^e_{ab}\, u_e \, , \qquad dC^e_{ab} = 0 \, ,$$

and the result follows.

Incidentally, we have also established the following result.

THEOREM 4-5.5. *The subset of all isovectors of an ideal* I *of* $\Lambda(E_n)$ *that are also characteristic vectors of* I *forms a Lie subalgebra of* $T(E_n)$.

A result that is often of great utility in actual calculations is obtained as follows. Direct application of the definition of the Lie derivative gives

$$\pounds_u(v \rfloor \omega) - \pounds_v(u \rfloor \omega) = u \rfloor d(v \rfloor \omega) - v \rfloor d(u \rfloor \omega)$$

$$+ d(u \rfloor v \rfloor \omega - v \rfloor u \rfloor \omega)$$

$$= u \rfloor d(v \rfloor \omega) - v \rfloor d(u \rfloor \omega) + 2\, d(u \rfloor v \rfloor \omega)$$

and since $u \rfloor v \rfloor \omega = - v \rfloor u \rfloor \omega$. On the other hand, the fact that the Lie derivative acts as a derivation on exterior products and on inner multiplications shows that

$$\pounds_u(v \rfloor \omega) - \pounds_v(u \rfloor \omega) = [u,v] \rfloor \omega + v \rfloor \pounds_u \omega$$

$$- [v,u] \rfloor \omega - u \rfloor \pounds_v \omega$$

$$= 2[u,v] \rfloor \omega + 2\, v \rfloor u \rfloor d\omega + v \rfloor d(u \rfloor \omega) - u \rfloor d(v \rfloor \omega)$$

since $[u,v] = - [v,u]$. When these two expressions are combined and the common factor, 2 , is eliminated, we obtain the following result.

LEMMA 4-5.4. *Any exterior form* ω *and any pair of vector fields* u,v *satisfy the identity*

$$(4\text{-}5.26) \quad [u,v] \rfloor \omega + v \rfloor u \rfloor d\omega = u \rfloor d(v \rfloor \omega) - v \rfloor d(u \rfloor \omega) + d(u \rfloor v \rfloor \omega) \, .$$

We have already encountered a direct application of this result in the proof of the Cartan Theorem 4-3.4 (compare (4-5.26) with (4-3.19)). A further application of this result obtains when ω is a 1-form η. In this case $u \rfloor v \rfloor \eta = 0$, so that

$$(4\text{-}5.27) \qquad v \rfloor u \rfloor d\eta = u(v \rfloor \eta) - v(u \rfloor \eta) - [u,v] \rfloor \eta$$

for any $\eta \in \Lambda^1(E_n)$. On the other hand, if $\omega = d\beta$ is any exact form, we have

$$(4\text{-}5.28) \qquad [u,v] \rfloor d\beta = u \rfloor d(v \rfloor d\beta) - v \rfloor d(u \rfloor d\beta) + d(u \rfloor v \rfloor d\beta)$$

and hence

$$(4\text{-}5.29) \qquad [u,v] \rfloor d\beta - u \rfloor d(v \rfloor d\beta) + v \rfloor d(u \rfloor d\beta)$$

is a closed form of the same degree as β. In fact, (4-5.26) shows that

$$(4\text{-}5.30) \qquad [u,v] \rfloor \omega + v \rfloor u \rfloor d\omega - u \rfloor d(v \rfloor \omega) + v \rfloor d(u \rfloor \omega)$$

is an exact form for any form ω and any pair of vector fields u, v.

Suppose that u and v are characteristic vector fields of the form ω of arbitrary degree $(u \rfloor \omega = 0,\ v \rfloor \omega = 0)$. In this event, (4-5.26) yields

$$(4\text{-}5.31) \qquad [u,v] \rfloor \omega = u \rfloor v \rfloor d\omega .$$

Thus, since the characteristic vector fields of any given ω form a subspace of $T(E_n)$, we have the following result.

LEMMA 4-5.5. *The characteristic vector fields of a given form* ω *form a Lie subalgebra of* $T(E_n)$ *if and only if* $u \rfloor v \rfloor d\omega = 0$ *for every pair of characteristic vectors* u *and* v.

The following result often proves to be useful in applications.

LEMMA 4-5.6. *Let* $\omega^1,\ldots,\omega^r$ *be* r *given* 1-*forms and let* I *be the ideal generated by* $\omega^1,\ldots,\omega^r$. *If* I *is closed,* $dI \subset I$, *then the characteristic subspace* $S(I)$ *of* I *is comprised of isovectors of* I.

Proof. We have already seen that necessary and sufficient conditions for $dI \subset I$ are satisfaction of $d\omega^a = \Gamma^a_b \wedge \omega^b$. Thus, if $v \in T(E_n)$ belongs to $S(I)$, then $v \rfloor d\omega^a = (v \rfloor \Gamma^a_b)\, \omega^b \in I$ and hence $\pounds_v \omega^a = v \rfloor d\omega^a + d(v \rfloor \omega^a) = v \rfloor d\omega^a \in I$. Since any element ρ of I has the form $\rho = \lambda_a \wedge \omega^a$, $\pounds_v \omega^a \in I$ implies $\pounds_v \rho = (\pounds_v \lambda_a) \wedge \omega^a + \lambda_a \wedge \pounds_v \omega^a$ belongs to I, hence v is an isovector of I and the result is established.

An examination of the proof of Lemma 4-5.6 clearly shows that isovectors of the closed ideal I need not be characteristic vectors of I, for $\pounds_v \omega^a = v \rfloor d\omega^a + d(v \rfloor \omega^a) = (v \rfloor \Gamma^a_b)\, \omega^b - \Gamma^a_b\, v \rfloor \omega^b + d(v \rfloor \omega^a)$ can belong to I without $v \rfloor \omega^a = 0$. In general, *the characteristic vectors of* I *form a proper subset of the isovectors of* I.

4-6. INTEGRATION OF EXTERIOR FORMS, STOKES' THEOREM AND THE DIVERGENCE THEOREM

Let Y_r be a closed, r-dimensional region of r-dimensional number space E_r and let ∂Y_r denote the boundary of Y_r. We know that the dimension of $\Lambda^r(E_r)$ is one, so that any element ω of $\Lambda^r(E_r)$ can be written uniquely as

$$\omega = w(x^j)\mu, \tag{4-6.1}$$

where

$$\mu = dx^1 \wedge dx^2 \wedge \ldots \wedge dx^r \tag{4-6.2}$$

is the natural volume element of E_r with respect to the $(x^1,\ldots,x^r)$ coordinate cover. Further, any two elements of

$\Lambda^r(E_r)$ can only differ in the choice of the function $w(x^j)$ and in the order in which $dx^1, dx^2, \ldots, dx^n$ are multiplied. Accordingly, the choice of the order given by the natural volume element μ fixes all simple elements of $\Lambda^r(E_r)$ to within a factor ± 1. Choice of the $+$ sign, as in (4-6.1), thus fixes things completely. This choice of the order in which the factors appear in the basis element μ is referred to as a choice of *orientation*, while the making of this choice is said to assign an orientation to Y_r. In 3 dimensions the choice $\mu = dx \wedge dy \wedge dz$ fixes the right-hand orientation on E_3, while the choice of $\eta = dy \wedge dx \wedge dz = -dx \wedge dy \wedge dz$ for the natural volume element would fix a left-hand orientation on E_3. If $\Phi: \grave{}E_r \longrightarrow E_r \mid x^i = \phi^i(\grave{}x^j)$ is an invertible map, then

$$\Phi^*\mu = \det(\partial\phi^i/\partial\grave{}x^j)\grave{}\mu \,, \qquad \grave{}\mu = d\grave{}x^1 \wedge \ldots \wedge d\grave{}x^r \,,$$

in which case Φ is said to be *orientation preserving* if $\det(\partial\phi^i/\partial\grave{}x^j) > 0$ and to be *orientation reversing* if $\det(\partial\phi^i/\partial\grave{}x^j) < 0$. This shows that the choice of orientation with respect to any one coordinate cover (x^i) fixes the orientation for all coordinate covers that can be obtained from (x^i) by invertible coordinate transformations.

The choice of orientation for r-forms on E_r also fixes the orientation of (r-1)-forms on E_r. We have seen that $\Lambda^{r-1}(E_r)$ has a basis of the form

(4-6.3) $$\mu_i = \partial_i \lrcorner \mu \,, \qquad i = 1, \ldots, r$$

such that

(4-6.4) $$dx^i \wedge \mu_j = \delta^i_j \mu \,.$$

This shows that

(4-6.5) $$dx^1 \wedge \mu_1 = dx^2 \wedge \mu_2 = \ldots = dx^r \wedge \mu_r = \mu \,,$$

and hence any change in orientation induces a like change in the signs of the basis elements $\{\mu_i\}$ of $\Lambda^{r-1}(E_r)$. We therefore say that the basis $\{\mu_i\}$ of $\Lambda^{r-1}(E_r)$ assigns an *orientation to* $\Lambda^{r-1}(E_r)$ of *equal parity.*

Let

$$\omega = w(x^j)\mu \tag{4-6.6}$$

be a given element of $\Lambda^r(E_r)$. We define the integral of over the set Y_r by

$$\int_{Y_r} \omega = \int_{Y_r} w(x^j)\,dx^1 dx^2 \ldots dx^r , \tag{4-6.7}$$

where the integral on the right-hand side of (4-6.7) denotes r-dimensional Riemann integral over the region Y_r of r-dimensional Euclidean space. It is now a direct, but messy calculation to establish Stokes' theorem.

THEOREM 4-6.1. (Stokes) *If* $\omega \in \Lambda^{r-1}(E_r)$ *is defined over a closed, arcwise connected, simply connected, r-dimensional region* Y_r *of* E_r *and* Y_r *has a smooth boundary,* ∂Y_r *, then*

$$\int_{Y_r} d\omega = \int_{\partial Y_r} \omega , \tag{4-6.8}$$

when the orientation of ∂Y_r *is of equal parity with the orientation of* Y_r .

We will not give a proof of Stokes' theorem since there is a trivial reformulation that reduces Stokes' theorem to the divergence theorem, and the proof of the latter is simply a careful application of the integration by parts theorem.

THEOREM 4-6.2. *If* $\omega \in \Lambda^{r-1}$ *is defined over a closed, arcwise connected, simply connected, r-dimensional region* Y_r *of* E_r *with a smooth boundary* ∂Y_r *, then the resolution of* ω *on the basis* $\{\mu_i\}$ *of* $\Lambda^{r-1}(E_r)$ *by*

(4-6.9) $$\omega = w^i(x^j)\mu_i$$

reduces Stokes' theorem to the divergence theorem

(4-6.10) $$\int_{Y_r} \partial_i w^i(x^j)\mu = \int_{\partial Y_r} w^i(x^j)\mu_i \;.$$

Proof. We have already seen that (4-6.9) implies

(4-6.11) $$d\omega = \partial_i w^i(x^j)\mu \;.$$

Now, simply substitute (4-6.9) and (4-6.11) into (4-6.8) to obtain (4-6.10). Conversely, since any $\omega \in \Lambda^{r-1}(E_r)$ is uniquely represented by (4-6.9), a substitution of (4-6.9) and (4-6.11) into (4-6.10) yields (4-6.8).

This is all well and good for forms of degree r and degree $r-1$ on E_r, but what happens for forms of degree r in E_n with $r<n$? Let Ψ be the smooth map

$$\Psi : Y_r \subset E_r \xrightarrow{\text{onto}} \mathcal{Y}_r \subset E_n \mid x^i = \psi^i(u^1,\ldots,u^r), \quad \text{rank}(\partial\psi^i/\partial u^\alpha) = r$$

then Ψ^* acting on $\omega \in \Lambda^r(E_n)$ induces the r-form $\Psi^*\omega$ on the r-dimensional set Y_r in E_r. The map Ψ may also be viewed as inducing the coordinate cover $\{u^\alpha\}$ of E_r as surface coordinates of $\mathcal{Y}_r$, in which case the orientation $\mu(u^\alpha)$ of E_r induces an orientation on $\mathcal{Y}_r$. The integral of ω over $\mathcal{Y}_r$ in E_n can then be defined by

(4-6.12) $$\int_{\mathcal{Y}_r} \omega = \int_{Y_r} \Psi^*\omega \;.$$

Note that the integrand on the right-hand side does not vanish identically because $\text{rank}(\partial\psi^i/\partial u^\alpha) = r$ and hence

$$\int_{Y_r} \Psi^*\omega = \int_{Y_r} h(u^\alpha)\,du^1 \wedge du^2 \wedge \ldots \wedge du^r = \int_{Y_r} h(u^\alpha)\mu(u^\alpha)$$

is well defined. This definition, when combined with Theorem 4-6.1, gives the general version of Stokes' theorem.

THEOREM 4-6.3. *If* $\omega \in \Lambda^{r-1}(E_n)$, *then*

(4-6.13) $$\int_{Y_r} d\omega = \int_{\partial Y_r} \omega$$

provided Y_r *and* ∂Y_r *are such that there exists a map*

(4-6.14) $$\Psi : Y_r \subset E_r \xrightarrow{\text{onto}} Y_r \subset E_n \mid x^i = \psi^i(u^\alpha) , \ \text{rank}(\partial\psi^i/\partial u^\alpha) = r$$

for which Y_r *is a closed, arcwise connected, simply connected, r-dimensional region with a smooth boundary.*

A more useful way of writing this result is

(4-6.15) $$\int_{Y_r} \Psi^* d\omega = \int_{Y_r} d\Psi^*\omega = \int_{\partial Y_r} \Psi^*\omega$$

since $\int_{Y_r} \Psi^* d\omega$ and $\int_{\partial Y_r} \Psi^*\omega$ are the actual values involved.

CHAPTER V

ANTIEXACT DIFFERENTIAL FORMS AND LINEAR HOMOTOPY OPERATORS

5-1. PERSPECTIVE AND OVERVIEW

Anyone with a working knowledge of the exterior calculus of E. Cartan is well aware of the importance of the classes of closed and exact differential forms. In fact, it may be claimed that the study of the structure of the relations between these two classes of differential forms has led to a new plateau of understanding in several fundamental mathematical disciplines. There is an aspect of this problem that has been little touched, however. This is unfortunate, for, although this aspect is principally of a local nature and hence of no great modern significance, it is of considerable practical use and provides otherwise unavailable insights.

Let ω be a differential form on a starlike region S of an n-dimensional space. It is then well known [2,4] that one can define a homotopy operator H on forms on S for which $\omega = dH\omega + Hd\omega$ is identically satisfied on S. The usual case of this result is that it establishes the Poincaré lemma: any closed form on a starlike region is exact (i.e., $d\omega = 0$ implies $\omega = dH\omega$). There is significantly more information that is available, however. If ω is an arbitrary differential form on S, then $\omega = dH\omega + Hd\omega$ allows us to associate a unique exact form $\omega_e = dH\omega$ with ω. It is thus meaningful to refer to ω_e as the exact part of ω. One can then show that $\omega - \omega_e$ belongs to ker(H). We refer to $\omega_a = \omega - \omega_e = Hd\omega$ as the antiexact part of ω. The important thing about antiexact forms, in contrast with exact forms, is that they form a graded submodule, A(S), of the graded module $\Lambda(S)$ of exterior forms on S, and that H *inverts* d *on this submodule*. It is then clear that the decomposition $\omega = \omega_e + \omega_a$ and the invertability of d on A(S) allows the construction of general solutions of systems of exterior differential equation and a number of other useful results of both computational and structural natures. Thus, in perspective and overview, the substance of this Chapter is the study and application of the submodule A(S) of antiexact forms.

5-2. STARLIKE REGIONS

Let S be an open region of an n-dimensional space M. Such a region is said to be *starlike* with respect to one of its points $P_o \in S$ if the following conditions are met:

(i) S is contained in a coordinate neighborhood U of P_o,

(ii) the coordinate functions of U assign coordinates $\{x_o^i\}$ to the point P_o,

(iii) if P is any point of S and if the coordinate functions of U assign coordinates $\{x^i\}$ to P, then the set of points that are assigned the coordinates $\{x_o^i + \lambda(x^i - x_o^i)\}$ belongs to S for all $\lambda \in [0,1]$.

It is clear that the notion of a starlike region is a local one. Moreover, it is dependent upon the existence of a specific coordinate neighborhood with specific coordinate functions. We refer to the coordinate neighborhood U and the coordinate functions, that verify that S is a starlike, as a *preferred coordinate neighborhood* and *preferred coordinate functions* of S. For the time being, we assume that a starlike region S is referred to its preferred coordinate neighborhood with its preferred coordinate functions since the analysis will be of a local nature. Thus, if $P \in S$, we write $P:\{x^i\}$, where $\{x^i\}$ are the coordinates of P in the preferred coordinate system, and, of course, $P_o:\{x_o^i\}$. Thus, for now, we fix the coordinate cover of S. Later we will consider what happens when the coordinate cover is changed.

The local nature of starlike regions notwithstanding, there is certainly no scarcity of starlike regions. For instance, one can generate uncountably many that are starlike with respect to a given point, say P_o. Let $'S$ be a starlike region with respect to the point $'P_o$ of an n-dimensional space $'M$ and let K denote the collection of all regular maps of $'M$ to the n-dimensional space M that map $'P_o$ onto P_o. It is then clear that the image of $'S$ under any map $\Phi \in K$ yields a region in M that is starlike with respect to P_o. Further, the Φ-image of the preferred coordinate cover of $'S$ yields a preferred coordinate cover of $\Phi('S)$.

The fixed nature of the preferred coordinate cover of S allows us to take advantage of certain notational conveniences. If $P:\{x^i\}$ belongs to S, then each point with coordinates $\{x_o^i + \lambda(x^i - x_o^i)\}$ belongs to S. Thus, if we have an exterior differential form

$$(5\text{-}2.1) \qquad \omega = \omega_{i_1 \ldots i_k}(x^j)\, dx^{i_1} \wedge \ldots \wedge dx^{i_k}$$

defined on S, the exterior differential form

$$(5\text{-}2.2)\qquad \tilde{\omega}(\lambda) = \omega_{i_1\ldots i_k}(x_o^j + \lambda(x^j - x_o^j))\, dx^{i_1} \wedge \ldots \wedge dx^{i_k}$$

is well defined on S for every $\lambda \varepsilon [0,1]$ and $\tilde{\omega}(1) = \omega$. In particular, we have

$$(5\text{-}2.3)\qquad d(\tilde{\omega}(\lambda)) = \lambda(\widetilde{d\omega})(\lambda)$$

for every $\lambda \varepsilon [0,1]$, as follows directly from the definition of the exterior derivative.

A starlike region S also has a naturally associated vector field that has many of the properties of the classical "radius vector". This vector field is defined relative to the preferred coordinate system on S by

$$(5\text{-}2.4)\qquad X(x^i) = (x^i - x_o^i)\,\partial/\partial x^i = \frac{d}{d\lambda}\{x_o^i + \lambda(x^i - x_o^i)\}\,\partial/\partial x^i$$

and satisfies the relation

$$(5\text{-}2.5)\qquad \tilde{X}(\lambda) = X(x_o^i + \lambda(x^i - x_o^i)) = \lambda X$$

for all $\lambda \varepsilon [0,1]$.

5-3. THE HOMOTOPY OPERATOR H

Let ω be a differential form of degree k on a starlike region S. Since $\tilde{\omega}(\lambda)$ and X are well defined on S, the linear homotopy operator

$$(5\text{-}3.1)\qquad H\omega = \int_o^1 X \lrcorner\, \tilde{\omega}(\lambda)\lambda^{k-1} d\lambda$$

is well defined on S as a Riemann-Graves integral. H is thus well defined on $\Lambda(S)$. Its specific representation in terms of the preferred coordinate cover of S follows directly from (5-2.2) and (5-2.4):

$$(5\text{-}3.2)\qquad (H\omega)(x^i) = \int_o^1 X(x^i) \lrcorner\, \omega(x_o^i + \lambda(x^i - x_o^i))\lambda^{k-1} d\lambda\ .$$

Although some of the properties of H given in the following theorem are trivial consequences of others in the list, they are given explicitly in order to simplify proofs later on.

THEOREM 5-3.1. *The operator* H *is a linear operator on* $\Lambda(S)$ *with the following properties:*

H_1: H *maps* $\Lambda^k(S)$ *into* $\Lambda^{k-1}(S)$ *for* $k\geq 1$ *and maps* $\Lambda^o(S)$ *into zero,*

H_2: $dH + Hd =$ *identity for* $k\geq 1$, *and* $(Hdf)(x^i) = f(x^i) - f(x_o^i)$ *for* $k=0$

H_3: $(HH\omega)(x^i) = 0$, $(H\omega)(x_o^i) = 0$,

H_4: $HdH = H$, $dHd = d$,

H_5: $HdHd = Hd$, $dHdH = dH$, $(dH)(Hd) = 0$, $(Hd)(dH) = 0$,

H_6: $X\rfloor H = 0$, $HX\rfloor = 0$.

Proof. The linearity of H on $\Lambda(S)$ is clear from the defining equation (5-3.1). Property H_1 also follows directly from (5-3.1) since $X\rfloor$ maps $\Lambda^k(S)$ into $\Lambda^{k-1}(S)$ for $k\geq 1$ and maps $\Lambda^o(S)$ into the zero function on S ; When (5-3.1) is used to evaluate $H\omega$ and $Hd\omega$ for $\omega \in \Lambda^k(S)$, we obtain

$$(5\text{-}3.3)\qquad (dH + Hd)\omega = \int_o^1 \{d(X\rfloor\tilde{\omega}(\lambda))\lambda^{k-1} + X\rfloor\widetilde{(d\omega)}(\lambda)\lambda^k\}d\lambda .$$

However, (5-2.3) gives $\widetilde{(d\omega)}(\lambda) = \lambda^{-1}d(\tilde{\omega}(\lambda))$, and (5-3.3) can be written as

$$(5\text{-}3.4)\qquad (Hd + dH)\omega = \int_o^1 L_\lambda\tilde{\omega}(\lambda)\, d\lambda ,$$

where the linear operator L_λ is defined on $\Lambda(S)$ for all $\lambda \in [0,1]$ by

$$(5\text{-}3.5)\qquad L_\lambda\tilde{\omega}(\lambda) = \{d(X\rfloor\tilde{\omega}(\lambda)) + X\rfloor d(\tilde{\omega}(\lambda))\}\lambda^{k-1} .$$

It might appear that there is some difficulty with (5-3.5) if $k=0$. This is not the case, however, for $X\rfloor\tilde{f}(\lambda) = 0$ and $X\rfloor d(\tilde{f}(\lambda)) = \lambda\widetilde{(df)}(\lambda)$ for $f \in \Lambda^o(S)$. Now, (5-3.5) and the definition of the Lie derivative give $\lambda^{1-k} L_\lambda \tilde{\omega}(\lambda) = \pounds_X \tilde{\omega}(\lambda)$. Thus, since $\pounds$ is a derivation on Λ , $\alpha \in \Lambda^k(S)$, $\beta \in \Lambda^m(S)$ imply

$$\lambda^{1-k-m} L_\lambda \widetilde{(\alpha \wedge \beta)}(\lambda) = \lambda^{1-k-m} L_\lambda (\tilde{\alpha}(\lambda) \wedge \tilde{\beta}(\lambda))$$

$$= \lambda^{1-k} L_\beta \tilde{\alpha}(\lambda) \wedge \tilde{\beta}(\lambda) + \tilde{\alpha}(\lambda) \wedge \lambda^{1-m} L_\lambda \tilde{\beta}(\lambda) ,$$

so that

$$L_\lambda \widetilde{(\alpha \wedge \beta)}(\lambda) = L_\lambda \tilde{\alpha}(\lambda) \wedge \lambda^m \tilde{\beta}(\lambda) + \lambda^k \tilde{\alpha}(\lambda) \wedge L_\lambda \tilde{\beta}(\lambda) . \tag{5-3.6}$$

Accordingly, since any element of $\Lambda(S)$ can be written as a sum of simple monomials and an elementary calculation gives $L_\lambda \tilde{\alpha}(\lambda) = \frac{d}{d\lambda}(\lambda \tilde{\alpha}(\lambda))$ for $\alpha \in \Lambda^1(S)$, we obtain the following evaluation of $L_\lambda \tilde{\omega}(\lambda)$ for $\omega \in \Lambda^k(S)$:

$$L_\lambda \tilde{\omega}(\lambda) = \frac{d}{d\lambda}(\lambda^k \tilde{\omega}(\lambda)) . \tag{5-3.7}$$

A substitution of this result back into (5-3.4) gives $(Hd + dH)\omega = \int_0^1 \frac{d}{d\lambda}(\lambda^k \tilde{\omega}(\lambda)) d\lambda$. Thus, since $\tilde{\omega}(1) = \omega$, property H_2 is established. Since $X(x_o^i) = 0$ from (5-2.4), (5-3.2) yields $(H\omega)(x_o^i) = 0$. It also follows from (5-2.5) and (5-3.1) that

$$HH\omega = \int_0^1 X \rfloor \int_0^1 \widetilde{X \rfloor \tilde{\omega}(\lambda) \lambda^{k-1} d\lambda}\ (\mu) \mu^{k-2} d\mu$$

$$= \int_0^1 \int_0^1 X \rfloor \tilde{X}(\mu) \rfloor \widetilde{(\tilde{\omega}(\lambda))}(\mu) \lambda^{k-1} \mu^{k-2} d\lambda d\mu$$

$$= \int_0^1 \int_0^1 X \rfloor \mu X \rfloor \widetilde{\tilde{\omega}(\lambda)}(\mu) \lambda^{k-1} \mu^{k-2} d\lambda d\mu = 0 ,$$

which establishes H_3 . Property H_4 follows directly by allowing H and d to act on $dH + Hd$ = identity and using H_3 ; i.e., $H = H(\text{identity}) = H(dH + Hd) = HdH + HHd = HdH$. Allowing H and d to act on properties H_4 establishes H_5 . Properties

H_6 are established in exactly the same way as that used to establish H_3 ; i.e., (5-2.5) can be used to obtain a factor of the form $X \rfloor X \rfloor$ under the integral sign, which, of course, vanishes.

The operator H on a starlike region is of particular importance in many applications. It is not an unfamiliar operator in the exterior calculus, for it is customarily constructed in order to establish the Poincaré lemma; *every closed form on* S *is exact.* This result follows directly from H_2 since ω closed implies $\omega = dH\omega + Hd\omega = dH\omega$. In this vein, it may be remarked that the proof of property H_2 given above is somewhat simpler than given in most texts [2,4], for it requires explicit calculation only at the point where $L_\lambda \tilde{\omega}(\lambda) = \frac{d}{d\lambda}(\lambda\tilde{\omega}(\lambda))$ is established for 1-forms. There is, however, significantly more information that can be gleaned from properties H_1 through H_6 , and it is these additional properties that will prove to be instrumental in this monograph.

We consider a few elementary examples as an assist to the reader. Take 2-dimensional Euclidean space with its global Cartesian coordinate cover $\{x,y\}$ and let the 1-form ω be defined by

$$(5\text{-}3.8) \qquad \omega = y\, U(3-x)\, dx \ ,$$

where $U(z) = \{1 \text{ for } z > 0 \ , \ 0 \text{ for } z < 0\}$ is the unit function. Since $d\omega = U(3-x)\, dy \wedge dx$, the definition of H given by (5-3.1) yields

$$(5\text{-}3.9) \qquad H\omega = \int_0^1 x\, \lambda y\, U(3-\lambda x) d\lambda = \begin{cases} xy/2 & \text{for } x < 3 \\ 3^2 y/2x & \text{for } x > 3 \ , \end{cases}$$

$$(5\text{-}3.10) \qquad Hd\omega = \int_0^1 U(3-\lambda x)\lambda(ydx - xdy)d\lambda =$$

$$= \quad (ydx-xdy) \begin{cases} 1/2 & \text{for} \quad x < 3 \\ 3^2/2x^2 & \text{for} \quad x > 3 \; . \end{cases}$$

Thus, although both ω and $d\omega$ vanish for $x > 3$, $H\omega$ and $Hd\omega$ are nonzero for $x > 3$, $y \neq 0$. In fact, this "smearing out" that results from the action of H is characteristic of this operator. It now remains only to check that $dH + Hd$ = identity. However, (5-3.9) yields

$$(5\text{-}3.11) \qquad dH\omega \quad = \quad \frac{1}{2} \begin{cases} ydx+xdy & \text{for} \quad x < 3 \\ 3^2(-ydx+xdy)/x^2 & \text{for} \quad x > 3 \; , \end{cases}$$

so that (5-3.10) and (5-3.11) thus yield

$$(5\text{-}3.12) \qquad dH\omega + Hd\omega \quad = \quad \begin{cases} ydx & \text{for} \quad x < 3 \\ 0 & \text{for} \quad x > 3 \; . \end{cases}$$

For an example of an exact form, we take

$$(5\text{-}3.13) \qquad \omega \quad = \quad x^r \; U(3-x) \; dx \; , \qquad r > 0 \; ,$$

in which case (5-3.1) gives,

$$(5\text{-}3.14) \qquad H\omega \quad = \quad \frac{1}{r+1} \begin{cases} x^{r+1} & \text{for} \quad x < 3 \\ 3^{r+1} & \text{for} \quad x > 3 \; . \end{cases}$$

Notice that in both examples, $H\omega$ is continuous at $x = 3$ even though ω has a jump discontinuity there. It is clear from $dH + Hd$ = identity, that the action of H on a form will result in a form with an improved continuity class, as indeed it must.

5-4. THE EXACT PART OF A FORM AND THE VECTOR SUBSPACE OF EXACT FORMS

Let $E^k(S)$ denote the collection of all exact elements of $\Lambda^k(S)$. Since the set $E^k(S)$ is closed under addition and multiplication by numbers, but is not closed under multiplication by functions, $E^k(S)$ forms a linear subspace of $\Lambda^k(S)$ but not a submodule of $\Lambda^k(S)$. Further, $E^0(S)$ is empty. Since the exterior product of two exact forms is again an exact form, we can construct the graded algebra $E(S)$ of exact forms on S, and $E(S)$ is a subspace of $\Lambda(S)$ but not a submodule of $\Lambda(S)$. In fact, it is precisely because $E^0(S)$ is empty that $E(S)$ is not a submodule of $\Lambda(S)$. The following result is now an almost immediate consequence of properties H_1 through H_6.

LEMMA 5-4.1. *The operator* dH *maps* $E^k(S)$ *onto* $E^k(S)$ *and* $\Lambda(S)$ *onto* $E(S)$.

Proof. Since $dH\omega$ is exact for any $\omega \in \Lambda^k(S)$, dH maps $\Lambda^k(S)$ into $E^k(S)$. The result then follows upon showing that this mapping is onto. We do this with the following lemma that is of importance in its own right.

LEMMA 5-4.2. *The operator* d *is the inverse of the operator* H *when* H *is restricted to* $E^k(S)$.

Proof. Let ω be an arbitrary element of $E^k(S)$, then $\omega = d\alpha$ for some $\alpha \in \Lambda^{k-1}(S)$, and the restriction of H to $E^k(S)$ yields quantities of the form $H\omega = Hd\alpha$. Allowing d to act on both sides of this relation yields $dH\omega = dHd\alpha = d\alpha = \omega$ when H_4 is used. Thus dH restricted to $E^k(S)$ is the identity and the result follows. It also follows that dH is an onto map of $\Lambda^k(S)$ to $E^k(S)$ and hence Lemma 5-4.1 is established.

We now go back to property H_2 and note that any $\omega \in \Lambda^k(S)$ satisfies

$$(5\text{-}4.1) \qquad \omega = dH\omega + Hd\omega .$$

Since dH is the identity map of $E^k(S)$ and maps $\Lambda^k(S)$ onto $E^k(S)$, every element ω of $\Lambda^k(S)$ has a uniquely associated exact part, $dH\omega$. This motivates the following definition.

Definition 5-4.1. Let $\omega \in \Lambda^k(S)$ with $k \geq 1$. The element of

$E^k(S)$, defined by

(5-4.2) $$\omega_e = dH\omega ,$$

is the *exact part* of the form ω . Elements of $\Lambda^o(S)$ have no exact part.

If we use property H_2 , then for any $\omega \in \Lambda^k(S)$ we have $\omega = dH\omega + Hd\omega$, and Definition 5-4.1 yields $\omega = \omega_e + Hd\omega$. Accordingly, allowing H to act on both sides of this relation and making use of property H_3 shows that $H(\omega-\omega_e) = 0$. We thus have the following result that acts as motivation for the considerations of the next Section.

LEMMA 5-4.3. *Let* ω_e *be the exact part of a form* ω *, then* $\omega-\omega_e$ *belongs to the kernel of the linear operator* H .

5-5. THE MODULE OF ANTIEXACT FORMS

The results of Lemma 5-4.3 and the fact that any ω can be written as $\omega = \omega_e + Hd\omega$, by H_2 , suggests that we study quantities of the form $Hd\alpha$. In this regard, the following definition is the natural complement of Definition 5-4.1.

Definition 5-5.1. Let $\omega \in \Lambda^k(S)$ for any $k \geq 0$. The element

(5-5.1) $$\omega_a = Hd\omega = \omega - \omega_e$$

of $\Lambda^k(S)$ is the *antiexact* part of ω .

The collection of all antiexact elements of $\Lambda^k(S)$ is denoted by $A^k(S)$, for $k \geq 1$. Since $E^o(S)$ is empty and any form satisfies $\omega = dH\omega + Hd\omega = \omega_e + \omega_a$, we agree that any scalar function on S is its own antiexact part:

(5-5.2) $$A^o(S) = \Lambda^o(S) .$$

It follows immediately from H_3 and H_6 that the antiexact part of any $\omega \varepsilon \Lambda^k(S)$, for $k \geq 1$, satisfies

(5-5.3) $\quad X \rfloor \omega_a = 0$, $\omega_a(x_o^i) = 0$.

Conversely, if $\alpha \varepsilon \Lambda^k(S)$ satisfies $X \rfloor \alpha = 0$, define an element ω of $\Lambda^k(S)$ by $\omega = d\beta + \alpha$ for $\beta \varepsilon \Lambda^{k-1}(S)$. Since $\widetilde{(X \rfloor \alpha)}(\lambda) = \tilde{X}(\lambda) \rfloor \tilde{\alpha}(\lambda) = \lambda X \rfloor \tilde{\alpha}(\lambda)$, (5-3.1) shows that $H\alpha = 0$. Allowing H to act on both sides of $\omega = d\beta + \alpha$ then gives $H\omega = Hd\beta$ and we see that $\omega_e = dH\omega = dHd\beta = d\beta$ by H_4 . This in turn gives $\omega = \omega_e + \alpha$, from which we conclude that $\alpha = \omega_a$. However, if $\alpha = \omega_a$, then $\alpha = Hd\omega$ and we must have $\alpha(x_o^i) = 0$ for $k>1$. These considerations establish the following lemma.

LEMMA 5-5.1. $A^k(S) = \{\alpha \varepsilon \Lambda^k(S) | X \rfloor \alpha = 0 , \alpha(x_o^i) = 0$ *for* $k>1\}$.

LEMMA 5-5.2. *The operator* Hd *maps* $\Lambda^k(S)$ *onto* $A^k(S)$ *and* $A^n(S) = 0$ *for* S *of dimension* n .

Proof. That Hd is into $A^k(S)$ follows directly from $\omega_a = Hd\omega$. On the other hand, $Hd\omega_a = HdHd\omega = Hd\omega = \omega_a$, by H_5 , and hence Hd is onto $A^k(S)$. The latter result follows on noting that $\Lambda^k \xrightarrow{d} \Lambda^{k+1} \xrightarrow{H} \Lambda^k$ and $\Lambda^{n+1}(S) = 0$ for S of dimension n .

Definition 5-5.2. Any form $\alpha \varepsilon \Lambda^k(S)$ that satisfies $X \rfloor \alpha = 0$ and $\alpha(x_o^i) = 0$ for $k \geq 1$ is an *antiexact* form. $A^k(S)$ is the vector space of antiexact forms of degree k on S , and the antiexact part of any form is an antiexact form.

We are now in possession of the necessary groundwork in order to establish the theorem underlying this study.

THEOREM 5-5.1. *Antiexact forms possess the following properties:*

A_1: $A^k(S) \subset \ker(H)$ *for all* k ,

A_2: $\alpha \varepsilon A^k(S)$, $\beta \varepsilon A^m(S)$ *implies* $\alpha \wedge \beta \varepsilon A^{k+m}(S)$

A_3: $A^k(S)$ *is a* C^∞*-module for* $k \geq 1$,

A_4: H *is the inverse of* d *on* $A^k(S)$ *for* $k \geq 1$.

Proof. Property A_1 follows directly from Lemma 5-4.3 since any ω can be written as $\omega = \omega_e + \omega_a$ and $\omega - \omega_e$ belongs to ker(H). Clearly, A_2 holds for elements of $A^o(S) = \Lambda^o(S)$, and hence it suffices to establish A_2 for $\max(k,m) \geq 1$. Under the hypotheses of A_2 , we know that $\alpha \wedge \beta \in \Lambda^{k+m}(S)$. Since $\max(k,m) \geq 1$, Lemma 5-5.1 shows that $(\alpha \wedge \beta)(x_o^i) = 0$ because at least one of the factors vanishes at $\{x_o^i\}$. Further, $X \rfloor (\alpha \wedge \beta) = (X \rfloor \alpha) \wedge \beta + (-1)^k \alpha \wedge (X \rfloor \beta) = 0$ since $X \rfloor \alpha = 0$ and $X \rfloor \beta = 0$ by Lemma 5-5.1 Thus $X \rfloor (\alpha \wedge \beta) = 0$, $(\alpha \wedge \beta)(x_o^i) = 0$ and Lemma 5-5.1 implies $\alpha \wedge \beta \in A^{k+m}(S)$. A_3 then follows directly from A_2 since the set $A^k(S)$ is closed under addition and under exterior multiplication by all elements of $A^o(S) = \Lambda^o(S)$. Property H_2 gives, for $\omega \in \Lambda^k(S)$, $k \geq 1$, $\omega = Hd\omega + dH\omega$. However, A_1 shows that $H\omega = 0$ for $\omega \in A^k(S)$, and hence $\omega = Hd\omega$ for $\omega \in A^k(S)$ with $k \geq 1$.

Remark. For forms of degree zero, property H_2 gives $f(x^i) = f(x_o^i) + (Hdf)(x^i)$. Accordingly, the operator H can be used to invert the operator d on $A^k(S)$ for all values of k .

COROLLARY 5-5.1. *Let* A(S) *be the graded associative algebra of antiexact forms with the operations of addition and exterior multiplication taken from* $\Lambda(S)$, *then* A(S) *is a subalgebra of* $\Lambda(S)$ *that is given by* $A(S) = Hd(\Lambda(S))$, *and* A *is the* Hd-*projection of* Λ .

Proof. The result is a direct consequence of Theorem 5-5.1 and Lemma 5-5.2: the latter showing that Hd maps $\Lambda(S)$ onto A(S) . That Hd is a projection follows from property H_5 and Lemma 5-5.2.

COROLLARY 5-5.2. *The linear operator* H *maps* $\Lambda^{k+1}(S)$ *onto* $A^k(S)$ *for* $k \geq 1$ *and hence we have* $A^k(S) = H(\Lambda^{k+1}(S))$ *for* $k \geq 1$.

Proof. If $\omega \in \Lambda^{k+1}(S)$, then $H\omega \in \Lambda^k(S)$ by H_1 . Now, $(H\omega)(x_o^i) = 0$ by H_3 and $X \rfloor H\omega = 0$ by H_6 . Thus, $H\omega \in A^k(S)$ by Lemma

5-5.1. Conversely, if $\alpha \in A^k(S)$, then $\alpha = Hd\alpha$ by A_4 . There thus exists a $\beta = d\alpha \in \Lambda^{k+1}$ such that $\alpha = H\beta$ and $H(\Lambda^{k+1}(S))$ covers $A^k(S)$. In contrast with Corollary 5-5.1, we note that H is not a projection, for property H_3 gives $HH\omega = 0$ for all $\omega \in \Lambda(S)$.

The fact that $A(S)$ is a submodule of $\Lambda(S)$ is central in the applicability of antiexact exterior forms. This fact seems surprising on first reading since Corollary 5-5.1 shows that there exists a $\gamma \in \Lambda(E_n)$ for each pair α, β of $\Lambda(E_n)$ such that $Hd\gamma = Hd\alpha \wedge Hd\beta$ while it is an elementary exercise to see that $H(d\alpha \wedge d\beta) = Hd(\alpha \wedge d\beta) \neq Hd\alpha \wedge Hd\beta$. In fact, the definition of the operator H shows that $H(\rho \wedge \eta)$ is not expressible in terms of $H\rho$, $H\eta$ and the operations $(\wedge, +, d)$. The following Lemma provides the resolution of this seeming paradox.

<u>LEMMA 5-5.3</u>. *If* $\alpha \in A^k(S)$, *then there exists an* $\hat{\alpha} \in \Lambda^{k+1}(S)$ *such that*

$$\alpha = X \rfloor \hat{\alpha} . \tag{5-5.4}$$

<u>Proof</u>. Corollary 5-5.2 shows that there exists a $\beta \in \Lambda^{k+1}(S)$ for a given $\alpha \in A^k(S)$ such that $\alpha = H\beta$. Since β is the sum of the basis elements of $\Lambda^{k+1}(S)$ with coefficients from $\Lambda^o(S)$, both $\Lambda^{k+1}(S)$ and $A^k(S)$ are closed under addition, and $H(\rho+\eta) = H\rho + H\eta$, it is sufficient to establish the result for a monomial of the form $f(x^j)\, dx^{i_1} \wedge dx^{i_2} \wedge \ldots \wedge dx^{i_{k+1}}$. We now simply observe that

$$H\left(f(x^j)\, dx^{i_1} \wedge \ldots \wedge dx^{i_{k+1}}\right)$$

$$= \int_0^1 \lambda^k f(x_0^j + \lambda(x^j - x_0^j))\, d\lambda\; X \rfloor (dx^{i_1} \wedge \ldots \wedge dx^{i_{k+1}})$$

$$= X \rfloor (\hat{f}(x^j)\, dx^{i_1} \wedge \ldots \wedge dx^{i_{k+1}}) ,$$

where we set

$$\hat{f}(x^j) = \int_0^1 \lambda^k f(x_o^j + \lambda(x^j - x_o^j))\, d\lambda .$$

Thus, since (5-5.4) gives $X \rfloor \alpha = 0$ and $\alpha|_{x=x_o} = 0$ because $X|_{x=x_o} = 0$, the result is established.

If α and β are antiexact, Lemma 5-5.3 gives

$$\alpha = X \rfloor \hat{\alpha} , \quad \beta = X \rfloor \hat{\beta}$$

and hence

$$\alpha \wedge \beta = (X \rfloor \hat{\alpha}) \wedge (X \rfloor \hat{\beta}) = X \rfloor (\hat{\alpha} \wedge X \rfloor \hat{\beta}) .$$

Thus, since

$$Hd(X \rfloor (\hat{\alpha} \wedge X \rfloor \hat{\beta}) = X \rfloor (\hat{\alpha} \wedge X \rfloor \hat{\beta}) - dH(X \rfloor (\hat{\alpha} \wedge X \rfloor \hat{\beta}) = X \rfloor (\hat{\alpha} \wedge X \rfloor \hat{\beta}) ,$$

$$Hd(X \rfloor \hat{\alpha}) = X \rfloor \hat{\alpha} - dH(X \rfloor \hat{\alpha}) = X \rfloor \hat{\alpha} ,$$

we have

$$\alpha \wedge \beta = Hd(X \rfloor \hat{\alpha}) \wedge Hd(X \rfloor \hat{\beta}) = Hd(X \rfloor (\hat{\alpha} \wedge X \rfloor \hat{\beta})) = Hd\gamma$$

with $\gamma = X \rfloor (\hat{\alpha} \wedge X \rfloor \hat{\beta}) \varepsilon A(S)$. Further if ρ and η are elements of $\Lambda(S)$, Lemma 5-5.3 gives

$$H\rho = X \rfloor \alpha , \quad H\eta = X \rfloor \beta$$

for some α and $\beta \varepsilon \Lambda(S)$, in which case we have

$$H\rho \wedge H\eta = (X \rfloor \alpha) \wedge (X \rfloor \beta) = X \rfloor (\alpha \wedge X \rfloor \beta) = H\gamma .$$

Since H annihilates the antiexact part of γ it is sufficient to exhibit an exact γ for which the above relation holds:

$$\gamma = d(X \rfloor (\alpha \wedge X \rfloor \beta)) .$$

5-6. REPRESENTATIONS

Property H_2 has shown us that any $\omega \varepsilon \Lambda^k(S)$ with $k \geq 1$ can be written in the form $\omega = dH\omega + Hd\omega$. Thus, since $dH\omega = \omega_e \varepsilon E^k(S)$ and $Hd\omega = \omega_a \varepsilon A^k(S)$, we obtain the representation $\omega = \alpha + \beta$ with $\alpha \varepsilon E^k(S)$ and $\beta \varepsilon A^k(S)$. It would thus appear that we have to work with the two sets $E^k(S) = dH(\Lambda^k(S))$ and $A^k(S) = Hd(\Lambda^k(S))$ in order to represent any $\omega \varepsilon \Lambda^k(S)$. It turns out, however, that any element of $\Lambda^k(S)$ can be reconstructed from elements of $A(S)$ alone, as we now proceed to show.

THEOREM 5-6.1. *Any* $\omega \varepsilon \Lambda^k(S)$, $k \geq 1$, *has the unique representation* $\omega = d\alpha_1 + \alpha_2$ *under the conditions* $\alpha_1 \varepsilon A^{k-1}(S)$, $\alpha_2 \varepsilon A^k(s)$. *If these conditions are satisfied, then* $\alpha_2 = H(d\omega)$ *is unique, while* $\alpha_1 = H(\omega)$ *is unique for* $k>1$ *and* $\alpha_1 = H(\omega) +$ *constant for* $k=1$.

Proof. Property H_3 gives $\omega = dH\omega + Hd\omega$ for any $\omega \varepsilon \Lambda^k(S)$ with $k \geq 1$. Corollary 5-5.2 shows that $H\omega \varepsilon A^{k-1}(S)$ and $Hd\omega \varepsilon A^k(S)$. Thus, $\omega \varepsilon \Lambda^k(S)$ admits the representation $\omega = d\alpha_1 + \alpha_2$ with $\alpha_1 = H\omega \varepsilon A^{k-1}(S)$ and $\alpha_2 = Hd\omega \varepsilon A^k(S)$. It thus remains only to establish uniqueness under the conditions $\alpha_1 \varepsilon A^{k-1}(S)$, $\alpha_2 \varepsilon A^k(S)$. Suppose, therefore, that $\omega = d\alpha_1 + \alpha_2 = d\beta_1 + \beta_2$ with α_1 and β_1 in $A^{k-1}(S)$ and α_2 and β_2 in $A^k(S)$. This gives $d(\alpha_1 - \beta_1) = \beta_2 - \alpha_2$ and $0 = d(\beta_2 - \alpha_2)$. Since H inverts d on $A^{k+1}(S)$, we have $\beta_2 - \alpha_2 = Hd(\beta_2 - \alpha_2) = 0$. Thus, $d(\alpha_1 - \beta_1) = \beta_2 - \alpha_2 = 0$, and the Poincaré lemma yields $\alpha_1 - \beta_1 = d\mu$ for some $\mu \varepsilon \Lambda^{k-2}(S)$. Since $\alpha_1 - \beta_1 \subset \ker(H)$ by A_1 , allowing H to act on both sides of $\alpha_1 - \beta_1 = d\mu$ gives $0 = Hd\mu$. However, $\mu = dH\mu + Hd\mu$ by H_2 , and so we conclude that $\mu = dH\mu$. This in turn gives $\alpha_1 - \beta_1 = d^2 H\mu = 0$, and uniqueness is established.

This result can be paraphrased in the following direct form:

$$\Lambda^k(S) = d(A^{k-1}(S)) + A^k(S) , \quad k \geq 1 . \tag{5-6.1}$$

Similarly, H_2 can be used to obtain the direct sum representation

$$(5\text{-}6.2) \qquad \Lambda^k(S) \;=\; E^k(S) \oplus A^k(S) \;,$$

although this will turn out to be of significantly less use than (5-6.1). In this context, it is of interest to note that Corollary 5-5.2 follows from (5-6.1) by direct calculation:

$$H(\Lambda^{k+1}(S)) \;=\; H(dA^k(S) + A^{k+1}(S)) \;=\; Hd(A^k(S)) \;=\; A^k(S)$$

since H inverts d on $A^k(S)$. The following chart summarizes these findings in a convenient format:

$$\begin{array}{ccccccc}
 & & & \Lambda(S) & & & \\
 & & \swarrow^{Hd} & & \searrow^{dH} & & \\
0 & \xleftarrow{H} & A(S) & \overset{d}{\underset{H}{\rightleftarrows}} & E(S) & \xrightarrow{d} & 0 \\
 & & \downarrow{\scriptstyle id} & & \downarrow{\scriptstyle id} & & \\
 & & A(S) & \oplus & E(S) & = & \Lambda(S) \;.
\end{array}$$

COROLLARY 5-6.1. *The decomposition*

$$(5\text{-}6.3) \qquad \omega \;=\; \omega_e + \omega_a \;, \quad \omega_e \;=\; dH(\omega) \;, \quad \omega_a \;=\; H(d\omega)$$

of any form into the sum of an exact part and an antiexact part is unique.

Proof. The result follows directly from Theorem 5-6.1 for $\omega = d\alpha_1 + \alpha_2$ with $\alpha_1 = H(\omega)$, $\alpha_2 = H(d\omega)$ yields $\omega_e = d\alpha_1 = dH(\omega)$ and $\alpha_2 \in A(S)$.

COROLLARY 5-6.2. *An exterior form vanishes if and only if its exact part and its antiexact part vanish separately.*

Proof. Clearly, $\omega_e = 0$, $\omega_a = 0$ give $\omega = \omega_e + \omega_a = 0$.

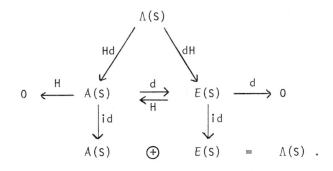

Conversely, $\omega = 0$ implies $H(\omega) = 0$ and $H(d\omega) = 0$ so that $\omega_e = dH(\omega) = 0$ and $\omega_a = H(d\omega) = 0$.

COROLLARY 5-6.3. *We have*

$$E^k(S) \cap A^k(S) = \text{0-element of } \Lambda^k(S) . \tag{5-6.4}$$

Proof. If $\omega = dH(\omega)$ then $\omega_a = Hd^2H(\omega) = 0$, and if $\omega = H(d\omega)$ then $\omega_e = dH^2(d\omega) = 0$. Thus, $E^k(S)$ and $A^k(S)$ have only the 0-element of $\Lambda^k(S)$ in common.

Consider the case where

$$\omega = 2x\,dx - y^2 dz$$

on E_3 and take $\{x_o^i\} = \{0,0,0\}$. We then have

$$H(\omega) = \int_o^1 (2\lambda x^2 - y^2\lambda^2 z)\,d\lambda = x^2 - \frac{1}{3}y^2 z ,$$

with $H(\omega) \varepsilon A^o(E_3)$ and

$$\omega_e = dH(\omega) = d(x^2 - \frac{1}{3}y^2 z) .$$

Use of this result and $\omega_a = \omega - \omega_e$ yields

$$\omega_a = \frac{2}{3}y\,(z\,dy - y\,dz)$$

with $\omega_a \varepsilon A^1(E_3)$. Thus, the representation theorem 5-6.1 gives

$$\omega = d(x^2 - \frac{1}{3}y^2 z) + \frac{2}{3}y\,(z\,dy - y\,dz) .$$

Corollary 5-6.1 and properties A_2 through A_4 lead to a direct generalization of the process $\int_o^z f(y)dy$ of definite integration to forms and antiexact forms. In order to see this, we note that (5-6.3) yields

$$\omega(x) = d(H\omega)(x) + H(d\omega)(x) ,$$

for any $\omega \varepsilon \Lambda(S)$, and the term $d(H\ )(x)$ plays the role of a "constant of integration" since $d(d(H\omega)(x)) \equiv 0$. There is also a direct analogue of *integration of parts* as the following lemma shows.

LEMMA 5-6.1. *If* $\alpha \in A^k(S)$ *and* $\beta \in A^m(S)$, *then*

$$H(d\alpha \wedge \beta) = \alpha \wedge \beta + (-1)^{k+1} H(\alpha \wedge d\beta) . \tag{5-6.5}$$

Proof. Since $\alpha \in A^k(S)$, $\beta \in A^m(S)$, we have $\alpha \wedge \beta \in A^{k+m}(S)$ by A_2 . Property A_4 thus gives $\alpha \wedge \beta = Hd(\alpha \wedge \beta)$, and this leads directly to (5-6.5).

As one might anticipate, use of H to invert d on $A(S)$, and the integration of parts formula (5-6.5) provide a natural and direct access to the study of exterior differential equations. We take up this topic beginning at Section 9.

5-7. CHANGE OF CENTER

If S is a starlike region with respect to the point $P \in S$, it is convenient to refer to P as a *center* of S . The homotopy operator H that is constructed on S by use of the center P will now be referred to as the homotopy operator with center P . Suppose that S is starlike with respect to several points P_1, P_2, etc. We can then construct homotopy operators H_1, H_2, ... , on S with centers P_1, P_2, ... , respectively. The question thus arises as to the relation between homotopy operators on S with different centers.

Let P_1 and P_2 be two centers of a starlike region S , and let H_1 and H_2 be the homotopy operators on S with centers P_1 and P_2 , respectively. If ω is any element of $\Lambda^k(S)$, property H_2 gives

$$\omega = dH_1\omega + H_1 d\omega = dH_2\omega + H_2 d\omega ,$$

and we obtain the relations

$$(5\text{-}7.1) \qquad d(H_1 - H_2)\omega = (H_2 - H_1)d\omega ,$$

$$(5\text{-}7.2) \qquad d(H_2 - H_1)(d\omega) = 0 .$$

Since the Poincaré lemma holds in S , (5-7.2) yields

$$(5\text{-}7.3) \qquad (H_2 - H_1)d\omega = d\beta$$

for some $\beta \varepsilon \Lambda^{k-1}(S)$. If we restrict β to belong to $A_2^{k-1}(S)$, then allowing H_2 to act on both sides of (5-7.3) yields $-H_2H_1d\omega = H_2d\beta = \beta$ since H_2 inverts d on $A_2(S)$, when this result together with (5-7.3) is substituted back into (5-7.2), we obtain $d(H_1 - H_2)\omega = d\beta$, and the Poincaré lemma gives

$$(5\text{-}7.4) \qquad (H_1 - H_2)\omega = \beta + d\alpha$$

for some $\alpha \varepsilon \Lambda^{k-2}(S)$. Restricting α to $A_2^{k-2}(S)$, application of H_2 to both sides of (5-7.4) gives $H_2H_1\omega = \alpha$. We thus have

$$(5\text{-}7.5) \qquad H_1\omega = H_2\omega + \beta + d\alpha$$

with

$$(5\text{-}7.6) \qquad \beta = - H_2H_1d\omega \varepsilon A_2^{k-1}(S) , \quad \alpha = H_2H_1\omega \varepsilon A_2^{k-2}(S) .$$

It now only remains to show that the restriction of α and β to $A_2(S)$ is inconsequential. For this, it is sufficient to substitute (5-7.6) and (5-7.5) and to use the fact that $dH_2 = -H_2d + id$ to obtain

$$\begin{aligned} H_2\omega + \beta + d\alpha &= H_2\omega - H_2H_1d\omega + dH_2H_1\omega \\ &= H_2\omega - H_2H_1d\omega + H_1\omega - H_2dH_1\omega = H_1\omega + H_2(\omega - H_1d\omega - dH_1\omega) \\ &= H_1\omega . \end{aligned}$$

These considerations establish the following lemma.

LEMMA 5-7.1. *Let* S *be a starlike region and let* H_1 *and* H_2 *be two homotopy operators on* S *with centers* P_1 *and* P_2, *respectively. The operators* H_1 *and* H_2 *then stand in the relation*

$$H_1\omega = H_2\omega + \beta + d\alpha ,$$

with $\beta = -H_2H_1d\omega$ *and* α *belonging to* $A_2(S)$, *and* α=constant *if* $\omega \varepsilon \Lambda^1(S)$, $\alpha = H_2H_1\omega$ *if* $\omega \varepsilon \Lambda^k(S)$, $k>1$.

This result is important, for it allows us to shift from any one center of S to any other center of S without essential change in any of the results. Granted, for $\omega = dH_1\omega + H_1d\omega$ and $\omega = dH_2\omega + H_2d\omega$, the exact element $dH_1\omega$ changes to the exact element $dH_2\omega$ and the antiexact element $H_1d\omega$ changes to the antiexact element $H_2d\omega$ on change from the center P_1 to the center P_2 , but the representation in terms of elements of $A_1(S)$ goes over exactly into the same representation in terms of elements of $A_2(S)$ and H_2 inverts d on $A_2(S)$ for exactly the same reasons that H_1 inverts d on $A_1(S)$.

For example, with $\omega = 2x\,dx - y^2dz$, we can construct H_1 with center $(0,0,0)$ and H_2 with center $(1,0,1)$. We then have

$$H_1\omega = x^2 - \frac{1}{3}y^2z , \quad H_2\omega = x^2 - \frac{1}{3}y^2z - 1 + \frac{1}{3}y^2 .$$

However, $H_1d\omega = \frac{2}{3}y\,(z\,dy - y\,dz)$, and hence

$$\beta = -H_2H_1\omega = -\frac{1}{3}y^2 .$$

A combination of these results shows that

$$H_1\omega = H_2\omega + \beta + 1 ,$$

and hence $\alpha = 1 =$ constant is admissible since $\omega \varepsilon \Lambda^1(E_3)$.

5-8. BEHAVIOR UNDER MAPPINGS

Let S be a starlike region of an m-dimensional space M with center P and with homotopy operator H with center P. Let Φ be a differentiable mapping from M to a space N of dimension $n \geq m$, that is regular on S. Finally, let $\Lambda(\Phi(S))$ denote the restriction of $\Lambda(N)$ to the range of Φ in the sense that $\Lambda(\Phi(S)) = \overset{-1}{\Phi}{}^*\Lambda(S)$, where $\overset{-1}{\Phi}$ is the relation inverse of Φ restricted to the range of Φ. If M and N have the same dimension m, then $\overset{-1}{\Phi}$ is then the mapping inverse of Φ since Φ is assumed to be regular on S (i.e., Φ^* maps at least one nonzero m-form onto a nonzero m-form).

If α is any k-form on $\Phi(S)$, then $\Phi^*\alpha$ is a k-form on S and $H\Phi^*\alpha$ belongs to $A(S)$. Since $\overset{-1}{\Phi}{}^*$ maps elements of $\Lambda(S)$ onto elements of $\Lambda(\Phi(S))$, the quantity $\overset{-1}{\Phi}{}^*H\Phi^*\alpha$ belongs to $\Lambda(\Phi(S))$. Thus, if we restrict α to $\Lambda(\Phi(S))$, then $\overset{-1}{\Phi}{}^*H\Phi^* = H^*$ induces a linear map of $\Lambda(\Phi(S))$ to $\Lambda(\Phi(S))$. The linear operator H^* obtained in this way is referred to as the *homotopy operator induced by* Φ on $\Lambda(\Phi(S))$.

THEOREM 5-8.1. *Let* Φ *be a map of a space* M *into a space* N *with* $\dim(M) \leq \dim(N)$ *and let* Φ *be regular on a starlike region* S *of* M *with center* P *and homotopy operator* H. *Then* $\Phi(S)$ *is a starlike region in* $\Phi(M)$ *with center* $\Phi(P)$, Φ *induces the homotopy operator* H^* *on* $\Lambda(\Phi(S))$ *that is defined by*

$$(5\text{-}8.1) \qquad \Phi^*H^* = H\Phi^*$$

and

$$(5\text{-}8.2) \qquad dH^* + H^*d = \text{identity}$$

on $\Lambda(\Phi(S))$.

Proof. Since Φ is regular on S, $\Phi(S)$ is a regular region of N of dimension equal to that of M. The preferred coordinate system on S can thus be lifted by Φ to a coordinate cover of $\Phi(S)$ from which we conclude that $\Phi(S)$ is starlike with center $\Phi(P)$. If we cut down $\Lambda(N)$ to $\Lambda(\Phi(S))$, then the invertibility of Φ on $\Phi(S)$ yields a well defined map $\overset{-1}{\Phi}{}^*$ of $\Lambda(S)$ onto $\Lambda(\Phi(S))$. Thus, pulling (5-8.1) back to $\Lambda(\Phi(S))$ by $\overset{-1}{\Phi}{}^*$ yields the operator $H^* = \overset{-1}{\Phi}{}^* H \Phi^*$ that we obtained previously. It thus remains only to verify the relation (5-8.2). If $\alpha \varepsilon \Lambda(\Phi(S))$, then $\beta = \Phi^*\alpha \varepsilon \Lambda(S)$ and H_2 gives $\Phi^*\alpha = dH\beta + Hd\beta = dH\Phi^*\alpha + Hd\Phi^*\alpha = dH\Phi^*\alpha + H\Phi^*d\alpha$. Thus (5-8.1) yields $\Phi^*\alpha = \Phi^*(dH^*\alpha + H^*d\alpha)$. Since all forms that occur belong to $\Lambda(\Phi(S))$, action by $\overset{-1}{\Phi}{}^*$ yields (5-8.2).

This theorem is of particular importance if we allow Φ to be a map of M to M. In this event, every image of a starlike region S in M under a regular map of M to M is a starlike region and Φ induces a homotopy operator H^* on the image of S. Since such maps Φ can also be thought of as inducing changes in the coordinate cover of the underlying coordinate neighborhood U that contains S if Φ is regular on U, the relation $\Phi^*H^* = H\Phi^*$ gives a complete accounting of how homotopy operators behave under coordinate maps. It is also clear that $H_2 = \overset{-1}{\Phi}{}^* H_1 \Phi^*$ yields the change of center formula if $\Phi: S \to S$ is the translation that maps the center P_1 onto the center P_2.

Consider the map

$$\Phi: E_2 \to E_3 \mid x = 1 + \alpha , \quad y = \alpha + e^{-\beta} , \quad z = e^{\beta} ,$$

then $\Phi(E_2)$ is the surface $y = x + \frac{1}{z} - 1$, $z > 0$ in E_3 since $\overset{-1}{\Phi}(\Phi(E_2))$ is given by $\alpha = x - 1$, $\beta = \ell n\ z$. If $\omega = 2x\ dx - y\ dz$, then

$$\Phi^*\omega = 2(1+\alpha)d\alpha - (\alpha e^{\beta}+1)d\beta$$

is well defined on E_2. Let H be the homotopy operator on E_2 with center $(0,0)$, then

$$H\Phi^*\omega = \int_0^1 \{2(1+\lambda\alpha)\alpha - (\lambda\alpha e^{\lambda\beta}+1)\beta\}d\lambda$$

$$= \alpha(2+\alpha) - \frac{\alpha}{\beta}(\beta e^{\beta} - e^{\beta} + 1) - \beta .$$

However, ω restricted to $\Phi(E_2)$ (i.e., $y = x + \frac{1}{z} - 1$) gives

$$\omega\Big|_{\Phi(E_2)} = 2x\,dx - (x + \frac{1}{z} - 1)\,dz$$

while

$$H^*\omega\Big|_{\Phi(E_2)} = \overset{-1}{\Phi}{}^*(H\Phi^*\omega) = (x-1)(x+1) - \ell n\, z$$

$$- \frac{(x-1)}{\ell n\, z}\{z\,\ell n\, z - z + 1\} .$$

We note, in this regard, that $\Phi:(0,0) \longrightarrow (1,2,1)$, so that the center of H^* on $y = x + \frac{1}{z} - 1$, $z > 0$ is $(1,2,1)$.

5-9. AN INTRODUCTORY PROBLEM

The purpose of this section is to show how the homotopy operator, H, and the unique decomposition of forms into sums of exact and antiexact parts allow us to obtain explicit solutions of exterior differential equations. To this end, we confine our attention to the simplest problem in exterior differential equations: find all k-forms Ω on S that verify

$$d\Omega = \Gamma \wedge \Omega + \Sigma$$

for $\Gamma \varepsilon \Lambda^1(S)$ and $\Sigma \varepsilon \Lambda^{k+1}(S)$.

Clearly, Γ and Σ can not be assigned arbitrarily if there is to be a solution of $d\Omega = \Gamma \wedge \Omega + \Sigma$, for the closure condition, $d^2\Omega = 0$, of this equation must be satisfied. Exterior differentiation and elimination give us the complete resulting system

(5-9.1) $$d\Omega = \Gamma \wedge \Omega + \Sigma ,$$

$$(5\text{-}9.2) \qquad d\Sigma = \Gamma \wedge \Sigma - \Theta \wedge \Omega ,$$

$$(5\text{-}9.3) \qquad d\Gamma = \Theta , \quad d\Theta = 0 .$$

We thus have the problem of finding all $\Omega \varepsilon \Lambda^k(S)$, $\Gamma \varepsilon \Lambda^1(S)$, $\Sigma \varepsilon \Lambda^{k+1}(S)$, $\Theta \varepsilon \Lambda^2(S)$ that satisfy the system (5-9.1)-(5-9.3).

The system consisting of the two exterior equations (5-9.3) is particularly simple, so we start with it. Since the Poincaré lemma holds on S , $d\Theta = 0$ integrates to give

$$(5\text{-}9.4) \qquad \Theta = d\theta$$

and property H_2 and the first of (5-9.3) give

$$(5\text{-}9.5) \qquad \theta = H(\Theta) = H(d\Gamma) ,$$

so that $\theta \varepsilon A^1(S)$. Use of property H_2 and the first of (5-9.3) shows that $\Gamma = dH(\Gamma) + H(d\Gamma) = dH(\Gamma) + \theta$. We thus have

$$(5\text{-}9.6) \qquad \Gamma = d\gamma + \theta ,$$

with

$$(5\text{-}9.7) \qquad \gamma = H(\Gamma) , \quad \theta = H(d\Gamma)$$

belonging to $A(S)$.

We now turn to the system (5-9.1), (5-9.2) and make the similarity transformations

$$(5\text{-}9.8) \qquad \Omega = e^{\lambda}\omega , \quad \Sigma = e^{\lambda}\sigma$$

with $\lambda \varepsilon \Lambda^o(S)$. When (5-9.6) is used, this gives

$$d\omega = (d\gamma - d\lambda + \theta) \wedge \omega + \sigma ,$$

$$d\sigma = (d\gamma - d\lambda + \theta) \wedge \sigma - d\theta \wedge \omega .$$

If λ is given by

(5-9.9) $\lambda = \gamma = H(\Gamma)$,

then the system reduces to

$$d\omega = \theta \wedge \omega + \sigma , \quad d\sigma = \theta \wedge \sigma - d\theta \wedge \omega .$$

The further substitution

(5-9.10) $\rho = \theta \wedge \omega$,

yields the final reduced system

(5-9.11) $d\omega = \rho + \sigma , \quad d\sigma = - d\rho$

since $d\rho = d\theta \wedge \omega - \theta \wedge d\omega = d\theta \wedge \omega - \theta \wedge (\theta \wedge \omega + \sigma) = d\theta \wedge \omega - \theta \wedge \sigma$.

The integration of the system (5-9.11) is direct since property H_2 gives $\sigma = dH(\sigma) + H(d\sigma)$ and $\omega = dH(\omega) + H(d\omega)$. Thus $\sigma = dH(\sigma) - H(d\rho) = dH(\sigma) - \rho + dH(\rho)$, since $H(d\rho) + dH(\rho) = \rho$, and $\omega = dH(\omega) + H(\rho + \sigma)$. Accordingly, if we define $\eta \varepsilon A^k(S)$ and $\phi \varepsilon A^{k-1}(S)$ by

(5-9.12) $\eta = H(\sigma) , \quad \phi = H(\omega)$,

we obtain

(5-9.13) $\sigma = d\eta - \rho + dH(\rho) , \quad \omega = d\phi + H(\rho) + \eta$

because $H(\sigma) = H(d\eta - \rho + dH(\rho)) = H(d\eta - H(d\rho)) = Hd\eta = \eta$, where the last equality follows from the fact that H inverts d on $A(S)$ and $\eta \varepsilon A^k(S)$.

It now remains only to eliminate the subsidiary form ρ . Substituting the second of (5-9.13) into (5-9.10) gives us

$$\rho = \theta \wedge d\phi + \theta \wedge H(\rho) + \theta \wedge \eta ,$$

and hence

$$H(\rho) = H(\theta \wedge d\phi + \theta \wedge H(\rho) + \theta \wedge \eta) .$$

However, $\theta \in A(S)$, $\eta \in A(S)$ imply $\theta \wedge H(\rho) \in A(S)$, $\theta \wedge \eta \in A(S)$ by property A_2 , and hence property A_1 gives $H(\theta \wedge H(\rho) + \theta \wedge \eta) = 0$. We thus obtain

$$\rho = \theta \wedge d\phi + \theta \wedge H(\theta \wedge d\phi) + \theta \wedge \eta , \quad H(\rho) = H(\theta \wedge d\phi) ,$$

and (5-9.13) give

$$\omega = d\phi + \eta + H(\theta \wedge d\phi) ,$$

$$\sigma = d\eta - \theta \wedge \eta - \theta \wedge H(\theta \wedge d\phi) - Hd(\theta \wedge d\phi) .$$

A straightforward combination of the various substitutions given above establishes the following result.

THEOREM 5-9.1. *The general solution of the differential system*

(5-9.14) $$d\Omega = \Gamma \wedge \Omega + \Sigma , \quad d\Sigma = \Gamma \wedge \Sigma - \Theta \wedge \Omega ,$$

(5-9.15) $$d\Gamma = \Theta , \quad d\Theta = 0 ,$$

on a starlike region S *is given in terms of the antiexact forms*

(5-9.16) $$\gamma = H(\Gamma), \quad \phi = H(e^{-\gamma}\Omega), \quad \eta = H(e^{-\gamma}\Sigma), \quad \theta = H(d\Gamma)$$

by

(5-9.17) $$\Omega = e^{\gamma}\{d\phi + \eta + H(\theta \wedge d\phi)\} ,$$

(5-9.18) $$\Sigma = e^{\gamma}\{d\eta - \theta \wedge \eta - \theta \wedge H(\theta \wedge d\phi) - Hd(\theta \wedge d\phi)\} ,$$

(5-9.19) $$\Gamma = d\gamma + \theta , \quad \Theta = d\theta .$$

It is of interest to inquire as to what amount of information is required in order to determine the antiexact (k-1)-form ϕ . Since ϕ is given by the second of (5-9.16), (5-3.1) and (5-2.5) yield

$$\phi = \int_0^1 X \rfloor e^{-\tilde{\gamma}(\lambda)} \tilde{\Omega}(\lambda)\ \lambda^{k-1}\ d\lambda$$

$$= \int_0^1 e^{-\tilde{\gamma}(\lambda)} \widetilde{(X \rfloor \Omega)}(\lambda)\ \lambda^{k-2}\ d\lambda\ .$$

It thus follows that specification of the (k-1)-form

$$\xi = X \rfloor \Omega$$

on S served to determine ϕ uniquely; that is

$$\phi = \int_0^1 e^{-\tilde{\gamma}(\lambda)} \tilde{\xi}(\lambda)\ \lambda^{k-2}\ d\lambda\ .$$

Let us now go back to the staring point, namely

$$\text{(5-9.20)} \qquad d\Omega = \Gamma \wedge \Omega + \Sigma\ .$$

Theorem 5-9.1 shows that for given $\Gamma = d\gamma + \theta$, $\gamma = H(\Gamma)$, $\theta = H(d\Gamma)$, we can determine families of quantities Ω and Σ that satisfy (5-9.20). There is, however, a significant simplification that can be effected by noting that the sum $\Gamma \wedge \Omega + \Sigma$ does not determine either Γ or Σ uniquely. Thus, if we substitute $\Gamma = d\gamma + \theta$ into (5-9.20), we obtain

$$\text{(5-9.21)} \qquad d\Omega = d\gamma \wedge \Omega + \Sigma'$$

with

$$\text{(5-9.22)} \qquad \Sigma' = \Sigma + \theta \wedge \Omega\ .$$

Now, (5-9.17) and (5-9.18) show that (5-9.22) yields

$$\text{(5-9.23)} \qquad \Sigma' = e^{\gamma} d(\eta + H(\theta \wedge d\phi))\ ,$$

and hence

$$\text{(5-9.24)} \qquad d\Sigma' = d\gamma \wedge \Sigma'\ .$$

The original system

$$d\Omega = \Gamma \wedge \Omega + \Sigma , \quad d\Sigma = \Gamma \wedge \Sigma - \Theta \wedge \Omega ,$$

$$d\Gamma = \Theta , \quad d\Theta = 0 ,$$

is thus equivalent to the system

$$d\Omega = d\gamma \wedge \Omega + \Sigma' , \quad d\Sigma' = d\gamma \wedge \Sigma' , \quad \gamma = H(\Gamma) .$$

Application of Theorem 5-9.1 to this equivalent system gives the following result.

THEOREM 5-9.2. *Any system of exterior equations*

(5-9.25) $$d\Omega = \Gamma \wedge \Omega + \Sigma , \quad d\Sigma = \Gamma \wedge \Sigma - \Theta \wedge \Omega ,$$

(5-9.26) $$d\Gamma = \Theta , \quad d\Theta = 0 ,$$

on S *is equivalent to the system*

(5-9.27) $$d\Omega = d\gamma \wedge \Omega + \Sigma' , \quad d\Sigma' = d\gamma \wedge \Sigma' , \quad \gamma = H(\Gamma)$$

and this system has the general solution

(5-9.28) $$\Omega = e^{\gamma}(d\phi + \eta) , \quad \Sigma' = e^{\gamma} d\eta$$

with $\Sigma' = \Sigma + H(d\Gamma) \wedge \Omega$

(5-9.29) $$\phi = H(e^{-\gamma}\Omega) , \quad \eta = H(e^{-\gamma}\Sigma') = Hd(e^{-\gamma}\Omega)$$

belonging to A(S) .

It may have seemed to be a waste of time to solve the original system, when the equivalent system (5-9.27) is so much simpler. There are instances, however, in which the solution to the original system is required in terms of the original forms Γ and Σ . This is particularly true when dealing with several Ω's, as in the case in later sections.

A class of exterior forms that figures heavily in applications consists of those for which $d\Omega = \Gamma \wedge \Omega$.

Definition 5-9.1. An exterior form Ω of degree k is said to be *recursive* with *coefficient form* Γ if Ω satisfies

$$(5\text{-}9.30) \qquad d\Omega = \Gamma \wedge \Omega , \quad \Theta \wedge \Omega = 0 , \quad d\Gamma = \Theta .$$

Theorem 5-9.1 then leads to the following immediate conclusion.

THEOREM 5-9.3. *If* Ω *is a recursive exterior form with coefficient form* Γ *then*

$$(5\text{-}9.31) \qquad \Omega = e^{\gamma}\{d\phi + H(\theta \wedge d\phi)\}$$

with

$$(5\text{-}9.32) \qquad \gamma = H(\Gamma) , \quad \theta = H(d\Gamma) , \quad \phi = H(e^{-\gamma}\Omega) ,$$

and θ *must satisfy*

$$(5\text{-}9.33) \qquad d\theta \wedge \{d\phi + H(\theta \wedge d\phi)\} = 0 .$$

This result is awkward, in view of the condition (5-9.33) that results from the requirement $\Theta \wedge \Omega = d\Gamma \wedge \omega = 0$. This motivates consideration of a special class of recursive forms.

Definition 5-9.2. A recursive form Ω is said to be *gradient recursive* if the coefficient Γ is exact.

Theorem 5-9.3 then gives the following immediate result.

THEOREM 5-9.4. *If* Ω *is a gradient recursive* k-*form with coefficient form* $d\gamma$, *then*

$$(5\text{-}9.34) \qquad \Omega = e^{\gamma}\, d\phi$$

with $\phi = H(e^{-\gamma}\Omega)$.

We observe from this result that $d\phi = e^{-\gamma}\Omega$, from which we have the following conclusion.

COROLLARY 5-9.1. *A k-form Ω is gradient recursive if and only if there exists a scalar-valued function γ such that $e^{-\gamma}\Omega$ is exact.*

Suppose that we are given a $\Gamma \varepsilon \Lambda^1(S)$ and a $\Sigma \varepsilon \Lambda^{k+1}(S)$ and we wish to determine all $\Omega \varepsilon \Lambda^k(S)$ such that

$$(5\text{-}9.35) \qquad d\Omega = \Gamma \wedge \Omega + \Sigma \; , \quad \Gamma \neq 0 \; .$$

It may seem, from Theorem 5-9.1, that we have to check that $d\Sigma = \Gamma \wedge \Sigma - \Theta \wedge \Omega$ is satisfied in order to make use of the solution given by (5-9.17) through (5-9.19). This is not the case as we now show. Since $\Gamma \neq 0$ is given, the first of (5-9.16) yields $\gamma = H(\Gamma)$, and (5-9.19) gives $\theta = \Gamma - d\gamma$. The third of (5-9.16) then gives $\eta = H(e^{-\gamma}\Sigma)$. If we now use these γ, θ and η in (5-9.17) through (5-9.19), we obtain a set Ω, Σ, Γ, Θ, for each choice of the (k-1)-form ϕ, that identically satisfies the system (5-9.14), (5-9.15). This is easily verified by a direct calculation based on (5-9.17)-(5-9.19). In effect, what happens is that (5-9.17)-(5-9.19) constitutes a solution of (5-9.35) that satisfies the second half of (5-9.14), namely $\Theta \wedge \Omega = \Gamma \wedge \Sigma - d\Sigma$, as a constraint.

5-10. CANONICAL FORMS FOR UNCLOSED 2-FORMS

We have seen that the classic theorem of Darboux provides a canonical form for any closed 2-form (see also [1,4,6]). Specifically, let $\pi(x)$ be a closed 2-form on an open region S of a manifold M, and define the *rank* of $\pi(x)$ on S to be the even integer $2p$ such that $(\pi(x))^{(p)} \neq 0$ for some x in S and $(\pi(x))^{(p+1)} = 0$ for all x in S. Here, $(\pi(x))^{(q)}$ denotes the qth exterior power of $\pi(x)$. The points $x \varepsilon S$ for which $(\pi(x))^{(p)} \neq 0$ are *regular points* of $\pi(x)$, while those points for which $(\pi(x))^{(p)} = 0$ are referred to as *critical points* of $\pi(x)$. The Darboux theorem establishes the existence of $2p$

scalar-valued functions $u_i(x)$, $i = 1,\ldots,2p$, that are linearly independent at all regular points of $\pi(x)$ and such

$$\pi(x) = \sum_{j=1}^{p} du_j(x) \wedge du_{j+p}(x) = d\left(\sum_{j=1}^{p} u_j(x)\, du_{j+p}(x)\right)$$

on S .

The question naturally arises as to what happens when the given 2-form is not closed. An answer to this question can be provided almost directly by use of Theorem 5-9.1. Let Ω be the given 2-form and assume that $d\Omega \neq 0$ on S . Starting with this given Ω , we can always construct the differential system

$$\text{(5-10.1)}\qquad \begin{aligned} d\Omega &= \Gamma \wedge \Omega + \Sigma\ , \quad d\Sigma = \Gamma \wedge \Sigma - \Theta \wedge \Omega\ , \\ d\Gamma &= \Theta\ , \quad d\Theta = 0\ , \end{aligned}$$

and thereby determine a 1-form Γ and the 3-form Σ , although not necessarily uniquely. Theorem 5-9.1 then gives

$$\text{(5-10.2)}\qquad \Omega = e^{\gamma}(d\phi + \eta + H(\theta \wedge d\phi))$$

where $\gamma \in A^0(S)$, $\phi \in A^1(S)$, $\eta \in A^2(S)$, $\theta \in A^1(S)$ are determined by

$$\text{(5-10.3)}\qquad \gamma = H(\Gamma),\quad \phi = H(e^{-\gamma}\Omega),\quad \eta = H(e^{-\gamma}\Sigma),\quad \theta = H(d\Gamma)\ .$$

Now, $d\phi$ is a closed 2-form to which the Darboux theorem is applicable, and we have the following result.

<u>THEOREM 5-10.1</u>. *Let* Ω *be a given 2-form on* S *with the differential system*

$$\text{(5-10.4)}\qquad d\Omega = \Gamma \wedge \Omega + \Sigma\ , \quad d\Sigma = \Gamma \wedge \Sigma - \Theta \wedge \Omega\ ,$$

and $d\Gamma = \Theta$, $d\Theta = 0$. *The 2-form* Ω *admits the canonical representation*

$$\text{(5-10.5)}\qquad \Omega = e^{\gamma}(\pi + \eta + H(\theta \wedge \pi))\ ,$$

with

$$(5\text{-}10.6)\qquad \gamma = H(\Gamma)\ ,\quad \theta = H(d\Gamma)\ ,\quad \eta = H(e^{-\gamma}\Sigma)\ ,$$

and

$$(5\text{-}10.7)\qquad \pi = \sum_{j=1}^{p} du_j \wedge du_{j+p}$$

where 2p *is the rank of* $dH(e^{-\gamma}\Omega)$ *on* S . *Further, the* 2p *functions* $\{u_i\}$ *are linearly independent on all regular points of* $dH(e^{-\gamma}\Omega)$.

Further results can be obtained from Theorem 5-10.1 by particularization of the structures of Γ and Σ .

COROLLARY 5-10.1. *If* Ω *is a 2-form such that*

$$(5\text{-}10.8)\qquad d\Omega = d\gamma \wedge \Omega + \Sigma\ ,$$

then

$$(5\text{-}10.9)\qquad \Omega = e^{\gamma}(\pi + \eta)\ ,\quad \Sigma = e^{\gamma}\, d\eta$$

with

$$(5\text{-}10.10)\qquad \eta = H(e^{-\gamma}\Sigma)$$

and

$$(5\text{-}10.11)\qquad \pi = \sum_{j=1}^{p} du_j \wedge du_{j+p}\ ,$$

where 2p *is the rank of* $dH(e^{-\gamma}\Omega)$ *on* S . *Further,* Σ *is a gradient recursive 3-form with coefficient form* $d\gamma$.

For example, the general solution of

$$(5\text{-}10.12)\qquad d\Omega = d(3x^2y^3) \wedge \Omega + y^2\, dx \wedge dy \wedge dz\ ,\quad \Omega \in \Lambda^2(E_3)$$

can be obtained through the identification $\Gamma = d(3x^2y^3)$, $\Sigma = y^2\, dx \wedge dy \wedge dz$. The considerations at the end of Section 10 yield the general solution

$$(5\text{-}10.13)\quad \Omega = e^{3x^2y^3}\left\{d\phi + \frac{1-e^{-3x^2y^3}}{15x^2y}\,(x\ dy\wedge dz - y\,dx\wedge dz + z\ dx\wedge dy)\right\}$$

for any $\phi \in \Lambda^1(E_3)$. However, $d\phi$ is a closed 2-form so that we may always write

$$d\phi = du(x,y,z)\wedge dv(x,y,z)$$

for any two scalar-valued functions $u(x,y,z)$ and $v(x,y,z)$.

5-11. DIFFERENTIAL SYSTEMS OF DEGREE k AND CLASS r

The next several sections study problems associated with solving simultaneous systems of exterior differential equations. The purpose of this section is to obtain a formulation of these problems.

A significant simplification obtains through introduction of matrix notation. Let $M_{r,s}(\Lambda^k)$ denote the collection of all r-by-s matrices of elements of Λ^k . If $\underset{\sim}{\Gamma} = ((\Gamma^p_q)) \in M_{a,b}(\Lambda^k)$ and $\underset{\sim}{\Omega} = ((\Omega^r_s)) \in M_{b,c}(\Lambda^m)$, then $\underset{\sim}{\Sigma} = \underset{\sim}{\Gamma}\wedge\underset{\sim}{\Omega} = ((\Sigma^p_s)) \in M_{a,c}(\Lambda^{k+m})$ is defined by

$$\Sigma^p_s = \sum_{q=1}^{b} \Gamma^p_q \wedge \Omega^q_s \ ;$$

that is, the matrix product with exterior multiplication. It follows readily from this definition that

$$(\underset{\sim}{\Gamma}\wedge\underset{\sim}{\Omega})^T = (-1)^{km}\,(\underset{\sim}{\Omega})^T\wedge(\underset{\sim}{\Gamma})^T \ ,$$

where the superior T denotes transpose and $\underset{\sim}{\Gamma}\in M_{a,b}(\Lambda^k)$, $\underset{\sim}{\Omega}\in M_{b,c}(\Lambda^m)$.

<u>Definition 5-11.1.</u> A collection of forms $\{\underset{\sim}{\Omega}, \underset{\sim}{\Gamma}, \underset{\sim}{\Sigma}\}$, with $\underset{\sim}{\Omega}\in M_{r,1}(\Lambda^k)$, $\underset{\sim}{\Gamma}\in M_{r,r}(\Lambda^1)$, $\underset{\sim}{\Sigma}\in M_{r,1}(\Lambda^{k+1})$ is said to form a *differential system of degree* k *and class* r on S if

(5-11.1) $$d\underset{\sim}{\Omega} = \underset{\sim}{\Gamma} \wedge \underset{\sim}{\Omega} + \underset{\sim}{\Sigma}$$

is satisfied throughout S .

An arbitrary collection $\{\underset{\sim}{\Omega}, \underset{\sim}{\Gamma}, \underset{\sim}{\Sigma}\}$ can not form a differential system on S , for the closure of the system (5-11.1) must likewise be satisfied throughout S . Exterior differentiation of (5-11.1) gives $\underset{\sim}{0} = d\underset{\sim}{\Gamma} \wedge \underset{\sim}{\Omega} - \underset{\sim}{\Gamma} \wedge (\underset{\sim}{\Gamma} \wedge \underset{\sim}{\Omega} + \underset{\sim}{\Sigma}) + d\underset{\sim}{\Sigma} = (d\underset{\sim}{\Gamma} - \underset{\sim}{\Gamma} \wedge \underset{\sim}{\Gamma}) \wedge \underset{\sim}{\Omega} - \underset{\sim}{\Gamma} \wedge \underset{\sim}{\Sigma} + d\underset{\sim}{\Sigma}$. Thus, if we define $\underset{\sim}{\Theta} \varepsilon M_{r,r}(\Lambda^2)$ by $d\underset{\sim}{\Gamma} = \underset{\sim}{\Gamma} \wedge \underset{\sim}{\Gamma} + \underset{\sim}{\Theta}$, we obtain the conditions $d\underset{\sim}{\Sigma} = \underset{\sim}{\Gamma} \wedge \underset{\sim}{\Sigma} - \underset{\sim}{\Theta} \wedge \underset{\sim}{\Omega}$. The subsidiary equations $d\underset{\sim}{\Gamma} = \underset{\sim}{\Gamma} \wedge \underset{\sim}{\Gamma} + \underset{\sim}{\Theta}$ yield the further closure conditions $d\underset{\sim}{\Theta} = \underset{\sim}{\Gamma} \wedge \underset{\sim}{\Theta} - \underset{\sim}{\Theta} \wedge \underset{\sim}{\Gamma}$. Since there are no further closure conditions, we have the following result.

THEOREM 5-11.1. *A collection* $\underset{\sim}{\Omega} \varepsilon M_{r,1}(\Lambda^k)$, $\underset{\sim}{\Gamma} \varepsilon M_{r,r}(\Lambda^1)$, $\underset{\sim}{\Sigma} \varepsilon M_{r,1}(\Lambda^{k+1})$ *forms a differential system of degree* k *and class* r *on* S *if and only if*

(5-11.2) $$d\underset{\sim}{\Omega} = \underset{\sim}{\Gamma} \wedge \underset{\sim}{\Omega} + \underset{\sim}{\Sigma}$$

(5-11.3) $$d\underset{\sim}{\Sigma} = \underset{\sim}{\Gamma} \wedge \underset{\sim}{\Sigma} - \underset{\sim}{\Theta} \wedge \underset{\sim}{\Omega}$$

(5-11.4) $$d\underset{\sim}{\Gamma} = \underset{\sim}{\Gamma} \wedge \underset{\sim}{\Gamma} + \underset{\sim}{\Theta}$$

(5-11.5) $$d\underset{\sim}{\Theta} = \underset{\sim}{\Gamma} \wedge \underset{\sim}{\Theta} - \underset{\sim}{\Theta} \wedge \underset{\sim}{\Gamma}$$

are satisfied throughout S *for some* $\underset{\sim}{\Theta} \varepsilon M_{r,r}(\Lambda^2)$.

For $r = n = \dim(S)$ and $k = 1$, the equations (5-11.2) through (5-11.5) are just the equations of structure of E. Cartan and their closure equations if the n entries of $\underset{\sim}{\Omega}$ are linearly independent. Further, if $r = n$, the system (5-11.4), (5-11.5) constitutes the second half of the Cartan structure equations irrespective of the value of k . Differential systems are thus of intrinsic interest in regard to the structure of the region S of the manifold. On the other hand, we take particular note of the fact that a differential system makes no demands concerning the linear independence of the entries that comprise $\underset{\sim}{\Omega}$ or the degree of these entries. The study of differential systems of degree k will thus afford a foundation for a structural analysis of the region S relative to its forms of degree k , for general k . Differential systems are also of importance in the study of the intrinsic structure of a collection of forms of equal degree, for any such collection of forms can be used to construct the matrix $\underset{\sim}{\Omega}$ and exterior differentiation then leads to a differential system.

The analogy with the Cartan structure equations, is, however, a useful one in the general context of differential systems. We accordingly give the following definitions in relation to equations (5-11.2) through (5-11.5) which we refer to as the "differential system (*)".

Definition 5-11.2. The entries of $\underset{\sim}{\Gamma}$ are the *connection* 1-forms of the differential system (*) and the equations

$$d\underset{\sim}{\Gamma} = \underset{\sim}{\Gamma} \wedge \underset{\sim}{\Gamma} + \underset{\sim}{\Theta}$$

are the *connection equations*.

Definition 5-11.3. The entries of $\underset{\sim}{\Theta}$ are the *curvature* 2-forms of the differential system (*) and the equations

$$d\underset{\sim}{\Theta} = \underset{\sim}{\Gamma} \wedge \underset{\sim}{\Theta} - \underset{\sim}{\Theta} \wedge \underset{\sim}{\Gamma}$$

are the *curvature equations*.

Definition 5-11.4. The entries of $\underset{\sim}{\Sigma}$ are the *torsion* (k+1)-forms of the differential system (*) and the equations

$$d\underset{\sim}{\Sigma} = \underset{\sim}{\Gamma} \wedge \underset{\sim}{\Sigma} - \underset{\sim}{\Theta} \wedge \underset{\sim}{\Omega}$$

are the *torsion equations*.

5-12. INTEGRATION OF THE CONNECTION EQUATIONS

We start the analysis of the differential system (*) by studying the connection equation

$$(5\text{-}12.1) \qquad d\underset{\sim}{\Gamma} = \underset{\sim}{\Gamma} \wedge \underset{\sim}{\Gamma} + \underset{\sim}{\Theta} .$$

Use of the representation Theorem 5-6.1 allows us to write

$$(5\text{-}12.2) \qquad \underset{\sim}{\Gamma} = \underset{\sim}{\Gamma}_e + \underset{\sim}{\Gamma}_a , \quad \underset{\sim}{\Gamma}_e = d\underset{\sim}{\gamma} ,$$

where

$$(5\text{-}12.3) \qquad \underset{\sim}{\gamma} = H(\underset{\sim}{\Gamma})$$

belongs to $A^0(S) = \Lambda^0(S)$ with $\underset{\sim}{\gamma}(x_0^i) = \underset{\sim}{0}$. A substitution of (5-12.2) into (5-12.1) gives us

$$(5\text{-}12.4) \qquad d\underset{\sim}{\Gamma}_a = d\underset{\sim}{\gamma} \wedge d\underset{\sim}{\gamma} + d\underset{\sim}{\gamma} \wedge \underset{\sim}{\Gamma}_a + \underset{\sim}{\Gamma}_a \wedge d\underset{\sim}{\gamma} + \underset{\sim}{\Theta} .$$

If $\underset{\sim}{A} \in M_{r,r}(\Lambda^0)$ is a nonsingular matrix on S , we can change variables by the substitution

$$(5\text{-}12.5) \qquad \underset{\sim}{\Gamma}_a = \underset{\sim}{A}\, \bar{\underset{\sim}{\Gamma}}\, \underset{\sim}{A}^{-1} , \quad \underset{\sim}{\Theta} = \underset{\sim}{A}\, \bar{\underset{\sim}{\Theta}}\, \underset{\sim}{A}^{-1}$$

with $\bar{\underset{\sim}{\Gamma}} \in M_{r,r}(A^1)$, due to the module property of antiexact forms. In this event, (5-12.4) becomes

$$(5\text{-}12.6) \qquad d\bar{\underset{\sim}{\Gamma}} = \underset{\sim}{A}^{-1}(d\underset{\sim}{\gamma}\, \underset{\sim}{A} - d\underset{\sim}{A}) \wedge \bar{\underset{\sim}{\Gamma}} + \bar{\underset{\sim}{\Gamma}} \wedge \underset{\sim}{A}^{-1}(d\underset{\sim}{\gamma}\, \underset{\sim}{A} - d\underset{\sim}{A})$$
$$+ \underset{\sim}{A}^{-1} d\underset{\sim}{\gamma} \wedge d\underset{\sim}{\gamma}\, \underset{\sim}{A} + \bar{\underset{\sim}{\Theta}} .$$

Since $\bar{\underset{\sim}{\Gamma}} \in M_{r,r}(A^1)$, $\bar{\underset{\sim}{\Gamma}} = H(d\bar{\underset{\sim}{\Gamma}})$ by property A_4 . A significant simplification can thus be achieved in the system (5-12.6) if we can choose $\underset{\sim}{A}$ so that $d\underset{\sim}{\gamma}\, \underset{\sim}{A} - d\underset{\sim}{A}$ belongs to $M_{r,r}(A^1)$, for in that event, $(d\underset{\sim}{\gamma}\, \underset{\sim}{A} - d\underset{\sim}{A}) \wedge \bar{\underset{\sim}{\Gamma}}$ and $\bar{\underset{\sim}{\Gamma}} \wedge (d\underset{\sim}{\gamma}\, \underset{\sim}{A} - d\underset{\sim}{A})$ belong to $M_{r,r}(A^2)$ by property A_2 . We therefore consider the system

$$(5\text{-}12.7) \qquad d\underset{\sim}{A} = d\underset{\sim}{\gamma}\,\underset{\sim}{A} - \underset{\sim}{A}\,\underset{\sim}{\mu}\,, \quad \underset{\sim}{\mu} \in M_{r,r}(A^1)\,.$$

Exterior differentiation of (5-12.7) yields the closure conditions

$$(5\text{-}12.8) \qquad d\underset{\sim}{\mu} = -\underset{\sim}{A}^{-1}\,d\underset{\sim}{\gamma}\wedge d\underset{\sim}{\gamma}\,\underset{\sim}{A} + \underset{\sim}{\mu}\wedge\underset{\sim}{\mu}\,.$$

Since $\underset{\sim}{\mu} \in M_{r,r}(A^1)$, we have $\underset{\sim}{\mu} = H(d\underset{\sim}{\mu})$ and $H(\underset{\sim}{\mu}\wedge\underset{\sim}{\mu}) = \underset{\sim}{0}$. Thus, (5-12.8) yields

$$(5\text{-}12.9) \qquad \underset{\sim}{\mu} = -H(\underset{\sim}{A}^{-1}\,d\underset{\sim}{\gamma}\wedge d\underset{\sim}{\gamma}\,\underset{\sim}{A})\,.$$

When this is substituted into (5-12.7), we accordingly obtain

$$(5\text{-}12.10) \qquad d\underset{\sim}{A} = d\underset{\sim}{\gamma}\,\underset{\sim}{A} + \underset{\sim}{A}\,H(\underset{\sim}{A}^{-1}\,d\underset{\sim}{\gamma}\wedge d\underset{\sim}{\gamma}\,\underset{\sim}{A})$$

and (5-12.6), (5-12.7) and (5-12.8) yield

$$(5\text{-}12.11) \qquad d\underset{\sim}{\bar{\Gamma}} = \underset{\sim}{\mu}\wedge\underset{\sim}{\bar{\Gamma}} + \underset{\sim}{\bar{\Gamma}}\wedge\underset{\sim}{\mu} - d\underset{\sim}{\mu} + \underset{\sim}{\mu}\wedge\underset{\sim}{\mu} + \underset{\sim}{\bar{\Theta}}\,.$$

If we now use the facts that H inverts d on $A(S)$ and that every element of $A(S)$ belongs to the kernel of H, (5-12.11) yields $\underset{\sim}{\bar{\Gamma}} = -\underset{\sim}{\mu} + H(\underset{\sim}{\bar{\Theta}})$. Thus, (5-12.2), (5-12.5) and (5-12.11) give $\underset{\sim}{\Gamma} = d\underset{\sim}{\gamma} + \underset{\sim}{A}\,\underset{\sim}{\bar{\Gamma}}\,\underset{\sim}{A}^{-1} = d\underset{\sim}{\gamma} - \underset{\sim}{A}\,\underset{\sim}{\mu}\,\underset{\sim}{A}^{-1} + \underset{\sim}{A}\,H(\underset{\sim}{\bar{\Theta}})\,\underset{\sim}{A}^{-1}$. An elimination of the term $\underset{\sim}{A}\,\underset{\sim}{\mu}$ between this result and equation (5-12.7) then results in $\underset{\sim}{\Gamma} = (d\underset{\sim}{A} + \underset{\sim}{A}\,H(\underset{\sim}{\bar{\Theta}}))\underset{\sim}{A}^{-1}$. Since $\underset{\sim}{A} \in M_{r,r}(A^0)$, $\underset{\sim}{A} = \underset{\sim}{A}(x_0^i) + H(d\underset{\sim}{A})$, and (5-12.10) yields $\underset{\sim}{A} = \underset{\sim}{A}_0 + H(d\underset{\sim}{\gamma}\,\underset{\sim}{A})$. We have thus established the following results.

<u>THEOREM 5-12.1</u>. *The general solution of the connection equations,*

$$d\underset{\sim}{\Gamma} = \underset{\sim}{\Gamma}\wedge\underset{\sim}{\Gamma} + \underset{\sim}{\Theta}\,,$$

is given in terms of $\underset{\sim}{\Theta}$ *and* $\underset{\sim}{\gamma} = H(\underset{\sim}{\Gamma})$ *by*

$$(5\text{-}12.12) \qquad \underset{\sim}{\Gamma} = (d\underset{\sim}{A} + \underset{\sim}{A}H(\underset{\sim}{\bar{\Theta}}))\,\underset{\sim}{A}^{-1}\,, \quad \underset{\sim}{\bar{\Theta}} = \underset{\sim}{A}^{-1}\,\underset{\sim}{\Theta}\,\underset{\sim}{A}$$

in the region $D \subset S$ *where the matrix integral equation*

$$(5\text{-}12.13) \qquad \underset{\sim}{A} = \underset{\sim o}{A} + H(d\underset{\sim}{\gamma}\, \underset{\sim}{A}) , \quad \underset{\sim}{\gamma} = H(\underset{\sim}{\Gamma})$$

possesses a nonsingular solution.

5-13. THE ATTITUDE MATRIX: EXISTENCE AND UNIQUENESS

The results established in Theorem 5-12.1 show that everything hinges on the question of the solvability of the matrix integral equation $\underset{\sim}{A} = \underset{\sim o}{A} + H(d\underset{\sim}{\gamma}\, \underset{\sim}{A})$ and the domain D over which the solution is nonsingular. We give the following definition in analogy with the underlying theory of the Cartan structure equations.

<u>Definition 5-13.1.</u> If a matrix $\underset{\sim}{A} \in M_{r,r}(\Lambda^o)$ satisfies the matrix integral equation

$$(5\text{-}13.1) \qquad \underset{\sim}{A}(x) = \underset{\sim}{A}(x_o) + H(d\underset{\sim}{\gamma}\, \underset{\sim}{A})(x) , \quad \underset{\sim}{\gamma} = H(\underset{\sim}{\Gamma})$$

on a domain $D \subset S$ and $\det(\underset{\sim}{A}) \neq 0$ throughout D, then $\underset{\sim}{A}$ is an *attitude* matrix of the connection $\underset{\sim}{\Gamma}$ and D is a *domain of regularity* of $\underset{\sim}{A}$.

The first thing to do is to establish existence and uniqueness of the solution of (5-13.1) for given $\underset{\sim}{A}(x_o) = \underset{\sim o}{A}$. To this end, introduce the following uniform norm of indexed functions on a set U by

$$\|\{P^{j\dots}_{i\dots}\}\|_U = \sum_{i\dots} \sum_{j\dots} \sup_{x \in \bar{U}} |P^{j\dots}_{i\dots}(x)|$$

where $\bar{U}$ denotes the closure of U, so that

$$\|\underset{\sim}{A}\|_U = \sum_{i,j} \sup_{x \in \bar{U}} |A^i_j(x)| .$$

If we introduce the open ball $B_b = \{x \mid \max_i |x^i - x^i_o| < b\}$, then

$$G(\underset{\sim o}{A};R) = \{\underset{\sim}{A} \in M_{r,r}(\Lambda^o) \mid \|\underset{\sim}{A} - \underset{\sim o}{A}\|_{\bar{B}_b} \le R\}$$

is a closed subset of the space of r^2-tuples of elements of

$\Lambda^o(B_b)$. $G(\underset{\sim}{A}_o;R)$ is thus a closed and complete subset of the Banach space $B(r^2;\bar{B}_b)$ of r^2-tuples of continuous functions on $\bar{B}_b$ with the uniform norm $||\underset{\sim}{A}||_{\bar{B}_b}$. Define the operator J and L on $B(r^2;\bar{B}_b)$ by

$$(5\text{-}13.2)\qquad J\underset{\sim}{A} = \underset{\sim}{A}_o + L\underset{\sim}{A}\ ,\quad L\underset{\sim}{A} = H(d\underset{\sim}{\gamma}\ \underset{\sim}{A})\ ,$$

then L is a linear operator on $B(r^2;\bar{B}_b)$ and both J and L map $B(r^2;\bar{B}_b)$ into itself because $d\underset{\sim}{\gamma}\,\varepsilon\,\Lambda^1(S)$ so that its entries are 1-forms with continuous coefficients.

Since $\underset{\sim}{\gamma} = H(\underset{\sim}{\Gamma})$ is given, we can write

$$d\underset{\sim}{\gamma} = ((\eta^i_{j\ell}(x)dx^\ell))\ ,$$

in which case (5-13.2) and (5-3.2) give

$$(5\text{-}13.3)\quad LA^i_j(x^m) = \int_0^1 (x^\ell - x^\ell_o)\eta^i_{k\ell}(x^m_o+\lambda(x^m-x^m_o))A^k_j(x^m_o+\lambda(x^m-x^m_o))d\lambda\ .$$

A straightforward sequence of manipulations yields

$$||L\underset{\sim}{A}||_{\bar{B}_b} \leq nb \sum_{i,k,\ell} \max_{x\varepsilon\bar{B}_b} \int_0^1 |\eta^i_{k\ell}(x^m_o+\lambda(x^m-x^m_o))|d\lambda\ ||\underset{\sim}{A}||_{\bar{B}_b}\ ,$$

and hence, if U is any closed set in M that contains $\bar{B}_b$,

$$(5\text{-}13.4)\qquad ||L\underset{\sim}{A}||_{\bar{B}_b} \leq nb||\{\eta^i_{k\ell}\}||_U\ ||\underset{\sim}{A}||_{\bar{B}_b}\ .$$

Now, (5-13.2) gives $J\underset{\sim}{C} - J\underset{\sim}{A} = L(\underset{\sim}{C}-\underset{\sim}{A})$, and hence (5-13.4) yields

$$||J\underset{\sim}{C} - J\underset{\sim}{A}||_{\bar{B}_b} \leq nb||\{\eta^i_{k\ell}\}||_U\ ||\underset{\sim}{C}-\underset{\sim}{A}||_{\bar{B}_b}\ .$$

J is thus a contraction mapping on $B(r^2;\bar{B}_b)$ provided b is chosen so that $nb||\{\eta^i_{k\ell}\}||_U < 1$. For simplicity, we choose

$$(5\text{-}13.5)\qquad b^{-1} = 2n||\{\eta^i_{k\ell}\}||_U\ ,$$

so that

$$(5\text{-}13.6)\qquad ||J\underset{\sim}{C} - J\underset{\sim}{A}||_{\bar{B}_b} \leq c\ ||\underset{\sim}{C} - \underset{\sim}{A}||_{\bar{B}_b}\ ,\quad c = \frac{1}{2}\ ,$$

and the contraction constant is $1/2$. Further, (5-13.2), (5-13.4) and (5-13.5) combine to give

$$(5\text{-}13.7) \qquad ||J\underset{\sim}{A}_o - \underset{\sim}{A}_o||_{\bar{B}_b} = ||L\underset{\sim}{A}_o||_{\bar{B}_b} \leq \frac{1}{2}||\underset{\sim}{A}_o||_{\bar{B}_b} .$$

Thus, J moves the center $\underset{\sim}{A}_o$ of the closed ball $G(\underset{\sim}{A}_o;R)$ a distance at most $(1-c)R = R/2$ provided R is chosen so that $||\underset{\sim}{A}_o|| \leq R$. We therefore elect the option in the choice of R to take

$$(5\text{-}13.8) \qquad R = ||\underset{\sim}{A}_o||_{\bar{B}_b} .$$

Combining the above results, we have shown that J is a map of the Banach space $B(r^2;\bar{B}_b)$ that is a contraction with contraction constant $c = 1/2$ for $b^{-1} = 2n||\{\eta^i_{k\ell}\}||_U$ and moves the center of the closed ball $G(\underset{\sim}{A}_o, ||\underset{\sim}{A}_o||_{\bar{B}_b})$ a distance at most $(1-c)||\underset{\sim}{A}_o||_{\bar{B}_b} = \frac{1}{2}||\underset{\sim}{A}_o||_{\bar{B}_b}$. The Banach fixed point theorem then gives the following results.

LEMMA 5-13.1. *Let* U *be a given closed set in* M *that contains the starlike region* S. *The matrix integral equation* $\underset{\sim}{A} = \underset{\sim}{A}_o + H(d\underset{\sim}{\gamma}\,\underset{\sim}{A})$ *has a unique solution in the closed ball* $\bar{B}_b$, *for*

$$(5\text{-}13.9) \qquad b^{-1} = 2n||\{\eta^i_{k\ell}\}||_U , \quad d\underset{\sim}{\gamma} = ((\eta^i_{k\ell}\,dx^\ell))$$

and given $\underset{\sim}{A}_o$, *and this solution satisfies*

$$(5\text{-}13.10) \qquad ||\underset{\sim}{A} - \underset{\sim}{A}_o||_{\bar{B}_b} \leq ||\underset{\sim}{A}_o||_{\bar{B}_b} , \quad \underset{\sim}{A}(x^m_o) = \underset{\sim}{A}_o .$$

It is now a trivial matter to extend this solution throughout a starlike region S. Since S is starlike, any point $P:\{x^i\}$ in S can be connected to the point $P_o:\{x^i_o\}$ by the curve $u^i(\lambda) = x^i_o - \lambda(x^i - x^i_o)$ for $0 \leq \lambda \leq 1$. Further, this curve is of finite length since S is contained in a coordinate neighborhood D of P_o. A finite sequence of balls B_b can thus be constructed so that the center of each ball lies on the curve

$\{u^i(\lambda)\}$ and the center of each ball is contained in the previous ball of the sequence. Now, each of these balls is a starlike region whose center can be taken as the center of the ball. We can thus construct a homotopy operator on each ball with center at the center of the ball. Lemma 5-13.1 then implies the existence and uniqueness of the solution to $\underset{\sim}{A} = \underset{\sim}{A}_o + H(d\underset{\sim}{\gamma}\,\underset{\sim}{A})$ on each of these because b does not change with the location of the center of the ball, by (5-13.9), for a given $\underset{\sim}{A}_o$ on each of these balls. If we then choose $\underset{\sim}{A}_o$ of the ball in question to be equal to the values of the matrix $\underset{\sim}{A}$ from the previous ball that obtain from evaluation at the center of the ball in question, Lemma 5-7.1 on the change of centers then shows that the matrix $\underset{\sim}{A}$ can then be extended uniquely over such a finite sequence of balls because (5-13.10) gives $||\underset{\sim}{A}|| \leq 2^N ||\underset{\sim}{A}_o||$ for the Nth ball of such a sequence. Thus, the solution can be continued throughout S provided only that S be contained in some sufficiently large but compact set U of M .

<u>LEMMA 5-13.2</u>. *If* S *is a starlike region that is contained in a sufficiently large compact set* U *of* M , *then the matrix integral equation* $\underset{\sim}{A} = \underset{\sim}{A}_o + H(d\underset{\sim}{\gamma}\,\underset{\sim}{A})$ *has a unique solution on* S *that satisfies* $\underset{\sim}{A}(x_o^m) = \underset{\sim}{A}_o$.

This result provides the basis for the following existence and uniqueness theorem.

<u>THEOREM 5-13.1</u>. *Let* $\underset{\sim}{A}_o$ *be any constant element of* $M_{r,r}(\Lambda^o)$ *such that* $\det(\underset{\sim}{A}_o) \neq 0$. *Every connection matrix* $\underset{\sim}{\Gamma}$ *has a unique attitude matrix* $\underset{\sim}{A}$ *such that*

$$\text{(5-13.11)} \qquad \det(\underset{\sim}{A}(x)) = \det(\underset{\sim}{A}_o) \exp(\operatorname{tr} \underset{\sim}{\gamma}) ,$$

and the domain of regularity of $\underset{\sim}{A}$ *contains the starlike region* S .

<u>Proof</u>. Lemma 5-13.2 establishes the existence and uniqueness of

the matrix $\underset{\sim}{A}$, for given $\underset{\sim}{A}_o$, over the set U that contains S. Definition 5-13.1 then shows that this $\underset{\sim}{A}$ is an attitude matrix of $\underset{\sim}{\Gamma}$ with domain of regularity S provided we show that $\det(\underset{\sim}{A}) \neq 0$ on S. Let a denote the determinant of $\underset{\sim}{A}$ on U. Since $\underset{\sim}{A}$ satisfies $d\underset{\sim}{A} = d\underset{\sim}{\gamma}\, \underset{\sim}{A} - \underset{\sim}{A}\underset{\sim}{\mu}$ on U, we have

$$da = a[\mathrm{tr}(d\underset{\sim}{\gamma}) - \mathrm{tr}(\underset{\sim}{\mu})] ,$$

where $\mathrm{tr}(\cdot)$ denotes the trace of the corresponding matrix. If we define the function b by $b = a\exp(-\mathrm{tr}(\underset{\sim}{\gamma}))$, then b satisfies $db = -b\,\mathrm{tr}(\underset{\sim}{\mu})$. Now, $\underset{\sim}{\mu} \in M_{r,r}(\Lambda^1)$ implies $\mathrm{tr}(\underset{\sim}{\mu}) \in \Lambda^1$, and hence $b \in \Lambda^0$ implies that $b = b_o + Hdb = b_o + H(-b\,\mathrm{tr}(\underset{\sim}{\mu})) = b_o$. Accordingly, we obtain $a = b_o \exp(\mathrm{tr}(\underset{\sim}{\gamma}))$. Now $\underset{\sim}{\gamma} = H(\underset{\sim}{\Gamma})$ so that $\underset{\sim}{\gamma}(x_o) = \underset{\sim}{0}$, by property H_3, and hence $a = \det(\underset{\sim}{A}) = \det(\underset{\sim}{A}_o)\exp(\mathrm{tr}(\underset{\sim}{\gamma}))$; that is, $\det(\underset{\sim}{A}) \neq 0$ throughout S.

5-14. PROPERTIES OF ATTITUDE MATRICES

The results established in Sections 12 and 13 show that an attitude matrix, $\underset{\sim}{A}$, of a connection matrix $\underset{\sim}{\Gamma}$, satisfies

$$(5\text{-}14.1) \qquad d\underset{\sim}{A} = d\underset{\sim}{\gamma}\,\underset{\sim}{A} - \underset{\sim}{A}\,\underset{\sim}{\mu}$$

with $\underset{\sim}{\mu} \in M_{r,r}(\Lambda^1)$ and

$$(5\text{-}14.2) \qquad \underset{\sim}{\gamma} = H(\underset{\sim}{\Gamma}) ,$$

in which case

$$(5\text{-}14.3) \qquad \underset{\sim}{\Gamma} = (d\underset{\sim}{A} + \underset{\sim}{A}H(\bar{\underset{\sim}{\Theta}}))\underset{\sim}{A}^{-1} , \quad \bar{\underset{\sim}{\Theta}} = \underset{\sim}{A}^{-1}\,\underset{\sim}{\Theta}\,\underset{\sim}{A}$$

and $\underset{\sim}{A}$ is uniquely determined by

$$(5\text{-}14.4) \qquad \underset{\sim}{A} = \underset{\sim}{A}_o + H(d\underset{\sim}{\gamma}\,\underset{\sim}{A})$$

for a given, constant $\underset{\sim}{A}_o$ such that $\det(\underset{\sim}{A}_o) \neq 0$.

If we set $\underset{\sim}{A} = \underset{\sim}{B}\,\underset{\sim}{A}_o$, then (5-14.4) gives $\underset{\sim}{B}\,\underset{\sim}{A}_o = \underset{\sim}{A}_o +$

$H(d\gamma\, B\, A_o)$. Accordingly, since A_o is a constant matrix, $H(d\gamma\, B\, A_o) = H(d\gamma\, B) A_o$, we obtain

$$(5\text{-}14.5) \qquad B = E + H(d\gamma\, B) ,$$

where E is the r-by-r identity matrix. Further, if $A = B\, A_o$ is substituted into (5-14.3), we obtain

$$\Gamma = (dB\, A_o + B\, A_o\, H(A_o^{-1} B^{-1} \Theta\, B\, A_o)) A_o^{-1} B^{-1} = (dB + B\, H(B^{-1} \Theta\, B)) B^{-1} .$$

Accordingly, the connection Γ is independent of the choice of the constant matrix A_o such that $\det(A_o) \neq 0$, and we may thus assign A_o without loss of generality.

LEMMA 5-14.1. *A one-to-one correspondence can be established between connection matrices* Γ *and attitude matrices* A *by the assignment* $A_o = E$, *and* Γ *is invariant under the transformation* $A \longrightarrow A\, C_o$ *for any constant matrix* C_o *such that* $\det(C_o) \neq 0$. The consequence of this lemma is that it allows us to replace equations (5-14.4) by the equations

$$A = E + H(d\gamma\, A)$$

throughout the remainder of this discussion.

We now turn to the study of the relations between A and γ . Any given γ can always be written as

$$(5\text{-}14.6) \qquad \gamma = \frac{p}{n} E + R$$

with

$$(5\text{-}14.7) \qquad p = tr(\gamma), \quad tr(R) = 0 .$$

Since $\gamma = H(\Gamma)$, $\gamma(x_o^m) = 0$, and hence

$$(5\text{-}14.8) \qquad p(x_o^m) = 0 , \quad R(x_o^m) = 0 .$$

Now, set $A = e^{p/n} B$, in which case (5-14.8) yields $A_o = E = B_o$.

When this is substituted into (5-14.1), we obtain $d\underset{\sim}{B} = d\underset{\sim}{R}\,\underset{\sim}{B} - \underset{\sim}{B}\,\underset{\sim}{\mu}$, and hence $\underset{\sim}{B}$ satisfies $\underset{\sim}{B} = \underset{\sim}{E} + H(d\underset{\sim}{R}\,\underset{\sim}{B})$. Thus, since (5-13.11) and (5-14.7) give $\det(\underset{\sim}{B}(x)) = \det(\underset{\sim}{B}_o)\exp(\mathrm{tr}\,\underset{\sim}{R}) = 1$, we have established the following result.

LEMMA 5-14.2. *The decomposition*

(5-14.9) $$H(\underset{\sim}{\Gamma}) = \frac{p}{n}\underset{\sim}{E} + \underset{\sim}{R}\,, \quad p = H(\mathrm{tr}\,\underset{\sim}{\Gamma})\,, \quad \mathrm{tr}\,\underset{\sim}{R} = 0\,,$$

yields the attitude matrix

(5-14.10) $$\underset{\sim}{A} = \exp(p/n)\,\underset{\sim}{B}$$

with

(5-14.11) $$d\underset{\sim}{B} = d\underset{\sim}{R}\,\underset{\sim}{B} - \underset{\sim}{B}\,\underset{\sim}{\mu}\,, \quad \underset{\sim}{B} = \underset{\sim}{E} + H(d\underset{\sim}{R}\,\underset{\sim}{B})\,,$$

and $\underset{\sim}{B}$ *is a unitary matrix such that*

(5-14.12) $$\det(\underset{\sim}{B}) = 1\,.$$

The results established thus far can be placed in a more useful context in the following manner. The attitude matrix of any connection matrix is a nonsingular matrix valued function of position on S that is generated by $\underset{\sim}{A} = \underset{\sim}{E} + H(d\underset{\sim}{\gamma}\,\underset{\sim}{A})$, $\underset{\sim}{\gamma} = H(\underset{\sim}{\Gamma})$. Thus, if we take the standard representation of $GL(r,\mathbb{R})$ in terms of nonsingular r-by-r matrices, we may view $\underset{\sim}{A}$ as a mapping ρ_γ of S into $GL(r,\mathbb{R})$:

$$\rho_\gamma : S \longrightarrow GL(r,\mathbb{R}) \mid P_o \longrightarrow \underset{\sim}{E}\,.$$

The latter restriction follows from the condition $\underset{\sim}{A}(x_o^m) = \underset{\sim}{E}$. Now, define the maps h_p and u_R by

$$h_p : S \longrightarrow \exp(p/n)\underset{\sim}{E} \mid p(x_o^m) = 0\,,$$

$$u_R : S \longrightarrow SL(r,\mathbb{R}) \mid P_o \longrightarrow \underset{\sim}{E}\,,$$

so that h_p is a mapping into the homothetic matrices and u_R is a mapping into the unitary group. Lemmas 5-14.1 and 5-14.2 can thus be combined to yield the following results.

LEMMA 5-14.3. *The unique attitude matrix,* $\underset{\sim}{A}$, *of a connection matrix,* $\underset{\sim}{\Gamma}$, *is a map*

$$(5\text{-}14.13) \qquad \rho_\gamma : S \longrightarrow GL(r,\mathbb{R}) \mid P_o \longrightarrow \underset{\sim}{E}$$

that is generated by $\underset{\sim}{\gamma} = H(\underset{\sim}{\Gamma})$ *from the matrix integral equation* $\underset{\sim}{A} = \underset{\sim}{E} + H(d\underset{\sim}{\gamma}\,\underset{\sim}{A})$. *The decomposition*

$$(5\text{-}14.14) \qquad \underset{\sim}{\gamma} = \frac{p}{n}\underset{\sim}{E} + \underset{\sim}{R} , \quad p = H(\operatorname{tr} \underset{\sim}{\Gamma}) , \quad \operatorname{tr} \underset{\sim}{R} = 0$$

yields the decomposition $\rho_\gamma = h_p \times u_R$ *in terms of the matrix product, where*

$$(5\text{-}14.15) \qquad h_p : S \longrightarrow \exp(p/n)\,\underset{\sim}{E} \mid p(x_o^m) = 0 ,$$

$$(5\text{-}14.16) \qquad u_R : S \longrightarrow SL(r,\mathbb{R}) \mid P_o \longrightarrow \underset{\sim}{E} ,$$

and u_R *is generated by* $\underset{\sim}{R}$ *from the matrix integral equation* $\underset{\sim}{B} = \underset{\sim}{E} + H(d\underset{\sim}{R}\,\underset{\sim}{B})$.

This result suggests that the attitude matrix of a connection $\underset{\sim}{\Gamma}$ belongs to a matrix group if $\underset{\sim}{\gamma} = H(\underset{\sim}{\Gamma})$ is an infinitesimal generating matrix of that matrix group. This is indeed the case, as we now show.

Let $\underset{\sim}{J}$ be a nonsingluar constant element of $M_{r,r}$. There are basically two kinds of elementary matrix subgroups of $GL(r,\mathbb{R})$; those that satisfy $\underset{\sim}{F}^T\underset{\sim}{J}\underset{\sim}{F} = \underset{\sim}{J}$ and those that satisfy $\underset{\sim}{G}^{-1}\underset{\sim}{J}\underset{\sim}{G} = \underset{\sim}{J}$. In the first instance, the infinitesimal generating matrices $\underset{\sim}{f}$ satisfy $\underset{\sim}{f}^T\underset{\sim}{J} + \underset{\sim}{J}\,\underset{\sim}{f} = \underset{\sim}{0}$, and, in the second instance, the infinitesimal generating matrices satisfy $-\underset{\sim}{g}\,\underset{\sim}{J} + \underset{\sim}{J}\,\underset{\sim}{g} = \underset{\sim}{0}$.

We first suppose that the matrix $\underset{\sim}{R}$ in (5-14.14) satisfies

$$(5\text{-}14.17) \qquad \underset{\sim}{R}^T\underset{\sim}{J} + \underset{\sim}{J}\,\underset{\sim}{R} = \underset{\sim}{0} ,$$

in which case $tr(\underset{\sim}{R}) = 0$, $\underset{\sim}{B} = \underset{\sim}{E} + H(d\underset{\sim}{R}\,\underset{\sim}{B})$, and

$$(5\text{-}14.18) \qquad d\underset{\sim}{B} = d\underset{\sim}{R}\,\underset{\sim}{B} - \underset{\sim}{B}\,\underset{\sim}{\mu} .$$

Since (5-14.18) implies

$$d\underset{\sim}{B}^T = \underset{\sim}{B}^T\, d\underset{\sim}{R}^T - \underset{\sim}{\mu}^T\, \underset{\sim}{B}^T ,$$

it follows that

$$d(\underset{\sim}{B}^T\underset{\sim}{J}\underset{\sim}{B}) = d\underset{\sim}{B}^T\underset{\sim}{J}\underset{\sim}{B} + \underset{\sim}{B}^T\underset{\sim}{J}d\underset{\sim}{B} = \underset{\sim}{B}^T(d\underset{\sim}{R}^T\underset{\sim}{J} + \underset{\sim}{J}d\underset{\sim}{R})\underset{\sim}{B} - \underset{\sim}{\mu}^T\underset{\sim}{B}^T\underset{\sim}{J}\underset{\sim}{B} - \underset{\sim}{B}^T\underset{\sim}{J}\underset{\sim}{B}\underset{\sim}{\mu}$$

$$= -\underset{\sim}{\mu}^T\underset{\sim}{B}^T\underset{\sim}{J}\underset{\sim}{B} - \underset{\sim}{B}^T\underset{\sim}{J}\underset{\sim}{B}\underset{\sim}{\mu}$$

when (5-14.17) is used. Since $\underset{\sim}{\mu} \in M_{r,r}(A^1)$, the module property of antiexact forms shows that $\underset{\sim}{\mu}^T\underset{\sim}{B}^T\underset{\sim}{J}\underset{\sim}{B} + \underset{\sim}{B}^T\underset{\sim}{J}\underset{\sim}{B}\underset{\sim}{\mu}$ belongs to $ker(H)$. Thus, allowing H to act on both sides of this relation, property H_2 and $\underset{\sim}{B}_o = \underset{\sim}{E}$ yield

$$\underset{\sim}{B}^T\underset{\sim}{J}\underset{\sim}{B} = \underset{\sim}{B}_o^T\underset{\sim}{J}\underset{\sim}{B}_o = \underset{\sim}{J} .$$

LEMMA 5-14.5. *If the matrix* $\underset{\sim}{R}$ *in* (5-14.14) *satisfies*

$$(5\text{-}14.19) \qquad \underset{\sim}{R}^T\underset{\sim}{J} + \underset{\sim}{J}\underset{\sim}{R} = \underset{\sim}{0} ,$$

so that $\underset{\sim}{R}$ *is an infinitesimal generator of the matrix subgroup* $U(\underset{\sim}{J};r,\mathbb{R})$ *of the unitary group whose elements satisfy* $\underset{\sim}{F}^T\underset{\sim}{J}\underset{\sim}{F} = \underset{\sim}{J}$ *for* $\underset{\sim}{J}$ *a constant, nonsingular, r-by-r matrix, then the solution of* $\underset{\sim}{B} = \underset{\sim}{E} + H(d\underset{\sim}{R}\,\underset{\sim}{B})$ *belongs to* $U(\underset{\sim}{J};r,\mathbb{R})$ *at every point of* S *and* $\underset{\sim}{R}$ *generates the map*

$$(5\text{-}14.20) \qquad u_R(\underset{\sim}{J}) : S \longrightarrow U(\underset{\sim}{J};r,\mathbb{R}) \mid P_o \longrightarrow \underset{\sim}{E} .$$

COROLLARY 5-14.1. *If* $\underset{\sim}{R}$ *satisfies* $\underset{\sim}{R}^T + \underset{\sim}{R} = \underset{\sim}{0}$, *then* $\underset{\sim}{B}$ *is a proper orthogonal matrix at every point of* S .

COROLLARY 5-14.2. *If* $\underset{\sim}{R}$ *satisfies* $\underset{\sim}{R}^T\underset{\sim}{I} + \underset{\sim}{I}\underset{\sim}{R} = \underset{\sim}{0}$, *where*

$r = 2q$ *and* $\underset{\sim}{I}$ *is the fundamental sympletic matrix* $\begin{pmatrix} \underset{\sim}{0} & \underset{\sim}{E} \\ -\underset{\sim}{E} & \underset{\sim}{0} \end{pmatrix}$, *then* $\underset{\sim}{B}$ *is a proper symplectic matrix at every point of* S.

We now turn to the second alternative where $-\underset{\sim}{g}\underset{\sim}{J} + \underset{\sim}{J}\underset{\sim}{g} = \underset{\sim}{0}$. Since this equation places no constraint on the value of $tr(\underset{\sim}{g})$, we consider the situation wherein $-\underset{\sim}{\gamma}\underset{\sim}{J} + \underset{\sim}{J}\underset{\sim}{\gamma} = \underset{\sim}{0}$ for given, constant, nonsingular $\underset{\sim}{J}$. If we note that $d\underset{\sim}{A}^{-1} = -\underset{\sim}{A}^{-1} d\underset{\sim}{A}\, \underset{\sim}{A}^{-1}$, it follows that $d(\underset{\sim}{A}^{-1}\underset{\sim}{J}\underset{\sim}{A}) = \underset{\sim}{A}^{-1}(-d\underset{\sim}{R}\,\underset{\sim}{J} + \underset{\sim}{J}d\underset{\sim}{R})\underset{\sim}{A} + \underset{\sim}{\mu}\underset{\sim}{A}^{-1}\underset{\sim}{J}\underset{\sim}{A} - \underset{\sim}{A}^{-1}\underset{\sim}{J}\underset{\sim}{A}\underset{\sim}{\mu} = \underset{\sim}{\mu}\underset{\sim}{A}^{-1}\underset{\sim}{J}\underset{\sim}{A} - \underset{\sim}{A}^{-1}\underset{\sim}{J}\underset{\sim}{A}\underset{\sim}{\mu}$. An integration by means of the operator H then yields $\underset{\sim}{A}^{-1}\underset{\sim}{J}\underset{\sim}{A} = \underset{\sim}{A}_o^{-1}\underset{\sim}{J}\underset{\sim}{A}_o = \underset{\sim}{J}$.

LEMMA 5-14.6. *If* $\underset{\sim}{\gamma}$ *satisfies*

$$(5\text{-}14.21) \qquad -\underset{\sim}{\gamma}\underset{\sim}{J} + \underset{\sim}{J}\underset{\sim}{\gamma} = \underset{\sim}{0},$$

so that $\underset{\sim}{\gamma}$ *is an infinitesimal generator of the matrix subgroup* $JL(r,\mathbb{R})$ *of* $GL(r,\mathbb{R})$ *whose elements satisfy* $\underset{\sim}{G}^{-1}\underset{\sim}{J}\underset{\sim}{G} = \underset{\sim}{J}$, *for* $\underset{\sim}{J}$ *a constant, nonsingular, r-by-r matrix, then the solution of* $\underset{\sim}{A} = \underset{\sim}{E} + H(d\underset{\sim}{\gamma}\,\underset{\sim}{A})$ *belongs to* $JL(r,\mathbb{R})$ *at every point of* S *and* $\underset{\sim}{\gamma}$ *generates the map*

$$(5\text{-}14.22) \qquad J_{\gamma}: S \longrightarrow JL(r,\mathbb{R}) \mid P_o \longrightarrow \underset{\sim}{E}.$$

COROLLARY 5-14.3. *If* $\underset{\sim}{\gamma}$ *commutes with* $\underset{\sim}{J}$, *then* $\underset{\sim}{A}$ *commutes with* $\underset{\sim}{J}$.

The above results give a fairly complete characterization of the relation between $\underset{\sim}{\gamma}$ and $\underset{\sim}{A}$. In general, unless $\underset{\sim}{\gamma}$ or the trace-free part of $\underset{\sim}{\gamma}$ satisfies a relation of the form $\underset{\sim}{\gamma}^T\underset{\sim}{J} + \underset{\sim}{J}\underset{\sim}{\gamma} = \underset{\sim}{0}$ or $-\underset{\sim}{\gamma}\underset{\sim}{J} + \underset{\sim}{J}\underset{\sim}{\gamma} = \underset{\sim}{0}$, for some nonsingular $\underset{\sim}{J}$, all that can be concluded is that $\underset{\sim}{A}$ belongs to $GL(r,\mathbb{R})$ at every point of S.

LEMMA 5-14.7. *The matrix integral equation*

$$(5\text{-}14.23) \qquad \underset{\sim}{A} = \underset{\sim}{A}_o + H(d\underset{\sim}{\gamma}\,\underset{\sim}{A}), \quad \underset{\sim}{\gamma} = H(\underset{\sim}{\Gamma})$$

is equivalent to the matrix integral equation

(5-14.24) $\quad \underset{\sim}{A} = \underset{\sim}{A}_o + H(\underset{\sim}{\Gamma}\,\underset{\sim}{A})$.

Proof. This result follows directly from the fact that $\underset{\sim}{\Gamma} = d\underset{\sim}{\gamma} + H(d\underset{\sim}{\Gamma})$ and hence $H(\underset{\sim}{\Gamma}\,\underset{\sim}{A}) = H(d\underset{\sim}{\gamma}\,\underset{\sim}{A} + H(d\underset{\sim}{\Gamma})\,\underset{\sim}{A}) = H(d\underset{\sim}{\gamma}\,\underset{\sim}{A})$ since $H(d\underset{\sim}{\Gamma})\,\underset{\sim}{A} \varepsilon A^1(S)$ and hence belongs to the kernel of H .

LEMMA 5-14.8. *If* $\underset{\sim}{\gamma} = H(\underset{\sim}{\Gamma}) = \underset{\sim}{u} + \underset{\sim}{v}$, *then the attitude matrix,* $\underset{\sim}{A}$, *of* $\underset{\sim}{\Gamma}$ *is given by*

(5-14.25) $\quad \underset{\sim}{A} = \underset{\sim}{B}\,\underset{\sim}{C}$

where $\underset{\sim}{B}$ *and* $\underset{\sim}{C}$ *satisfy the matrix integral equations*

(5-14.26) $\quad \underset{\sim}{B} = \underset{\sim}{B}_o + H(d\underset{\sim}{u}\,\underset{\sim}{B})$,

(5-14.27) $\quad \underset{\sim}{C} = \underset{\sim}{C}_o + H(\underset{\sim}{B}^{-1}\,d\underset{\sim}{v}\,\underset{\sim}{B}\,\underset{\sim}{C})$

with $\underset{\sim}{A}_o = \underset{\sim}{B}_o\,\underset{\sim}{C}_o$.

Proof. Under the hypotheses, (5-14.1) yields $d\underset{\sim}{A} = (d\underset{\sim}{u} + d\underset{\sim}{v})\underset{\sim}{A} - \underset{\sim}{A}\underset{\sim}{\mu}$, and the substitution $\underset{\sim}{A} = \underset{\sim}{B}\,\underset{\sim}{C}$ yields $d\underset{\sim}{B} + \underset{\sim}{B}\,d\underset{\sim}{C}\,\underset{\sim}{C}^{-1} = d\underset{\sim}{u}\,\underset{\sim}{B} + d\underset{\sim}{v}\,\underset{\sim}{B} - \underset{\sim}{B}\,\underset{\sim}{C}\,\underset{\sim}{\mu}\,\underset{\sim}{C}^{-1}$. Now, with $\underset{\sim}{B}$ satisfying (5-14.26), we have $d\underset{\sim}{B} = d\underset{\sim}{u}\,\underset{\sim}{B} - \underset{\sim}{B}\,\underset{\sim}{\mu}_B$, and hence $\underset{\sim}{C}$ must satisfy $d\underset{\sim}{C} = (\underset{\sim}{B}^{-1}\,d\underset{\sim}{v}\,\underset{\sim}{B})\underset{\sim}{C} - (\underset{\sim}{C}\,\underset{\sim}{\mu} - \underset{\sim}{\mu}_B\,\underset{\sim}{C}^{-1})$. Thus, since $(\underset{\sim}{C}\,\underset{\sim}{\mu} - \underset{\sim}{\mu}_B\,\underset{\sim}{C}^{-1}) \varepsilon A^1(S)$, and hence to $\ker(H)$, $\underset{\sim}{C} = \underset{\sim}{C}_o + H(d\underset{\sim}{C})$ yields (5-14.27).

As an example, consider the case where $r = 2 < n$ and $\underset{\sim}{\gamma}$ is given by

$$\underset{\sim}{\gamma} = \begin{pmatrix} (u+v)/2 & f(v) \\ g(v) & (u-v)/2 \end{pmatrix} ,$$

where u and v are C^1 functions of $\{x^i\}$ that are functionally independent and

$$f(v) = c_1 + \int_o^v h(t)\,e^t\,dt ,$$

$$g(v) = c_2 - \int_0^v h(t)\, e^{-t}\, dt \, .$$

It is clear from (5-12.9) that significant simplification occurs if $d\underset{\sim}{\gamma} \wedge d\underset{\sim}{\gamma} = \underset{\sim}{0}$, as is the case here, for then $\underset{\sim}{\mu} = \underset{\sim}{0}$ and $\underset{\sim}{A}$ must satisfy the linear equation $d\underset{\sim}{A} = d\underset{\sim}{\gamma}\, \underset{\sim}{A}$. A straightforward, but somewhat lengthy calculation shows that

$$\underset{\sim}{A} = e^{(u+v)/2} \begin{pmatrix} \cos\beta & \sin\beta \\ -e^{-v}\sin\beta & e^{-v}\cos\beta \end{pmatrix}$$

with

$$\beta(v) = \int_0^v h(t)\, dt \, ,$$

where we have chosen the constants of integration such that $\underset{\sim}{A} = \underset{\sim}{E}$ at $u = v = 0$. In this example, $\underset{\sim}{\Gamma}$ is given by

$$\underset{\sim}{\Gamma} = \underset{\sim}{\Gamma}_a + \begin{pmatrix} (du+dv)/2 & h(v)e^{v}dv \\ -h(v)e^{-v}dv & (du-dv)/2 \end{pmatrix} .$$

5-15. INTEGRATION OF THE CURVATURE EQUATIONS

Theorem 5-12.1 gave the general solution of the connection equations as

(5-15.1) $$\underset{\sim}{\Gamma} = (d\underset{\sim}{A} + \underset{\sim}{A}\, H(\bar{\underset{\sim}{\Theta}}))\underset{\sim}{A}^{-1}$$

where $\underset{\sim}{A}$ is a solution of

(5-15.2) $$\underset{\sim}{A} = \underset{\sim}{A}_o + H(d\underset{\sim}{\gamma}\, \underset{\sim}{A}) \, , \quad \underset{\sim}{\gamma} = H(\underset{\sim}{\Gamma})$$

and $\bar{\underset{\sim}{\Theta}}$ is defined in terms of $\underset{\sim}{\Theta}$ and $\underset{\sim}{A}$ by

(5-15.3) $$\bar{\underset{\sim}{\Theta}} = \underset{\sim}{A}^{-1}\underset{\sim}{\Theta}\, \underset{\sim}{A} \, .$$

The quantities $\underset{\sim}{\Theta}$ still remain to be determined, however, for they must satisfy the curvature equations

$$(5\text{-}15.4)\qquad d\underset{\sim}{\Theta} = \underset{\sim}{\Gamma} \wedge \underset{\sim}{\Theta} - \underset{\sim}{\Theta} \wedge \underset{\sim}{\Gamma} .$$

We now proceed to solve (5-15.4).

Under the substitution $\underset{\sim}{\Theta} = \underset{\sim}{A}\, \underset{\sim}{\bar{\Theta}}\, \underset{\sim}{A}^{-1}$ that is the inverse of (5-15.3), the system (5-15.4) becomes

$$(5\text{-}15.5)\qquad d\underset{\sim}{\bar{\Theta}} = -\underset{\sim}{A}^{-1} d\underset{\sim}{A} \wedge \underset{\sim}{\bar{\Theta}} + \underset{\sim}{\bar{\Theta}} \wedge \underset{\sim}{A}^{-1} d\underset{\sim}{A} + \underset{\sim}{A}^{-1} (\underset{\sim}{\Gamma} \wedge \underset{\sim}{\Theta} - \underset{\sim}{\Theta} \wedge \underset{\sim}{\Gamma}) \underset{\sim}{A}$$

$$= H(\underset{\sim}{\bar{\Theta}}) \wedge \underset{\sim}{\bar{\Theta}} - \underset{\sim}{\bar{\Theta}} \wedge H(\underset{\sim}{\bar{\Theta}}) ,$$

where the last equality follows from (5-15.1). We now use the decomposition of forms into exact and antiexact parts to write

$$(5\text{-}15.6)\qquad \underset{\sim}{\bar{\Theta}} = d\underset{\sim}{\theta} + \underset{\sim}{\bar{\Theta}}_a , \quad \underset{\sim}{\theta} = H(\underset{\sim}{\bar{\Theta}}) \in A^1(S) ,$$

in which case (5-15.5) becomes

$$(5\text{-}15.7)\qquad d\underset{\sim}{\bar{\Theta}}_a = -d(\underset{\sim}{\theta} \wedge \underset{\sim}{\theta}) + \underset{\sim}{\theta} \wedge \underset{\sim}{\bar{\Theta}}_a - \underset{\sim}{\bar{\Theta}}_a \wedge \underset{\sim}{\theta} .$$

Since $\underset{\sim}{\theta} \wedge \underset{\sim}{\bar{\Theta}}_a$ and $\underset{\sim}{\bar{\Theta}}_a \wedge \underset{\sim}{\theta}$ belong to $A^3(S)$ by property A_2, and hence to the kernel of H, while H inverts d on $A(S)$ by property A_4, application of H to (5-15.7) yields

$$(5\text{-}15.8)\qquad \underset{\sim}{\bar{\Theta}}_a = -\underset{\sim}{\theta} \wedge \underset{\sim}{\theta} .$$

Accordingly, (5-15.6) yields

$$(5\text{-}15.9)\qquad \underset{\sim}{\bar{\Theta}} = d\underset{\sim}{\theta} - \underset{\sim}{\theta} \wedge \underset{\sim}{\theta} .$$

Combination of the above results now yields the following theorem.

THEOREM 5-15.1. *The general solution of the connection and the curvature equations of a differential system is given by*

$$(5\text{-}15.10)\qquad \underset{\sim}{\Gamma} = (d\underset{\sim}{A} + \underset{\sim}{A}\,\underset{\sim}{\theta})\underset{\sim}{A}^{-1} ,$$

$$(5\text{-}15.11)\qquad \underset{\sim}{\Theta} = \underset{\sim}{A}(d\underset{\sim}{\theta} - \underset{\sim}{\theta} \wedge \underset{\sim}{\theta})\underset{\sim}{A}^{-1} ,$$

where the matrix $\underset{\sim}{A}$ *and the matrix* $\underset{\sim}{\theta}$ *of antiexact* 1-*forms are given by*

$$(5\text{-}15.12)\qquad \underset{\sim}{A} = \underset{\sim o}{A} + H(d\underset{\sim}{\gamma}\,\underset{\sim}{A}),\quad \underset{\sim}{\gamma} = H(\underset{\sim}{\Gamma}),\quad \det(\underset{\sim o}{A}) \neq 0,$$

$$(5\text{-}15.13)\qquad \underset{\sim}{\theta} = H(\underset{\sim}{A}^{-1}\,\underset{\sim}{\Theta}\,\underset{\sim}{A}),$$

and $\underset{\sim}{\Gamma}$ *and* $\underset{\sim}{\Theta}$ *are independent of the choice of* $\underset{\sim o}{A}$.

It is clear from these results that $\underset{\sim}{\Gamma}$ determines $\underset{\sim}{\Theta}$ uniquely, for $\underset{\sim}{\Theta} = d\underset{\sim}{\Gamma} - \underset{\sim}{\Gamma}\wedge\underset{\sim}{\Gamma}$ from the connection equations and (5-15.10) gives $\underset{\sim}{\theta} = \underset{\sim}{A}^{-1}(\underset{\sim}{\Gamma}\,\underset{\sim}{A} - d\underset{\sim}{A})$. On the other hand, a given curvature matrix $\underset{\sim}{\Theta}$ can result from many connections.

As an example, suppose that $\underset{\sim}{U}\in\Lambda^k_{2,1}$ is to satisfy $d\underset{\sim}{U} = \underset{\sim}{\Gamma}\wedge\underset{\sim}{U}$ with

$$\underset{\sim}{\Gamma} = \begin{pmatrix} 0 & a \\ b & 0 \end{pmatrix} d\xi,\quad \xi\in\Lambda^0_{1,1},$$

then $d\underset{\sim}{\Gamma} = 0$, $\underset{\sim}{\Gamma}\wedge\underset{\sim}{\Gamma} = 0$, and hence $\underset{\sim}{\Theta} = \underset{\sim}{0}$, $\underset{\sim}{\Sigma} = \underset{\sim}{0}$. In this case

$$\underset{\sim}{A} = \exp\left[\begin{pmatrix} 0 & a \\ b & 0 \end{pmatrix}\xi\right]$$

satisfies $\underset{\sim}{A} = \underset{\sim}{E} + H(\underset{\sim}{\Gamma}\underset{\sim}{A})$; that is

$$\underset{\sim}{\Gamma} = (d\underset{\sim}{A})\underset{\sim}{A}^{-1}.$$

Thus

$$\underset{\sim}{U} = \exp\left[\begin{pmatrix} 0 & a \\ b & 0 \end{pmatrix}\xi\right] d\underset{\sim}{\phi}$$

with $\underset{\sim}{\phi}\in\Lambda^{k-1}_{2,1}$; i.e.,

$$\underset{\sim}{U} = \begin{pmatrix} \cosh(\xi\sqrt{ab}) & \dfrac{a}{\sqrt{ab}}\sinh(\xi\sqrt{ab}) \\ \dfrac{b}{\sqrt{ab}}\sinh(\xi\sqrt{ab}) & \cosh(\xi\sqrt{ab}) \end{pmatrix} d\underset{\sim}{\phi}.$$

5-16. INTEGRATION OF A DIFFERENTIAL SYSTEM

Now that we have solved the connection equations and the curvature equations, so as to obtain

$$(5\text{-}16.1)\qquad \underset{\sim}{\Gamma} = (d\underset{\sim}{A} + \underset{\sim}{A}\,\underset{\sim}{\theta})\underset{\sim}{A}^{-1}, \quad \underset{\sim}{A} = \underset{\sim o}{A} + H(\underset{\sim}{\Gamma}\underset{\sim}{A}),$$

$$(5\text{-}16.2)\qquad \underset{\sim}{\Theta} = \underset{\sim}{A}(d\underset{\sim}{\theta} - \underset{\sim}{\theta}\wedge\underset{\sim}{\theta})\underset{\sim}{A}^{-1}, \quad \underset{\sim}{\theta} = H(\underset{\sim}{A}^{-1}\underset{\sim}{\Theta}\,\underset{\sim}{A}),$$

the task outstanding consists of integrating the remaining equations

$$(5\text{-}16.3)\qquad d\underset{\sim}{\Omega} = \underset{\sim}{\Gamma}\wedge\underset{\sim}{\Omega} + \underset{\sim}{\Sigma}, \quad d\underset{\sim}{\Sigma} = \underset{\sim}{\Gamma}\wedge\underset{\sim}{\Sigma} - \underset{\sim}{\Theta}\wedge\underset{\sim}{\Omega}$$

of the differential system (5-11.2)-(5-11.5). Under the substitution

$$(5\text{-}16.4)\qquad \underset{\sim}{\Omega} = \underset{\sim}{A}\,\underset{\sim}{\omega}, \quad \underset{\sim}{\Sigma} = \underset{\sim}{A}\,\underset{\sim}{\sigma},$$

equations (5-16.1) through (5-16.3) combine to yield

$$(5\text{-}16.5)\qquad d\underset{\sim}{\omega} = \underset{\sim}{\theta}\wedge\underset{\sim}{\omega} + \underset{\sim}{\sigma}, \quad d\underset{\sim}{\sigma} = \underset{\sim}{\theta}\wedge\underset{\sim}{\sigma} - (d\underset{\sim}{\theta} - \underset{\sim}{\theta}\wedge\underset{\sim}{\theta})\wedge\underset{\sim}{\omega}.$$

If we put

$$(5\text{-}16.6)\qquad \underset{\sim}{\rho} = \underset{\sim}{\theta}\wedge\underset{\sim}{\omega},$$

and note that this implies $d\underset{\sim}{\rho} = d\underset{\sim}{\theta}\wedge\underset{\sim}{\omega} - \underset{\sim}{\theta}\wedge(\underset{\sim}{\theta}\wedge\underset{\sim}{\omega} + \underset{\sim}{\sigma})$, we arrive at the system

$$(5\text{-}16.7)\qquad d\underset{\sim}{\omega} = \underset{\sim}{\rho} + \underset{\sim}{\sigma}, \quad d\underset{\sim}{\sigma} = -\,d\underset{\sim}{\rho}.$$

We now set

$$(5\text{-}16.8)\qquad \underset{\sim}{\sigma} = d\underset{\sim}{\eta} + \underset{\sim a}{\sigma}, \quad \underset{\sim}{\eta} = H(\underset{\sim}{\sigma})\,\varepsilon\,A^{1}(S)$$

$$(5\text{-}16.9)\qquad \underset{\sim}{\omega} = d\underset{\sim}{\phi} + \underset{\sim a}{\omega}, \quad \underset{\sim}{\phi} = H(\underset{\sim}{\omega})\,\varepsilon\,A^{0}(S)$$

and obtain

(5-16.10) $$d\underset{\sim}{\omega}_a = \underset{\sim}{\rho} + d\underset{\sim}{\eta} + \underset{\sim}{\sigma}_a , \quad d\underset{\sim}{\sigma}_a = - d\underset{\sim}{\rho} .$$

Since H inverts d on $A(S)$, we obtain $\underset{\sim}{\omega}_a = H(\underset{\sim}{\rho} + d\underset{\sim}{\eta} + \underset{\sim}{\sigma}_a) = H(\underset{\sim}{\rho}) + \underset{\sim}{\eta}$ and $\underset{\sim}{\sigma}_a = - H(d\underset{\sim}{\rho}) = dH(\underset{\sim}{\rho}) - \underset{\sim}{\rho}$ because $dH + Hd =$ identity. Equations (5-16.8) and (5-16.9) thus yield

(5-16.11) $$\underset{\sim}{\omega} = d\underset{\sim}{\phi} + H(\underset{\sim}{\rho}) + \underset{\sim}{\eta} , \quad \underset{\sim}{\sigma} = d\underset{\sim}{\eta} + dH(\underset{\sim}{\rho}) - \underset{\sim}{\rho} .$$

It now remains only to eliminate the subsidiary variable $\underset{\sim}{\rho}$. A direct combination of (5-16.6) and (5-16.11) yield

$$\underset{\sim}{\rho} = \underset{\sim}{\theta} \wedge d\underset{\sim}{\phi} + \underset{\sim}{\theta} \wedge H(\underset{\sim}{\rho}) + \underset{\sim}{\theta} \wedge \underset{\sim}{\eta}$$

so that

$$H(\underset{\sim}{\rho}) = H(\underset{\sim}{\theta} \wedge d\underset{\sim}{\phi} + \underset{\sim}{\theta} \wedge H(\underset{\sim}{\rho}) + \underset{\sim}{\theta} \wedge \underset{\sim}{\eta}) = H(\underset{\sim}{\theta} \wedge d\underset{\sim}{\phi})$$

because $\underset{\sim}{\theta} \wedge H(\underset{\sim}{\rho})$ and $\underset{\sim}{\theta} \wedge \underset{\sim}{\eta}$ belong to $A(S)$. Thus, we obtain

$$\underset{\sim}{\rho} = \underset{\sim}{\theta} \wedge d\underset{\sim}{\phi} + \underset{\sim}{\theta} \wedge H(\underset{\sim}{\theta} \wedge d\underset{\sim}{\phi}) + \underset{\sim}{\theta} \wedge \underset{\sim}{\eta} , \quad H(\underset{\sim}{\rho}) = H(\underset{\sim}{\theta} \wedge d\underset{\sim}{\phi})$$

and hence

$$\underset{\sim}{\omega} = d\underset{\sim}{\phi} + \underset{\sim}{\eta} + H(\underset{\sim}{\theta} \wedge d\underset{\sim}{\phi}) ,$$

$$\underset{\sim}{\sigma} = d\underset{\sim}{\eta} + dH(\underset{\sim}{\theta} \wedge d\underset{\sim}{\phi}) - \underset{\sim}{\theta} \wedge d\underset{\sim}{\phi} - \underset{\sim}{\theta} \wedge H(\underset{\sim}{\theta} \wedge d\underset{\sim}{\phi}) - \underset{\sim}{\theta} \wedge \underset{\sim}{\eta} .$$

A combination of the above substitutions now establishes the following result.

THEOREM 5-16.1. *The general solution of the differential system*

(5-16.12) $$d\underset{\sim}{\Omega} = \underset{\sim}{\Gamma} \wedge \underset{\sim}{\Omega} + \underset{\sim}{\Sigma} , \quad d\underset{\sim}{\Sigma} = \underset{\sim}{\Gamma} \wedge \underset{\sim}{\Sigma} - \underset{\sim}{\Theta} \wedge \underset{\sim}{\Omega} ,$$

(5-16.13) $$d\underset{\sim}{\Gamma} = \underset{\sim}{\Gamma} \wedge \underset{\sim}{\Gamma} + \underset{\sim}{\Theta} , \quad d\underset{\sim}{\Theta} = \underset{\sim}{\Gamma} \wedge \underset{\sim}{\Theta} - \underset{\sim}{\Theta} \wedge \underset{\sim}{\Gamma}$$

is given in terms of the matrices of antiexact forms

$$(5\text{-}16.14)\qquad \underset{\sim}{\phi} = H(\underset{\sim}{A}^{-1}\underset{\sim}{\Omega})$$

$$(5\text{-}16.15)\qquad \underset{\sim}{\eta} = H(\underset{\sim}{A}^{-1}\underset{\sim}{\Sigma})$$

$$(5\text{-}16.16)\qquad \underset{\sim}{\theta} = H(\underset{\sim}{A}^{-1}\underset{\sim}{\Theta}\ \underset{\sim}{A})$$

and the attitude matrix $\underset{\sim}{A}$, *satisfying*

$$(5\text{-}16.17)\qquad \underset{\sim}{A} = \underset{\sim}{A}_o + H(d\underset{\sim}{\gamma}\ \underset{\sim}{A}) ,\quad \underset{\sim}{\gamma} = H(\underset{\sim}{\Gamma}) ,\quad \det(\underset{\sim}{A}_o) \neq 0 ,$$

by

$$(5\text{-}16.18)\qquad \underset{\sim}{\Omega} = \underset{\sim}{A}\{d\underset{\sim}{\phi} + \underset{\sim}{\eta} + H(\underset{\sim}{\theta} \wedge d\underset{\sim}{\phi})\} ,$$

$$(5\text{-}16.19)\qquad \underset{\sim}{\Sigma} = \underset{\sim}{A}\{d\underset{\sim}{\eta} - \underset{\sim}{\theta} \wedge \underset{\sim}{\eta} - H(d\underset{\sim}{\theta} \wedge d\underset{\sim}{\phi}) - \underset{\sim}{\theta} \wedge H(\underset{\sim}{\theta} \wedge d\underset{\sim}{\phi})\} ,$$

$$(5\text{-}16.20)\qquad \underset{\sim}{\Gamma} = (d\underset{\sim}{A} + \underset{\sim}{A}\underset{\sim}{\theta})\underset{\sim}{A}^{-1} ,$$

$$(5\text{-}16.21)\qquad \underset{\sim}{\Theta} = \underset{\sim}{A}(d\underset{\sim}{\theta} - \underset{\sim}{\theta} \wedge \underset{\sim}{\theta})\underset{\sim}{A}^{-1} .$$

The following corollary is not immediate.

COROLLARY 5-16.1. *We have*

$$(5\text{-}16.22)\qquad (\underset{\sim}{A}^{-1}\underset{\sim}{\Omega})_e = d\underset{\sim}{\phi} ,\quad (\underset{\sim}{A}^{-1}\underset{\sim}{\Omega})_a = \underset{\sim}{\eta} + H(\underset{\sim}{\theta} \wedge d\underset{\sim}{\phi}) ,$$

$$(5\text{-}16.23)\qquad (\underset{\sim}{A}^{-1}\underset{\sim}{\Sigma})_e = d\underset{\sim}{\eta},\quad (\underset{\sim}{A}^{-1}\underset{\sim}{\Sigma})_a = -\underset{\sim}{\theta} \wedge \underset{\sim}{\eta} - H(\underset{\sim}{\theta} \wedge d\underset{\sim}{\phi}) - \underset{\sim}{\theta} \wedge H(\underset{\sim}{\theta} \wedge d\underset{\sim}{\phi}) ,$$

$$(5\text{-}16.24)\qquad (\underset{\sim}{\Gamma}\underset{\sim}{A})_e = d\underset{\sim}{A} ,\quad (\underset{\sim}{\Gamma}\underset{\sim}{A})_a = \underset{\sim}{A}\underset{\sim}{\theta} ,$$

$$(5\text{-}16.25)\qquad (\underset{\sim}{A}^{-1}\underset{\sim}{\Theta}\ \underset{\sim}{A})_e = d\underset{\sim}{\theta} ,\quad (\underset{\sim}{A}^{-1}\underset{\sim}{\Theta}\ \underset{\sim}{A})_a = -\underset{\sim}{\theta} \wedge \underset{\sim}{\theta} .$$

5-17. EQUIVALENT DIFFERENTIAL SYSTEMS

If we go back to Definition 5-11.1, we see that the general differential system

$$(5\text{-}17.1) \qquad d\underset{\sim}{\Omega} = \underset{\sim}{\Gamma} \wedge \underset{\sim}{\Omega} + \underset{\sim}{\Sigma} \,, \quad d\underset{\sim}{\Sigma} = \underset{\sim}{\Gamma} \wedge \underset{\sim}{\Sigma} - \underset{\sim}{\Theta} \wedge \underset{\sim}{\Omega}$$

$$(5\text{-}17.2) \qquad d\underset{\sim}{\Gamma} = \underset{\sim}{\Gamma} \wedge \underset{\sim}{\Gamma} + \underset{\sim}{\Theta} \,, \quad d\underset{\sim}{\Theta} = \underset{\sim}{\Gamma} \wedge \underset{\sim}{\Theta} - \underset{\sim}{\Theta} \wedge \underset{\sim}{\Gamma}$$

arises from the system of equations

$$(5\text{-}17.3) \qquad d\underset{\sim}{\Omega} = \underset{\sim}{\Gamma} \wedge \underset{\sim}{\Omega} + \underset{\sim}{\Sigma}$$

by adjoining to this system the equations that obtain from it by demanding closure. Now, for given $\underset{\sim}{\Omega}$, the quantities $\underset{\sim}{\Gamma}$ and $\underset{\sim}{\Sigma}$ that occur in (5-17.3) are not uniquely determined. This leads to the following natural definition of equivalence.

<u>Definition 5-17.1</u>. Two differential systems,

$$d\underset{\sim}{\Omega} = \underset{\sim}{\Gamma} \wedge \underset{\sim}{\Omega} + \underset{\sim}{\Sigma} \,, \quad d\underset{\sim}{\Sigma} = \underset{\sim}{\Gamma} \wedge \underset{\sim}{\Sigma} - \underset{\sim}{\Theta} \wedge \underset{\sim}{\Omega} \,,$$

$$d\underset{\sim}{\Gamma} = \underset{\sim}{\Gamma} \wedge \underset{\sim}{\Gamma} + \underset{\sim}{\Theta} \,, \quad d\underset{\sim}{\Theta} = \underset{\sim}{\Gamma} \wedge \underset{\sim}{\Theta} - \underset{\sim}{\Theta} \wedge \underset{\sim}{\Gamma} \,,$$

and

$$d\underset{\sim}{\bar{\Omega}} = \underset{\sim}{\bar{\Gamma}} \wedge \underset{\sim}{\bar{\Omega}} + \underset{\sim}{\bar{\Sigma}} \,, \quad d\underset{\sim}{\bar{\Sigma}} = \underset{\sim}{\bar{\Gamma}} \wedge \underset{\sim}{\bar{\Sigma}} - \underset{\sim}{\bar{\Theta}} \wedge \underset{\sim}{\bar{\Omega}} \,,$$

$$d\underset{\sim}{\bar{\Omega}} = \underset{\sim}{\bar{\Gamma}} \wedge \underset{\sim}{\bar{\Gamma}} + \underset{\sim}{\bar{\Theta}} \,, \quad d\underset{\sim}{\bar{\Theta}} = \underset{\sim}{\bar{\Gamma}} \wedge \underset{\sim}{\bar{\Theta}} - \underset{\sim}{\bar{\Theta}} \wedge \underset{\sim}{\bar{\Gamma}}$$

are said to be *equivalent* on S if and only if $\underset{\sim}{\bar{\Omega}} = \underset{\sim}{\Omega}$ at all points of S.

There are obviously many differential systems that are equivalent to a given differential system. They are obtained by the requirement that

$$d\underset{\sim}{\Omega} = \underset{\sim}{\Gamma} \wedge \underset{\sim}{\Omega} + \underset{\sim}{\Sigma} = \underset{\sim}{\bar{\Gamma}} \wedge \underset{\sim}{\Omega} + \underset{\sim}{\bar{\Sigma}} \,,$$

from which we obtain

$$(5\text{-}17.4) \qquad \underset{\sim}{\bar{\Sigma}} = \underset{\sim}{\Sigma} + (\underset{\sim}{\Gamma} - \underset{\sim}{\bar{\Gamma}}) \wedge \underset{\sim}{\Omega} \,.$$

Theorem 5-16.1 can then be used to obtain the solutions of the two equivalent systems whereby the various antiexact forms that determine these solutions can be related. The results are rather complex and not altogether illuminating, with the exception of one very important case.

Let us start with a given differential system with connection matrix $\underset{\sim}{\Gamma}$, in which case,

$$(5\text{-}17.5)\qquad \underset{\sim}{\Gamma} = (d\underset{\sim}{A} + \underset{\sim}{A}\,\underset{\sim}{\theta})\underset{\sim}{A}^{-1}, \quad \underset{\sim}{A} = \underset{\sim}{A}_o + H(d\underset{\sim}{\gamma}\,\underset{\sim}{A}), \quad \underset{\sim}{\gamma} = H(\underset{\sim}{\Gamma}).$$

If we now take

$$(5\text{-}17.6)\qquad \bar{\underset{\sim}{\Gamma}} = d\underset{\sim}{A}\,\underset{\sim}{A}^{-1}$$

then

$$(5\text{-}17.7)\qquad \bar{\underset{\sim}{\gamma}} = H(\bar{\underset{\sim}{\Gamma}}) = H(d\underset{\sim}{A}\,\underset{\sim}{A}^{-1}) = H(d\underset{\sim}{A}\,\underset{\sim}{A}^{-1} + \underset{\sim}{A}\,\underset{\sim}{\theta}\,\underset{\sim}{A}^{-1}) = \underset{\sim}{\gamma}$$

and the attitude matrix $\bar{\underset{\sim}{A}}$ of the connection $\bar{\underset{\sim}{\Gamma}}$ can be taken to be the same as the attitude matrix $\underset{\sim}{A}$ of the connection $\underset{\sim}{\Gamma}$, for $\underset{\sim}{A}$ and $\bar{\underset{\sim}{A}}$ satisfy the same matrix integral equation $\underset{\sim}{A} = \underset{\sim}{A}_o + H(d\underset{\sim}{\gamma}\,\underset{\sim}{A})$. In this event, (5-17.4) yields $\bar{\underset{\sim}{\Sigma}} = \underset{\sim}{\Sigma} + (\underset{\sim}{\Gamma} - \bar{\underset{\sim}{\Gamma}}) \wedge \underset{\sim}{\Omega}$ and hence (5-16.18), (5-16.19), (5-16.20) and (5-17.6) yield

$$(5\text{-}17.8)\qquad \bar{\underset{\sim}{\Sigma}} = \underset{\sim}{A}\{d\underset{\sim}{\eta} - \underset{\sim}{\theta}\wedge\underset{\sim}{\eta} - H(d\underset{\sim}{\theta}\wedge d\underset{\sim}{\phi}) - \underset{\sim}{\theta}\wedge H(\underset{\sim}{\theta}\wedge d\underset{\sim}{\phi}) + \underset{\sim}{\theta}\wedge(d\underset{\sim}{\phi} + \underset{\sim}{\eta} + H(\underset{\sim}{\theta}\wedge d\underset{\sim}{\phi}))\}$$

$$= \underset{\sim}{A}\{d\underset{\sim}{\eta} - H(d\underset{\sim}{\theta}\wedge d\underset{\sim}{\phi}) + \underset{\sim}{\theta}\wedge d\underset{\sim}{\phi}\} = \underset{\sim}{A}d(\underset{\sim}{\eta} + H(\underset{\sim}{\theta}\wedge d\underset{\sim}{\phi})) .$$

Accordingly, (5-16.14) and (5-16.15) yield

$$(5\text{-}17.9)\qquad \bar{\underset{\sim}{\phi}} = H(\underset{\sim}{A}^{-1}\bar{\underset{\sim}{\Omega}}) = H(\underset{\sim}{A}^{-1}\underset{\sim}{\Omega}) = \underset{\sim}{\phi},$$

$$(5\text{-}17.10)\qquad \bar{\underset{\sim}{\eta}} = H(\underset{\sim}{A}^{-1}\bar{\underset{\sim}{\Sigma}}) = Hd(\underset{\sim}{\eta} + H(\underset{\sim}{\theta}\wedge d\underset{\sim}{\phi})) = \underset{\sim}{\eta} + H(\underset{\sim}{\theta}\wedge d\underset{\sim}{\phi}),$$

while $d\bar{\underset{\sim}{\Gamma}} = d\bar{\underset{\sim}{\Gamma}}\wedge d\bar{\underset{\sim}{\Gamma}} + \bar{\underset{\sim}{\Theta}}$ and (5-17.6) give

$$(5\text{-}17.11)\qquad \bar{\underset{\sim}{\Theta}} = \underset{\sim}{0}, \quad \bar{\underset{\sim}{\theta}} = \underset{\sim}{0} .$$

We have thus established the following result.

THEOREM 5-17.1. *Any differential system*

$$d\underset{\sim}{\Omega} = \underset{\sim}{\Gamma} \wedge \underset{\sim}{\Omega} + \underset{\sim}{\Sigma} , \quad d\underset{\sim}{\Sigma} = \underset{\sim}{\Gamma} \wedge \underset{\sim}{\Sigma} - \underset{\sim}{\Theta} \wedge \underset{\sim}{\Omega} \tag{5-17.12}$$

$$d\underset{\sim}{\Gamma} = \underset{\sim}{\Gamma} \wedge \underset{\sim}{\Gamma} + \underset{\sim}{\Theta} , \quad d\underset{\sim}{\Theta} = \underset{\sim}{\Gamma} \wedge \underset{\sim}{\Theta} - \underset{\sim}{\Theta} \wedge \underset{\sim}{\Gamma} \tag{5-17.13}$$

is equivalent to a curvature-free differential system

$$d\underset{\sim}{\Omega} = \underset{\sim}{\bar{\Gamma}} \wedge \underset{\sim}{\Omega} + \underset{\sim}{\bar{\Sigma}} , \quad d\underset{\sim}{\bar{\Sigma}} = \underset{\sim}{\bar{\Gamma}} \wedge \underset{\sim}{\bar{\Sigma}} , \tag{5-17.14}$$

$$d\underset{\sim}{\bar{\Gamma}} = \underset{\sim}{\bar{\Gamma}} \wedge \underset{\sim}{\bar{\Gamma}} , \quad \underset{\sim}{\bar{\Theta}} = \underset{\sim}{0} \tag{5-17.15}$$

whose general solution is given by

$$\underset{\sim}{\Omega} = \underset{\sim}{A}(d\underset{\sim}{\phi} + \underset{\sim}{\bar{\eta}}) , \tag{5-17.16}$$

$$\underset{\sim}{\bar{\Sigma}} = \underset{\sim}{A}\, d\underset{\sim}{\bar{\eta}} , \tag{5-17.17}$$

$$\underset{\sim}{\bar{\Gamma}} = d\underset{\sim}{A}\, \underset{\sim}{A}^{-1} , \tag{5-17.18}$$

with

$$\underset{\sim}{A} = \underset{\sim}{A}_o + H(\underset{\sim}{\bar{\Gamma}}\underset{\sim}{A}) = \underset{\sim}{A}_o + H(\underset{\sim}{\Gamma}\underset{\sim}{A}) , \quad \det(\underset{\sim}{A}_o) \neq 0 , \tag{5-17.19}$$

$$\underset{\sim}{\bar{\eta}} = H(\underset{\sim}{A}^{-1}\underset{\sim}{\bar{\Sigma}}) , \quad \underset{\sim}{\phi} = H(\underset{\sim}{A}^{-1}\underset{\sim}{\Omega}) , \tag{5-17.20}$$

and $\underset{\sim}{\bar{\Sigma}} = \underset{\sim}{\Sigma} + (\underset{\sim}{\Gamma} - \underset{\sim}{\bar{\Gamma}}) \wedge \underset{\sim}{\Omega}$, $\underset{\sim}{\bar{\eta}} = \underset{\sim}{\eta} + H(\underset{\sim}{\theta} \wedge d\underset{\sim}{\phi})$.

The above result is, to some extent, a case of "robbing Peter to pay Paul", for we have simply taken away from the $\underset{\sim}{\Gamma} \wedge \underset{\sim}{\Omega}$ and included them in the $\underset{\sim}{\Sigma}$. From this point of view, obtaining an equivalent system that leaves both $\underset{\sim}{\Omega}$ and $\underset{\sim}{\Sigma}$ unchanged would be most useful. This turns out to be possible if $\underset{\sim}{\Omega}$ is such that there exists a nonzero $\underset{\sim}{P} \in M_{r,r}(\Lambda^1)$ such that

$$\underset{\sim}{P} \wedge \underset{\sim}{\Omega} = \underset{\sim}{0} . \tag{5-17.21}$$

Under these circumstances, $\underset{\sim}{\Gamma}\wedge\underset{\sim}{\Omega} = (\underset{\sim}{\Gamma} + \underset{\sim}{P})\wedge\underset{\sim}{\Omega}$ and exterior differentiation of (5-17.21) yields

$$(5\text{-}17.22)\qquad d\underset{\sim}{P}\wedge\underset{\sim}{\Omega} = \underset{\sim}{P}\wedge\underset{\sim}{\Gamma}\wedge\underset{\sim}{\Omega} - \underset{\sim}{P}\wedge\underset{\sim}{\Sigma}$$

because $d\underset{\sim}{\Omega} = \underset{\sim}{\Gamma}\wedge\underset{\sim}{\Omega} + \underset{\sim}{\Sigma}$. If we set $\hat{\underset{\sim}{\Gamma}} = \underset{\sim}{\Gamma} + \underset{\sim}{P}$, the connection equation for $\hat{\underset{\sim}{\Gamma}}$ gives $\hat{\underset{\sim}{\Theta}} = \underset{\sim}{\Theta} + d\underset{\sim}{P} - \underset{\sim}{\Gamma}\wedge\underset{\sim}{P} - \underset{\sim}{P}\wedge\underset{\sim}{\Gamma} - \underset{\sim}{P}\wedge\underset{\sim}{P}$, from which we can use (5-17.22) to conclude that

$$(5\text{-}17.23)\qquad \hat{\underset{\sim}{\Theta}}\wedge\underset{\sim}{\Omega} = \underset{\sim}{\Theta}\wedge\underset{\sim}{\Omega} + \underset{\sim}{P}\wedge\underset{\sim}{\Sigma}\ .$$

The torsion equation, $d\underset{\sim}{\Sigma} = \underset{\sim}{\Gamma}\wedge\underset{\sim}{\Sigma} - \underset{\sim}{\Theta}\wedge\underset{\sim}{\Omega}$ thus yields

$$(5\text{-}17.24)\qquad d\underset{\sim}{\Sigma} = \hat{\underset{\sim}{\Gamma}}\wedge\underset{\sim}{\Sigma} - \hat{\underset{\sim}{\Theta}}\wedge\underset{\sim}{\Omega}\ .$$

LEMMA 5-17.1. *If there exists a nonzero* $\underset{\sim}{P}\in M_{r,r}(\Lambda^1)$ *such that*

$$(5\text{-}17.25)\qquad \underset{\sim}{P}\wedge\underset{\sim}{\Omega} = \underset{\sim}{0}\ ,$$

then the differential system

$$d\underset{\sim}{\Omega} = \underset{\sim}{\Gamma}\wedge\underset{\sim}{\Omega} + \underset{\sim}{\Sigma}\ ,\qquad d\underset{\sim}{\Sigma} = \underset{\sim}{\Gamma}\wedge\underset{\sim}{\Sigma} - \underset{\sim}{\Theta}\wedge\underset{\sim}{\Omega}$$

$$d\underset{\sim}{\Gamma} = \underset{\sim}{\Gamma}\wedge\underset{\sim}{\Gamma} + \underset{\sim}{\Theta}\ ,\qquad d\underset{\sim}{\Theta} = \underset{\sim}{\Gamma}\wedge\underset{\sim}{\Theta} - \underset{\sim}{\Theta}\wedge\underset{\sim}{\Gamma}$$

is equivalent to the differential system

$$(5\text{-}17.26)\qquad d\underset{\sim}{\Omega} = \hat{\underset{\sim}{\Gamma}}\wedge\underset{\sim}{\Omega} + \underset{\sim}{\Sigma}\ ,\qquad d\underset{\sim}{\Sigma} = \hat{\underset{\sim}{\Gamma}}\wedge\underset{\sim}{\Sigma} - \hat{\underset{\sim}{\Theta}}\wedge\underset{\sim}{\Omega}\ ,$$

$$(5\text{-}17.27)\qquad d\hat{\underset{\sim}{\Gamma}} = \hat{\underset{\sim}{\Gamma}}\wedge\hat{\underset{\sim}{\Gamma}} + \hat{\underset{\sim}{\Theta}}\ ,\qquad d\hat{\underset{\sim}{\Theta}} = \hat{\underset{\sim}{\Gamma}}\wedge\hat{\underset{\sim}{\Theta}} - \hat{\underset{\sim}{\Theta}}\wedge\hat{\underset{\sim}{\Gamma}}$$

with

$$(5\text{-}17.28)\qquad \hat{\underset{\sim}{\Gamma}} = \underset{\sim}{\Gamma} + \underset{\sim}{P}\ .$$

Let $\underset{\sim}{A}$ denote the attitude matrix of the given connection $\underset{\sim}{\Gamma}$. We then have

$$(5\text{-}17.29) \qquad \underset{\sim}{\Gamma} = (d\underset{\sim}{A} + \underset{\sim}{A}\ \underset{\sim}{\theta})\underset{\sim}{A}^{-1} .$$

If $\underset{\sim}{P}$ satisfies the hypothesis of Lemma 5-17.1, we may write, without loss of generality,

$$(5\text{-}17.30) \qquad \underset{\sim}{P} = \underset{\sim}{A}(d\underset{\sim}{B} + \underset{\sim}{B}\ \underset{\sim}{\pi})\underset{\sim}{B}^{-1}\underset{\sim}{A}^{-1} ,$$

where $\underset{\sim}{\pi} \in M_{r,r}(A^1)$, as suggested by Lemma 5-14.8. It then follows immediately from (5-17.28) through (5-17.30) that

$$(5\text{-}17.31) \qquad \hat{\underset{\sim}{\Gamma}} = (d\hat{\underset{\sim}{A}} + \hat{\underset{\sim}{A}}\ \hat{\underset{\sim}{\theta}})\hat{\underset{\sim}{A}}^{-1}$$

with

$$(5\text{-}17.32) \qquad \hat{\underset{\sim}{A}} = \underset{\sim}{A}\ \underset{\sim}{B} ,$$

$$(5\text{-}17.33) \qquad \hat{\underset{\sim}{\theta}} = \underset{\sim}{B}^{-1}\ \underset{\sim}{\theta}\ \underset{\sim}{B} + \underset{\sim}{\pi} .$$

Thus, with

$$(5\text{-}17.34) \qquad \underset{\sim}{\Omega} = \underset{\sim}{A}\ \underset{\sim}{B}\ \hat{\underset{\sim}{\omega}} , \quad \underset{\sim}{\Sigma} = \underset{\sim}{A}\ \underset{\sim}{B}\ \hat{\underset{\sim}{\sigma}} ,$$

(5-17.25), (5-17.30) and (5-17.34) yield $\underset{\sim}{B}^{-1}d\underset{\sim}{B} \wedge \underset{\sim}{B}^{-1}\underset{\sim}{A}^{-1}\underset{\sim}{\Omega} = -\underset{\sim}{\pi} \wedge \underset{\sim}{B}^{-1}\underset{\sim}{A}^{-1}\underset{\sim}{\Omega}$, and we obtain

$$(5\text{-}17.35) \qquad \hat{\underset{\sim}{\theta}} \wedge \hat{\underset{\sim}{\omega}} = \underset{\sim}{B}^{-1}(\underset{\sim}{\theta}\ \underset{\sim}{B} - d\underset{\sim}{B}) \wedge \hat{\underset{\sim}{\omega}} .$$

Thus, $\hat{\underset{\sim}{\omega}}$ and $\hat{\underset{\sim}{\sigma}}$ satisfy

$$(5\text{-}17.36) \qquad d\hat{\underset{\sim}{\omega}} = \underset{\sim}{B}^{-1}(\underset{\sim}{\theta}\ \underset{\sim}{B} - d\underset{\sim}{B}) \wedge \hat{\underset{\sim}{\omega}} + \hat{\underset{\sim}{\sigma}}, \quad d\hat{\underset{\sim}{\sigma}} = -d(\underset{\sim}{B}^{-1}(\underset{\sim}{\theta}\ \underset{\sim}{B} - d\underset{\sim}{B}) \wedge \hat{\underset{\sim}{\omega}}) .$$

In the case where $k = 1$ and $\hat{\underset{\sim}{\sigma}} = \underset{\sim}{0}$ $(\underset{\sim}{\Sigma}=\underset{\sim}{0})$, a significant simplification occurs because (5-17.36) then implies that $\underset{\sim}{B}$ can be chosen so that $(\underset{\sim}{\theta}\ \underset{\sim}{B} - d\underset{\sim}{B}) \wedge \hat{\underset{\sim}{\omega}} = 0$. However, $\hat{\underset{\sim}{\sigma}} = \underset{\sim}{0}$ implies that $\underset{\sim}{\Omega}$ is recursive; that is $d\underset{\sim}{\Omega} = \underset{\sim}{\Gamma} \wedge \underset{\sim}{\Omega}$, and we obtain the Frobenius theorem: $\underset{\sim}{\Omega} = \underset{\sim}{A}\underset{\sim}{B}\ d\underset{\sim}{\psi}$.

LEMMA 5-17.2. *If* $k = 1$ *and* $\underset{\sim}{\Omega}$ *is recursive, then* $\underset{\sim}{\Omega}$ *is gradient-recursive:* $\underset{\sim}{d\Omega} = d(\underset{\sim\sim}{AB}) \wedge \underset{\sim}{\Omega}$.

5-18. n-FORMS AND INTEGRATION

Let S be a starlike region of an n-dimensional space M with center P_o and let $\{x^i\}$ be a preferred coordinate cover of S for which $P_o:\{x_o^i = 0\}$. The natural volume element (n-form) of S with respect to the coordinate cover $\{x^i\}$ is denoted by

$$(5\text{-}18.1) \qquad \mu = dx^1 \wedge dx^2 \wedge \ldots \wedge dx^n .$$

Now, μ is closed, and hence Theorem 5-6.1 gives the unique representation

$$(5\text{-}18.2) \qquad \mu = dH(\mu) .$$

It also follows from (5-16.1) that $\tilde{\mu}(\lambda) = \mu$, and hence (5-3.1) shows that

$$H(\mu) = \int_o^1 X \rfloor \tilde{\mu}(\lambda)\lambda^{n-1} d\lambda = \int_o^1 \lambda^{n-1} d\lambda \; X \rfloor \mu = \frac{1}{n} X \rfloor \mu .$$

However

$$X \rfloor \mu = \sum_{i=1}^{n} (-1)^{i+1} x^i dx^1 \wedge \ldots \wedge dx^{i-1} \wedge dx^{i+1} \wedge \ldots \wedge dx^n$$

which leads to the following natural system of n linearly independent (n-1)-forms

$$(5\text{-}18.3) \qquad \mu_i = (-1)^{i+1} dx^1 \wedge \ldots \wedge dx^{i-1} \wedge dx^{i+1} \wedge \ldots \wedge dx^n ,$$

$$i = 1, \ldots, n .$$

It thus follows that $\rho = H(\mu) = \frac{1}{n} x^i \mu_i$. Further, since $\rho \varepsilon A^{n-1}(S)$, we have $X \rfloor \rho = 0$ and $\rho(x_o^i) = 0$. These considerations establish the following results.

LEMMA 5-18.1. *Let* S *be a starlike region of an* n-*dimen-*

sional space whose center P_o *has the coordinates* $x_o^i = 0$ *in a preferred coordinate cover* $\{x^i\}$ *of* S , *and let* μ *be the natural volume element of* S *with respect to* $\{x^i\}$. *There exists a unique* (n-1)-*form*

$$(5\text{-}18.4) \qquad \rho = \frac{1}{n} x^i \mu_i$$

on S *such that*

$$(5\text{-}18.5) \qquad \mu = d\rho ,$$

$$(5\text{-}18.6) \qquad X \rfloor \rho = 0 , \quad \rho(x_o^i) = 0 .$$

The quantities μ_i are of an obvious utility in view of their properties which we list in the following lemma.

LEMMA 5-18.2. *The* n *linearly independent* (n-1)-*forms*

$$(5\text{-}18.7) \qquad \mu_i = (-1)^{i+1} dx^i \wedge \ldots \wedge dx^{i-1} \wedge dx^{i+1} \wedge \ldots \wedge dx^n$$

have the following properties:

$$(5\text{-}18.8) \qquad d\mu_i = 0 ,$$

$$(5\text{-}18.9) \qquad dx^i \wedge \mu_j = \delta^i_j \mu ,$$

$$(5\text{-}18.10) \qquad d(F^i \mu_i) = (\partial_i F^i)\mu$$

for $F^i(x) \in \Lambda^o(S)$

$$(5\text{-}18.11) \qquad v \rfloor \mu = v^i \mu_i$$

for any $v = v^i \partial/\partial x^i \in T(S)$, *and any* (n-1)-*form* σ *on* S *can be written uniquely as* $\sigma = \sigma^i \mu_i$ *with* $\sigma^i \in \Lambda^o(S)$.

Proof. The result (5-18.8) follows directly from (5-18.7), as does (5-18.9). Exterior differentiation of $F^i \mu_i$ with $F^i \in \Lambda^o(S)$ and use of (5-18.8), (5-18.9) give $d(F^i \mu_i) =$

$(dF^i) \wedge \mu_i = (\partial_j F^i dx^j) \wedge \mu_i = (\partial_i F^i)\mu$, and (5-18.10) is established. The final result, (5-18.11), follows directly from the properties of inner multiplication and (5-18.7).

THEOREM 5-18.1. *Let* S *be a starlike region of an* n-*dimensional space whose center* P_o *has coordinates* $x_o^i = 0$ *in a preferred coordinate cover* $\{x^i\}$ *of* S , *and let*

$$(5\text{-}18.12) \qquad \omega(x) = f(x)\mu$$

be any n-*form on* S , *where* μ *is the natural volume element of* S *with respect to the coordinate cover* $\{x^i\}$. *There exists a unique, antiexact,* (n-1)-*form*

$$(5\text{-}18.13) \qquad \rho = F^i \mu_i$$

on S *such that*

$$(5\text{-}18.14) \qquad \omega = d\rho ,$$

$$(5\text{-}18.15) \qquad X \rfloor \rho = 0 , \quad \rho(x_o^i) = 0 ,$$

and the functions $F^i(x)$ *are given by*

$$(5\text{-}18.16) \qquad F^i = x^i h(f)$$

where

$$(5\text{-}18.17) \qquad h(f)(x) = \int_0^1 \lambda^{n-1} \tilde{f}(\lambda)d\lambda$$

satisfies

$$(5\text{-}18.18) \qquad n\, h(f) + x^i \partial_i h(f) = f$$

identically on S .

Proof. Since ω is an n-form on an n-dimensional region, $d\omega = 0$, and Theorem 5-6.1 gives the unique representation $\omega = dH(\omega)$. We set $\rho = H(\omega)$, in which case ρ satisfies (5-18.4) and

(5-18.5). Now, $\rho = H(\omega) = H(f\ \mu) = \int_0^1 X \rfloor \widetilde{(f\ \mu)}(\lambda)\lambda^{n-1}\, d\lambda =$ $X \rfloor \mu \int_0^1 \tilde{f}(\lambda)\lambda^{n-1}\, d\lambda = h(f) X \rfloor \mu$ since $\mu(\lambda) = \mu$. We now use (5-18.11) to obtain $X \rfloor \mu = x^i\ \mu_i$. Thus, $\rho = F^i\ \mu_i$, where $F^i = x^i\ h(f)$. This establishes (5-18.13) through (5-18.17). Finally, we have $\omega = f\ \mu = d\rho = d(h(f)x^i\ \mu_i) = d(h(f)x^i) \wedge \mu_i =$ $\partial_j(h(f)x^i)dx^j \wedge \mu_i = (x^i\ \partial_i h(f) + n\ h(f))\mu$, where we have used (5-18.8) and (5-18.9). Thus, since μ constitutes a bases for $\Lambda^n(S)$, we obtain the identical satisfaction of (5-18.18) on S.

<u>COROLLARY 5-18.1</u>. *There exists one and only one collection of* n *functions* $F^i(x)$ *on* S *of the form* $u(x)x^i$ *such that*

(5-18.19) $$\partial_i F^i = f\ ,$$

for given f. *The function* $u(x)$ *is given by* $u(x) = h(f)(x)$.

<u>Proof</u>. Given any function $f(x)$ on S, we construct the n-form $f\mu$. The result then follows directly from Theorem 5-18.1 since $f\mu = d(F^i\ \mu_i) = (\partial_i F^i)\mu$, and $F^i = h(f)x^i$ is unique due to the uniqueness of ρ.

<u>COROLLARY 5-18.2</u>. *Let* $\omega = f\ \mu$ *be a given* n-*form on* S. *There exists one and only one vector field* F *on* S *that is proportional to* X *such that*

(5-18.20) $$\omega = f\mu = d(F \rfloor \mu) = \mathfrak{L}_F \mu$$

and this F *is given by*

(5-18.21) $$F = h(f)X\ .$$

<u>Proof</u>. Theorem 5-18.1 gives $\omega = f\ \mu = d\rho$ with $\rho = h(f)x^i\ \mu_i =$ $(h(f)x^i\ \partial/\partial x^i) \rfloor \mu$, where the last equality obtains from (5-18.11). However, it follows trivially that $\omega = f\ \mu = d(\rho + d\phi) = d\eta$ where ρ is the antiexact part of η that is unique, while $d\phi$ is the exact part of η and hence an arbitrary exact (n-1)-form.

Now, if $d\phi = g(x)X \rfloor \mu = g(x)x^i \mu_i$, then we must have $0 = d^2\phi = d(gx^i \mu_i) = (x^i \partial_i g + n\, g)\mu$, in which case $g(x)$ must be a function that is homogeneous of degree $-n$ or else vanish throughout S . Since the origin of coordinates is contained in S , $g(x)$ can not be homogeneous of degree $-n$ on S for it would then be discontinuous on S . Thus, $d\phi = g(x)X \rfloor \mu$ can hold for continuous $g(x)$ on S only if $g(x) = 0$, and the result is established.

THEOREM 5-18.2. *Let* B_n *be an open, connected, arc-wise simply connected set whose closure,* $\bar{B}_n$ *, is compact and contained in a starlike region* S *of an n-dimensional space. Let the boundary of* $\bar{B}_n$ *be locally smooth with the exception of a finite number of edges and vertices and let* x^i *be a preferred coordinate system on* S *such that the center of* S *has coordinates* $x^i = 0$. *There exists a unique, antiexact,* $(n-1)$*-form* ρ *on* S *, for given* $\omega \in \Lambda^n(S)$ *such that*

$$\int_{\bar{B}_n} \omega = \int_{\partial \bar{B}_n} \rho \tag{5-18.22}$$

$$X \rfloor \rho = 0 , \quad \rho(0) = 0 . \tag{5-18.23}$$

This form ρ *is determined by*

$$\rho = F^i \mu_i , \quad F^i = h(f)x^i \tag{5-13.24}$$

for $\omega = f\,\mu$.

Proof. Under the given hypotheses, Theorem 5-18.1 gives $\omega = d\rho$ where ρ verifies (5-18.23), (5-18.24) for $\omega = f\,\mu$, and is unique. The result then follows directly from Stokes' theorem [3,5]. In effect, Theorem 5-18.2 reduces Stokes' theorem to the divergence theorem, for we obviously have

$$\int_{\bar{B}_n} \omega = \int_{\bar{B}_n} f\,\mu = \int_{\bar{B}_n} (\partial_i F^i)\mu = \int_{\partial \bar{B}_n} F^i \pi_i . \tag{5-18.25}$$

5-19. k-FORMS AND INTEGRATION OVER k-DIMENSIONAL DOMAINS

Let ω be a k-form that is defined on a region R of an n-dimensional space, M_n, and let Φ be a regular map from a starlike region S of a k-dimensional space, M_k, into a k-dimensional set $\Phi(S)$ that is contained in R. In this event, $\Phi^*\omega$ is a k-form on the k-dimensional region S of M_k to which we can apply the results of the previous section. To this end, let $\{y^i\}$ be a preferred coordinate system on S such that the center of S has coordinates $y^i = 0$, and let H be the homotopy operator on S with respect to this preferred system of coordinates. It then follows immediately from Theorem 5-18.1 that there exists a unique $\rho \in \Lambda^{k-1}(S)$ such that

$$\Phi^*\omega = d\rho = dH(\Phi^*\omega) , \tag{5-19.1}$$

$$Y \rfloor \rho = 0 , \quad \rho(0) = 0 , \tag{5-19.2}$$

where $Y = y^i \, \partial/\partial y^i$. Now that we know that

$$\rho = H(\Phi^*\omega) \tag{5-19.3}$$

we can use $\overset{-1}{\Phi}{}^*$ to map ρ onto a (k-1)-form restricted on $\Phi(S)$. Likewise, using $\overset{-1}{\Phi}{}^*$ to map $\Phi^*\omega$ onto the restriction of ω to $\Phi(S)$, (5-19.1) yields the relation

$$\overset{-1}{\Phi}{}^*(\Phi^*\omega) = \overset{-1}{\Phi}{}^*(dH\Phi^*\omega) = d\overset{-1}{\Phi}{}^*H\Phi^*\omega \tag{5-19.4}$$

on $\Phi(S)$. However, Theorem 5-8.1 gives us the operator $H^* = \overset{-1}{\Phi}{}^*H\Phi^*$ on forms restricted to $\Phi(S)$.

<u>LEMMA 5-19.1</u>. *Let ω be a k-form on a region R of an n-dimensional space and let Φ be a regular mapping of a starlike region S of a k-dimensional set into R. There exists a unique (k-1)-form ρ on $\Phi(S)$ such that*

$$\omega = d\rho \tag{5-19.5}$$

holds upon restriction to $\Phi(S)$ *and*

$$(5\text{-}19.6) \qquad \rho = H^{*}(\omega) ,$$

$$(5\text{-}19.7) \qquad V \rfloor \Phi^{*}\rho = 0 , \quad \Phi^{*}\rho(0) = 0 .$$

This allows us to establish the following result.

THEOREM 5-19.1. *Let the hypotheses of Lemma 15-19.1 be satisfied and let* B_k *be an open, simply connected, arc-wise simply connected subset of* S *whose closure* $\bar{B}_k$ *is contained in* S . *Let* ω *be a* k*-form of an* n*-dimensional space* M *with* $n \geq k$ *and let* $\bar{\mathcal{B}}_k = \Phi(\bar{B}_k)$. *The following integration formula then obtains:*

$$(5\text{-}19.8) \qquad \int_{\bar{\mathcal{B}}_k} \omega = \int_{\bar{B}_k} \Phi^{*}\omega = \int_{\bar{B}_k} d\rho = \int_{\partial\bar{B}_k} \rho = \int_{\partial\bar{\mathcal{B}}_k} \overset{-1}{\Phi}{}^{*}\rho = \int_{\partial\bar{\mathcal{B}}_k} H^{*}(\omega) .$$

This result shows that any integral of any form that is well defined over a region $\mathcal{B}$ can be written uniquely as an integral over the boundary of $\mathcal{B}$ provided only that the boundary of $\mathcal{B}$ is sufficiently smooth that Stokes' theorem may be applied. Accordingly, any well defined k-measure can be evaluated in terms of a (k-1)-measure on the boundary of the set of definition under appropriate smoothness conditions on the measure and the boundary of the set of definition.

Careful note must be taken of the fact that (5-19.8) involves the operator H^{*} rather than the operator H . There is a significant difference, for H is defined for artibrary forms on the manifold, while H^{*} is defined only for forms restricted to $\bar{\mathcal{B}}_k = \Phi(\bar{B}_k)$. This is clear from the derivation, for it is only upon restriction to $\bar{\mathcal{B}}_k$ that a k-form becomes the $\overset{-1}{\Phi}{}^{*}$-image of a k-form on a k-dimensional space, and hence is exact. On the other hand, an arbitrary k-form on an n-dimensional space need not be exact. In fact, we have

$$(5\text{-}19.9) \qquad \omega = dH\omega + Hd\omega ,$$

as a consequence of property H_2. Application of Stokes' theorem to (5-19.9) yields

$$(5\text{-}19.10) \qquad \int_{\bar{B}_k} \omega = \int_{\partial\bar{B}_k} H\omega + \int_{\bar{B}_k} Hd\omega .$$

A comparison of (5-19.8) and (5-19.10) shows that

$$(5\text{-}19.11) \qquad \int_{\partial\bar{B}_k} H^*(\omega) = \int_{\partial\bar{B}_k} H\omega + \int_{\bar{B}_k} Hd\omega ,$$

which possibly gives the most self-evident distinction between the operators H and H^* in integral relations.

5-20. TOP DOWN GENERATION OF BASES

Calculation of the homotopy operator's application to forms of degree n, $n-1$, $n-2$, ..., are quite tedious if we use the standard bases for Λ^n, Λ^{n-1}, ... It was shown in Section 3-5, however, that there were alternative bases for Λ^n, Λ^{n-1}, ... that are generated from the top down and that result in certain definite computational efficiencies. It is therefore useful to return to this topic in somewhat more detail here.

Let U be a coordinate neighborhood of an n-dimensional space with the associated coordinate functions $\{x^i\}$. The natural volume element of U and the natural basis elements of $T(U)$ are denoted by $\mu = dx^1 \wedge dx^2 \wedge \ldots \wedge dx^n$ and $\{t_i\} = \{\partial/\partial x^1, \partial/\partial x^2, \ldots, \partial/\partial x^n\}$, respectively. Define the collection of (n-1)-forms $\{\mu_i\}$ by

$$(5\text{-}20.1) \qquad \mu_i = t_i \rfloor \mu , \quad i = 1,\ldots,n .$$

If $v = v^i t_i$ is any vector field on U, then $v \rfloor \mu = v^i \mu_i$, from which it follows that the μ_i's defined by (5-20.1) agree with those defined in Section 5-18 if the starlike region S of Section 5-18 is contained in U and the coordinate functions are chosen so that the center of S has coordinates $x^i = 0$, $i = 1,\ldots,n$. The quantities μ_i defined by (5-20.1) are, however, defined for any coordinate cover of any coordinate neighborhood. Thus, the definition given by (5-20.1) is universal, as opposed to the definition given in Section 5-18 that was dependent upon a preferred coordinate cover of a starlike region.

The definition given by (5-20.1) has the further feature that it allows definition of forms of decreasing degree by recursion:

(5-20.2) $$\mu_{i_1\ldots i_k} = t_{i_1} \lrcorner \mu_{i_2\ldots i_k} .$$

We thus have

$$\mu_{i_1\ldots i_k} = t_{i_1} \lrcorner t_{i_2} \lrcorner \cdots \lrcorner t_{i_k} \lrcorner \mu .$$

These forms have very useful properties, as the following Lemma shows.

<u>LEMMA 5-20.1</u>. *Let* U *be a coordinate neighborhood of an* n-*dimensional space with coordinate functions* $\{x^i\}$ *and let* $\mu = dx^1 \wedge dx^2 \wedge \ldots \wedge dx^n$ *and* $\{t_i = \partial/\partial x^i,\ i = 1,\ldots,n\}$ *be the natural volume element of* U *and the natural basis for* T(U) , *respectively. The sequence of forms of decreasing degree, defined by*

(5-20.3) $$\mu_i = t_i \lrcorner \mu , \quad \mu_{i_1\ldots i_k} = t_{i_1} \lrcorner \mu_{i_2\ldots i_k} ,$$

has the following properties: (1) $\{\mu_{i_1\ldots i_k}\}$ *belongs to* Λ^{n-k} *and is completely antisymmetric in the indices* $i_1,\ldots,i_k$; (2) *the set* $\{\mu_{i_1\ldots i_k} \mid i_1 < i_2 < \ldots < i_k\}$ *forms a basis for* Λ^{n-k} ; (3) *we have*

(5-20.4) $$d\mu_{i_1\ldots i_k} = 0 ,$$

(5-20.5) $$dx^j \wedge \mu_{i_1 \ldots i_k} = k\, \delta^j_{[i_1} \mu_{i_2 \ldots i_k]} ,$$

where the square brackets signify complete antisymmetrization (i.e., all even permutations minus all odd permutations).

Proof. Properties (1) and (2) follow immediately from the definition (5-20.3). It thus remains only to establish (5-20.4) and (5-20.5), Now $d\mu_i = d(t_i \rfloor \mu) = d(t_i \rfloor \mu) + t_i \rfloor d\mu = \pounds_{t_i} \mu = 0$, since $d\mu = 0$ and μ and t_i are both constant fields. Similarly $d\mu_{ij} = d(t_i \rfloor \mu_j) + t_i \rfloor d\mu_j$, since $d\mu_j = 0$, and hence $d\mu_{ij} = \pounds_{t_i} \mu_j = \pounds_{t_i}(t_j \rfloor \mu) = [t_i, t_j] \rfloor \mu + t_j \rfloor \pounds_{t_i} \mu = 0$. Writing this result as $d\mu_{ij} = \pounds_{t_i} \mu_j = 0$, (5-20.4) follows by recursion. If v is any vector field and ω and Ω are forms with $\omega \varepsilon \Lambda^k$, $v \rfloor (\omega \wedge \Omega) = (v \rfloor \omega) \wedge \Omega + (-1)^k \omega \wedge (v \rfloor \Omega)$, so that $\omega \wedge (v \rfloor \Omega) = (-1)^k v \rfloor (\omega \wedge \Omega) + (-1)^{k+1} (v \rfloor \omega) \wedge \Omega$. Thus, with $\omega = dx^j$, $v = t_{i_1}$, $\Omega = \mu_{i_2 \ldots i_k}$, we have

$$dx^j \wedge \mu_{i_1 \ldots i_k} = dx^j \wedge (t_{i_1} \rfloor \mu_{i_2 \ldots i_k})$$

$$= - t_{i_1} \rfloor (dx^j \wedge \mu_{i_2 \ldots i_k}) + \delta^j_{i_1} \mu_{i_2 \ldots i_k} .$$

Since this is recursive, we need only note that $dx^j \wedge \mu_i = \delta^j_i \mu$, for $k = 1$, in order to obtain (5-20.5) by recursion and antisymmetrization.

It is clear from these results that we can confine our attention to the cases $k = 1$ and $k = 2$, for everything then follows by recursion. Let F be an arbitrary (n-1)-form (i.e., $k = 1$). Since $\{\mu_i\}$ is a basis for Λ^{n-1}, F can be written uniquely as

(5-20.6) $$F = F^i \mu_i .$$

The question that arises as to the properties of $\{F^i\}$. We first note that $\mu_i = t_i \lrcorner \mu$ implies that a coordinate map $\{x^i\} \longrightarrow \{{}^\backprime x^i\}$ induces the map

$$(5\text{-}20.7) \qquad {}^\backprime\mu_j = \Delta({}^\backprime x, x) \frac{\partial x^i}{\partial {}^\backprime x^j} \mu_i ,$$

where $\Delta({}^\backprime x,x) = \det(\partial {}^\backprime x^i/\partial x^j)$ is the Jacobian of the map. Thus, since $F = F^i \mu_i = {}^\backprime F^i \, {}^\backprime\mu_i$, we obtain

$$(5\text{-}20.8) \qquad {}^\backprime F^i = \Delta({}^\backprime x,x)^{-1} \frac{\partial {}^\backprime x^i}{\partial x^j} F^j ;$$

that is, $\{F^i\}$ are the components of a contravariant vector density. This gives us the correspondence

$$(5\text{-}20.9) \qquad F = F^i \mu_i \longrightarrow \hat{F} = F^i t_i ,$$

where $\hat{F}$ is the corresponding vector density (i.e., a quantity with the transformation law $\hat{F} = F^i t_i$, ${}^\backprime\hat{F} = {}^\backprime F^i \, {}^\backprime t_i$, ${}^\backprime F = \Delta({}^\backprime x,x)^{-1} F$). Conversely, let $\hat{v} = \hat{V}^i t_i$ be an arbitrary vector density, then $\hat{v} \lrcorner \mu$ belongs to Λ^{n-1} and (5-20.3) gives

$$(5\text{-}20.10) \qquad \hat{v} \lrcorner \mu = \hat{V}^i t_i \lrcorner \mu = \hat{V}^i \mu_i ,$$

and hence the correspondence (5-20.9) is 1-to-1.

<u>LEMMA 5-20.2</u>. *Any (n-1)-form F is represented in terms of the basis $\{\mu_i\}$ by the components of a vector density and*

$$(5\text{-}20.11) \qquad F = F^i \mu_i , \quad \hat{F} = F^i t_i$$

establishes a 1-to-1 correspondence between (n-1)-forms and vector densitites.

An identical argument can be made for the case $k = 2$.

<u>LEMMA 5-20.3</u>. *Any (n-2)-form F is represented in terms of the basis $\{\mu_{ij}\}$ by the components of a bivector density and*

$$(5\text{-}20.12) \qquad F = F^{ij} \mu_{ij} , \quad \hat{F} = F^{ij} t_i \wedge t_j$$

establishes a 1-to-1 correspondence between (n-2)-*forms and bi-vector densities.*

The following result now follows directly by use of (5-20.4), (5-20.5), (5-20.10) and (5-20.11).

LEMMA 5-20.4. *Use of the bases* $\{\mu_i\}$ *and* $\{\mu_{ij}\}$ *for* Λ^{n-1} *and* Λ^{n-2}, *respectively, yield*

(5-20.13) $$dF = d(F^i \mu_i) = (\partial_i F^i)\mu ,$$

(5-20.14) $$dG = d(G^{ij} \mu_{ij}) = 2(\partial_i G^{ij}) \mu_j .$$

The result for arbitrary k should now be clear. Use of the basis $\{\mu_{i_1 \ldots i_k}\}$ leads to a representation of any (n-k)-form in terms of the entries of a k-vector density and establishes a 1-to-1 correspondence between (n-k)-forms and k-vector densities: $F \rightleftarrows \hat{F}$ with

(5-20.15) $$F = F^{i_1 \ldots i_k} \mu_{i_1 \ldots i_k} , \quad \hat{F} = F^{i_1 \ldots i_k} t_{i_1} \wedge \ldots \wedge t_{i_k} .$$

We also have

(5-20.16) $$dF = k(\partial_{i_1} F^{i_1 \ldots i_k}) \mu_{i_2 \ldots i_k} .$$

It is thus clear that use of the bases $\{\mu_i\}$, $\{\mu_{ij}\}$, etc. provide all of the necessary structure needed in order to obtain the "divergence like" quantities $\partial_{i_1} F^{i_1 i_2 \ldots i_k}$ and the 1-to-1 correspondences between (n-k)-forms and k-vector densities without introducing an innerproduct structure into Λ and T or the corresponding Hodge star operator.

There is another obvious correspondence that obtains by use of the top down and the bottom up bases to represent the same form. We illustrate this in the case n = 3 , since its generalization to arbitrary n is simple but cumbersome to write. We obtain the results immediately by simply writing

$$(5\text{-}20.17)\quad \begin{aligned} f_{ijk}\, dx^i \wedge dx^j \wedge dx^k &= F\,\mu\ , \\ f_{ij}\, dx^i \wedge dx^j &= F^i\,\mu_i\ , \\ f_i\, dx^i &= F^{ij}\,\mu_{ij}\ , \\ f &= F^{ijk}\,\mu_{ijk}\ . \end{aligned}$$

This gives the 1-to-1 reciprocal maps Γ and γ ,

$$(5\text{-}20.18)\quad \Gamma\{f_{ijk}\} = \{F\},\ \Gamma\{f_{ij}\} = \{F^i\},\ \Gamma\{f_i\} = \{F^{ij}\},\ \Gamma\{f\} = \{F^{ijk}\}\ ,$$

$$(5\text{-}20.19)\quad \gamma\{F\} = \{f_{ijk}\},\ \gamma\{F^i\} = \{f_{ij}\},\ \gamma\{F^{ij}\} = \{f_i\},\ \gamma\{F^{ijk}\} = \{f\}\ ,$$

that must be used whenever top down and bottom up representations are used simultaneously.

A specific application of these results obtains when we are faced with computing $H\omega$ when the degree of ω is large. If $\omega \varepsilon \Lambda^{n-k}$, we may write

$$(5\text{-}20.20)\quad \omega = F^{i_1 \dots i_k}\,\mu_{i_1 \dots i_k}\ .$$

In this event, we obtain

$$\begin{aligned} H\omega &= \int_0^1 X \rfloor\, (\tilde{F}^{i_1 \dots i_k})(\lambda)\ \mu_{i_1 \dots i_k}\ \lambda^{n-k-1}\, d\lambda \\ &= x^j \left(\int_0^1 (\tilde{F}^{i_1 \dots i_k})(\lambda)\ \lambda^{n-k-1}\, d\lambda \right) \mu_{j i_1 \dots i_k} \end{aligned}$$

since the μ's are constant-valued forms. Thus, if we define the operator h_k by

$$(5\text{-}20.21)\quad h_k(F^{i_1 \dots i_k}) = \int_0^1 (\tilde{F}^{i_1 \dots i_k})(\lambda)\ \lambda^{n-k-1}\, d\lambda\ ,$$

we obtain the result that

$$(5\text{-}20.22)\quad H\omega = x^j\, h_k(F^{i_1 \dots i_k})\ \mu_{j i_1 \dots i_k}\ .$$

Thus, in particular, for $k = 1$, we have $\omega = F^i \mu_i$

$$(5\text{-}20.23) \qquad H\omega = x^j h_1(F^i) \mu_{ji} .$$

Clearly, the operator h, that is defined by (5-18.17), is the operator h_o in the sequence (5-20.21). The results of Section 5-18 can thus be extended to forms of arbitrary degree. Further, since $\omega = F^i \mu_i = dH(F^i \mu_i) + Hd(F^i \mu_i) = dH(F^i \mu_i) + H(\partial_j F^j \mu)$, the operators h_o and h_1 satisfy the identity

$$(5\text{-}20.24) \qquad F^i = (n-1) h_1(F^i) + x^j \partial_j h_1(F^i) + x^i\{h_o(\partial_j F^j) - \partial_j h_1(F^j)\} .$$

Similar identities can be obtained for h_o, h_1, and h_2 by starting with $\omega = F^{ij} \mu_{ij}$. We leave the construction of these identities to the interested reader.

5-21. SOLUTIONS OF MAXWELL'S EQUATIONS IN THE PRESENCE OF ELECTRIC AND MAGNETIC CHARGES

One of the classic applications of the exterior calculus is in the area of electrodynamics, for, as is well known, Maxwell's equations in vacuum can be written as $df = 0$, $d^*f = 0$, where $f \in \Lambda^2(M_4)$ is the 2-form of the electromagnetic field and $^*f$ is the metric dual of f. Significantly more information can be obtained, however, by use of the results given in previous sections. In fact, general solutions of Maxwell's equations can be obtained when there are both electric and magnetic charges present. A magnetic charge is what is commonly referred to as a magnetic monopole.

Let M_4 be a flat, 4-dimensional space with a global coordinate cover $\{x^\alpha\}$; $x^1 = x$, $x^2 = y$, $x^3 = z$, $x^4 = t$ where x, y, z are the standard Cartesian coordinates of 3-dimensional Euclidean space and t = time. The 4-vector density field of free electric current is denoted by

$$(5\text{-}21.1)\qquad \mathcal{J} = J^{\alpha}\, t_{\alpha} ,$$

where $t_{\alpha} = \partial/\partial x^{\alpha}$ and $J^4 = q$ is the density of free electric charge. Similarly, the 4-vector pseudodensity field of free magnetic current is denoted by

$$(5\text{-}21.2)\qquad \mathcal{G} = G^{\alpha}\, t_{\alpha} ,$$

where $G^4 = g$ is the pseudodensity of magnetic charge.

Let $\mu = dx^1 \wedge dx^2 \wedge dx^3 \wedge dx^4$ be the natural volume element of M_4. The laws of conservation of free electric charge and of conservation of free magnetic charge take the forms

$$(5\text{-}21.3)\qquad d(\mathcal{J} \rfloor \mu) = d(J^{\alpha}\, \mu_{\alpha}) = (\partial_{\alpha} J^{\alpha})\mu = 0 ,$$

and

$$(5\text{-}21.4)\qquad d(\mathcal{G} \rfloor \mu) = d(G^{\alpha}\, \mu_{\alpha}) = (\partial_{\alpha} G^{\alpha})\mu = 0 ,$$

respectively. The string of equalities of (5-21.3) and (5-21.4) are obtained from the definition of the $\mu_{\alpha} = t_{\alpha} \rfloor \mu$ and Theorem 5-20.1. Since the 3-forms $J^{\alpha}\, \mu_{\alpha}$ and $G^{\alpha}\, \mu_{\alpha}$ are closed, by (5-21.3) and (5-21.4), and M_4 is globally starlike, there exist 2-forms $H^{\alpha\beta}\, \mu_{\alpha\beta}$ and $F^{\alpha\beta}\, \mu_{\alpha\beta}$ on M_4 such that

$$(5\text{-}21.5)\qquad J^{\alpha}\, \mu_{\alpha} = d(H^{\alpha\beta}\, \mu_{\alpha\beta}) , \quad G^{\alpha}\, \mu_{\alpha} = d(F^{\alpha\beta}\, \mu_{\alpha\beta}) ,$$

which are equivalent to

$$(5\text{-}21.6)\qquad J^{\alpha} = 2\, \partial_{\beta} H^{\beta\alpha} , \quad G^{\alpha} = 2\, \partial_{\beta} F^{\beta\alpha}$$

by Theorem 5-20.1. It is then a trivial matter to see that

$$H^{\beta\alpha} = - H^{\alpha\beta} ,$$

$$(5\text{-}21.7)\qquad 2((H^{\alpha\beta})) = \begin{pmatrix} 0 & -H_z & H_y & D_x \\ & 0 & -H_x & D_y \\ & & 0 & D_z \\ & & & 0 \end{pmatrix},$$

$$(5\text{-}21.8)\qquad 2((F^{\alpha\beta})) = \begin{pmatrix} 0 & E_z & -E_y & B_x \\ & 0 & E_x & B_y \\ & & 0 & B_z \\ & & & 0 \end{pmatrix},\quad F^{\beta\alpha} = -F^{\alpha\beta},$$

replicate the full system of Maxwell's equations

$$(5\text{-}21.9)\qquad \vec{\nabla}\times\vec{H} - \partial_t\vec{D} = \vec{J},\quad \vec{\nabla}\cdot\vec{D} = q,$$

$$(5\text{-}21.10)\qquad -\vec{\nabla}\times\vec{E} - \partial_t\vec{B} = \vec{G},\quad \vec{\nabla}\cdot\vec{B} = g.$$

All that we have really done so far, is to show that Maxwell's equations are a direct consequence of the conservation of free electric charge and free magnetic charge. There is significantly more that can be gleaned, however. Let H be the homotopy operator on M_4 with the origin of coordinates as the center. Thus

$$(5\text{-}21.11)\qquad \mathcal{X} = x^\alpha\, t_\alpha\,.$$

It then follows directly from (5-21.5), (5-21.6) and $dH + Hd =$ identity, that (5-21.5) and (5-21.6) are identically satisfied by

$$H^{\alpha\beta}\,\mu_{\alpha\beta} = dK + H(J^\alpha\,\mu_\alpha)\,,$$

$$F^{\alpha\beta}\,\mu_{\alpha\beta} = dA + H(G^\alpha\,\mu_\alpha)\,,$$

where K and A are arbitrary 1-forms on M_4. This establishes the following result.

THEOREM 5-21.1. *Maxwell's equations,* $J^\alpha\,\mu_\alpha = d(H^{\alpha\beta}\,\mu_{\alpha\beta})$, $G^\alpha\,\mu_\alpha = d(F^{\alpha\beta}\,\mu_{\alpha\beta})$, *are identically satisfied on* M_4 *by*

$$(5\text{-}21.12) \qquad \mathcal{H} = H^{\alpha\beta}\,\mu_{\alpha\beta} = dK + H(J^{\alpha}\,\mu_{\alpha}) \,,$$

$$(5\text{-}21.13) \qquad \mathcal{F} = F^{\alpha\beta}\,\mu_{\alpha\beta} = dA + H(G^{\alpha}\,\mu_{\alpha}) \,,$$

for any 1-forms K *and* A *on* M_4 .

We first note that the absence of magnetic charges yields $F^{\alpha\beta}\,\mu_{\alpha\beta} = dA$, so that A is the 1-form of the standard 4-vector potential of the classic electromagnetic field. The results have been obtained without the introduction of a metric into M_4 (an inner product on $T(M_4)$). Accordingly, it is preferable to refer to A as the 1-form potential. Further, we also have the 1-form potential K that is associated with the structure of the electric current field. In fact, the results given in the above theorem appear to be new, both from the complete first integral (5-21.12) of the electric current equations, and the existence of a single-valued potential 1-form for the $(F^{\alpha\beta}\,\mu_{\alpha\beta})$-field in the presence of magnetic charges. In any event, it is now clear from whence the actual equations that determine the electromagnetic field arise. Certainly it is not from Maxwell's equations, for we have obtained their general solution. Rather, the electromagnetic field is determined solely from its constitutive relations $C(H^{\alpha\beta};F^{\alpha\beta},\dots) = 0$ whereby the fields $H^{\alpha\beta}\,\mu_{\alpha\beta}$ and $F^{\alpha\beta}\,\mu_{\alpha\beta}$ are related; that is (5-21.12) and (5-21.13) must be substituted into $C(H^{\alpha\beta};F^{\alpha\beta};\dots) = 0$ in order to obtain the equations that A and K must satisfy. We further note that

$$(5\text{-}21.14) \qquad \mathcal{J} \rfloor (F^{\alpha\beta}\,\mu_{\alpha\beta}) = \mathcal{W}_q$$

$$(5\text{-}21.15) \qquad \mathcal{G} \rfloor (H^{\alpha\beta}\,\mu_{\alpha\beta}) = \mathcal{W}_g$$

give the Lorentz force 1-form $\mathcal{W}_q$ that acts on the electric charges and the 1-form of force $\mathcal{W}_g$ that acts on the magnetic charges. These 1-forms of forces satisfy the conditions

$$(5\text{-}21.16) \qquad J\rfloor W_q = 0\,, \quad G\rfloor W_g = 0$$

identically, so that we secure the standard requirements of a 4-dimensional force distributions.

Every theory that is considered respectable by physicists is derivable from a variational principle. Accordingly, let R_4 be the closure of an open, arcwise connected, simply connected point set in M_4 and consider the "action" functional

$$(5\text{-}21.17) \qquad S(R_4) = \int_{R_4} \{dA \wedge H + dK \wedge F - A \wedge (J\rfloor \mu) - K \wedge (G\rfloor \mu) + H(G\rfloor \mu) \wedge H + H(J\rfloor \mu) \wedge F - H \wedge F\}\,.$$

The functional derivative of $S(R_4)$ that is generated by

$$A \longrightarrow A + \sigma_1 \delta A, \quad K \longrightarrow K + \sigma_2 \delta K, \quad H \longrightarrow H + \sigma_3 \delta H, \quad F \longrightarrow F + \sigma_4 \delta F$$

is easily seen to be given by

$$\begin{aligned}(5\text{-}21.18) \qquad \delta S(R_4) = \int_{R_4} \{&\sigma_1 [d(\delta A) \wedge H - \delta A \wedge (J\rfloor \mu)] \\ &+ \sigma_2 [d(\delta K) \wedge F - \delta K \wedge (G\rfloor \mu)] \\ &+ \sigma_3 [dA + H(G\rfloor \mu) - F] \wedge \delta H \\ &+ \sigma_4 [dK + H(J\rfloor \mu) - H] \wedge \delta F\}.\end{aligned}$$

However, $d(\delta A) \wedge H = d(\delta A \wedge H) + \delta A \wedge dH$, $d(\delta K) \wedge F = d(\delta K \wedge F) + \delta K \wedge dF$, and (5-21.18) becomes

$$\begin{aligned}(5\text{-}21.19) \qquad \delta S(R_4) = \int_{R_4} \{&\sigma_1 \delta A \wedge (dH - J\rfloor \mu) + \sigma_2 \delta K \wedge (dF - G\rfloor \mu) \\ &+ \sigma_3 \delta H \wedge (dA + H(G\rfloor \mu) - F) \\ &+ \sigma_4 \delta F \wedge (dK + H(J\rfloor \mu) - H)\} \\ &+ \int_{\partial R_4} \{\delta A \wedge H + \delta K \wedge F\}\,.\end{aligned}$$

Accordingly, the fundamental lemma of the calculus of variations shows that $S(R_4)$ is stationary relative to all variations of A and K than vanish on ∂R_4 and relative to all variations of F and H if and only if

$$\text{(5-21.20)} \qquad \frac{\delta S}{\delta A} = 0 : \quad dH - J \rfloor \mu = 0 , \qquad \frac{\delta S}{\delta K} = 0 : \quad dF - G \rfloor \mu = 0 ,$$

$$\text{(5-21.21)} \qquad \frac{\delta S}{\delta H} = 0 : \quad dA + H(G \rfloor \mu) - F = 0 , \qquad \frac{\delta S}{\delta F} = 0 : \quad dK + H(J \rfloor \mu) - H = 0 .$$

This variational principle thus yields both the field equations, (5-21.20) and the first integrals of the field equations, (5-21.21).

The electric and magnetic currents have been held fixed during the variational process that yields the field equations and their first integrals. The question naturally arises as to what obtains upon variation of the current distributions. The part of the action integral (5-21.17) that involves the electric and magnetic currents is given by

$$\text{(5-21.22)} \qquad S_c(R_4) = \int_{R_4} \{-A \wedge (J \rfloor \mu) - K \wedge (G \rfloor \mu) + H \wedge H(G \rfloor \mu) + F \wedge H(J \rfloor \mu)\} .$$

The variations induced in these terms by $J \longrightarrow J + \sigma_5 \delta J$, $G \longrightarrow G + \sigma_6 \delta G$ are quite easily computed. They give

$$\text{(5-21.23)} \qquad \delta S_c(R_4) = \int_{R_4} \{\sigma_5[-A \wedge (\delta J \rfloor \mu) + F \wedge H(\delta J \rfloor \mu)] + \sigma_6[-K \wedge (\delta G \rfloor \mu) + H \wedge H(\delta G \rfloor \mu)]\} .$$

An application of the fundamental lemma of the calculus of variations can not be made until the primatives δJ and δG occur as simple factors. We are thus faced with the problem of rewriting the terms $F \wedge H(\delta J \rfloor \mu)$ and $H \wedge H(\delta J \rfloor \mu)$ in terms of equivalent expressions that remove the operator H from acting on δJ and δG. Since H is a linear operator and $F \wedge H(\delta J \rfloor \mu)$ is a 4-form on M_4, H possesses a formal adjoint H^+ that satisfies

$$\int_{R_4} \{F \wedge H(\delta J \rfloor \mu) - H^+(F) \wedge (\delta J \rfloor \mu)\} = 0 .$$

In this event, (5-21.23) becomes

$$(5\text{-}21.24) \qquad \delta S_c(R_4) = \int_{R_4} \{\sigma_5[-A + H^+(F)] \wedge (\delta J \rfloor \mu) + \sigma_6[-K + H^+(H)] \wedge (\delta G \rfloor \mu)\} .$$

This formal variational process will lead to the forces that act on the electric and magnetic current distributions provided the variations of J and G are constrained so as to preserve the statements of conservation of electric and magnetic charge [31]. Clearly, this process hinges on the construction of the formal adjoint of the operator H, which we take up in the next Section.

5-22. THE ADJOINT OPERATOR H^+

We saw in the last Section that the adjoint, H^+, of the homotopy operator H arises naturally in certain problems of the calculus of variations, and is, in fact, an operator whose properties and intrinsic utility place it on an equal footing with the operator H. We therefore make a detailed study of H^+ in this Section.

Definition 5-22.1. Let S be a starlike region of an n-dimensional manifold, let $\bar{S}$ be the closure of S, and let $\partial\bar{S}$ denote the boundary of $\bar{S}$. Let L be a linear operator that maps $\Lambda(S)$ into $\Lambda(S)$. If a linear operator L^+ exists that maps $\Lambda(S)$ into $\Lambda(S)$ and verifies

$$(5\text{-}22.1) \qquad \int_{\bar{S}} \{\alpha \wedge L(\beta) - L^{+}(\alpha) \wedge \beta\} = \int_{\partial\bar{S}} P(\alpha,\beta) ,$$

for all α and β such that $\alpha \wedge L(\beta) \in \Lambda^n(S)$, then L^+ is said to be the *generalized adjoint* of L with bilinear concomitant $P(\alpha,\beta)$. If $P(\alpha,\beta) = 0$, then L^+ is said to be the *adjoint* of L .

We first establish a simple working lemma that will simplify calculation of the adjoint of H .

LEMMA 5-22.1. *Let* $\alpha(x) \in \Lambda^k(S)$, $\beta(x) \in \Lambda^{n-k+1}(S)$ *and let* v *be any vector field on* S , *then*

$$(5\text{-}22.2) \qquad \alpha \wedge (v \rfloor \beta) = (-1)^{k+1} (v \rfloor \alpha) \wedge \beta .$$

Proof. Under the hypotheses concerning the degrees of α and β , $\alpha \wedge \beta = 0$. Thus, $0 = v \rfloor (\alpha \wedge \beta) = (v \rfloor \alpha) \wedge \beta + (-1)^k \alpha \wedge (v \rfloor \beta)$, from which the result follows directly.

Let $\alpha(x)$ and $\beta(x)$ be arbitrary forms on S such that $\alpha \in \Lambda^k(S)$, $\beta \in \Lambda^{n-k+1}(S)$. It then follows directly from (5-2.2) and (5-3.1) that

$$(\alpha \wedge H(\beta))(x) = \int_0^1 \alpha(x) \wedge \left(X(x) \rfloor \beta(\lambda x)\right) \lambda^{n-k} \, d\lambda ,$$

and hence Lemma 5-22.1 yields

$$(5\text{-}22.3) \qquad (\alpha \wedge H(\beta))(x) = (-1)^{k+1} \int_0^1 \left(X(x) \rfloor \alpha(x)\right) \wedge \beta(\lambda x) \, \lambda^{n-k} \, d\lambda .$$

Accordingly, we have

$$(5\text{-}22.4) \qquad \int_{\bar{S}} \alpha \wedge H(\beta) = (-1)^{k+1} \int_{\bar{S}} \int_0^1 \left(X(x) \rfloor \alpha(x)\right) \wedge \beta(\lambda x) \, \lambda^{n-k} \, d\lambda .$$

Now, $(X(x) \rfloor \alpha(x)) \wedge \beta(\lambda x) \in \Lambda^n(S)$ and hence it is a multiple of μ . We may therefore write

$$(X(x) \rfloor \alpha(x)) \wedge \beta(\lambda x) = \widehat{X(x) \rfloor \alpha(x)} \vee \widehat{\beta(\lambda x)} \; \mu$$

with the obvious symbolic meaning for the quantities on the right-hand side of this expression that multiply μ. Now, introduce n-dimensional spherical coordinates with P_o as the center. This gives $\mu = r^{n-1}\, dr \wedge d\Omega$, where $d\Omega$ is the element of n-dimensional solid angle. It then follows directly that (5-22.4) assumes the following equivalent form

$$(5\text{-}22.5)\qquad \int_{\bar{S}} \alpha \wedge H(\beta)$$
$$= (-1)^{k+1} \int_{\odot_n} \int_o^{R(\Omega)} \int_o^1 \widehat{X(r\ldots) \rfloor \alpha(r\ldots)} \vee \widehat{\beta(\lambda r\ldots)}\, \lambda^{n-k}\, d\lambda\, r^{n-1}\, dr\, d\Omega ,$$

where the dots signify dependence on the n-1 angular coordinates.

Since all (n-1) angles are held constant during the λ and r integrations, our considerations can be simplified by introducing

$$(5\text{-}22.6)\qquad \xi(\Omega)$$
$$= (-1)^{k+1} \int_o^{R(\Omega)} \int_o^1 \widehat{X(r\ldots) \rfloor \alpha(r\ldots)} \vee \widehat{\beta(\lambda r\ldots)}\, \lambda^{n-k}\, d\lambda\, r^{n-1}\, dr$$

so that

$$(5\text{-}22.7)\qquad \int_{\bar{S}} \alpha \wedge H(\beta) = \int_{\odot_n} \xi(\Omega)\, d\Omega .$$

Again, since all angles are held fixed in (5-22.6), we may interchange the order of the r and the λ integrations to obtain

$$(5\text{-}22.8)\qquad \xi(\Omega)$$
$$= (-1)^{k+1} \int_o^1 \int_o^{R(\Omega)} \widehat{X(r\ldots) \rfloor \alpha(r\ldots)} \vee \widehat{\beta(\lambda r\ldots)}\, \lambda^{n-k}\, r^{n-1}\, dr\, d\lambda .$$

Since this is an iterated integral, we may make the change of variables $\lambda r = u$, $r = u\lambda^{-1}$, $dr = \lambda^{-1}\, du$ for the r-wise integration with λ held fixed. This gives us the expression

(5-22.9) $\xi(\Omega)$

$$= (-1)^{k+1}\int_0^1\int_0^{\lambda R(\Omega)} \widehat{X(\tfrac{u}{\lambda}\ldots) \rfloor \alpha(\tfrac{u}{\lambda}\ldots)} \vee \widehat{\beta(u\ldots)}\lambda^{-k}\, u^{n-1}\, du\, d\lambda\ .$$

Thus, if we interchange the order of integration in this iterated integral and then set $u = r$, we obtain

(5-22.10) $\xi(\Omega)$

$$= (-1)^{k+1}\int_0^{R(\Omega)}\int_{\frac{r}{R(\Omega)}}^1 \widehat{X(\tfrac{r}{\lambda}\ldots) \rfloor \alpha(\tfrac{r}{\lambda}\ldots)} \vee \widehat{\beta(r\ldots)}\, r^{n-1}\, \lambda^{-k}\, d\lambda\, dr$$

$$= \int_0^{R(\Omega)} \left\{(-1)^{k+1}\int_{\frac{r}{R(\Omega)}}^1 \widehat{X(\tfrac{r}{\lambda}\ldots) \rfloor \alpha(\tfrac{r}{\lambda}\ldots)}\lambda^{-k}\, d\lambda\right\} \vee \widehat{\beta(r\ldots)}\, r^{n-1}\, dr\ .$$

The lower limit $r\, R(\Omega)^{-1}$ in the λ-wise integral can be changed to zero if we introduce a function $\hat{e}(\frac{r}{\lambda},\Omega)$ such that

$$(5\text{-}22.11)\qquad \hat{e}(\tfrac{r}{\lambda},\Omega) = \begin{cases} 1 & \text{for}\quad r/\lambda \geq r/R(\Omega) \\ 0 & \text{for}\quad r/\lambda < r/R(\Omega)\ . \end{cases}$$

We thus obtain

$$(5\text{-}22.12)\qquad \xi(\Omega) = \int_0^{R(\Omega)} \left\{(-1)^{k+1}\int_0^1 \hat{e}(\tfrac{r}{\lambda},\Omega)\widehat{X(\tfrac{r}{\lambda}\ldots) \rfloor \alpha(\tfrac{r}{\lambda}\ldots)}\lambda^{-k}\, d\lambda\right\} \vee \widehat{\beta(r\ldots)}\, r^{n-1}\, dr\ .$$

It is now a simple matter to substitute (5-22.12) back into (5-22.7) and then change back to the original coordinate cover $\{x^i\}$. When this is done, we obtain

$$(5\text{-}22.13)\qquad \int_{\bar{S}} \alpha \wedge H(\beta) = \int_{\bar{S}} \left\{(-1)^{k+1}\int_0^1 e(\tfrac{x}{\lambda})X(\tfrac{x}{\lambda}) \rfloor \alpha(\tfrac{x}{\lambda})\lambda^{-k}\, d\lambda\right\} \wedge \beta(x)\ ,$$

where

$$(5\text{-}22.14)\qquad e\left(\frac{x}{\lambda}\right) = \begin{cases} 1 & \text{for } \frac{x}{\lambda} \in \bar{S} \\ 0 & \text{for } \frac{x}{\lambda} \notin \bar{S} \,. \end{cases}$$

However, (5-2.5) gives $X(\frac{x}{\lambda}) = \lambda^{-1} X(x)$, and hence the notation introduced by (5-2.2) yields

$$(5\text{-}22.15)\qquad \int_{\bar{S}} \alpha \wedge H(\beta)$$

$$= \int_{\bar{S}} \left\{ (-1)^{k+1} \int_0^1 e(\tfrac{x}{\lambda}) X \rfloor \tilde{\alpha}(\tfrac{1}{\lambda}) \lambda^{-(k+1)} \, d\lambda \right\} \wedge \beta$$

$$= \int_{\bar{S}} \left\{ (-1)^{k+1} \int_1^{\infty} X \rfloor (e\alpha)(\lambda) \lambda^{k-1} \, d\lambda \right\} \wedge \beta \,,$$

where the latter equality results from the transformations $\lambda \longrightarrow \nu^{-1}$, $\nu \longrightarrow \lambda$. Since α and β are arbitrary forms such that $\alpha \wedge H(\beta) \in \Lambda^n(S)$, use of Definition 5-22.1 gives the following result.

<u>THEOREM 5-22.1</u>. *Let* S *be a starlike region of an* n*-dimensional space with center* P_o *, let* x^i *be a preferred coordinate cover of* S *and let* $\bar{S}$ *denote the closure of* S *relative to the Euclidean topology induced by identifying the points of* S *in the preferred coordinate cover* x^i *with the points in* E_n *with the same coordinate values. The homotopy operator* H *with center* P_o *has an adjoint operator* H^+ *that is defined for any* $\alpha(x) \in \Lambda^k(S)$ *by*

$$(5\text{-}22.16)\qquad H^+(\alpha) = (-1)^{k+1} \int_1^{\infty} X \rfloor \widetilde{(e\alpha)}(\lambda) \lambda^{k-1} \, d\lambda$$

where e *is the characteristic function of* $\bar{S}$ *. Thus,*

$$(5\text{-}23.17)\qquad H^+(\alpha)(x) = (-1)^{k+1} \int_1^{\infty} e(\lambda x) X(x) \rfloor \alpha(\lambda x) \lambda^{k-1} \, d\lambda$$

with

$$(5\text{-}22.18)\qquad e(\lambda x) = \begin{cases} 1 & \text{for} \quad \lambda x \in \bar{S} \\ 0 & \text{for} \quad \lambda x \notin \bar{S} . \end{cases}$$

COROLLARY 5-22.1. *The operator* H^+ *is a linear operator that maps* $\Lambda^k(S)$ *into* $A^{k-1}(S)$.

Proof. That H^+ maps $\Lambda^k(S)$ into $\Lambda^{k-1}(S)$ follows directly from (5-22.16) or (5-22.17), and the linearity of H^+ is likewise immediate. Since $X(x)$ does not depend on λ , it can be taken outside the integral in (5-22.17) so as to obtain

$$(5\text{-}22.19)\qquad H^+(\alpha)(x) = (-1)^{k+1} X(x) \lrcorner \int_1^\infty e(\lambda x)\alpha(\lambda x)\lambda^{k-1}\, d\lambda .$$

Thus, if $\{x_o^i\}$ denote the coordinates of the center of S , we have

$$(5\text{-}22.20)\qquad X(x) \lrcorner H^+(\alpha)(x) = 0 , \quad H^+(\alpha)(x_o) = 0 ,$$

where the last equality follows from $X(x_o) = 0$ and the fact that the integral is finite due to the continuity of α and the presence of the characteristic function e of $\bar{S}$. Thus, $H^+(\alpha) \in A^{k-1}(S)$ by Definition 5-5.2.

Since the image of any α under H^+ is antiexact, we have the following immediate result.

COROLLARY 5-22.1. *The operator* H^+ *satisfies*

$$(5\text{-}22.21)\qquad HH^+\alpha \equiv 0 , \quad H^+H\alpha \equiv 0 .$$

Let α and β be forms such that $\alpha \wedge \beta \in \Lambda^n(S)$. If we use property H_2 , it follows that $\alpha \wedge \beta = \alpha \wedge dH(\beta) + \alpha \wedge H(d\beta)$, and hence

$$(5\text{-}22.22)\qquad \int_{\bar{S}} \alpha \wedge \beta = \int_{\bar{S}} \alpha \wedge dH(\beta) + \int_{\bar{S}} \alpha \wedge H(d\beta) .$$

Now, let $\alpha \in \Lambda^k(S)$. In this event, $\alpha \wedge dH(\beta) = (-1)^k d(\alpha \wedge H(\beta)) +$

$(-1)^{k+1}\, d\alpha \wedge H(\beta)$, so that

$$\int_{\bar{S}} \alpha \wedge dH(\beta) = (-1)^k \int_{\partial\bar{S}} \alpha \wedge H(\beta) + (-1)^{k+1} \int d\alpha \wedge H(\beta)$$

$$= (-1)^k \int_{\partial\bar{S}} \alpha \wedge H(\beta) + (-1)^{k+1} \int H^+(d\alpha) \wedge \beta \ .$$

Similarly,

$$\int_{\bar{S}} \alpha \wedge H(d\beta) = \int_{\bar{S}} H^+(\alpha) \wedge d\beta$$

$$= (-1)^{k-1} \int_{\partial\bar{S}} H^+(\alpha) \wedge \beta + (-1)^k \int dH^+(\alpha) \wedge \beta \ .$$

A combination of these equalities thus yields

$$\text{(5-22.23)} \qquad \int_{\bar{S}} \alpha \wedge \beta = \int_{\bar{S}} \alpha \wedge (dH + Hd)(\beta)$$

$$= \int_{\bar{S}} \{(-1)^k (dH^+ - H^+ d)(\alpha)\} \wedge \beta$$

$$+ \int_{\partial\bar{S}} (-1)^k \{\alpha \wedge H(\beta) - H^+(\alpha) \wedge \beta\} \ .$$

Definition 5-22.1 thus yields the following result.

COROLLARY 5-22.3. *The generalized adjoint of* $dH + Hd$ *is given, for any* $\alpha \in \Lambda^k(S)$, *by*

$$\text{(5-22.24)} \qquad (dH + Hd)^+(\alpha) = (-1)^k (dH^+ - H^+ d)(\alpha)$$

with the bilinear concomitant

$$\text{(5-22.25)} \qquad P(\alpha,\beta) = (-1)^k \{\alpha \wedge H(\beta) - H^+(\alpha) \wedge \beta\}$$

for $\alpha \wedge \beta \in \Lambda^n(S)$.

This result may seem strange on first examination, for $dH + Hd =$ identity on S . However, it is easily seen that

$$(5\text{-}22.26) \qquad \int_{\bar{S}} \alpha \wedge d\beta = \int_{\bar{S}} \{(-1)^{k+1} d\alpha\} \wedge \beta + \int_{\partial\bar{S}} (-1)^k \alpha \wedge \beta$$

for any α and β such that $\alpha \wedge d\beta \in \Lambda^n(S)$. The occurrence of a nonzero bilinear concomitant thus comes about from the presence of the operator d ; that is, the generalized adjoint of d is given by $(-1)^{k+1} d$, for any $\alpha \in \Lambda^k(S)$, and the bilinear concomitant is $(-1)^k \alpha \wedge \beta$.

5-23. VARIATIONS AND STATIONARITY

The study of the adjoint operator, H^+ , was motivated by the particular variational problem considered in Section 5-21. It would therefore seem appropriate to examine a general class of variational problems that involve differential forms and the action of H on these differential forms. Since the action of H on a differential form introduces a strong nonlocality, these considerations will do more than just reproduce the known classical results.

Since we are concerned only with C^∞-structures, the following lemma is obvious.

LEMMA 5-23.1. *Let* α *be a* k-*form and let* $\Omega(\alpha,\ldots)$ *be an* n-*form on an* n-*dimensional space that is constructed from the* k-*form* α *and a finite number of other forms and vectors. There exists a unique* (n-k)-*form* $\delta\Omega/\delta\alpha$ *such that*

$$(5\text{-}23.1) \qquad \left\{\frac{d}{d\varepsilon}\,\Omega(\alpha+\varepsilon\eta,\ldots)\right\}\Big|_{\varepsilon=0} = \frac{\delta\Omega}{\delta\alpha}(\alpha,\ldots) \wedge \eta$$

is satisfied identically in η *for all* $\eta \in \Lambda^k$.

For example, suppose that v is a given vector field, $\alpha \in \Lambda^1$, $f(\gamma)$ is an infinitely differentiable function of the argument γ , and $\Omega(\alpha,\ldots) = f(v \rfloor \alpha)\omega$ where ω is an n-form. We then have

$$\left\{\frac{d}{d\varepsilon}\,\Omega(\alpha+\varepsilon\eta)\right\}\Big|_{\varepsilon=0} = f'(v \rfloor \alpha)(v \rfloor \eta)\omega$$

for all $\eta \varepsilon \Lambda^1$. However, $\omega \wedge \eta \varepsilon \Lambda^{n+1}$, so that $0 = v \rfloor (\omega \wedge \eta) = (v \rfloor \omega) \wedge \eta + (-1)^n \omega (v \rfloor \eta)$, and hence $(v \rfloor \eta)\omega = (-1)^{n+1}(v \rfloor \omega) \wedge \eta$. This gives

$$\left\{\frac{d}{d\varepsilon}\,\Omega(\alpha+\varepsilon\eta)\right\}\Bigg|_{\varepsilon=0} = (-1)^{n+1} f'(v \rfloor \alpha)(v \rfloor \omega) \wedge \eta ,$$

and hence $\delta\Omega/\delta\alpha = (-1)^{n+1} f'(v \rfloor \alpha)(v \rfloor \omega)$.

Definition 5-23.1. Let α be a k-form on a starlike region S of an n-dimensional space and let $\Omega(\alpha, d\alpha, H\alpha, Hd\alpha, \ldots)$ be a specific n-form that is constructed from α, $d\alpha$, $H\alpha$, $Hd\alpha$ and a finite number of given forms and vector fields on S; that is, Ω is a map from $\Lambda^k(S) \times E^{k+1}(S) \times A^{k-1}(S) \times A^k(S)$ into $\Lambda^n(S)$. The map

$$I[\alpha] = \int_{\bar{S}} \Omega(\alpha, d\alpha, H\alpha, Hd\alpha, \ldots)$$

of $\Lambda^k(S)$ into the real line is the *action functional* generated by the given map Ω.

Definition 5-23.2. An element α of $\Lambda^k(S)$ is said to *stationarize* the action functional $I[\alpha]$ if and only if

$$(5\text{-}23.2.)\qquad \delta_\eta I[\alpha] \overset{\text{def}}{=} \left\{\frac{d}{d\varepsilon}\, I[\alpha+\varepsilon\eta]\right\}\Bigg|_{\varepsilon=0} = 0$$

for all k-forms η that vanish on $\partial\bar{S}$.

It is clear from the above that (5-23.2) is equivalent to the standard stationarity condition

$$(5\text{-}23.3.)\qquad I[\alpha+\varepsilon\eta] = I[\alpha] + o(\varepsilon)$$

of the calculus of variations, where, of course, $o(\varepsilon)$ is computed with respect to the uniform norm on the real line.

We now proceed to the calculation. If α is to render $I[\alpha]$ stationary in value, then (5-23.2) must hold for all $\eta \varepsilon \Lambda^k(S)$ that vanish on the boundary of $\bar{S}$. Thus, since

$$I[\alpha + \varepsilon\eta] = \int_{\bar{S}} \Omega(\alpha + \varepsilon\eta,\ d\alpha + \varepsilon d\eta,\ H\alpha + \varepsilon H\eta,\ Hd\alpha + \varepsilon Hd\eta,\ \ldots)\ ,$$

due to the linearity of the operators H and d, use of Lemma 5-23.1 gives us

$$(5\text{-}23.4) \qquad \delta_\eta I[\alpha] = \int_{\bar{S}} \left\{ \frac{\delta\Omega}{\delta\alpha} \wedge \eta + \frac{\delta\Omega}{\delta d\alpha} \wedge d\eta + \frac{\delta\Omega}{\delta H\alpha} \wedge H\eta + \frac{\delta\Omega}{\delta Hd\alpha} \wedge Hd\eta \right\} .$$

It now follows directly from Definition 5-22.1 and Theorem 5-22.1 that

$$(5\text{-}23.5) \qquad \delta_\eta I[\alpha] = \int_{\bar{S}} \left\{ \left(\frac{\delta\Omega}{\delta\alpha} + H^+\!\left(\frac{\delta\Omega}{\delta H\alpha}\right)\right) \wedge \eta + \left(\frac{\delta\Omega}{\delta d\alpha} + H^+\!\left(\frac{\delta\Omega}{\delta Hd\alpha}\right)\right) \wedge d\eta \right\} .$$

Now, $\xi = \delta\Omega/\delta d\alpha + H^+(\delta\Omega/\delta Hd\alpha)$ is an $(n-k-1)$-form, and hence $\xi \wedge d\eta = (-1)^{n-k-1} d(\xi \wedge \eta) + (-1)^{n-k} d\xi \wedge \eta$. Thus (5-23.5) and Stokes' theorem yield

$$(5\text{-}23.6) \qquad \delta_\eta I[\alpha] = (-1)^{n-k-1} \int_{\partial\bar{S}} \left(\frac{\delta\Omega}{\delta d\alpha} + H^+\!\left(\frac{\delta\Omega}{\delta Hd\alpha}\right)\right) \wedge \eta$$
$$+ \int_{\bar{S}} \left\{ \frac{\delta\Omega}{\delta\alpha} + H^+\!\left(\frac{\delta\Omega}{\delta H\alpha}\right) + (-1)^{n-k} d\left(\frac{\delta\Omega}{\delta d\alpha} + H^+\!\left(\frac{\delta\Omega}{\delta Hd\alpha}\right)\right)\right\} \wedge \eta .$$

Since (5-23.6) need vanish only for all η that vanish on $\partial\bar{S}$, the fundamental lemma of the calculus of variations leads directly to the following results.

THEOREM 5-23.1. *The element* α *of* $\Lambda^k(S)$ *stationarizes the action functional*

$$(5\text{-}23.7) \qquad I[\alpha] = \int_{\bar{S}} \Omega(\alpha,\ d\alpha,\ H\alpha,\ Hd\alpha,\ \ldots)$$

that is generated by the given map Ω *if and only if* α *satisfies*

$$(5\text{-}23.8) \qquad \frac{\delta\Omega}{\delta\alpha} + H^+\!\left(\frac{\delta\Omega}{\delta H\alpha}\right) = (-1)^{n-k+1} d\left\{ \frac{\delta\Omega}{\delta d\alpha} + H^+\!\left(\frac{\delta\Omega}{\delta Hd\alpha}\right)\right\} .$$

throughout S .

There are obvious generalizations that can be obtained. For instance, an integral over $\partial\bar{S}$ could be included in the definition of $I[\alpha]$, in which case we could eliminate the requirement that η vanish on $\partial\bar{S}$ provided α satisfies the natural boundary conditions that are thereby induced. In fact, it follows directly from (5-23.6) that η need not vanish on $\partial\bar{S}$ if α is such that $\{\delta\Omega/\delta d\alpha + H^{+}(\delta\Omega/\delta H d\alpha)\}$ vanish on $\partial\bar{S}$. We leave these aspects of the general stationarization problem to the interested reader.

REFERENCES

1. Ślebodziński, W.: *Exterior Forms and Their Applications* (Polish Scientific Publishers, Warsaw, 1970).

2. Flanders, H.: *Differential Forms* (Academic Press, New York, 1963).

3. Cartan, E.: *Les systèmes differentials extérieurs et leurs applications géométriques* (Hermann, Paris, 1945).

4. Cartan, H.: *Differential Forms* (Houghton Mifflin Co., Boston, 1970).

5. Lovelock, D. and H. Rund: *Tensors, Differential Forms, and Variational Principles* (John Wiley, New York, 1975).

6. Sternberg, S.: *Lectures on Differential Geometry* (Prentice-Hall, Englewood Cliffs, 1964).

7. Edelen, D. G. B.: *Lagrangian Mechanics of Nonconservative Nonholonomic Systems* (Noordhoff, Leyden, 1977).

8. Harrison, B. K. and F. B. Estabrook: Geometric Approach to Invariance Groups and Solutions of Partial Differential Systems, *J. Math. Phys. 12* (1971), 653.

9. Estabrook, F. B.: Some Old and New Techniques for the Practical Use of Exterior Differential Forms, in *Bäcklund Transformations, the Inverse Scattering Method, Solutions, and their Applications* (*Lecture Notes in Mathematics No. 515*, Springer, Berlin, 1976).

10. Ovsjannikov, L. V.: *Gruppovye svostva differentsialny uravneni* (U.S.S.R. Academy of Science, Novosibirsk, 1962) (*Group Properties of Differential Equations*, G. W. Bluman, translation, 1967).

11. Bluman, G. W. and J. D. Cole: *Similarity Methods for Differential Equations* (Springer, Berlin, 1974).

12. Duff, G. F. D.: *Partial Differential Equations* (University of Toronto Press, Toronto, 1956).

13. Forsythe, A. R.: *Theory of Differential Equations* (Dover, New York, 1959), Vol. V, 315-343.

14. Forsythe, A. R.: *Theory of Differential Equations* (Dover, New York, 1959), Vol. VI, 425-455.

15. Walquist, H. D. and F. B. Estabrook: Prolongation structures of nonlinear evolution equations, *J. Math. Phys.* *16* (1975), 1.

16. Corones, J.: Solitons and simple pseudopotentials, *J. Math. Phys.* *17* (1976), 765.

17. Estabrook, F. B. and H. D. Wahlquist: Prolongation structures of nonlinear evolution equations, II, *J. Math. Phys.* *17* (1976), 1293.

18. Morris, H. C.: Prolongation structures and nonlinear evolution equations in two spatial dimensions, *J. Math. Phys.* *17* (1976), 1870.

19. Morris, H. C.: Prolongation structures and a generalized inverse scattering problem, *J. Math. Phys.* *17* (1976), 1867.

20. Corones, J.: Solitons, pseudopotentials, and certain Lie algebras, *J. Math. Phys.* *18* (1977), 163.

21. Morris, H. C.: A prolongation structure for the AKNS system and its generalization, *J. Math. Phys.* *18* (1977), 533.

22. Morris, H. C.: Prolongation structures and nonlinear evolution equations in two spatial dimensions, II. A generalized nonlinear Schrödinger equation, *J. Math. Phys.* *18* (1977), 285.

23. Herman, R.: *The Geometry of Non-Linear Differential Equations Bäcklund Transformations, and Solitons* (Math. Sci. Press, Brookline, Ma. 1976).

24. Morse, P. M. and H. Feshbach: *Methods of Theoretical Physics, Part I* (McGraw-Hill, New York, 1953).

25. Edelen, D. G. B.: *Nonlocal Variations and Local Invariance of Fields* (American Elsevier, New York, 1969).

26. Rund, H.: "Variational problems and Bäcklund transformations associated with the sine-Gordon and Korteweg-Devries equations and their extensions," in *Bäcklund Transformation, the Inverse Scattering Method, Solitons, and their Applications*, ed. by R. M. Miura, New York, Springer, 1976, p. 199-226 (Lecture Notes in Mathematics No. 515).

27. Goldschmidt, H. and S. Sternberg: *Ann. Inst. Fourier* *23*, *1* (1973), 203.

28. Rund, H.: The Hamilton-Jacobi theory of the geodesic fields of Carathéodory in the calculus of variations of multiple integrals (The Greek Mathematical Society, C. Carathéodory Symposium, September 3-7, 1973).

29. Dedecker, P.: On the generalization of symplectic geometry to multiple integrals in the calculus of variations, in *Differential Geometric Methods in Mathematical Physics* (Lecture Notes in Mathematics, Vol. 570, Springer, Berlin, 1977).

30. Carathéodory, C.: Uber die variationsrechnung bei mehrfachen integralen., *Acta Math.* (*Szeged*) *4* (1929), 193.

31. Edelen, D. G. B.: A metric free electrodynamics with electric and magnetic charges, *Ann. Phys.* (*N.Y.*) *112* (1978), 366.

INDEX